# 4th International Conference on Advanced Materials and Intelligent Manufacturing 2023

AA007112

Journal of Physics: Conference Series Volume 2720

Guangzhou, China
1-3 December 2023

ISBN: 978-1-7138-9277-9
ISSN: 1742-6588

**Printed from e-media with permission by:**

Curran Associates, Inc.
57 Morehouse Lane
Red Hook, NY 12571

proceedings.com

**Some format issues inherent in the e-media version may also appear in this print version.**



Institute of Physics
Dirac House, Temple Back
Bristol BS1 6BE UK

Phone: 44 1 17 929 7481
Fax: 44 1 17 920 0979

techtracking@iop.org

**Additional copies of this publication are available from:**

Curran Associates, Inc.
57 Morehouse Lane
Red Hook, NY 12571 USA
Phone: 845-758-0400
Fax: 845-758-2633
Email: curran@proceedings.com
Web: www.proceedings.com

# TABLE OF CONTENTS

# Preface

The 2023 4th International Conference on Advanced Materials and Intelligent Manufacturing (ICAMIM 2023) was held in Guangzhou, China during December 1st to 3rd, 2023. The series of ICAMIM conference was created with the aim to be an interdisciplinary incubator for the worldwide exchange of information and innovative ideas about the latest research results in the fields of advanced materials and intelligent manufacturing.

At ICAMIM 2023, we wished to bring together a broad community of researchers from different branches of the rapidly developing complex science to discuss the fundamental theoretical challenges as well as experimental and applied aspects of many practical problems. The agenda was conducted mainly based on three parts, covering keynote speeches, oral & poster presentations, and academic tour.

There were six keynote speeches and many regular oral talks in the morning and afternoon sessions. The topics of keynote speeches cover: a) Scientific Carbon Reduction, Green Development, Manufacturing Industry to Achieve the Goal of "Double Carbon" Efforts; b) ChatGPT with Intelligent Manufacturing System; c) Challenges Facing the Industrialization of Sodium-Ion Batteries; d) Microstructure Control and Application Properties of Ultra-High Strength and Toughness Steels; e) Double Growing Carbon Background Material Base Environmental Functional Materials Development Prospects; f) Current Status and Development of the New Materials Industry in Guangdong. These speeches allowed all the participants to interact with the speakers and people around and get some new insights from each other. An academic discussion part was also held, the topic of which was related to the recent and anticipated future progress in advanced materials and intelligent manufacturing, discussing a variety of questions and opinions important for the understanding of the technology and management methods of the industry itself.

The papers submitted to this Conference were carefully reviewed by expert referees, covering topics on: Composites, New feature Materials, Integrated Manufacturing Systems, Advanced Manufacturing Technology, Intelligent Mechatronics, etc. We are very grateful to the reviewers for their efficient and thoughtful actions. Various excellent papers were included here based on the referee reports.

We are thankful to the members of Technical Program Committee for their time and efforts in preparing the program of ICAMIM 2023. We also appreciate the behind-the-scenes work of the members of Organizing Committee in preparing the conference site, arrangements for excursions and accommodation, handling the financial support for participants, and so on. Finally, we acknowledge the publishing support from Journal of physics: Conference Series

The Committee of ICAMIM 2023

# Committee Member

**General Conference Chair**
Prof. Liejun Li, South China University of Technology, China
Prof. Ali Reza Kamali, University of Cambridge, UK

**Publication Chair**
Prof. Jixiang Gao, Guangdong Polytechnic Normal University, China

**Technical Program Committee Chair**
Prof. Tongguang Zhai, Shandong Jianzhu University, China

**Organizing Committee**
Prof. Jixiang Gao, Guangdong Polytechnic Normal University, China
Prof. Xuegong Yu, Zhejiang University, China
Prof. Zude Zhou, Wuhan University of Technology, China
Prof. Xin Chen, Xi'an Jiaotong University, China
Prof. Cong Wang, Beihang University, China
A. Prof. Kai Wu, Tongji University, China
Prof. Weihua Dan, Sichuan University, China
Prof. Sirong Yu, China University of Petroleum, China
Prof. Shilu Dai, South China University of Technology, China
Prof. Sheng-Rui Jian, I-Shou University, Taiwan, China
Prof. Mingyang Chen, Jiangsu University, China
Prof. Mingxin Huang, The University of Hong Kong, China
Prof. Feng Wen, Hainan University, China
Dr. Hao Peng, Yangtze Normal University, China
A. Prof. Kai Wu, Tongji University, China
Prof. Chuang Feng, Nanjing University of Technology, China
Prof. Zhu Weibing, Xihua University, China
Prof. Weihua Dan, Sichuan University, China

**Technical Program Committee**
Prof. Dr. Yasuto Hijikata, Saitama University, Japan
Prof. Eckart H. Meiburg, University of California, USA
Prof. DAVID BASSIR, French University of Technology, France
Prof. Volkan Kahya, Karadeniz Technical University, Turkey
Prof. David E. Esezobor, University of Lagos, Nigeria
A. Prof. Nour F. Attia, National Institute of Standards (NIS), Egypt
A. Prof. Yipin Wang, Chang'an University, China
A. Prof. Mohamed Ahmed Ramadan, Helwan University, Egypt
Dr. Prashant J. Patil, Abhinav Education Society's College of Engineering and Technology Wadwadi, India
Dr. Quah Hock Jin, University Sains Malaysia, Malaysia
Dr. Alexander Lukin, Western-Caucasus Research Center, Russian Federation
Dr. R. RANJITH KUMAR, Viyen Biotech LLP, India
Dr. Xiao Qi, Northeastern University, China
Dr. Yang Luo, Swiss Federal Laboratories, Switzerland
Dr. Hicham Bakkali, University of Cadiz, Spain
Dr. P K Dash, NMIT, India

# Peer Review Statement

All papers published in this volume have been reviewed through processes administered by the Editors. Reviews were conducted by expert referees to the professional and scientific standards expected of a proceedings journal published by IOP Publishing.

- **Type of peer review:** Single Anonymous
- **Conference submission management system:** Morressier
- **Number of submissions received:** 135
- **Number of submissions sent for review:** 108
- **Number of submissions accepted:** 68
- **Acceptance Rate (Submissions Accepted / Submissions Received × 100):** 50.4
- **Average number of reviews per paper:** 2
- **Total number of reviewers involved:** 14
- **Contact person for queries:**
  **Name:** Xuexia Ye
  **Email:** xx.ye@keoaeic.org
  **Affiliation:** AEIC Academic Exchange Information Centre

# Study on schottky barrier of Cu/Graphene/4H-SiC interface based on first principles

**Congcong Wang**[1, 2, 5], **Siyu Wu**[1, 3], **Zhenyu Jiang**[1, 4] **and Yingjie Huang**[1,3]

[1] Institute of High Energy Physics, Chinese Academy of Sciences, Beijing, 100049, China
[2] State Key Laboratory of Particle Detection and Electronics, Beijing, 100049, China
[3] Jilin University, Changchun, 130012, China
[4] Liaoning University, Shenyang, 110036, China

[5]Corresponding author's e-mail: wangcc@ihep.ac.cn

**Abstract.** High stability 4H-SiC ohmic contact is currently a key technical challenge that silicon carbide devices urgently need to overcome. In the paper, the interfacial structures, atomic interactions and Schottky barrier height (SBH) of Cu/Graphene/4H-SiC were studied using the first-principles method. According to research, the SBH for Cu/G/4H-SiC is lower than the SBH for Cu/4H-SiC. The reasons for this phenomenon mainly include the following: 1. The graphene C atoms saturate the dangling bonds on the 4H-SiC surface and the influence of the metal-induced-gap-states (MIGS) at the interface is decreased. 2. A new phase is formed by inserting graphene between the Cu and 4H-SiC have low work functions. 3. An interfacial-electric-dipole layer (IEDL) formed at the interface of 4H-SiC and graphene may also reduce the SBH. These results make them to be promising candidates for future radiation resistant electronics.

## 1. Introduction

Silicon carbide has potential to be used for fast particle detection in radiation environment because of its wider band gap and high electron mobility[1-3]. High stability ohmic contact is currently a key technical challenge that silicon carbide devices urgently need to overcome. The quality of ohmic contact performance directly affects the device performance[4]. It is difficult to directly form ohmic contact between metals and wide band gap SiC. Due to the electronic affinity and surface density of states constraints of silicon carbide, the ohmic contact resistivity formed by high-temperature annealing is high, with poor stability and low process repeatability. High temperature annealing will activate impurities in silicon carbide, and the activated impurities will cause electrical signal scattering, which limits the improvement of the signal noise ratio (SNR), charge collection, time resolution and operational stability [1].

Stefan Hertel found that graphene can provide perfect ohmic contact with SiC epitaxial layers[5]. Zheng Jiangshan et al. verified that applying voltage can achieve p-type ohmic contact at the metal/graphene/SiC interface[6].Ji Yuting and Jia Yuping confirmed that graphene, as an intermediate transition layer between metal and silicon carbide, can reduce the metal/4H-SiC interface barrier, form a good ohmic contact, improve charge collection efficiency and time resolution[7]. The graphene insertion layer can form carbon compounds at the interface between the metal and silicon carbide, thereby reducing the ohmic contact resistance. Graphene as an intermediate transition layer may be beneficial to reducing the SBH of the Cu/4H-SiC interface and improving the ohmic characteristics. However, the relevant mechanism is not yet known and there is a lack of corresponding theoretical basis.

In this paper, in order to explain the Schottky barrier reduction mechanism, we optimize the cell structure based on first principles, analyze the band gap, density of states (DOS) and differential charge density to study the SBH of Cu/Graphene/4H-SiC.

## 2. Calculation method

The paper builds a highly symmetrical, up-and-down rectangular structure model, in which there are two layers of Cu, monolayer graphene and four layers of 4H-SiC carbon atoms. In order to eliminate the influence of dangling bonds on the bottom, the C atoms on the SiC (000$\bar{1}$) surface are passivated with H atoms, and the unit cells are isolated with a 25Å vacuum layer to eliminate interactions caused by periodicity. The lattice mismatch is reasonable and the structure is stable. The lattice optimization values are in the table1:

**Table 1.** The optimization parameters.

| parameter name: | value |
|---|---|
| cutoff energy | 400eV |
| convergence tolerance of force | $<$ 0.03eV/Å |
| maximal displacement of atoms | $< 1\times10^{-3}$Å |
| total energy | $< 2.0\times10^{-6}$ eV/atom |
| internal stress | $<$ 0.05 GPa |
| SCF tolerance | $< 1.5\times10^{-6}$ eV/atom |

## 3. Analysis and discussion

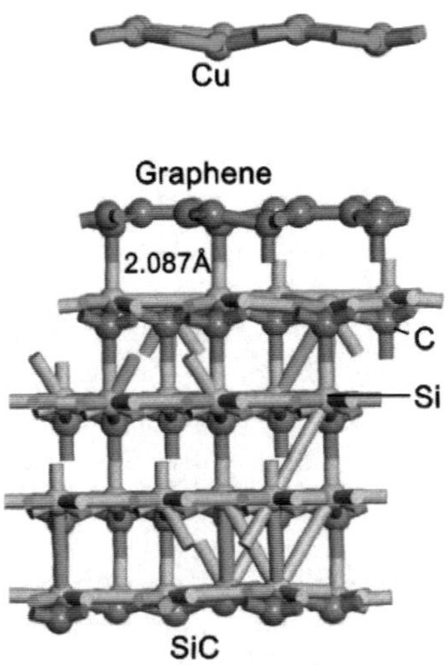


**Figure 1.** Cu/Graphene/4H-SiC crystal structure.

Figure 1 show that the optimized- graphene by the model does not a two-dimensional planar form. This demonstrates that the interaction between carbon and carbon has changed, and it is no longer $sp^2$ hybridization. The Si-C distance formed by C atoms and Si atoms is 2.087Å, which basically reaches the bonding distance. The optimized structure appears to be reconstructed, which is consistent with the research of Varchon F.[8]. Therefore, this model is proved to be reliable.

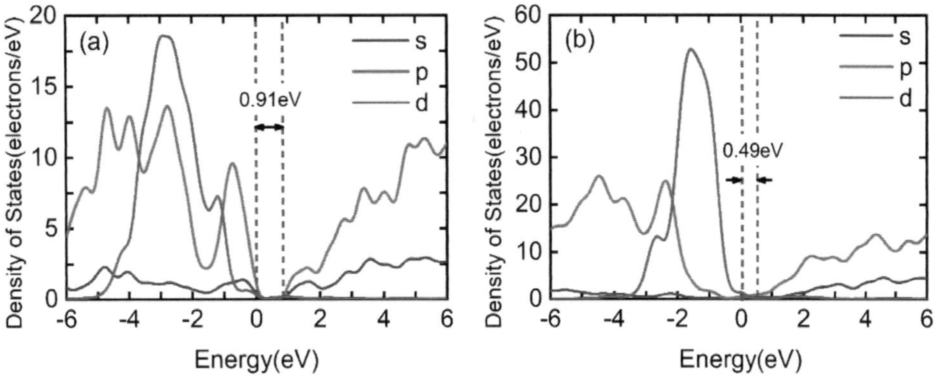


**Figure 2.** (a) Density of states for Cu/4H-SiC; (b) Density of states for Cu/Graphene/4H-SiC.

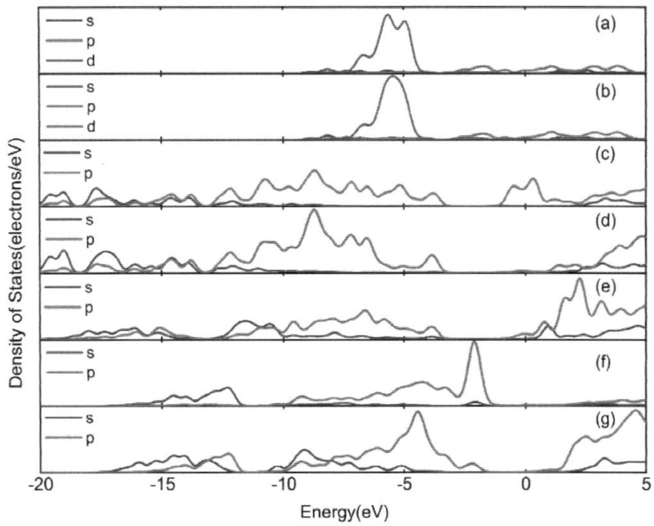


**Figure 3.** Density of states for Cu/Graphene/4H-SiC: (a) The Cu at the top; (b) The Cu near graphene; (c) The C near Cu in graphene;(d) The C near 4H-SiC in graphene; (e) C in 4H-SiC near graphene; (f) The bottom C in 4H-SiC; (g) The bottom Si in 4H-SiC.

Figure 3 show the density of states for Cu/Graphene/4H-SiC.The electronic structure of epitaxial growth monolayer graphene is obviously different from that of the free-state graphene. Epitaxially grown monolayer of graphene exhibit a large gap and a fixed Fermi level with a less dispersive electron state. For the Si plane, the Fermi level is near the conduction band, and for C plane, the Fermi level is in the band gap. These electron states are determined by the dangling bonds on the surface. That is, the Si surface is determined by the Si dangling bonds, and the C surface is determined by the C dangling bonds. Compared with the density of states for Si atoms near the Fermi level ($E_f$) in the lower layer of 4H-SiC, the DOS of Si atoms near graphite rarefaction increases. The decrease of electronic states near the Fermi level in the DOS of graphene C atoms near 4H-SiC (0001). This indicates that the Si at the 4H-SiC (0001) interface reacts with the C atom in graphene, and the electronic states interact and overlap with each other. The DOS of the C atom near the metal Cu in graphite has little change compared to the DOS of the Cu atom near the graphite. The DOS of the two layers of metal at the top did not change significantly, indicating that there is no reaction between Cu and graphite. In order to visually display the gain and loss of electrons at the interface, the (a) and (b) interfaces of the Cu/Graphene/4H-SiC atomic layer are taken for differential charge density. The Figure 4 shows that there is no electron overlap between graphene and Cu, further proving that Cu does not interact with graphene. The mismatch between graphene and silicon carbide lattice and the polarization of the 4H-SiC itself result in charge transfer between graphene and substrate. Electrons transfer from Si atoms on the surface of 4H-SiC to graphene C atoms. The charge depletion region formed on the Si surface and the charge accumulation region formed on the C surface jointly form the interface electric field. This results in the formation of interface electric dipoles, and the presence of this electric field may also reduce the Schottky barrier to some extent. According to the bonding mode, it can be inferred that the two forms a covalent bond, which saturates the surface overhang bond of 4H-SiC[9]. This to some extent reduces the intrinsic interface states of 4H-SiC, which is beneficial for reducing the Schottky barrier[10]. This proves that graphite rarefaction interacts with Si atoms in 4H-SiC, but does not interact with metal Cu.

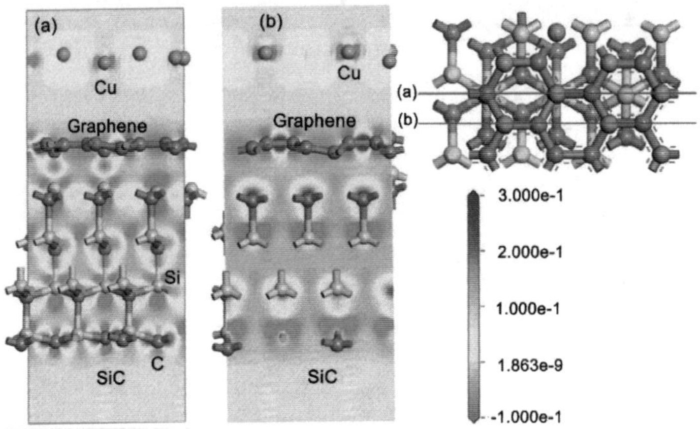


**Figure 4.** Cu/Graphene/4H-SiC: (a) a section differential charge density; (b) b section differential charge density.

Figure 5 show the DOS and PDOS of atoms in each layer for Cu/Graphene/4H-SiC. We can see the electronic states in the forbidden band of 4H-SiC gradually increase from the bottom layer to the interface layer. Interface states near the Fermi level arise from scratch, among which electronic states are the most obvious, and these interface states rapidly decay toward the interior. In the interior (second, third and fourth layers) there are no impurities and deformation, but electronic states still appear. The intrinsic interface states may cause partial pinning of the Fermi level. The influence of self-induced energy gap states and the overlap of electronic states caused by interactions at the interface lead to the appearance of a very high peak near the Fermi level of the first 4H-SiC layer.

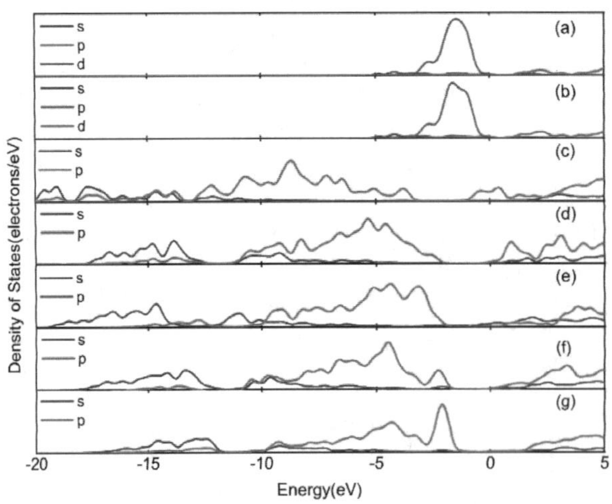


**Figure 5.** The DOS and PDOS of atoms in each layer for Cu/Graphene/4H-SiC: (a) The Cu atom at the top; (b) The Cu atoms near graphene; (c) graphene; (d) First-layer for 4H-SiC; (e) Second-layer for 4H-SiC; (f) Third-layer for 4H-SiC; (g) Fourth-layer for 4H-SiC.

## 4. Conclusion and perspectives

In the paper, the interfacial structures, atomic interactions and Schottky barrier height (SBH) of Cu/Graphene 4H-SiC were studied based on the first-principles method. The results show that the SBH for Cu/G/4H-SiC is lower than the SBH for Cu/4H-SiC. There are several reasons for this phenomenon. Firstly, the graphene C atoms saturate the dangling bonds on the 4H-SiC surface and the influence of the metal-induced-gap-states (MIGS) at the interface is decreased, so the Schottky barrier is lowered. Secondly, a new phase is formed by inserting graphene between the Cu and 4H-SiC have low work functions. Finally, an interfacial-electric-dipole layer (IEDL) formed at the interface of 4H-SiC and graphene may also reduce the SBH. These results make them to be promising candidates for future radiation resistant electronics.

**Acknowledgments**
This work was supported by the National Natural Science Foundation (12305207) and the State Key Laboratory of Particle Detection and Electronics (SKLPDE-ZZ-202312)

**References**

[1] Nava F, Bertuccio G, Cavallini A and Vittone E 2008 *Measurement Science and Technology* 19 102001.
[2] Ruddy F H and Seidel J G 2007 *Nuclear Instruments and Methods in Physics Research Section B Beam Interactions with Materials and Atoms* 263 163.
[3] Haijuan F U, Wang L, Wang H, Cao X, Liu L, Fan G, Yugang M A, Xue H, Zhang Y and Xinrong H U 2019 *Nuclear Techniques* 42 020501.
[4] Xuliang D, Xianfei J, Dejun W and Lingqin H 2022 *Acta Phys. Sin.* 71 058102.
[5] Stefan H, Finkler A, Krieger M and Weber H B 2015 *Materials Science Forum* 821-823 933.
[6] Zheng J, Li E, Cui Z and Ma D 2020 *Physica E Low-dimensional Systems and Nanostructures* 124 114277.
[7] Jia Y, Sun X, Shi Z, Jiang K, Wu T, Liang H, Cui X, Lü W and Li D 2021 *Diamond and Related Materials* 115 1.
[8] Varchon F, Feng R, Hass J, Li X, Nguyen B N, Naud C, Mallet P, Veuillen J Y, Berger C and Conrad E H 2007 *Phys Rev Lett* 99 126805.1.
[9] Kozlov S M, Viñes F, Nilius N, Shaikhutdinov S and Neyman K M 2012 *The Journal of Physical Chemistry Letters* 3 1956.
[10] Jing W, Meicheng L, Gan W, Yonghua L, Jianping Y, Zhanxu L, Donghui L and Xuekang C 2000 *Semiconductor Optoelectronics* 21 261.

# Study on road performance of asphalt mixture based on POE/PA6 modifier

**Liujie Guo[1], Zongjun Pan[1,2], Changxiaoxi Liu[1], Yongcai Yang[1], and Yanhe Sui[1]**

[1]RoadMainT Co., Ltd., Beijing, 100095, China

[2] Corresponding author's e-mail: panzongjun@roadmaint.com

**Abstract:** A POE/PA6 modifier with increased reactivity was prepared using the twin-screw extrusion granulation process and added to the asphalt mixture. The effect of the POE/PA6 modifier on high-temperature stability, low-temperature anti-cracking, and water stability of asphalt mixture was studied. The results show that compared with the asphalt mixture mixed with recycled PE, the asphalt mixture mixed with POE/PA6 has nearly four times the rutting resistance of the former at 70°C and 1.0 MPa, the fracture energy and flexibility index are twice the former at -10°C, and the maximum bending strain and water stability are also significantly improved. This modifier shows a good engineering application prospect in highway construction and maintenance under high temperatures and heavy load traffic environments.

## 1. Introduction

In recent years, through the large-scale application of polymer materials such as SBS and recycled PE in highway maintenance, the service life of ordinary national and provincial arterial highway asphalt pavement in China has been greatly extended, and the quality of highway maintenance has been significantly improved. However, with the rapid economic and social development, heavy highway traffic is becoming more common. The early damage phenomenon dominated by asphalt pavement ruts and cracks is still prominent [1], especially in the long longitudinal slope, heavy lanes, intersections, bus lanes, toll station squares, and other sections under heavy traffic conditions. Conventional treatment solutions are difficult to solve this industry problem fundamentally. Some areas try to use semi-flexible grouting road surfaces, epoxy asphalt, and other technical solutions but also face the problems of complex construction technology, long closed traffic time, and high project costs.

As an engineering plastic with excellent performance, PA6 has many applications, featuring high strength, wear, oil, and chemical resistance [2]. Still, it also has problems such as high brittleness at low temperatures and poor compatibility with non-polar materials [3]. Ethylene-octene copolymer (POE) is widely used in wire, cable, daily necessities, and other industries because of its high elasticity and good processing properties [4-6]. Toughening of PA6 with POE can improve the comprehensive properties of PA6 [7], but no relevant studies on POE/PA6 in asphalt mixtures have been seen. Therefore, the author prepared a POE/PA6 modifier with increased reactivity and added it to the asphalt mixture to investigate its influence on comprehensive road performance and provide information for asphalt pavement maintenance under heavy traffic.

## 2. Material preparation

### *2.1. Preparation of POE/PA6 modifier*

The PA6, capacity enhancing POE, antioxidant, and other additives used in this paper are commercially available products. Firstly, all kinds of raw materials are mixed and dispersed according to the set proportion, and the capacity enhancing POE occupies 25% of the mass of PA6. Then, the premix is extruded and granulated through the twin screw extruder, and the particle size of the material is kept below 5 mm, and it is used after drying.

### *2.2. Preparation of asphalt mixture*

The bitumen used in this experiment is grade A 70# petroleum bitumen, and the mineral material used is diabase of 10-15 mm and 5-10 mm produced in Shanxi, limestone of 0-5 mm, and mineral powder. The test results of raw materials all meet the requirements of relevant regulations. The grading composition and curve of the mineral material are shown in Table 1 and Figure 1.

**Table 1.** The pass rate of each sieve for AC-13 gradation.

| Sieve size (mm) | 16 | 13.2 | 9.5 | 4.75 | 2.36 | 1.18 | 0.6 | 0.3 | 0.15 | 0.075 |
|---|---|---|---|---|---|---|---|---|---|---|
| Upper limit | 100 | 100 | 85 | 68 | 50 | 38 | 28 | 20 | 15 | 8 |
| Lower limit | 100 | 90 | 68 | 38 | 24 | 15 | 10 | 7 | 5 | 4 |
| Synthetic gradation | 100 | 96 | 80.2 | 47.8 | 35.4 | 28.2 | 21 | 12.5 | 9.9 | 6.9 |

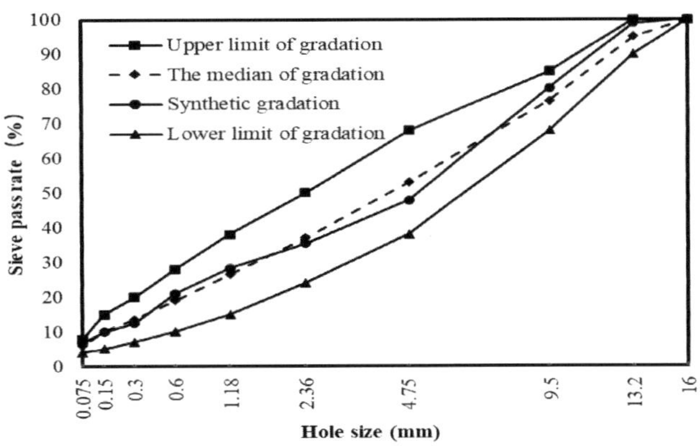


**Figure 1.** Mineral grade curve of asphalt mixture.

Marshall design method confirmed the optimum asphalt-stone ratio of the AC-13 asphalt mixture. The forming conditions of the specimen are as follows. Mixing temperature is 175°C-185°C, compaction molding temperature is 170°C-180°C, and double-sided co MPaction is 75 times. The mixing process involves adding the coarse and fine aggregates along with a PA6/POE modifier, which accounts for 0.5% of the total ore mass, into the mixing pot and mixing for 90 seconds. The measured petroleum asphalt mixing 90 s, and finally, the mineral powder mixing 90 s. The experimental results of asphalt mixture are shown in Table 2, and 5.4% is finally adopted as the optimal asphalt-stone ratio of the AC-13 asphalt mixture.

**Table 2.** Test results of AC-13 asphalt mixture.

| Asphalt-stone ratio (%) | Bulk volume relative density | Voidage (%) | Voids of mineral aggregate (%) | Asphalt saturation (%) | Marshall stability (kN) | Flow value (0.1 mm) |
|---|---|---|---|---|---|---|
| 4.5 | 2.422 | 7.5 | 17.6 | 57.4 | 17.2 | 3.8 |
| 5.0 | 2.441 | 5.9 | 17.3 | 66.2 | 19.1 | 4.2 |
| 5.5 | 2.483 | 3.6 | 16.3 | 77.9 | 21.6 | 4.5 |
| 6.0 | 2.468 | 2.7 | 17.2 | 84.4 | 18.5 | 4.9 |

## 3. Performance test

To fully appraise the road performance of asphalt mixture with POE/PA6, in this study, the performance of the mixture was compared with that of the recovered PE asphalt mixture with the same content and tested the stability, anti-cracking, and water stability of asphalt mixture according to the optimal asphalt-stone ratio determined above.

### *3.1. High-temperature stability*

Marshall stability and dynamic stability tests were carried out on the above two mixtures to characterize the anti-rutting performance of the asphalt mixture at high temperatures, and a specific experimental method was referred to the relevant requirements in JTG E20-2011. The dynamic stability test adopts two different health temperatures and test pressures, 60°C, 0.7 MPa, and 70°C, 1.0 MPa, respectively. The former simulates general temperature and traffic load conditions, while the latter simulates high temperature and overloading environment. The test results are shown in Table 3.

**Table 3.** Test results of stability and dynamic stability of different asphalt mixtures.

| Additive type | Marshall stability (kN) | Dynamic stability (Times/mm) | |
|---|---|---|---|
| | | 60°C, 0.7 MPa | 70°C, 1.0 MPa |
| No additive | 10.3 | 1153 | It cannot be tested |
| Recycled PE | 13.2 | 16235 | 4548 |
| POE/PA6 | 21.8 | 37820 | 17724 |

As seen from Table 3, compared with the matrix asphalt mixture, the stability and dynamic stability of the mixture after the addition of recycled PE have been greatly improved. The shear strength of the mixture has been significantly enhanced, while the addition of the POE/PA6 modifier has a more significant improvement in the stability and dynamic stability indicators. It is worth mentioning that the asphalt mixture with recycled PE shows excellent rutting resistance at 60°C and 0.7 MPa. Still, the performance decreases significantly at 70°C and 1.0 MPa. The matrix asphalt mixture cannot measure the test result because the result is too small, which is also the main reason it cannot meet the high temperature and heavy load environment. There are two possible reasons why the asphalt mixture with POE/PA6 modifier has excellent high-temperature rutting resistance. On the one hand, reactivity enhancement POE contains active groups such as epoxide, which can react chemically with the amine or imine group of the nylon molecular chain during the melt extrusion blending process. Thus, the interface binding force between the two phases is enhanced [8-9]. On the other hand, the epoxy group can also react with the hydroxyl and carboxyl groups in the asphalt and hydroxyl groups on the aggregate, resulting in the asphalt mixture forming a mixture with a certain spatial network structure. The possible chemical reaction is shown in Figure 2.

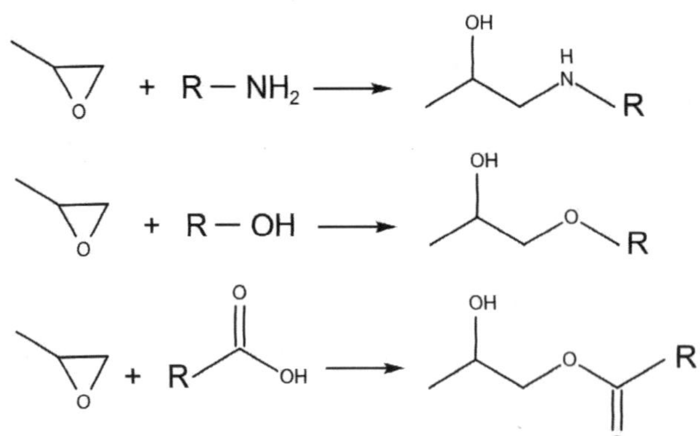


**Figure 2.** Schematic diagram of possible chemical reactions.

*3.2. Low-temperature anti-cracking*

The low-temperature anti-cracking of the mixture was evaluated by using a three-point beam test at -10°C and a semi-circular bending test at -10°C. The test methods were referred to the relevant provisions in JTG E20-2011 and AASHTO TP124-2016. The results of the three-point beam test are shown in Table 4.

**Table 4.** The results of the three-point beam test at -10°C.

| Additive type | Fracture deflection (mm) | Maximum flexural strain (με) |
|---|---|---|
| No additive | 0.389 | 2165 |
| Recycled PE | 0.361 | 1986 |
| POE/PA6 | 0.499 | 2676 |

It can be seen from the above table that the fracture deflection and the maximum flexural strain of the asphalt mixture show the same variation law. The greater the fracture deflection is, the greater the bending strain is. The PA6/POE modifier mixture showed the best fracture deflection and bending strain. In contrast, the mixture with recovered PE showed a decrease in fracture deflection and maximum flexural strain compared with the matrix asphalt mixture.

The semi-circular bending experimental data and curve of three different mixtures at -10°C are shown in Figure 3 and Table 5.

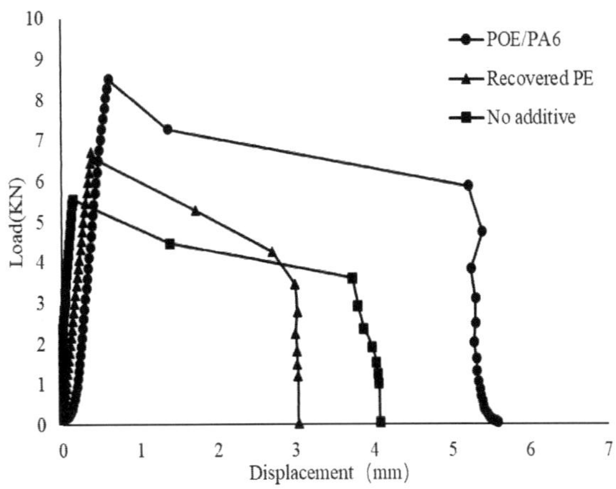


**Figure 3.** Semi-circle bending curve of mixture mixed with different additives.

**Table 5.** The results of the semi-circular bending test at -10°C.

| Additive type | Peak load (kN) | Peak displacement (mm) | Fracture energy (J/m$^2$) | Flexibility index |
|---|---|---|---|---|
| No additive | 5.54 | 0.148 | 5759 | 3.6 |
| Recycled PE | 6.70 | 0.383 | 5093 | 2.2 |
| PA6/POE | 8.51 | 0.610 | 11496 | 4.2 |

As can be seen from Figure 3 and Table 5, compared with the base asphalt mixture and the mixture mixed with recycled PE, the asphalt mixture mixed with the PA6/POE modifier shows excellent performance in terms of peak load, peak displacement, fracture energy, and flexibility index. Compared with the base asphalt mixture and the mixture mixed with recycled PE, the peak load of the asphalt mixture mixed with PA6/POE increased by 53.6% and 27%, respectively, because the addition of PA6/POE raised the flexural strength of the mixture. As several important indicators of fracture toughness, the increase of peak displacement, fracture energy, and flexibility index is more obvious, showing excellent low-temperature cracking resistance. It can be seen that POE with increased reactivity has a better toughening effect on PA6.

### *3.3. Water stability*

This paper used two indexes of soaking stability residual strength ratio and freeze-thaw splitting residual strength ratio of asphalt mixture to appraise the water stability of different mixture types. The former was obtained by placing asphalt mixture samples in a 60°C water tank for 48 h and testing the stability ratio before and after placement. The latter is obtained by adding a certain amount of water to the specimen, maintaining it at -18°C for 16 h, then immediately placing it in water at 60°C for 24 h, and testing the ratio of splitting strength. The specific test method is referred to the relevant provisions of JTG E20-2011, and the test results are shown in Figure 4.

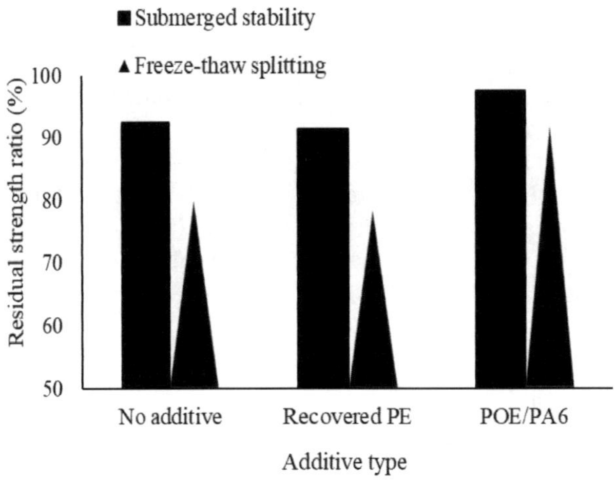


**Figure 4.** Water stability of asphalt mixtures with different additives.

The results in Figure 4 show that the residual stability of immersed water and the ratio of splitting strength of the three mixtures meet the technical requirement of the industry standard (no less than 80% and 75%, respectively) when compared with the first two mixtures, the water stability of the asphalt mixture mixed with PA6/POE modifier is significantly better, especially the ratio of splitting strength. The reason may be that a modifier of the reactive group and aggregate hydroxyl groups on the surface of the chemical reaction happens, greatly improving the interface adhesion of asphalt-aggregate and resistance to moisture damage to the asphalt mixture of the interface. Therefore, the mechanical strength of the asphalt mixture is not significantly reduced by water immersion, resulting in a greatly increased residual strength ratio.

## 4. Conclusion

(1) Due to the chemical reaction between the active group of POE and the amine or imide group of the PA6 molecular chain and the hydroxyl group on the surface of asphalt and aggregate, the shear strength of the mixture is greatly improved.

(2) The fracture energy and flexibility index of asphalt mixture mixed with POE/PA6 is significantly greater than that of base asphalt mixture and asphalt mixture mixed with recovered PE, and POE with increased reactivity has a better toughening effect on PA6.

(3) The active groups in the POE/PA6 modifier react chemically with the hydroxyl group on the aggregate surface, which greatly improves the interface bonding force of the asphalt aggregate, thus improving the water stability of the mixture.

(4) In summary, the addition of the POE/PA6 modifier to the asphalt mixture can effectively slow down the occurrence of ruts, cracks, pits, and other diseases on the asphalt pavement. It has a good engineering application prospect in the field of highway construction and maintenance under high temperatures and heavy load traffic environments.

Journal of Physics: Conference Series **2720** (2024) 012002 doi:10.1088/1742-6596/2720/1/012002

**References**

[1] Titin H. The frame work of statistical acceptance plan based on the permanent deformation potential of asphalt concrete [D]. Pennsylvania: the Pennsylvania State University, 2001.
[2] Ye X D, Huang W, Jia F, Gao R M, Qin Y and Chen C. Research progress on synthesis, modification technology and application of polyamide 6 [J]. Modern Chemical Industry, 2022, 42(10): 71–75.
[3] Yin X. Methods of toughening nylon and their actualities [J]. Plastics Additives, 2014(5): 8-10.
[4] Chen X H, Chen X T, Yin Y L, and Lu J. Preparation and flame-retardant properties of IFR/FeOOH/POE composites [J]. China Plastics Industry, 2013, 41(5): 73-77.
[5] Bazli L, Bagherian M H, Karrabi M, Abbassi-Sourki F, and Azizi H. Effect of starch ratio and compatibilization on the viscoelastic behavior of POE/starch blends [J]. Journal of Applied Polymer Science, 2020, 137(29): 48877.
[6] Li T, Ma L F, Bao R Y, Qi G Q, Yang W, Xie B H and Yang M B. A new approach to construct segregated structures in thermoplastic polyolefin elastomers towards improved conductive and mechanical properties [J]. Journal of Materials Chemistry A, 2015, 3(10): 5482-5490.
[7] Ibrahim B A, Kadum K M. Influence of polymer blending on mechanical and thermal properties [J]. Modern Applied Science, 2010, 4(9): 157-161.
[8] Wang J, Wan J, Wen S, Gong C L, Zheng G W and Liu H. Preparation of HDPE-g-GMA and application in improving the co MPatibility of HDPE/PA6 blends [J]. Modern Plastics Processing And Applications, 2018, 30(4): 38-41.
[9] Huang JJ, Keskkula H, Paul DR. Comparison of the toughening behavior of nylon 6 versus an amorphous polyamide using various maleated elastomers [J]. Polymer, 2006, 47(10): 639–651.

# Synthesis of g-$C_3N_4$/HC composite and its visible light catalytic performance

**Jinshan Liu[1], Yuwei Zhu[1], Yulan Ren[1,2],Heruo Li[1], Shuang Gao[1] and Jianfei Wu [1]**

[1]College of Chemistry and Chemical Engineering, Mudanjiang Normal University, Mudanjiang 157011 China

[2]Corresponding author's e-mail: rylmsy@163.com

**Abstract:** Thermal polymerization prepared g-$C_3N_4$/HC composite with good visible light catalytic ability using corn stover and urea as raw materials. The structure of the g-$C_3N_4$/HC composite was characterized by XRD, and it was determined that the composite was composed of g-$C_3N_4$ and HC. The morphology of the g-$C_3N_4$/HC composite was characterized by SEM. g-$C_3N_4$ is dispersed in sheet form on HC tubes with a regular porous structure. The degradation efficiency of RhB by g-$C_3N_4$/HC reached 99.14% after 70 min irradiation by a 70 W metal halide lamp. The cyclic degradation experiments proved that g-$C_3N_4$/HC has good reusability. Compared with g-$C_3N_4$, the visible light absorption capacity of g-$C_3N_4$/HC is enhanced, and the electron-hole recombination rate is also reduced, which results in the higher photocatalytic capacity of g-$C_3N_4$/HC. The free radical trapping experiments confirmed that the hole free radical and the hydroxyl free radical are the primary active substances in the photodegradation reaction.

## 1. Introduction

The discharge of various wastes causes the pollution of water bodies. A lot of research has been done to find solutions to pollution. In recent years, photocatalytic oxidation technology has attracted wide attention due to its features, such as effective utilization of sunlight, complete degradation, and no secondary pollution.

As a non-metallic semiconductor material, g-$C_3N_4$ has non-toxicity characteristics, good biocompatibility, good thermal stability, and cheap raw materials. g-$C_3N_4$ still has good stability when exposed to light in strong acid or base solutions due to the strong covalent bond between carbon and nitrogen atoms. However, g-$C_3N_4$ has some shortcomings, such as a small visible light response range and rapid recombination of photogenerated electrons with holes, which limit its application in the field of photocatalysis. Qi et al. [1] prepared g-$C_3N_4$/$Bi_2Fe_4O_9$ composite photocatalysts using a mechanical mixing method. After 240 min irradiation by 300 W xenon lamp, the degradation efficiency of RhB (10 mg/L) reached 87.59%. Xu et al. [2] prepared Ag/g-$C_3N_4$/$TiO_2$ ternary composite nanomaterials. After 110 min irradiation by 300 W xenon lamp, 10 mg/L RhB was almost completely degraded. Alsulmi et al. [3] constructed AgI/g-$C_3N_4$ heterojunctions containing g-$C_3N_4$ and silver iodide nanoparticles by acoustic chemistry, which showed 92% degradation of RhB after irradiation with a 300 W xenon lamp for 120 min.

Biochar takes agricultural waste as raw material. It has a low price, good pore structure, and rich functional groups. Liu et al. [4] studied the biochar modification by S, CuS, and MgO. CuS-Mg/S-BC was synthesized by the one-pot sulfurization method. After 120 min irradiation by a 150 W xenon lamp, the degradation efficiency of RhB reached 95.7%. However, the above research still has shortcomings, such as high energy consumption and long degradation time. In this paper, biochar was combined with g-$C_3N_4$ to enhance the photocatalytic degradation ability of the material.

## 2. Experiments

### *2.1. Preparation of biochar*
The corn stalk was dried naturally, and the powder was obtained after grinding and sifting. 0.5 g powder was placed in a beaker, mixed with 20 mL deionized water, and transferred to a 30 mL polytetrafluoroethylene lined high-pressure reactor, hydrothermal reaction at 180°C for 12 h. We cool and centrifuge to collect black solids. The product is washed, dried and ground into powder, recorded as HC.

### *2.2. Preparation of g-$C_3N_4$*
We weigh 10 g urea in a 30 ml crucible, put it into a Muffle furnace, heat it to 550°C at the speed of 5°C/min, hold it for 4 h, cool it to room temperature, and take it out. The product is recorded as g-$C_3N_4$.

### *2.3. Preparation of g-$C_3N_4$/HC*
A certain amount of biochar and 0.2 g g-$C_3N_4$ powder was evenly mixed in a crucible, calcined for 2 h at 550°C, cooled to room temperature, and collected to obtain HC/ g-$C_3N_4$.

### *2.4. Characterization*
The crystal structure of g-$C_3N_4$/HC was determined by XRD with Cu-kα ray. The wavelength was 0.154 nm, the scanning range of 2θ was 10 ~ 80°, the test voltage was 40 kV, and the step size was 0.02°. The morphology of g-$C_3N_4$/HC was observed by Hitachi 4800 SEM with an accelerated voltage of 20 kV. The light absorption ability of g-$C_3N_4$/HC was tested by UV-2600 UV-visible spectrophotometer (DRS) with an integrating sphere, and the scanning range was 220 ~ 850 nm. An F-7000 fluorescence spectrophotometer measured the fluorescence spectra (PL) of g-$C_3N_4$/HC.

### *2.5. Photocatalytic ability test*
The photocatalytic degradation ability of g-$C_3N_4$/HC to RhB solution with a concentration of 10 mg/L was tested using a metal halide lamp with a power of 70 W to simulate a visible light source. The procedure was as follows: In a clean beaker, we add 100 mL of RhB solution at 10 mg/L, then add 0.05 g g-$C_3N_4$/HC, and place the beaker on the magnetic stirrer. RhB reached adsorption-desorption equilibrium on the catalyst surface after magnetic stirring for 30 min under dark conditions. A 70 W metal halide lamp was turned on, and 4 mL liquid was taken into a centrifuge tube every 30 min during the illumination period and then placed into a high-speed centrifuge and centrifuged for 8 min. The absorbance of the supernatant at 554 nm was measured. The degradation efficiency of samples was calculated according to the formula [5].

## 3. Results and discussions

### *3.1. Composition and morphology analysis of samples*
The composition of g-$C_3N_4$/HC was analyzed by XRD. It can be seen from Figure 1 that the XRD curve of g-$C_3N_4$/HC has a characteristic wide peak at 22.38°, which belongs to biochar [6]. The diffraction peaks at 13.10° and 27.40° correspond to the (100) and (002) crystal faces of g-$C_3N_4$ (JCPDS 87-1526). XRD results showed that HC combined with g-$C_3N_4$ successfully.

The morphology of the sample was analyzed by SEM. Figure 2 (a) shows that biochar is a large tubular structure with many holes on its surface. Figure 2 (b) shows that g-$C_3N_4$ exists as granular aggregates. The SEM image of g-$C_3N_4$/HC (Figure 2 (c)) showed that there are many neatly arranged pores on the surface of the tubular biochar, and g-$C_3N_4$ is dispersed in sheets on the surface of the biochar, and no agglomeration phenomenon is observed.

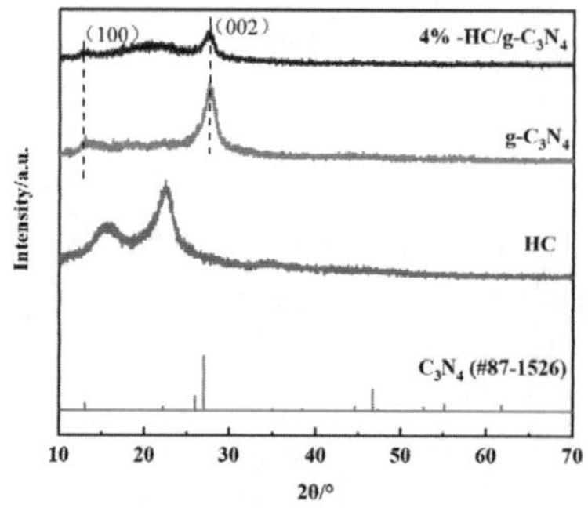


**Figure 1.** XRD pattern of the samples.

**Figure 2.** SEM images: HC (a), g-$C_3N_4$ (b) and g-$C_3N_4$/HC (c).

*3.2. Optical properties analysis of samples*

Figure 3 (a) shows the UV-VIS DRS spectra of the samples. The edge of the absorption band of pure phase g-$C_3N_4$ is about 456 nm, while the edge of the absorption band of the composite sample combined with biochar extends to 487 nm. It can be seen from the figure that the absorption intensity of the catalyst on the visible region is enhanced. The enhancement of visible light absorption capacity boosts the photodegradation ability of g-$C_3N_4$/HC.

Fluorescence spectra can illustrate the photogenerated electron and hole pair recombination rate of the catalyst. As can be seen from Figure 3 (b), the fluorescence peak intensity of g-$C_3N_4$/HC is significantly reduced, which is due to the recombination between HC and g-$C_3N_4$ so that the photogenerated electrons can be effectively transferred from g-$C_3N_4$ to HC, and the electron-hole separation rate is improved. At the same time, biochar can absorb part of the photoluminescence of g-$C_3N_4$ and generate the quenching effect. The increase in electron-hole separation rate further boosts the photocatalytic degradation ability of g-$C_3N_4$/HC.

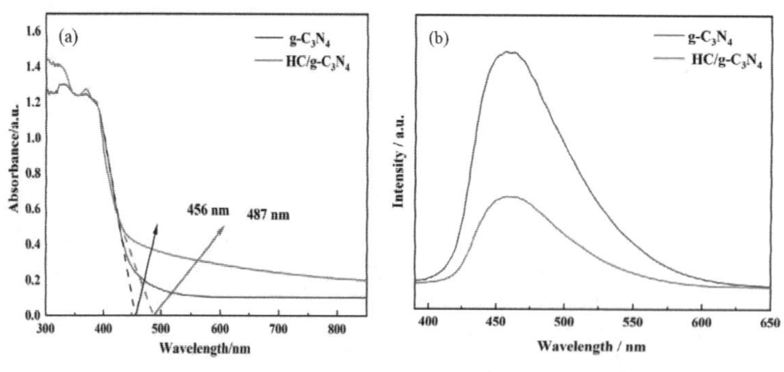


**Figure 3.** UV-VIS DRS spectra (a) and PL spectra (b).

### *3.3. Photocatalytic performance analysis of g-$C_3N_4$/HC*

Figure 4 (a) shows the effect of the amount of biochar on the photocatalytic degradation ability of composite samples. When the amount of biochar is 4%, the photocatalytic degradation ability of the composite is the best. After 70 minutes of exposure to a 70 W metal halide lamp, RhB was almost completely degraded. Figure 4 (b) shows the kinetic curve of the photocatalytic degradation process. The rate constant of RhB degradation by composite photocatalyst doped with 4% biochar is 10 times that of g-$C_3N_4$.

Figure 4 (c) shows that the degradation ability of g-$C_3N_4$/HC composite photocatalyst to RhB has almost no change after four cycles of experiments, indicating that g-$C_3N_4$/HC composite photocatalyst has good reusability. As shown in Figure 4 (d), the hole free radical and the hydroxyl free radical are the primary active substances in the photocatalytic reaction, $\cdot O_2^-$ plays a secondary role.

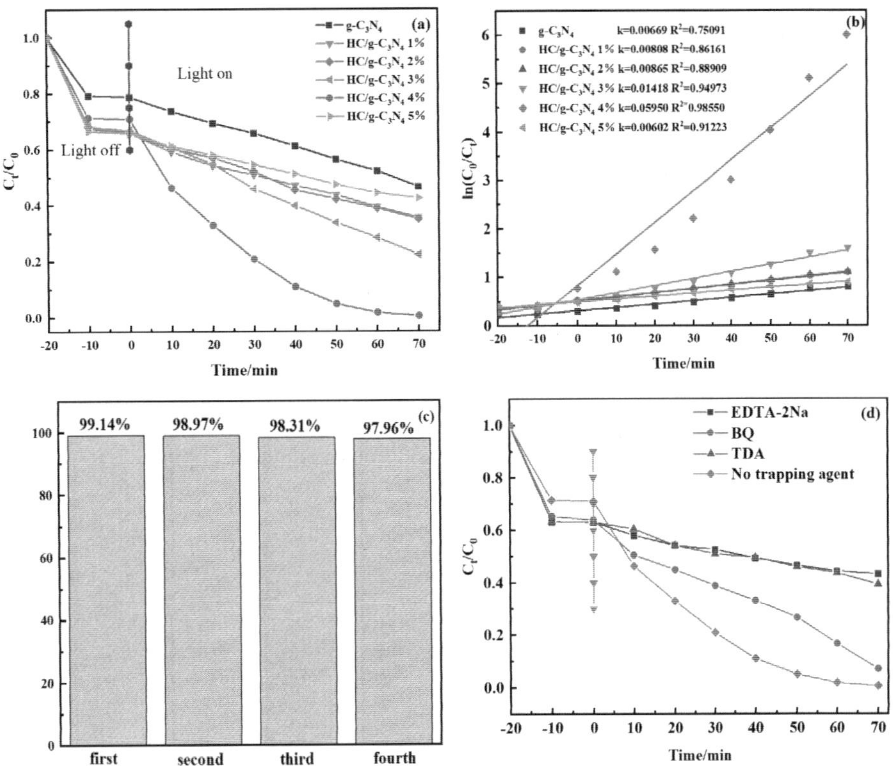


**Figure 4.** Photocatalytic degradation curve (a), kinetic curve (b), cyclic degradation experiment (c), and free radical capture experiment (d).

## 4. Conclusions

g-$C_3N_4$/HC photocatalyst was prepared by thermal polymerization. When the doping amount of biochar was 4%, the prepared g-$C_3N_4$/HC had good photocatalytic degradation performance for RhB. The simulated pollutants can be almost completely degraded after 70 minutes of exposure to 70 W metal halide lamps. Adding biochar enlarges the photoresponse range and increases the photogenerated carrier migration rate. Cyclic degradation experiments proved that the samples have good cyclic stability. The primary active substances in the photocatalytic reaction of the composite photocatalyst were $h^+$ and $\cdot OH$.

**Acknowledgments**
Thanks for the support of the College Student Innovation Project (No. 202210233006) and the Teaching Reform Project (No. SJGY20220617) of Heilongjiang Province (No. GP2022005).

**Reference**

[1] Qi S Y, Guan L. Zhang R Y, Wu S Q, and Zhang K Y. (2023) Study on the Photocatalytic Degradation of Rhodamine B by g-$C_3N_4$/$Bi_2Fe_4O_9$ Heterojunction Photocatalyst. *Journal of Inorganic and Organometallic Polymers and Materials*. 33: 3675-3683. https://doi.org/10.1007/s10904-023-02800-y.

[2] Xu J H and Ma S Y. (2023) High-efficiency photocatalytic degradation of RhB by protonation of g-$C_3N_4$ with Ag-loaded $TiO_2$ nanofibers. *Journal of Crystal Growth*. 617: 127290. https://doi.org/10.1016/j.jcrysgro.2023.127290.

[3] Alsulmi A, Alsulm I A, Alsulm M, Hesham M, Soltan A, Messih M F A and Ahmed M A. (2023) Sonochemical Fabrication of S-Scheme AgI/g-$C_3N_4$ Heterojunction for Efcient Photocatalytic Degradation of RhB Dye. Journal of Inorganic and Organometallic Polymers and Materials. *Journal of Inorganic and Organometallic Polymers and Materials*. 33(9). https://doi.org/10.1007/s10904-023-02854-y.

[4] Liu X K, Liu T T, Zhou Y L, Wen J, Dong J M, Chang C G and Zheng W X. (2023) Synergic effect of CuS and MgO for boosting adsorption-photocatalytic activity of S-doped biocha. *Journal of Physics and Chemistry of Solids*. 185: 111781. https://doi.org/10.1016/j.jpcs.2023.111781.

[5] Ren Y L, Li H R, Gao S, Zhu Y W, Liu X Y, and Wang X. (2023) Ag-doped $Ag_3PO_4$/$Bi_2SiO_5$ nanocomposite: controlled synthesis and good photocatalytic activity of a new ternary nanocomposite based on plasmon resonance effect. *Applied Physics A*. 129: 454. https://doi.org/10.1007/s00339-023-06714-x.

[6] Kumar P, Patel A K, Singhania R R, Chen C W, Saratale R G and Dong C D. (2023) Enhanced copper (II) bioremediation from wastewater using nano magnetite ($Fe_3O_4$) modified biochar of Ascophyllum nodosum. *Bioresource Technology*. 388: 129654. https://doi.org/10.1016/j.biortech.2023.129654.

# High-temperature stability, water stability, and ice melting properties of salt-storage bituminous mixture AC-13

**Huani Liu [1], Zhaoming Chen [1], Lidong Wu [1], Qingliang Wang [2], and Yanmin Wang [2,3]**

[1] Weihai Highway Survey & Design Institute Co., Ltd. Weihai, 264206 China
[2] School of Civil Engineering, Shandong Jiaotong University, Jinan, 250357 China

[3] Corresponding author's e-mail: ymwangsdu@139.com

**Abstract:** This study investigated the road performance and ice melting performance of the AC-13 bituminous mixture prepared with the equal volume replacement of mineral powder from a self-made salt-storage bitumen pavement additive. The results show that adding salt storage bitumen additive improves the bituminous mixture's high-temperature stability and reduces the bituminous mixture's water stability. When the replacement rates of the salt storage additive are 50%, 75%, and 100%, the Tensile Strength Ratio (TSR) decreases by 3.35%, 5.95%, and 8.11%, respectively. The dynamic stability (DS) also increases by 4.21%, 7.30%, and 5.62%, respectively. The melting ice rate and salt precipitation tests demonstrate that the salt storage bituminous mixture with a replacement rate of 100% achieves a 36% ice melting rate and 0.3% salt precipitation at -5°C, meeting the standard requirements.

## 1. Introduction

Salt-storage bituminous mixture is a mixture that uses salt-storage bitumen pavement additives to replace part or all the mineral filler in traditional bituminous mixtures. The salt-storage bituminous mixture has the function of melting ice and snow actively. There are essential differences between salt-storage additives and ordinary mineral fillers regarding chemical composition, density, etc. Correspondingly, salt-storage additives used in the salt-storage bituminous mixture, the road performance will also be different. Therefore, the variation of pavement performance of salt-storage bituminous mixture has always been the focus of research, and a series of achievements have been made.

Wu et al. [1] investigated the influence of MFL on bituminous mixtures, and the findings indicated that the inclusion of MFL leads to a decrease in the road performance of these mixtures., especially water stability. Sun [2] prepared a salt-storage bituminous mixture with a self-made slow-release salt-storage filler and found that the water damage resistance of the salt-storage bituminous mixture decreased. Liu [3] conducted an experiment where he substituted snow-melting and ice-suppressing materials for the filler in asphalt. The objective was to examine how this replacement affects the performance of bituminous mixtures under varying temperature conditions and their water stability and other road properties. Guo et al. [4] studied the effect of salt chemicals and high elastomer content on salt-treated bitumen wear layers. Various studies have shown that replacing mineral powder with environmentally friendly deicing materials can improve the high-temperature stability of salt-treated bituminous mixtures but adversely affects its water stability and low-temperature performance [5-8].

This study used an equal amount of self-made salt storage additive instead of mineral powder to prepare a bitumen salt storage mixture AC-13. Rutting tests evaluated the high-temperature stability of salt-treated bituminous mixtures. Marshall impregnation and freeze-thaw cycle gap tests were performed on bituminous mixtures. The melting ice rate and salt precipitation tests were conducted on the salt-storage bituminous mixture to directly evaluate the ice and snow melting effect of the salt-storage bituminous mixture.

## 2. Experiment

### *2.1. Materials*

SBS modified bitumen was used as the bitumen binder. Limestone mineral powder was used as filler. The salt-storage bitumen pavement additives were homemade. Their specific indicators are shown in Table 1.

Crushed limestone gravel was used as the coarse aggregate, and limestone chips were used as the fine aggregate in the bituminous mixture. All the aggregates were sieved to single particle size for later use.

**Table 1.** Indexes of salt-storage bitumen pavement additive.

| Items | Test value | Basic requirements |
|---|---|---|
| Density | 2.18 | ≥1.7 |
| Appearance | Earthy yellow, uniform color, no clumps | Uniform color, no clumps, and oil |
| Chloride content /% | 38.0 | ≥30 |
| Maximum particle size /mm | 0.15 | ≤0.3 |

### *2.2. Preparation of salt-storage bituminous mixture AC-13*

The mineral mixture used in the AC-13 bituminous mixture is continuously densely graded, and the gradation curve is shown in Figure 1. According to the Standard Test Method for Asphalt and bituminous mixtures for Road Construction (JTG E20-2011, China), the optimal asphalt aggregate content of AC-13 bituminous mixture was determined to be 5.0% through the Marshall test.

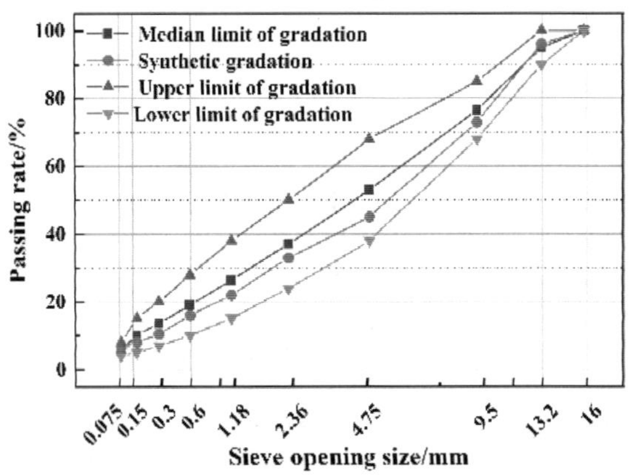


**Figure 1.** Grading curve of AC-13.

The salt-storage bituminous mixture was prepared by replacing the mineral powder with an equal volume of salt-storage additives (replacement rates are 50%, 75%, and 100%, respectively). The asphalt-aggregate ratio remained unchanged. The salt-storage additives must be mixed with the mineral powder and added to the mixing pot simultaneously.

### *2.3. Pavement performance tests of bituminous mixture AC-13*

According to the guidelines outlined in (JTG E20-2011, China), the laboratory performed rutting and bending tests on small beam specimens to assess their properties for highway engineering. To evaluate the water stability of the bituminous mixture, the immersion test and freeze-thaw splitting test were also carried out using the Marshall method.

*2.4. Test on thaw performance of salt storage bituminous mixture AC-13*

Using the standard reference "Deicing and thaw material employed in the bituminous mixture (JT/T 1210.2-2018, China)", we directly assess the ice melting effectiveness of the bituminous mixture stored with salt through experiments such as the ice melting rate test and salting out test. Meanwhile, the freezing point and conductivity tests indirectly indicate the ice melting and ice blocking capabilities of the bituminous mixture stored with salt.

**Figure 2.** Test process diagram of melting ice rate test.

The test process of the melting ice rate test is shown in Figure 2. To begin, we measured 40 mL of distilled water and placed it in a crystallization dish with a diameter of 100 mm. The dish was then placed in a refrigerator at -20°C to freeze the water into ice cubes for later use. The Marshall sample, measuring 101.6 × 63.5 mm in size, was allowed to cool to room temperature and placed inside an environmental box set to -5°C for 4 hours. Subsequently, the ice, known as the initial mass $m_0$, was removed from the freezer and promptly placed on top of the sample. After the sample was kept in an environmental chamber at -5°C for 2 hours, the ice was removed, and its remaining mass was determined as $m_2$. The percentage of the remaining mass of the ice to the initial mass is calculated according to Formula (1), that is, the ice melting rate.

$$\sigma = \frac{m_0 - m_2}{m_0} \times 100\% \quad (1)$$

Where σ is the ice melting rate (%). The initial mass of the ice is represented as $m_0$. The residual mass of the ice after being maintained at a temperature of -5°C for 2 hours is denoted as $m_2$.

Figure 3 illustrates the salt precipitation examination for bituminous mixtures, which aims to evaluate the salt discharge of the salt-storage bituminous mixture sample using the soaking technique. To conduct the test, the following procedures are undertaken: First, we take 1000 ml of distilled water, put it into a 2000 ml beaker, and keep the temperature at 23°C in a constant temperature water bath; Marshall test pieces of bituminous mixture after cooling and demoulded were slowly placed into beakers at room temperature for further heat preservation. The salinity concentration of the solution after soaking for 10 min and 24 h was measured with the salinity measurement module of the magnetic DDSJ-308F conductivity meter, and the salt precipitation of test pieces was calculated according to Formula (2).

$$C_S = C_{24} - C_0 \quad (2)$$

Where $C_S$ is the salt precipitation of test specimen (%); $C_{24}$ is the solution salt solubility of specimen at constant temperature for 24 h (%); $C_0$ is the solution salt solubility of specimens at constant temperature for 10 min (%).

**Figure 3.** Test process diagram of salt release test.

## 3. Results and discussion

Figure 4 shows the rutting test results of the salted bituminous mixture. Compared with ordinary AC-13 asphalt mixture, the dynamic stability (DS) of salt-storage asphalt mixture has somewhat improved. The DS of the bituminous mixture for salt storage had observed a rise of 4.21%, 7.30%, and 5.62% when the replacement rates were 50%, 75%, and 100%, respectively. This indicates that incorporating additives specifically for salt storage into the bitumen can improve the intercalation among aggregates, assuming the aggregate grading and oil-stone ratio remain constant, which is conducive to improving structural integrity. In addition, salt-storage asphalt pavement additives increase resistance to rut deformation at high temperatures and improve dynamic stability.

As can be observed from Figure 5(a), the immersion residual stability ($MS_0$) of salt-stored bituminous mixture specimens decreases with the increase of substitution rate. When the substitution rate is 100%, the $MS_0$ of the salted bituminous mixture drops from 90.3% of the ordinary AC-13 mixture to 86.05%. The $MS_0$ of the salted bituminous mixture samples with a 50% and 75% substitution rate decreased by 1.88% and 3.21%, respectively, but still met the specification requirements.

The freeze-thaw splitting test results of the salt-storage bituminous mixture shown in Figure 5(b) are similar to those of the immersion Marshall test. The strength ratio (TSR) of salt storage bituminous mixture decreases with the growth of the substitution ratio. The TSR of salt-preserving bituminous mixtures with replacement rates of 50%, 75%, and 100% was reduced by 3.35%, 5.95%, and 8.11%, respectively, and still met the lower limit requirements of the specification.

The above data analysis shows that the incorporation of salt-based bitumen pavement additives hinders the water stability of the mixture. Cracks can easily occur between aggregate and bitumen mortar in a freeze-thaw environment. After water intrusion, NaCl solution is formed with snow melt salt (NaCl) in salt storage bitumen pavement additive. Compared with water, NaCl solution is more likely to infiltrate the aggregate, so it is also more likely to penetrate the interface between bitumen and aggregate. This will reduce the bitumen aggregate adhesion and accelerate the peeling of the bitumen film on the aggregate surface [4, 8]. Additionally, submerging the salt-storage asphalt mix in water releases snowmelt salts. There are still gaps in the mix. After water enters the gap, it freezes and expands at low temperatures, accelerating water loss [9-13].

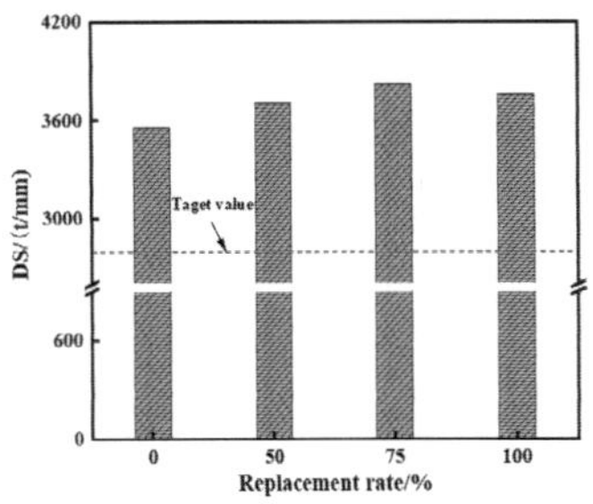


**Figure 4.** Results of the rutting test.

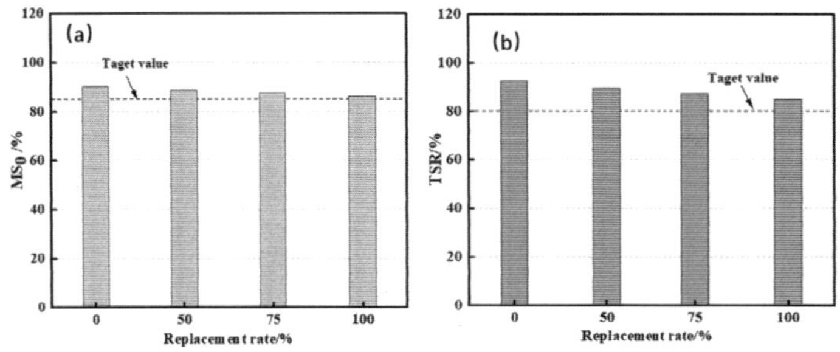


**Figure 5.** Results of the (a) Marshall immersion test (b) and the Freeze-thaw split test.

**Table 2.** Results of the melting ice rate of salt-storage bituminous mixture.

| Specimen number | $m_0$ (g) | $m_2$ (g) | σ (%) | Average value of σ (%) |
|---|---|---|---|---|
| 1 | 40.2 | 25.6 | 36.3 | 36 |
| 2 | 40.1 | 25.5 | 36.4 | |

**Table 3.** Results of the salt precipitation of salt-storage bituminous mixture.

| Specimen number | Test time | Saline concentration (%) | | | | $C_S$ (%) | |
|---|---|---|---|---|---|---|---|
| | | 1 | 2 | 3 | Average value | $C_S$ | The average value of $C_S$ |
| 1 | 10 min | 0.02 | 0.03 | 0.03 | 0.03 | 0.31 | 0.3 |
| | 24 h | 0.33 | 0.34 | 0.34 | 0.34 | | |
| 2 | 10 min | 0.03 | 0.02 | 0.02 | 0.02 | 0.32 | |
| | 24 h | 0.33 | 0.34 | 0.34 | 0.34 | | |

The melting rate test results of the salt storage bituminous mixture are shown in Table 2. The melting rate (σ) of the salt-storage bituminous mixture is calculated to be 36%, which is not less than 20% of the salt chemical materials in the JT/T 1210.2-2018 standard. This shows that the salt storage bituminous mixture with an exchange rate of 100% has good ice melting performance at -5°C.

The results of the salt precipitation test of the salt-storage bituminous mixture are shown in Table 3. It is calculated that the salt precipitation amount of the salt-storage bituminous mixture is 0.3%. According to the standard, the fluxing rate of salt chemical raw materials cannot be greater than 0.4%. The results show that within 48 hours, the snow melt salt precipitation in the bituminous mixture is very small. The short-term precipitation of snow melt salt can meet the requirements of melting snow in the early stage of winter snowfall and ensure that the long-term effect of snow melt and ice suppression in the later stage is not affected.

## 4. Conclusions

In this study, the salt-storage bituminous mixture AC-13 was prepared with the equal volume replacement of mineral powder from a self-made salt-storage bitumen pavement additive. We will explore the performance of salt-storage bituminous mixtures in terms of road performance and ice-melting capabilities. The key findings obtained are outlined below:

(1) Adding salt-storage bitumen pavement additive increases the dynamic stability and the resistance to rutting deformation. When the replacement rate is 75%, the ordinary AC-13 bituminous mixture exhibits a dynamic stability that is 7.30% lower compared to the salt-storage bituminous mixture.

(2) The addition of salt-storage bitumen pavement additives is not conducive to the water stability of the bituminous mixture. The $MS_0$ of immersion water and the TSR of the freeze-thaw splitting test decrease with the increase of the replacement rate.

(3) The melting performance of a salt-storage bituminous mixture with a replacement rate of 100% was directly evaluated by the melting ice rate test and salt precipitation test. The results show that σ is 36% and $C_S$ is 0.3% at -5°C, indicating that the ice melting effect of the bituminous mixture with salt storage meets the standard requirements and is highly efficient.

**References**

[1] WU S, YANG J, SUN X, et al. (2020) Preparation and characterization of anti-freezing asphalt pavement [J]. Construction and Building Materials, 236: 117579.
[2] SUN R. R., (2012) Research and development of slow-release salt-storage asphalt mixture and evaluation of its performance [D]. Harbin Institute of Technology.
[3] LIU Z., (2013) R&D and application of snow melting and ice suppression materials for asphalt pavement [D]. Chang'an University.
[4] Guo P., Feng Y. X., Meng X. C., et al., (2020) Microscopic Analysis of Salt Storage Snow Melting Deicing Agent and its Effect on Stability of Mixed Material [J]. Material Reports, 34(6): 62-65.
[5] LIU J., (2021) Study on ice melting performance of salinized asphalt mixture [D]. Chongqing Jiaotong University.
[6] LING L.S. (2022) Preparation of environmentally friendly de-icing filler and its properties of asphalt mortar and mixture [D]. Guangxi University. DOI: 10.27034/d.cnki.ggxiu.2022.000524.
[7] LIU Zhuang, SHA Aimin, JIANG Wei. Research Progress on Salt Storage Asphalt Pavement: Salinity Materials, Mixtures, Properties and Evaluation [J]. China Journal of Highway and Transport, 2019, 32(4): 18-31, 72.
[8] XIE X.B., (2017) Road performance and mesomechanical analysis of asphalt mixture under ultraviolet aging based on mortar characteristics [D]. Xi'an University of Architecture and Technology.
[9] YANG R. (2016) Study on water stability of salt-storage snowmelt and ice-blocking asphalt mixture [J]. Highway Transportation Science and Technology (Applied Technology Edition), 12(9): 68-70.
[10] WU S, YANG J, YANG R, et al. (2018) Investigation of microscopic air void structure of anti-freezing asphalt pavement with X-ray CT and MIP [J]. Construction and Building Materials, 178: 473-483. https://doi.org/10.1016/j.conbuildmat.2018.05.185.
[11] LUIS J. -G., ÁNGEL V.-Z., MIGUEL Á. C.-P., (2019) Sodium chloride effect in the mechanical properties of the bituminous mixtures [J]. Cold Regions Science and Technology,164, 102776. https://doi.org/10.1016/j.coldregions.2019.05.002.
[12] FELICE G., FILIPPO M., GIONANNI P., et al. (2012) Effectiveness of sodium chloride-based anti-icing filler in asphalt mixtures, Construction and Building Materials, 30: 174-179. https://doi.org/10.1016/j.conbuildmat.2011.12.036.
[13] XU, O., YANG X., XIANG S. et al. (2021) Migration characteristic and model of chloride ions for NaCl-based salt storage asphalt mixtures, Construction and Building Materials, 280: 122482. https://doi.org/10.1016/j.conbuildmat.2021.122482.

# Design of superhydrophobic coatings fabricated by spraying for anti-icing

**Lei Fan [1], Jinhang He [1], Bo Li [1], Jie Bai [1], Tao Zhu [2], Linbo Song [2], Xujiang Hua [2] and Yuan Yuan [2, 3]**

[1] Electric Power Research Institute of Guizhou Power Grid Co., Ltd., Guiyang 550002, China
[2] College of Materials Science and Engineering, Chongqing University, Chongqing 400044, China

[3]Corresponding author's E-mail: yuany@cqu.edu.cn.

**Abstract:** The icing of transmission lines has a serious impact on people's lives and disrupts the secure and steady functioning of the power grid, causing huge economic losses. To retard the icing of transmission line glass insulators, we prepared coatings on a glass slide by spraying using epoxy resin, fluorosilicone resin, and hydrophobic silicon dioxide ($SiO_2$). The study examined the microscopical morphology, wetting behaviour, and anti-icing and self-cleaning characteristics of coatings containing various $SiO_2$ mass fractions. The findings indicated that the $SiO_2$ mass fraction notably impacted the micro-nanostructure, anti-wettability, and anti-icing performance of coatings. In addition, the largest contact angle (168.2°), the smallest sliding angle (2.6°), and the longest freezing time (181.7 s) were measured for the superhydrophobic coating with $SiO_2$ mass fraction of 34%, which was attributed to the most uniform microstructure. The superhydrophobic coatings fabricated through spraying exhibited good anti-icing and self-cleaning properties. This facilitates the anti-icing and anti-fouling of transmission line glass insulators.

## 1. Introduction

Electric transmission lines running through high-altitude mountainous areas are especially prone to ice covering. The icing of the power grid causes power outages and serious accidents such as insulator flashover, line breakage, and tower collapse. This greatly affects the safe operation of grid transmission and distribution [1]. Therefore, numerous active and passive anti-icing techniques are proposed to solve the icing problem of transmission lines. Active anti-icing techniques include high-current ice melting, mechanical deicing, and manual deicing. Although the de-icing effect of these methods is significant, the de-icing process often consumes a lot of manpower, material, and financial resources. Taking advantage of the self-cleaning effect of lotus leaves, the coating technique is employed as a powerful passively anti-icing method. The superhydrophobic coatings have excellent water-repellent properties and exhibit good anti-icing performance, attracting widespread attention [2, 3].

Superhydrophobic coatings are often prepared by building micro-nanostructures on the material surface and treating it with low-surface-energy substances. Currently, the widely used methods for superhydrophobic surface preparation include chemical vapor deposition, etching, self-assembly, and magnetron sputtering. However, these methods are limited to preparing small flat coatings in the laboratory, making it difficult to satisfy the application of large-area coatings for complex structures in industrial environments. On the contrary, spraying is a convenient and economical method to prepare large-area superhydrophobic coatings on arbitrarily shaped substrates (such as transmission line glass insulators). Qin et al. [2] utilized a step-by-step spraying method to apply PTFE/epoxy superhydrophobic coatings onto the surface of aluminum plates. The experimental findings revealed that the superhydrophobic coatings prepared at low cost could exhibit extremely small ice adhesion strength (28 kPa). Pan et al. [3] combined spraying with hard mold plate hot pressing to prepare a carbon fiber/epoxy composite superhydrophobic coating, which exhibited a long ice-delay time and a bond strength of 50 kPa to ice. By one-step spraying, Zhang et al. [4] fabricated a multi-walled carbon nanotube superhydrophobic coating on the Q235 steel surface. It was found that the coating containing a 30% MWCNT mass fraction showed the most effective delayed icing properties. It is worth noting that the spraying methods currently used to prepare superhydrophobic coatings have problems with complex solution compositions, expensive solutes [4], and cumbersome processes [2, 3]. These limitations may hinder the potential use in the industry. Additionally, producing superhydrophobic anti-icing coatings using one-step spraying must be studied as a potential application for transmission line glass insulators.

In the present work, we fabricated an anti-icing coating using a simple spraying. Common epoxy resins, fluorosilicone resins, and hydrophobic silicon dioxide ($SiO_2$) were selected to fabricate coatings containing various $SiO_2$ mass fractions on the surface of glass slides using one-step spraying. This study analyzed the impact of $SiO_2$ content on the micro-nanostructure, anti-wetting, and anti-icing characteristics of coatings by utilizing scanning electron microscopy, a water contact angle measuring instrument, and a semiconductor cooling platform. The relationships between microstructure, hydrophobicity, and anti-icing properties were examined.

## 2. Experiment

### *2.1. Materials*

The modified bisphenol A epoxy resin (GCC135) was purchased from Shanghai Kunshan Lvxun Chemicals Industry Co., Ltd. The Fluorosilicone resin was supplied by Chengdu Aikeda Chemical Reagent Co., Ltd. Compared with other ceramics, $SiO_2$ nanoparticles have the advantages of low cost, easy surface modification and controllability, facilitating the preparation of superhydrophobic coatings. As a result, $SiO_2$ nanoparticles were chosen to prepare superhydrophobic anti-icing coatings. Shanghai Macklin Biochemical Co., Ltd produced the $SiO_2$ nanoparticles. The average size and specific surface area were 30 nm and 120 $m^2/g$, respectively. Absolute alcohol was supplied by Chongqing Chuandong Chemical (Group) Co., Ltd. Ethyl acetate was provided by Chengdu Kelong Chemical Co., Ltd. The glass slides (7101) were obtained from Jiangsu Feizhou Glass Plastic Co., Ltd.

### *2.2. Fabrication*

The masses of $SiO_2$ (0.5 g, 0.8 g, 1.1 g, 1.4 g, 1.7 g, and 2.0 g) were weighed with an electronic balance to prepare coatings with different $SiO_2$ mass fractions. Among them, the mass of epoxy resin was 2 g, the mass of fluorosilicone resin was 3 g, and the mass of ethyl acetate was 20 g. The epoxy resin to curing agent mass ratio was 10:3. The epoxy resin, fluorosilicone resin, $SiO_2$, and ethyl acetate were mixed in a container and agitated using magnetic stirring for 10 min, and then ultrasonically dispersed by an ultrasonic cleaner for 5 min. After ultrasonic dispersion, the curing agents were applied and agitated for 5 min to provide a uniform solution. The glass slides were washed by ultrasound for 5 min in absolute ethanol and deionized water before spraying. The mixing solution was transferred to a spray bottle. After that, the spray pistol was turned toward the glass slide for spraying. Subsequently, the bottom-up spraying was performed in an even manner. The nozzle of the spray gun had a diameter of 1.5 mm. The pressure and distance of spraying were 0.4 MPa and 25 cm, respectively. After spraying, the coating was left at room temperature for 30 min, immediately followed by overnight curing in an oven.

### *2.3. Characterization*

The micro-nanostructure of the coating was analyzed by scanning electron microscopy (SEM). The contact angle and sliding angle of coating at room temperature were obtained using a water contact angle meter. Approximately 6 μL of water droplets were utilized for testing. Five positions were chosen for the measurements. The average was calculated as the final test result. A homemade water droplet freezing instrument determined the freezing times of water droplets on the coating surface. Details of the testing process can be found in the published report [5]. Furthermore, self-cleaning tests were performed with Congo red simulated dust.

## 3. Results and discussion

### *3.1. Microscopic morphology*

The coatings with different $SiO_2$ mass fractions were fabricated on the surface of the glass slides by one-step spraying. SEM was utilized to investigate the microstructure of coatings containing various $SiO_2$ mass fractions, as demonstrated in Figure 1. It is obvious that the $SiO_2$ mass fraction significantly affects the micro-nanostructure on the coating surface. At a $SiO_2$ content of 10%, the coating surface is observed to be smoother, with a small amount of scattered particles and protrusions (Figure 1(a)). This is mainly owing to the low amount of $SiO_2$ addition. The $SiO_2$ nanoparticles are almost completely encapsulated by epoxy resin and fluorosilicone resin. With increasing $SiO_2$ content, the surface of coatings progressively turns rougher. Many micron-sized particles of varying sizes are observed in the low magnification images, and numerous granular nano-$SiO_2$ are found on the surface of epoxy and fluorosilicone resins in the high magnification images (Figures 1(b), (c), (d), (e), and (f)). In addition, it is noted that the coating surface has more large-sized pits and holes when the $SiO_2$ mass fraction is 16%, 22%, 28%, and 40% (low magnification images in Figures 1(b), (c), (d), and (f)). In contrast, the coating with a $SiO_2$ mass fraction of 34% has a flatter microstructure (Figure 1(e)). The surface is seen to be covered with a uniform micron-sized structure (peppered with nano-$SiO_2$).

### *3.2. Anti-wetting characteristics*

The contact angle is commonly employed to assess the hydrophobic properties of material surfaces, and the sliding angle is used to evaluate the rolling ability of water droplets. We determined the contact and sliding angles of glass and coatings, as shown in Figure 2. It can be found that the contact angle of coatings gradually improves with increasing the $SiO_2$ content (Figure 2(a)). As the $SiO_2$ content is raised to 22%, the contact angle of coatings is higher than 150°. Meanwhile, the water droplets present a rounded appearance on the surface. Compared to glass (69.2°), the coating with a $SiO_2$ content of 34% shows a contact angle as high as 168.2°. In contrast, the sliding angle of the coating exhibits a decreasing and then increasing trend as the $SiO_2$ content increases (Figure 2(b)). Notably, the glass and coatings containing $SiO_2$ contents of 10% and 16% have large sliding angles, and water droplets are easily adsorbed on the surface. However, the sliding angles of coatings with $SiO_2$ mass fractions of 34% and 40% are as low as 2.6° and 4.8°, respectively. At this point, the water droplet can roll away from the surface. It is well known that the contact angle of superhydrophobic coatings should be over 150°, and their sliding angle should be below 10°. This indicates that coatings with $SiO_2$ mass fractions of 34% and 40% have superhydrophobic surfaces and exhibit good superhydrophobic properties. On the other hand, it should be mentioned that the addition of excess $SiO_2$ results in the formation of large craters on the surface of the coating containing 40% $SiO_2$ (Figure 1(f)). This leads to a reduction in contact angle and an expansion in sliding angle.

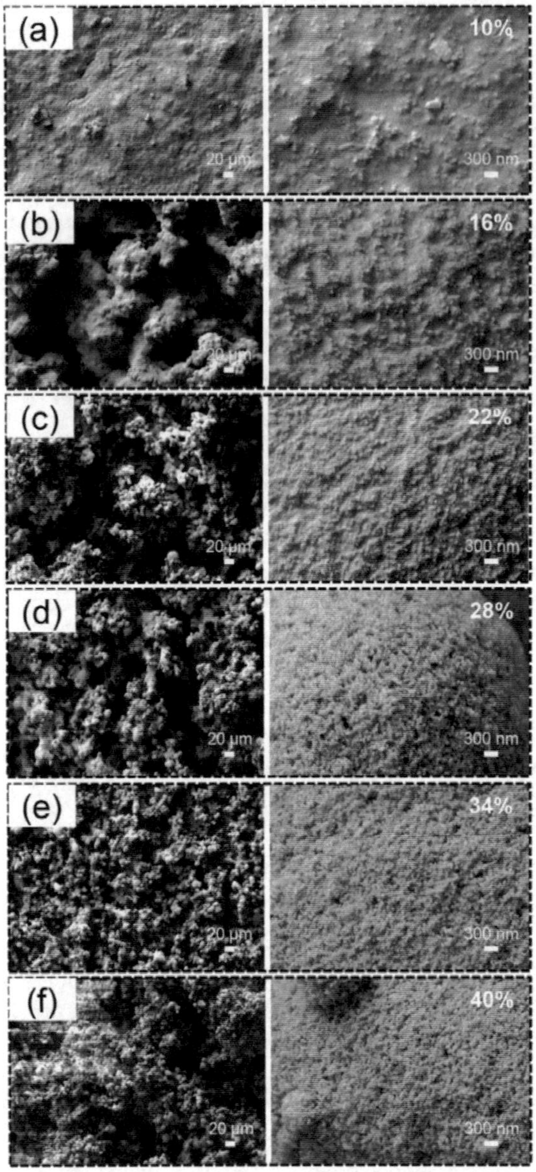


**Figure 1.** Micro-nanostructure of coatings containing various $SiO_2$ mass fractions.

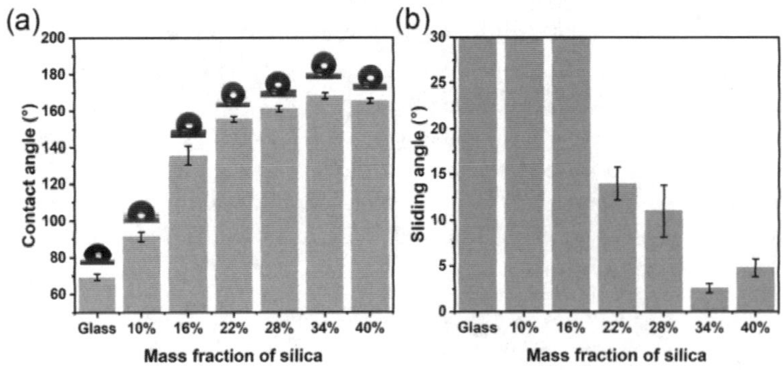


**Figure 2.** Wetting properties of glass and coatings.

*3.3. Anti-icing properties*

A semiconductor cooling platform performed the water droplet freezing tests to evaluate the impact of $SiO_2$ content on the anti-icing characteristics of coatings. The freezing processes of water droplets on glass and coatings containing various $SiO_2$ contents are shown in Figure 3. It is found that water droplet starts to nucleate from the bottom, and the ice crystals gradually expand to the whole droplet with time, which is a typical heterogeneous nucleation mode [1]. In addition, it can be noted that the water droplet lies flat on glass and coatings containing lower $SiO_2$ mass fractions (10% and 16%). The water droplets are completely frozen quickly, and the contact angle of a frozen water droplets (flat laying state) is small. With increasing $SiO_2$ content, the hydrophobic properties of the coating surface progressively improve. This leads to a longer water droplet freezing time, and the frozen water droplet (sphere-like state) still has a large contact angle.

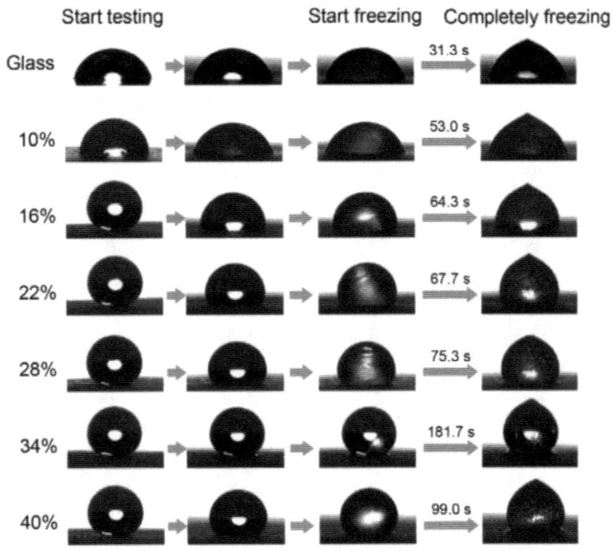


**Figure 3.** Freezing morphology of water droplets on glass and coatings.

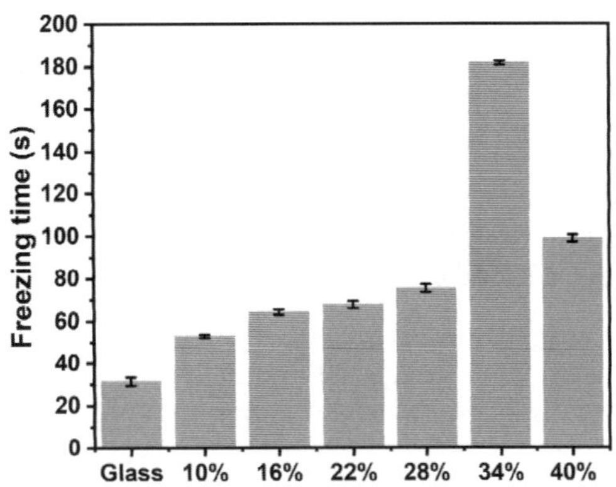


**Figure 4.** Freezing time of glass and coatings.

The freezing times of glass and coatings were extracted, as shown in Figure 4. Compared to the lowest freezing time (31.3 s) of glass, the freezing times of coatings are prolonged. Furthermore, the freezing time of coatings rises and falls as the $SiO_2$ mass fraction gradually increases. The superhydrophobic coating with a $SiO_2$ mass fraction of 34% shows the longest freezing time (181.7 s), approximately 6 times more than glass. We notice a strong correlation between the micro-nanostructure of coatings and the hydrophobic and anti-icing properties. As the $SiO_2$ content rises, the structure on the coating surface progressively gets rough (Figure 1). The hydrophobicity (Figure 2) and anti-icing properties (Figure 4) also gradually improve with the $SiO_2$ mass fraction. However, defects such as pits and holes in the rough structure (Figure 1(b), (c), (d), and (f)) cause a decrease in the hydrophobic and anti-icing characteristics of coatings. Therefore, the coating with a $SiO_2$ mass fraction of 34% displays the smoothest structure (Figure 1(e)), which leads to the best water repellences and anti-icing properties (freezing time of 181.7 s).

### *3.4. Self-cleaning properties*

The self-cleaning properties of glass and coatings were tested using Congo red powder to simulate pollutants, as shown in Figure 5. Obviously, for glass and coatings with a $SiO_2$ mass fraction of 10%, water drops from the dropper dissolve and accumulate with Congo red powder and cannot flow (Figures 5(a) and (b)). The primary reason is that the surface does not have a significant rough structure (Figure 1) and exhibits very poor hydrophobic properties (Figure 2). In contrast, the coating containing a 34% $SiO_2$ mass fraction exhibits a low sliding angle (Figure 2(b)) and good self-cleaning properties. This causes water droplets entrained with Congo red powder to run off the surface, resulting in a clean surface (Figure 5(c)).

**Figure 5.** Self-cleaning properties of glass and coatings.

Many studies have demonstrated that the rough structure on superhydrophobic surfaces contains numerous air cushions [1, 2, 4, 6, 7]. It dramatically minimizes the contact area, providing stability to the water droplet and creating a Cassie condition with exceptional hydrophobic properties [1]. Calculating the proportion of solid-liquid contact area of coatings can be performed using the Cassie model, assuming the intrinsic contact angle of coatings to be 108.2°. Among them, the solid-liquid contact area fraction for coating with a $SiO_2$ mass fraction of 34% is only 5.02%. This suggests that the superhydrophobic coatings fabricated in this study show outstanding hydrophobic properties. In this case, the water droplet on a superhydrophobic surface has a higher center of gravity, appearing globular to roll off the surface easily, thus avoiding the wetting of the coating surface and minimizing the likelihood of icing. Moreover, air is known to be a terrible heat conductor. Additionally, the very low solid-liquid contact area and substances with low surface energy on the surface of superhydrophobic coatings greatly reduce the heat transfer rate and considerably delay freezing. In summary, the coating containing 34% $SiO_2$ mass fraction not only exhibits the best hydrophobicity (Figure 2) and anti-icing properties (Figure 4) but also shows good self-cleaning properties (Figure 5), which facilitates the anti-icing and anti-fouling of transmission line insulators.

## 4. Conclusion

The impact of added $SiO_2$ content on the micro-nanostructure, anti-wetting, anti-icing, and self-cleaning characteristics of coatings fabricated by one-step spraying was studied. The uniformity of the rough structure strongly influences the anti-wettability and anti-icing characteristics of coatings. With the increase in $SiO_2$ content, the water-repellency and anti-icing characteristics of coatings tend to increase and decrease. The coating containing 34% $SiO_2$ mass fraction provides the smoothest micro-nanostructure so that the water droplet on the surface can maintain a stable Cassie state. It also exhibits good wetting resistance (contact and sliding angles of 168.2° and 2.6°), anti-icing (freezing time as long as 181.7 s), and self-cleaning performance. This provides some experimental references for the anti-icing and anti-fouling transmission line glass insulators.

## Acknowledgments

This work was supported by the Electric Power Research Institute of Guizhou Power Grid Co., Ltd., China (Contract No. 0666002022030103HX00003).

## References

[1] Olad, A., Maryami, F., Mirmohseni, A., Shayegani-Akmal, A.A. (2021) Potential of slippery liquid infused porous surface coatings as flashover inhibitors on porcelain insulators in icing, contaminated, and harsh environments. Prog Org Coat, 151: 106082. https://doi.org/10.1016/j.porgcoat.2020.106082.

[2] Qin, C., Mulroney, A.T., Gupta, M.C. (2020) Anti-icing epoxy resin surface modified by spray coating of PTFE Teflon particles for wind turbine blades. Mater Today Commun, 22: 100770. https://doi.org/10.1016/j.mtcomm.2019.100770.

[3] Pan, L., Xue, P.B., Wang, M.L., Wang, F., Guo, H.X., Yuan, X.S., Zhong, L., Yu, J. (2021) Novel superhydrophobic carbon fiber/epoxy composites with anti-icing properties. J Mater Res, 36: 1695–1704. https://doi.org/10.1557/s43578-021-00213-6.

[4] Zhang, F., Qian, H.C., Wang, L.T., Wang, Z., Du, C.W., Li, X.G., Zhang, D.W. (2018) Superhydrophobic carbon nanotubes/epoxy nanocomposite coating by facile one-step spraying. Surf Coat Tech, 341: 15–23. https://doi.org/10.1016/j.surfcoat.2018.01.045.

[5] Fan, L., Li, B., Wang, Y., He, J.H., Bai, J., Zhu, T., Yuan, Y. (2023) Superhydrophobic epoxy/fluorosilicone/PTFE coatings prepared by one-step spraying for enhanced anti-icing performance. Coatings, 13: 569. https://doi.org/10.3390/coatings13030569.

[6] Gaddam, A., Sharma, H., Karkantonis, T., Dimov, S. (2021) Anti-icing properties of femtosecond laser-induced nano and multiscale topographies. Appl Surf Sci, 552: 149443. https://doi.org/10.1016/j.apsusc.2021.149443.

[7] Mohammadian, B., Annavarapu, R.K., Raiyan, A., Nemani, S.K., Kim, S., Wang, M.H., Sojoudi, H. (2020) Delayed frost growth on nanoporous microstructured surfaces utilizing jumping and sweeping condensates. Langmuir, 36: 6635–6650. https://doi.org/10.1021/acs.langmuir.0c00413.

# Anti-icing application of superhydrophobic coating on glass insulator

**Lei Fan [1], Jinhang He [1], Bo Li [1], Mingyong Xia [2], Jian Liu [2], Tao Zhu [3], Yingying Zhao [3], Xujiang Hua [3], Yuan Yuan [3,4]**

[1]Electric Power Research Institute of Guizhou Power Grid Co., Ltd., Guiyang 550002, China
[2]Guizhou Chuangxing Electric Power Research Institute Co., Ltd., Guiyang 550081, China
[3]College of Materials Science and Engineering, Chongqing University, Chongqing 400044, China

[4]Corresponding author's e-mail: yuany@cqu.edu.cn.

**Abstract.** Insulators, as important components in transmission lines, are prone to disturb the safe running of the power system by the ice accumulation on surfaces. Traditional anti-icing coatings are difficult to practically apply to insulators. Here, superhydrophobic (SHP) coatings were fabricated on the glass insulator surface by spraying. We studied microstructure, wettability, water droplet bouncing behavior, and anti-glaze icing properties of SHP coating. The results demonstrated that SHP coatings had micro-nano rough structures. The excellent superhydrophobicity (contact angle of 165.2° and sliding angle of 3.7°) was achieved. The water droplet was easily adhered to the surface of the glass insulators. At the same time, individual water droplets could bounce away from the surface after impacting the SHP coating. In the glaze environment, water droplets sprayed onto the SHP coating merged with each other and slid off the surface. These significantly reduce the likelihood of freezing. Furthermore, the SHP coating could dramatically delay the glaze icing and decrease the icing area. The icing weight and icicle length were smaller than glass insulators. The SHP coatings prepared in this work display great potential for the anti-icing of glass insulators.

## 1. Introduction

Icing can easily cause accidents such as insulator flashover and even result in the loss of electrical insulation properties. These incidents negatively impact the stable functioning of power systems. A large number of scholars have carried out a lot of work for this purpose. Active deicing techniques such as mechanical deicing, current melting, and artificial deicing are used to solve the problem of ice accumulation on transmission lines. However, it should be pointed out that the active deicing technology has the defects of low efficiency and huge consumption, and it is difficult to realize the application on insulators [1]. In addition, the widely used water-repellent coatings (RTV and PRTV) have some antifouling effect, but their surfaces still suffer from the problem of ice cover.

In recent years, superhydrophobic (SHP) coating on the basis of the lotus leaf effect has shown outstanding anti-icing properties, becoming a hot research topic in the field [2]. In brief, by creating rough structures and modifying them with low-surface-energy substances, SHP coatings can be fabricated. Liao et al. [3] prepared ZnO/$SiO_2$/PTFE SHP coatings by RF magnetron sputtering. The contact angle reached a maximum of 167.2°, while the sliding angle was less than 1°. Within 120 min, only 17.9 % of the area (only the edge) of the film surface was frozen. Cao et al. [4] prepared metal-based SHP coatings by template and chemical vapor deposition. In an indoor environment with low temperature and high humidity, SHP coating displayed excellent anti-icing properties. The freezing time was significantly increased (about 300 s). Moreover, the ice adhesion strength (~14.8 kPa) decreased by 90% with respect to the substrate. Wan et al. [5] employed nanosecond laser processing to construct SHP surfaces on aluminum alloys. In the −15°C temperature environment, the SHP surface exhibited low adhesion (sliding angle below 6.3°). This essentially prevented ice formation, allowing water droplets to slide off the surface effortlessly. The delay time for icing measurement exceeded 700 s, causing a noteworthy decline in the icing amount. A durable SHP aluminum surface was fabricated through anodizing by Liu et al. [6]. The surface not only demonstrated excellent anti-frosting performance but also effectively retarded the glaze icing. In addition, due to the hard porous alumina and self-healing properties, the SHP aluminum surface maintained outstanding superhydrophobicity after 64 abrasion-damage cycles. Although the reported SHP surfaces have good anti-icing properties, most of the materials used above are metals [3–6], and there is a lack of relevant studies for glass insulators. On the other hand, the preparation of the aforementioned SHP coatings is limited by expensive and complicated equipment. The preparation process is cumbersome and only suitable for preliminary research in the laboratory, which makes it difficult to meet industrial mass production [4,5]. At this point, SHP anti-icing coatings can be conveniently and quickly prepared on specially shaped surfaces using simple and efficient spraying. Nevertheless, it is noteworthy that SHP coatings produced by spraying still focus on small areas of block specimens. Currently, there are limited reports on preparing SHP coatings by directly spraying them onto the glass insulator surface. The studies on water droplet bouncing and anti-glaze icing of SHP glass insulators are lacking.

In this work, SHP coatings were developed directly on the glass insulator surface by spraying. The micro-morphology and elemental content of SHP coatings were characterized with scanning electron microscopy and energy spectroscopy. A water contact angle meter and a high-speed camera were utilized to assess the wettability and water droplet bouncing properties of SHP coatings. Glaze icing tests were performed in an artificial climate chamber. The ice-covering morphology was captured with a cell phone camera. The glaze icing processes of SHP coatings were analyzed by measuring the icing weight and icicle length.

## 2. Experimental

### *2.1. Materials*
Modified Bisphenol A Epoxy Resin (GCC135) was supplied by China Kunshan Lvxun Chemicals Industry Co., Ltd. Polytetrafluoroethylene (PTFE) particles (average size of 200 nm) were bought from Shanghai Macklin Biochemical Co., Ltd. Fluorosilicone resin was provided by Chengdu Aikeda Chemical Reagent Co., Ltd. Ethyl acetate was offered by Chengdu Kelong Chemical Co., Ltd. Anhydrous ethanol was supplied by Chongqing Chuandong Chemical (Group) Co., Ltd. Ordinary glass insulators (FC70/146) were bought from Taobao.com. All reagents used in this work were analytical grade and applied directly without further treatment.

### *2.2. Preparation*
Ethyl acetate (25 g), fluorosilicone resin (3 g), and epoxy resin (2 g) were placed in a beaker. The mixture was magnetically agitated for 30 min. Next, PTFE powders (5 g) were applied, and magnetic stirring and ultrasonic shaking were conducted for 10 min. Afterward, epoxy resin hardener (0.6 g) was blended under magnetic agitation for 10 min to prepare a homogeneous solution. The mixture was placed in a spray bottle. The gun was targeted at the horizontally positioned glass insulator to spray. Before spraying, it was necessary to clean the glass insulators with anhydrous ethanol and deionized water and then dry them using hot air for later use. The nozzle of the spray gun had a diameter of 1.5 mm. The pressure and distance of spraying were 0.3 MPa and 15 cm, respectively. After spraying, the

glass insulator was left at room temperature for 30 min, immediately followed by overnight curing in a 70°C oven.

### *2.3. Characterization*
The microstructure and elemental composition were characterized by scanning electron microscopy (SEM, Zeiss Auriga) and energy spectroscopy (EDS). The three-dimensional structure and the roughness were evaluated by a laser confocal microscope (LEXT OLS4000). The contact angle and sliding angle were measured by a water contact angle meter (SINDIN SDC-100). Water droplets of 6 μL were used for testing, and three positions were selected for measurement and averaged. The water droplet bouncing behaviors were assessed by a high-speed camera (M220, Revealer). In addition, glaze icing tests were performed in the artificial climate chamber at three temperatures (−5°C, −8°C, and −12°C). The ambient humidity was 90%, and the wind speed was controlled at 1.2 m/s. Water droplets, measuring about 40 μm in diameter, were sprayed onto the glass insulator surface with a spray nozzle. The test lasted for 80 min. A cell phone camera was used to take pictures of the ice cover's appearance. After the test, the weight of the glass insulator was weighed, and the length of the icicle was measured.

## 3. Results and discussion

### *3.1. Macrophoto and microstructure*
SHP coating was prepared on the ordinary glass insulator surface by spraying, as displayed in Figure 1. It is observed that the glass insulator surface is relatively smooth. The contact angle is only 41.7°, and the sliding angle is over 90°. This makes it simpler for water droplets to stick to the surface and spread on the surface. In contrast, the SHP coating has a uniform white film layer distributed on the surface. The contact angle is significantly increased to 165.2°. At the same time, the sliding angle is also decreased to 3.7°. As a result, the surface of SHP coating displays outstanding water repellency and self-cleaning properties, allowing water droplets to quickly slide off the surface.

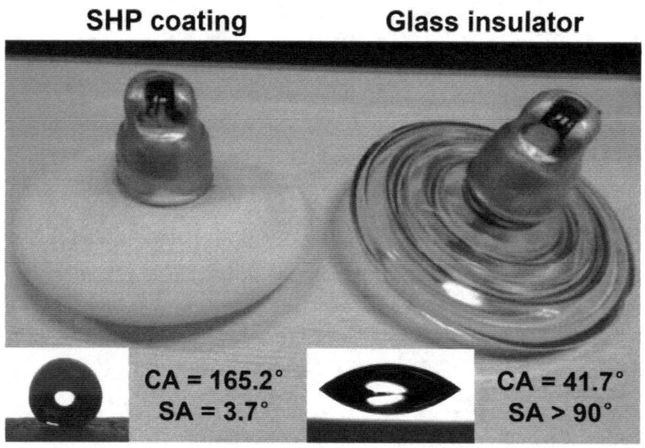


**Figure 1.** Macroscopic morphology and wettability of glass insulator and SHP insulator.

The microscopic morphology, elemental composition, and three-dimensional structure on the SHP coating surface were characterized, as presented in Figure 2. The SHP coating surface is filled with spherical PTFE nanoparticles, forming a typical micro-nano roughness structure. This meets the basic conditions for the construction of the SHP surface (Figure 2a). It is widely recognized that the presence of C−F groups results in low surface energy, which facilitates the achievement of superhydrophobicity. The F element with a content of up to 51.3% is detected on the SHP coating surface (Figure 2b), which originates from the fluorosilicone resin and PTFE particles. The large amount of C−F groups brought by the added fluorosilicone resin and PTFE particles effectively reduces the surface energy. Furthermore, the three-dimensional structure (Figure 2c) also verifies the establishment of a micro-nanoroughness structure with a surface roughness of 3.67 μm. Therefore, the SHP coating surface shows excellent hydrophobicity owing to the joint influence of micro-nanoroughness structures and low-surface-energy groups (Figure 1).

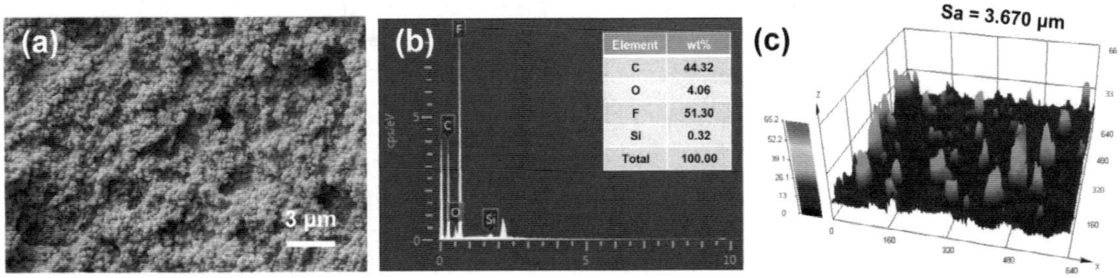


**Figure 2.** Microscopic morphology (a), elemental composition (b) and three-dimensional structure (c) of SHP coating.

*3.2. Water droplet bouncing behavior*

A large number of water droplet bouncing tests on flat plates have been reported. There is insufficient research on the bouncing behavior displayed by water droplets on glass insulators. Figure 3 exhibits the bouncing characteristics of single water droplets on glass and SHP insulator surfaces. During the test, a single water droplet is released from a distance of about 30 cm to collide with insulators. The observation reveals that individual water droplet impacting the glass insulator surface first spreads and then contract and finally spread flat on the surface. This is due to the large adhesion force on the glass insulator surface, which can easily absorb the water droplets. On the contrary, when a single water droplet hits the SHP insulator surface, the water droplet first rebounds significantly and then bounces off the surface. This suggests that there is a small adhesion force and energy loss between the SHP surface and the water droplet, enabling the water droplet to retain enough kinetic energy to rebound away from the surface. This is also attributed to the micro-nano-scale roughened structures (Figure 2) and excellent hydrophobicity (Figure 1) on the surface of the SHP insulator.

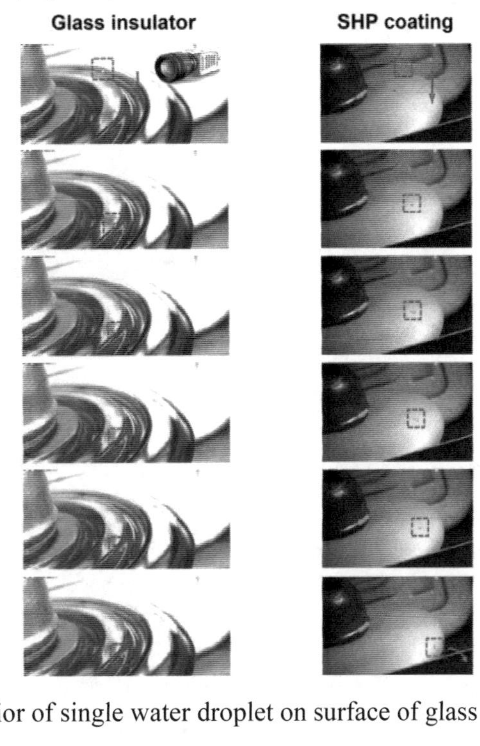


**Figure 3.** Bouncing behavior of single water droplet on surface of glass insulator and SHP insulator.

In order to simulate the glaze, the water droplet bouncing tests in a glaze-freezing environment were performed in an artificial climate chamber. Figure 4 illustrates the behaviors of water droplets bouncing on both glass and SHP insulators in the glaze icing environment. Different from a single water droplet, the spray nozzle forms small water droplets, similar to rain, that are blasted on the surface as the glaze freezes. Consistent with the bouncing behavior of a single water droplet, numerous small water droplets immediately adhere to the glass insulator and spread out on the surface upon impact. Subsequently, the small water droplets gradually expand to form a water film that completely covers the surface. In the case of the SHP insulator, it is evident that some of water droplets bounce off the surface immediately after impact, while others merge with each other and grow up before rolling away from the surface. As a result, the majority of water droplets in the glaze environment can roll off the surface of the SHP insulator. With excellent superhydrophobicity (Figures 1, 2, and 3), the SHP insulator surface retains a large dry area, which dramatically minimizes the likelihood of water droplet freezing and is conducive to the realization of glaze icing resistance of glass insulator.

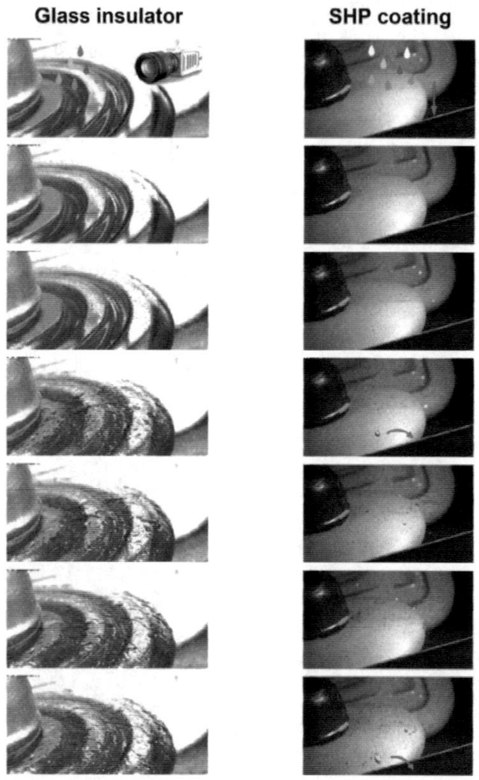


**Figure 4.** Bouncing behavior of water droplet on glass insulator and SHP insulator in glaze icing environment.

### *3.3. Anti-glaze icing performance*

To investigate the anti-glaze icing properties of the SHP insulator, the glaze icing tests under three low-temperature environments were performed in an artificial climate chamber. The ice-covering morphology is displayed in Figure 5. It is noted that as the ice-covering time extends, the glass insulator surface is first obscured by a film of apparent ice. Then, the ice grows downward along the ice layer to form a long icicle. The superior hydrophobicity (Figure 1) and water droplet bouncing properties (Figures 3 and 4) remarkably decrease the possibility of water droplet freezing [4,6]. A large number of small water droplets that do not have time to tumble away from the SHP insulator surface will freeze. However, the surface still maintains large un-iced areas, and the icicles formed at the edges are significantly reduced. In addition, the ambient temperature greatly affects the ice cover on the insulator surface. As the ambient temperature decreases, the icing on both glass and SHP insulator surfaces is gradually increasing. In particular, the icing on the glass insulator surface is more serious. At this point, the icing has completely wrapped around the insulator. The ice thickness and icicle length increase significantly. In three low-temperature environments, the surface of the SHP insulator can remain a certain dry area, and the ice-covered area and the length of the icicle are decreased.

**Figure 5.** Ice-covering morphology of glass insulator and SHP insulator in three low-temperature environments.

On the other hand, the icing weight and icicle length of the insulators were measured after the ice-covering test, as presented in Figure 6. As the ambient temperature decreases, the icing weight (Figure 6a) and icicle length (Figure 6b) of the glass and SHP insulators gradually increase, which agrees with the results in Figure 5. The icing weight of the SHP insulator drops by more than 35% compared to the glass insulator in the three low-temperature environments. It is noticed that the SHP insulator has no icicles at the edges in the low-temperature environment of −5°C (Figure 5). In contrast, the icicle length of the glass insulator has grown to 2.31 cm. As the ambient temperature decreases further, the icicle length of the SHP insulator increases rapidly due to localized areas of icing. However, SHP insulator still has significantly smaller icicle length than glass insulator. In summary, the prepared SHP insulator has excellent superhydrophobicity and water droplet bouncing performance, showing superior anti-glaze icing properties, and offers promising application on glass insulators for transmission lines.

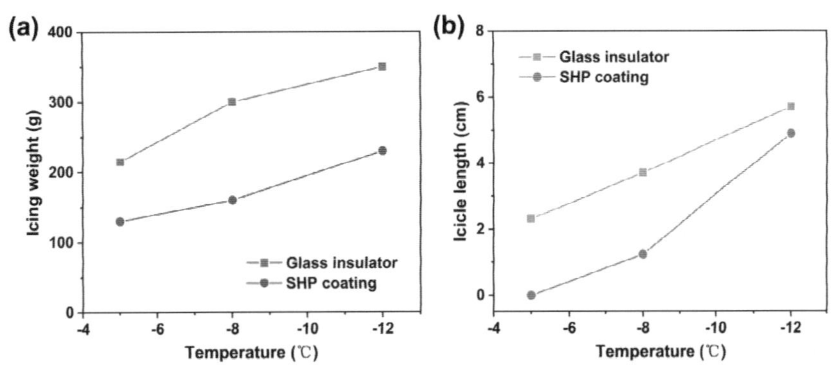


**Figure 6.** Icing weight (a) and icicle length (b) of glass insulator and SHP insulator.

## 4. Conclusion

A SHP coating was successfully fabricated on the glass insulator surface by spraying. Typical micro-nano structures and low-surface-energy nanoparticles were generated on the SHP coating surface. Contact angles of up to 165.2° and sliding angles down to 3.7° were measured. The excellent superhydrophobicity and low adhesion properties allowed a single water droplet to bounce away from the surface rapidly upon striking the SHP coating. In a glaze icing environment, numerous small water droplets gradually merged and rolled off the SHP coating surface. As a result, the majority of the area on the SHP insulator remained relatively dry compared to the extensive water layer present on the glass insulator surface. In three low-temperature environments, the surface of the glass insulator experienced more severe glaze icing. However, the icing area, icing weight, and icicle length on the SHP insulator surface were considerably reduced. The prepared SHP insulator shows good anti-glaze icing performance, which offers a valuable reference for applying anti-icing coatings to glass insulators in practical applications.

## Acknowledgments

This work was supported by the Electric Power Research Institute of Guizhou Power Grid Co., Ltd., China (Contract No. 0666002022030103HX00003).

## References

[1] Sarma, J., Zhang, L., Guo, Z.Q., Dai X.M. (2022) Sustainable icephobicity on durable quasi-liquid surface. Chem Eng J, 431: 133475. https://doi.org/10.1016/j.cej.2021.133475.

[2] Wang, L., Gong, Q.H., Zhan, S.H., Jiang, L., Zheng, Y.M. (2016) Robust Anti-Icing Performance of a Flexible Superhydrophobic Surface. Adv Mater, 28: 7729–7735. https://doi.org/10.1002/adma.201602480.

[3] Liao, R.J., Li, C., Yuan, Y., Duan, Y.Z., Zhuang A.Y. (2017) Anti-icing performance of ZnO/SiO2/PTFE sandwich-nanostructure superhydrophobic film on glass prepared via RF magnetron sputtering. Mater Lett, 206: 109–112. https://doi.org/10.1016/j.matlet.2017.06.127.

[4] Cao, Y.Z., Lu, Y.M., Liu, N., Li, Y., Wang, P.F., Dai, C.N., Wei Y. (2022) Multi-applicable, durable superhydrophobic anti-icing coating through template-method and chemical vapor deposition. Surf Interfaces, 32: 102100. https://doi.org/10.1016/j.surfin.2022.102100.

[5] Wan, Y.L., Yan, C.D., Yu, H.D., Wang B. (2021) Anti-Icing Performance of Superhydrophobic Surface with Square-Ring Structure Prepared by Nanosecond Laser. Adv Eng Mater, 23: 2100190. https://doi.org/10.1002/adem.202100190.

[6] Liu, G.Y., Yuan, Y., Liao, R.J., Xiang, H.Y., Wang, L., Yu Q., Zhang C. (2022) Robust and self-healing superhydrophobic aluminum surface with excellent anti-icing performance. Surf Interfaces, 28: 101588. https://doi.org/10.1016/j.surfin.2021.101588.

# Investigating the serve life of common earthed materials based on evaluations of the soil corrosion in offshore areas of East China

**Zhaomin Guo [1], Qi Shi [2], and Yejiao Wang [1, 3]**

[1]School of Mechanics and Engineering Science, Shanghai University, 99 Shangda Road, Shanghai 200444, China
[2]East Branch State Grid Corporation of China, 201 East Nanjing Road, Shanghai 200120, China

[3] Corresponding author's e-mail: yejiaowang@shu.edu.cn

**Abstract:** The serve life of common earthed materials in power systems is essentially affected by the soil environment, especially the potential corrosion problem of sodium chloride contaminated soil in offshore areas. This study measured the corrosion rates of three common earthed metal materials, Q235 carbon steel, pure Al, and H62 brass, in typical chloride saline soil using accelerated and field corrosion tests. The acceleration ratio and kinetic correlation coefficient under different acceleration conditions were calculated, and the laboratory and field corrosion time correlation models of common earthed metal materials were obtained. The results show a good correlation between soil accelerated and field corrosion tests. Under the same temperature and salinity, the correlation degree of Q235 carbon steel is larger when the water content is 30%. In comparison, the correlation degree of pure Al and H62 brass is larger when the water content is 20%. The laboratory and field corrosion time correlation models of Q235 carbon steel, pure Al, and H62 brass in chloride type saline soil environment are $T_1=18.32T_0$, $T_1=95.56T_0$, $T_1=15.78T_0$, which can effectively evaluate the serve life of common earthed metal materials in typical chloride type saline soil in the offshore area of East China.

## 1. Introduction

The grounding device in the power system plays a key role in ensuring the safe operation and maintenance of power equipment [1]. In recent years, the power system has continued to develop in the direction of higher voltage and larger capacity, and the requirements for the performance of grounding devices have gradually increased [2,3]. Based on satisfying the electrical performance, the anti-soil corrosion capacity of the grounding materials directly determines the operating life of the grounding device [4,5]. The soil in the eastern coastal areas of China can be classified as chloride-type coastal saline soil, rich in corrosive components such as chloride, sulfate, and magnesium salts [6-8]. This soil environment makes grounding metal materials more susceptible to corrosion damage, and the degree and speed of corrosion are usually significantly higher than those in inland areas [9,10]. Therefore, studying the service life of grounding metal materials in a chloride-type saline soil corrosion environment is significant.

In the field corrosion test, standard specimens are buried in a typical soil environment, and the corrosion mass loss of the specimens is tested according to a certain buried duration. This basic method can truly reflect the corrosion rate and performance evolution of metal materials in soil. However, the test data are precious due to the long time and high maintenance cost [11]. Compared with the field corrosion test, the accelerated corrosion test is an experimental method that can control the test conditions and accelerate the corrosion process through manual intervention. Its main purpose is to evaluate the corrosion trend of materials in specific media in a relatively short period [12]. The corrosion acceleration test is based on the premise that the results of the field corrosion test have a good correlation, and the acceleration test with a good correlation has more practical significance [13].

In this study, the corrosion rates of three grounding materials, Q235 carbon steel, pure Al, and H62 brass, in typical chloride-type saline soil in coastal areas of East China were measured based on the soil accelerated corrosion test and field corrosion test results. The laboratory and field corrosion time correlation models of three grounding materials in a chloride-type saline soil corrosion environment were obtained, which provided a theoretical basis for optimizing the depreciation strategy of metal equipment in electric power enterprises.

## 2. Materials and methods

### *2.1. Materials*

The test materials are Q235 carbon steel, pure Al, and H62 brass, and the sample size is 50 mm×25 mm ×2 mm, as shown in Figure 1. After the sample preparation is completed, the sample is ground, de-oiled, cleaned, and dried, weighed with an electronic balance with an accuracy of 0.1 mg, and placed in a dryer for use.

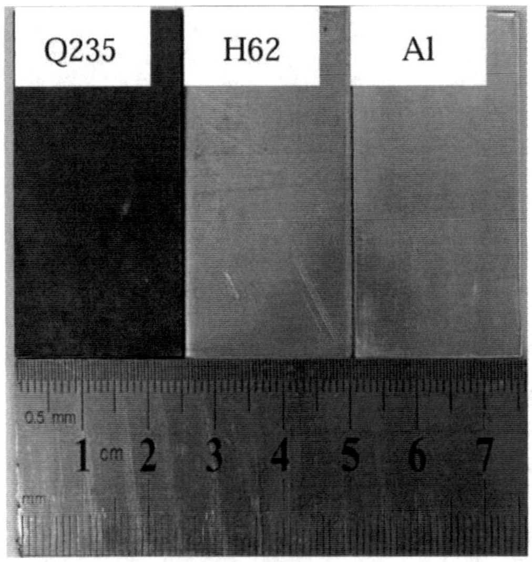


**Figure 1.** Metal samples.

The test soil was sampled from a site in Baoshan District, Shanghai. No gravel or other interference factors met the standard requirements of the China Soil Corrosion Test Network. The basic physical properties are shown in Table 1, and the soil sample can be classified as a low-liquid limit clay (CL).

**Table 1.** Basic physical properties of test soil.

| Parameters | Numerical |
|---|---|
| The moisture content of in-situ soil $w/\%$ | 26 |
| Liquid limit $w_L/\%$ | 37 |
| Plastic limit $w_P/\%$ | 21.59 |
| Plasticity index $I_P$ | 15.41 |
| $Cl^{-}/g \cdot kg^{-1}$ | 1.89 |
| $SO_4^{2-}/g \cdot kg^{-1}$ | 0.32 |
| Total salt content$/g \cdot kg^{-1}$ | 2.20 |

The configured test soil information is shown in Table 2, and the test soil is in Figure 2. The undisturbed soil samples were air-dried and ground in the natural environment and passed through a 0.83 mm sieve. They were placed in an oven at 110°C for 12 hours and then set aside. To simulate the corrosion of chloride-type saline soil in the coastal areas of East China, according to the 'Code for geotechnical engineering investigation', the NaCl analysis pure and deionized water were configured into 1.5%NaCl contaminated liquid for backup. The dry soil and NaCl contaminated solution were artificially mixed to prepare sodium chloride contaminated soil with 20% and 30% water content, respectively. In Table 2, the oxygen content in 1 # soil is higher than in 2 # soil.

**Table 2.** Test soil information.

| Number | 1#soil | 2#soil |
|---|---|---|
| Water content/% | 20 | 30 |

(a) (b)

**Figure 2.** Test soil: (a) 1#soil; (b) 2#soil.

*2.2. Methods*

Figure 3 shows the test process. The accelerated soil corrosion test was conducted in a programmable constant temperature and humidity test chamber. To accelerate metal corrosion in sodium chloride contaminated soil, the relative humidity of the test environment was controlled at 80%, and the temperature was set at 50°C. The metal samples were placed in parallel in the configured sodium chloride contaminated soil. At the same time, the distance from the soil in each direction of the test box was controlled to be 50 mm, and the sample spacing was set to be 25 mm. Three parallel samples were set respectively. Each test period was one month, and samples were taken every three days after the start of the test for weight loss measurement. During the test, the soil moisture content was measured and recorded every 24 hours, and deionized water was added regularly to supplement the water evaporated in the soil during the test.

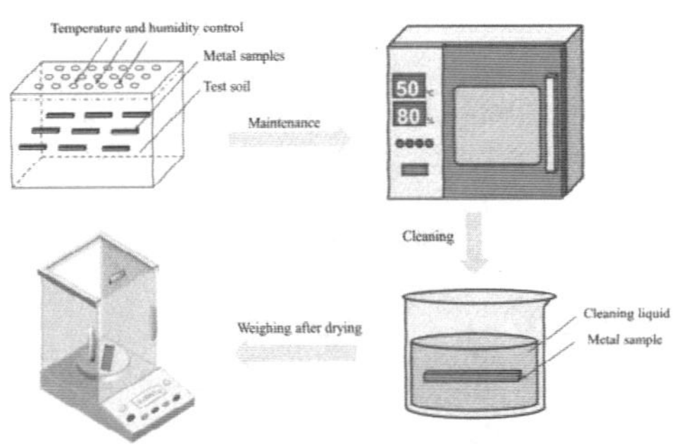


**Figure 3.** Test process.

After the corrosion test, the brush was used to remove the number of soil particles attached to the surface of the metal sample. Then, the metal sample was placed in the rust removal cleaning solution prepared according to the specifications. After the rust removal, the surface of the sample was cleaned with deionized water, dried, and weighed. The formula for the rust removal cleaning solution, cleaning time, and cleaning temperature of each metal material are shown in Table 3.

**Table 3.** Cleaning liquid formula and cleaning parameters.

| Test materials | Cleaning liquid formula | Cleaning time/min | Cleaning temperature/°C |
|---|---|---|---|
| Q235 carbon steel | 500 mL HCl ($\rho$= 1.19 g/mL), 3.5 g $C_6H_{12}N_4$, diluted with distilled water to 1000 mL | 10 | 20-25 |
| Pure Al | 50 mL $H_3PO_4$ ($\rho$=1.19 g/mL), 20 g $CrO_3$, diluted with distilled water to 1000 mL | 5 | 80 |
| H62 brass | 500 mL HCl ($\rho$ = 1.19 g/mL), diluted with distilled water to 1000 mL | 3 | 20-25 |

## 3. Results

### *3.1. Field corrosion test*

The field corrosion mass loss results of Q235 carbon steel, pure Al, and H62 brass in typical chloride-type saline soil are from the Dagang Material Corrosion Experimental Station of the National Material Environmental Corrosion Network. The specific corrosion weight loss values of the three earthed materials are shown in Table 4.

**Table 4.** Mass loss of earthed materials in the field corrosion test.

| Type of metal | Corrosion time/d | Mass loss per unit area/g·dm$^{-2}$ |
|---|---|---|
| Q235 carbon steel | 365 | 2.756 |
| | 730 | 3.408 |
| | 1095 | 3.968 |
| pure Al | 365 | 0.191 |
| | 730 | 0.324 |
| | 1095 | 0.453 |
| H62 brass | 365 | 1.010 |
| | 730 | 1.843 |
| | 1095 | 2.676 |

Table 4 shows that the corrosion rate of the three grounding materials in chloride-type saline soil is Q235 carbon steel > H62 brass > pure Al. In chloride-type saline soil, pure Al is relatively corrosion resistant, and the passivation film on the surface of Al is partially penetrated under the action of Cl-, forming pitting corrosion. H62 brass has obvious dezincification corrosion, and the corrosion rate is large. The corrosion rate of Q235 carbon steel is very high, above the national soil corrosion website standard (V) [14]. The poor corrosion resistance of Q235 carbon steel is mainly due to its main components, iron and carbon, and the lack of alloy elements to resist corrosion. Although Q235 carbon steel has high strength and low cost, it is susceptible to corrosion, especially in the corrosive environment of chloride-type saline soil with wet and high salt content.

The power function fitting of the corrosion mass loss of the three grounding materials in the field of 0-3a was carried out by using Equation (1). The fitting results are shown in Figure 4.

$$M = Dt^n \quad (1)$$

Where M is the corrosion mass loss per unit area corrosion (g·dm$^{-2}$). t is the corrosion time ($d$). D and n represent constants. The fitted D values of Q235 carbon steel, pure Al, and H62 brass are 0.384, 0.002 and 0.005, and the fitted n values of three grounding materials are 0.333, 0.793 and 0.894, respectively.

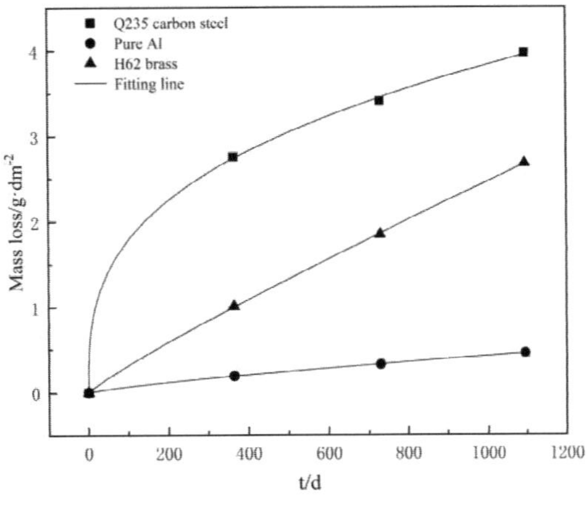


**Figure 4.** Mass loss curve of field corrosion test of grounding materials.

### *3.2. Accelerated corrosion test in the laboratory*

Figure 5 shows the corrosion mass loss kinetic curves of Q235 carbon steel, pure Al, and H62 brass under different accelerated conditions for one month. At the same temperature and salinity, with the decrease in water content, the corrosion mass loss of Q235 carbon steel and H62 brass increases by 3.0 and 8.1 times, respectively. In contrast, the corrosion mass loss of pure Al decreases by 1.6 times. The soil water content greatly affects the corrosion rate of Q235 carbon steel and H62 brass.

Soil is a multiphase system composed of solid, liquid, and gas. The physical and chemical properties of soil, such as Cl-content, water content, oxygen content, and temperature, affect soil corrosion. Under the same $Cl^-$content and temperature, the oxygen content of soil with 20% water content is higher than that of soil with 30% water content. Iron in Q235 carbon steel reacts with oxygen and water to form iron oxide, which is common in rust. The corrosion rate of Q235 carbon steel is greatly affected by oxygen content. H62 brass is mainly composed of copper and zinc. The addition of zinc helps to form an oxide film, which can form a protective layer on the surface and prevent further oxidation reactions. Although H62 brass has good corrosion resistance, it is still affected by oxygen content. Especially in the harsh chloride-type saline soil environment, the presence of oxygen and water will promote the oxidation reaction of zinc and copper to form an oxide layer, which affects the appearance and properties of brass to a certain extent. The surface of pure Al will quickly form a dense and uniform oxide film due to the reaction of alumina with oxygen to form alumina. Alumina is a very stable compound with good adhesion and corrosion resistance. This layer of oxide film is formed quickly in the air. Even if it is damaged, it can be regenerated quickly. This oxide layer prevents further oxidation reaction, thereby reducing the oxidation rate of pure Al, and the corrosion rate of pure Al in 20% moisture content soil with higher oxygen content is lower.

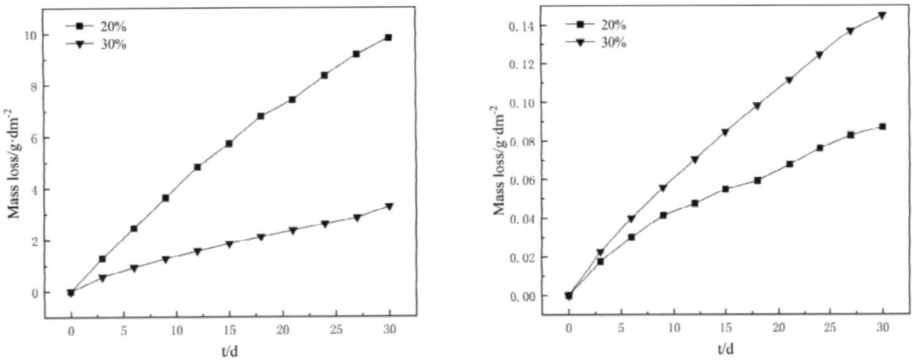


(a) (b)

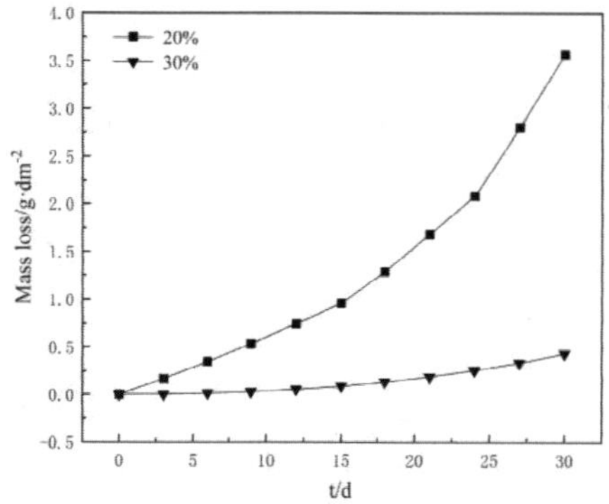


(c)

**Figure 5.** The mass loss curve of accelerated corrosion test of grounding materials (a) Q235 carbon steel; (b) Pure Al; (c) H62 brass.

## 4. Analysis and discussion

### *4.1. Acceleration ratio of accelerated corrosion test in laboratory*

The purpose of an accelerated corrosion test is to reduce the time required for the experiment to shorten the test cycle under the same corrosion degree. To measure the effect of the accelerated corrosion test directly, this paper uses Equation (2) to calculate the acceleration ratio [15].

$$a = \frac{t_0(M)}{t_1(M)} \tag{2}$$

Where a is the acceleration ratio when the corrosion mass loss reaches M, $t_0(M)$ is the test period when the corrosion mass loss reaches M in the field corrosion test, and $t_1(M)$ is the test period when the corrosion mass loss reaches M in the accelerated corrosion test. This formula uses the ratio of the non-accelerated test to the accelerated test cycle as the acceleration ratio of the accelerated corrosion test when the accelerated test and the non-accelerated test are at the same degree of corrosion mass loss.

The acceleration ratio is suitable not only for uniform corrosion but also for non-uniform corrosion and local corrosion. If the relationship between mass loss per unit area and time during the corrosion process satisfies the exponential law of Equation (1), namely:

Field corrosion test:

$$M_0 = D_0 t^{n_0} \tag{3}$$

Accelerated corrosion test in laboratory:

$$M_1 = D_1 t^{n_1} \tag{4}$$

We calculate the acceleration ratio with Equation (5):

$$a = \frac{D_1^{\frac{1}{n_1}}}{D_0^{\frac{1}{n_0}}} \cdot M^{\frac{1}{n_0} - \frac{1}{n_1}} \tag{5}$$

Figure 5 shows that for the accelerated experiment with a soil water content of 20, the corrosion kinetics of Q235 carbon steel and H62 brass changed greatly in one month, and the corrosion kinetics of pure Al changed greatly in the accelerated experiment with a soil water content of 30%. Therefore, the corrosion mass loss was fitted in two time periods of 0-15 d and 0-30 d, respectively, and the two cases of 0-15 d and 0-30 d were calculated when the acceleration ratio was calculated. Through

Equations (3)-(5), the corrosion mass loss of three grounding metal materials under different acceleration conditions was fitted, and the acceleration ratio was calculated under different acceleration conditions. The calculation results in Table 5 show that consistent with the change of accelerated corrosion mass loss, at the same temperature and salinity, Q235 carbon steel and H62 brass accelerated more in 20% water-bearing soil. In comparison, pure Al accelerated more in soil with 30% water content.

**Table 5.** $D_1$, $n_1$, M value of Q235 carbon steel, pure Al, and H62 brass in an accelerated corrosion test.

| Type of metal | Test soil | $D_1$ | $n_1$ | M /g·dm$^{-2}$ | a |
|---|---|---|---|---|---|
| Q235 carbon steel | 1#soil (0-15 d) | 0.474 | 0.925 | 5.733 | 226.486 |
| | 1#soil (0-30 d) | 0.598 | 0.828 | 9.806 | 574.026 |
| | 2#soil (0-30 d) | 0.973 | 0.358 | 3.291 | 21.124 |
| Pure Al | 1#soil (0-30 d) | 0.009 | 0.670 | 0.087 | 3.972 |
| | 2#soil (0-15 d) | 0.009 | 0.819 | 0.085 | 7.798 |
| | 2#soil (0-30 d) | 0.010 | 0.804 | 0.145 | 7.663 |
| H62 brass | 1#soil (0-15 d) | 0.047 | 1.112 | 0.959 | 23.752 |
| | 1#soil (0-30 d) | 0.007 | 1.832 | 3.569 | 50.621 |
| | 2#soil (0-15 d) | 0.001 | 2.289 | 0.435 | 4.890 |

### *4.2. Grey correlation method and life prediction*

The correlation of corrosion mass loss between the accelerated corrosion test in the laboratory and the field corrosion test is calculated by the grey correlation analysis method. Grey correlation analysis is based on the proximity of the curve shape of each factor sequence to analyse the development trend. It provides a quantitative tool for the development and change trend of the system, which is very suitable for the analysis of dynamic processes [16]. The calculation process of the correlation coefficient between the accelerated corrosion test in the laboratory and the field corrosion test results is as follows:

(1) The relationship model between corrosion mass loss and time is obtained by fitting the results of accelerated corrosion test in laboratory and field corrosion test, and the acceleration ratio is calculated by Equation (5).

(2) The time series $t_1$ of each accelerated corrosion test in the laboratory relative to the field corrosion test is calculated by Equation (6).

$$t_{1i} = \frac{t_{0i}}{a} \tag{6}$$

Where $t_{1i}$ is the i th data point in the accelerated corrosion test time series. $t_{0i}$ is the i th data point in the time series of field corrosion test. a is the acceleration ratio.

(3) The accelerated corrosion test in laboratory time series $t_1$ is substituted into the M-t model to calculate the corresponding corrosion mass loss sequence of the accelerated corrosion test.

(4) The correlation coefficient between the comparison sequence $x_i$ and the reference sequence $x_0$ at each moment can be calculated by Equation (7).

$$\xi_i(k) = \frac{\min_i \min_k |x_0(k) - x_i(k)| + \rho \cdot \max_i \max_k |x_0(k) - x_i(k)|}{|x_0(k) - x_i(k)| + \rho \cdot \max_i \max_k |x_0(k) - x_i(k)|} \tag{7}$$

Where $\rho$ is the resolution coefficient, which takes 0.5. $\min_i \min_k |x_0(k) - x_i(k)|$ is the minimum value of the absolute difference between the comparison sequence and the reference sequence. $\max_i \max_k |x_0(k) - x_i(k)|$ is the maximum value of absolute difference.

(5) The correlation coefficient $\gamma_i$ is calculated by Equation (8) as a quantitative representation of the degree of correlation between the comparison sequence and the reference sequence.

$$\gamma_i = \frac{1}{N}\sum_{k=1}^{N} \xi_i(k) \tag{8}$$

Based on the field corrosion test time of 365 d, 730 d, and 1095 d, the time series of the corresponding accelerated corrosion test in the laboratory is calculated by Equation (6), and the calculation results are shown in Table 6.

**Table 6.** Different experimental conditions correspond to the time series of field corrosion tests.

| Type of metal | Test soil | a | Time series corresponding to field corrosion tests (365 d, 730 d, 1095 d) |
|---|---|---|---|
| Q235 carbon steel | 1#soil (0-15 d) | 226.486 | 1.612, 3.223, 4.835 |
| | 1#soil (0-30 d) | 574.026 | 0.636, 1.272, 1.908 |
| | 2#soil (0-30 d) | 21.124 | 17.279, 34.558, 51.837 |
| pure Al | 1#soil (0-30 d) | 3.972 | 91.902, 183.803, 275.705 |
| | 2#soil (0-15 d) | 7.798 | 46.806, 93.612, 140.418 |
| | 2#soil (0-30 d) | 7.663 | 47.6321, 95.264, 142.896 |
| H62 brass | 1#soil (0-15 d) | 23.752 | 15.367, 30.735, 46.102 |
| | 1#soil (0-30 d) | 50.621 | 7.211, 14.421, 21.631 |
| | 2#soil (0-15 d) | 4.890 | 74.642, 149.285, 223.927 |

According to the time series under different accelerated conditions in Table 6 and the fitting results of their corrosion mass loss kinetics, the corrosion mass loss is calculated, and the corresponding comparison sequence is obtained. The data are normalized, and the correlation coefficients of accelerated tests of three kinds of grounding metal materials under different conditions are calculated by Equations (7) and (8). Table 7 shows that under the same temperature and salinity, the correlation coefficient of Q235 carbon steel in the accelerated corrosion test of higher water content soil is higher, reaching 0.940. The kinetic correlation coefficients of pure Al and H62 brass in the accelerated corrosion test of soil with low water content were higher, reaching 0.830 and 0.934, respectively.

**Table 7.** Dynamic correlation coefficients under different experimental conditions.

| Type of metal | Test soil | Reference and comparison series (365 d, 730 d, 1095 d)/$g\cdot dm^{-2}$ | Correlation coefficients |
|---|---|---|---|
| Q235 carbon steel | Field tests | 2.756, 3.408, 3.968 | |
| | 1#soil (0-15 d) | 0.736, 1.398, 2.035 | $\gamma_1 = 0.561$ |
| | 1#soil (0-30 d) | 0.411, 0.730, 1.021 | $\gamma_1 = 0.585$ |
| | 2#soil (0-30 d) | 2.701, 3.462, 4.004 | $\gamma_1 = 0.940$ |
| pure Al | Fiel d tests | 0.191, 0.324, 0.453 | |
| | 1#soil (0-30 d) | 0.198, 0.329, 0.448 | $\gamma_1 = 0.830$ |
| | 2#soil (0-15 d) | 0.252, 0.478, 0.695 | $\gamma_1 = 0.608$ |
| | 2#soil (0-30 d) | 0.218, 0.385, 0.537 | $\gamma_1 = 0.804$ |
| H62 brass | Field tests | 1.010, 1.843, 2.676 | |
| | 1#soil (0-15 d) | 0.981, 2.120, 3.329 | $\gamma_1 = 0.934$ |
| | 1#soil (0-30 d) | 0.251, 0.892, 1.876 | $\gamma_1 = 0.747$ |
| | 2#soil (0-15 d) | 3.442, 16.821, 42.549 | $\gamma_1 = 0.649$ |

According to the correlation analysis, the accelerated corrosion test in the laboratory can effectively simulate the corrosion of Q235 carbon steel, pure Al, and H62 brass in a chloride-type saline soil environment. By fitting the corrosion data with a high correlation coefficient, the corrosion weight loss function of Q235 carbon steel is $M = 0.973T^{0.358}$, the corrosion weight loss function of pure Al is $M = 0.009T^{0.670}$, and the corrosion weight loss function of H62 brass is $M = 0.047T^{1.112}$. The corrosion weight loss of the accelerated soil corrosion test is equal to the corrosion weight loss of the field corrosion test. The laboratory and field corrosion time-related time models of Q235 carbon steel, pure Al, and H62 brass in chloride type saline soil environment are $T_l = 18.32T_0$, $T_l = 95.56T_0$, and $T_l = 15.78T_0$ ($T_0$ is the field soil corrosion time, a; $T_l$ is the accelerated soil corrosion test in laboratory time, d).

The laboratory and field corrosion time correlation model can be used to predict the natural corrosion of Q235 carbon steel, pure Al, and H62 brass in a chloride-type saline soil environment. For example, the accelerated corrosion test of Q235 carbon steel in a chloride-type saline soil environment for 18.32

days can predict the corrosion weight loss kinetic characteristics of Q235 carbon steel in chloride-type saline soil environment in coastal areas of East China after 1 year, which has certain reference value for power enterprises.

## 5. Conclusion

(1) In this study, NaCl is added to the soil to simulate the corrosion environment of chloride-type saline soil in the coastal areas of East China. Controlling temperature, relative humidity, and salinity to accelerate the corrosion of metals in the soil can better simulate the corrosion weight loss characteristics of metal materials in saline soil environments.

(2) The water content of soil takes an essential role in the soil corrosion environment. In the correlation analysis of the accelerated corrosion test in the laboratory, under the same temperature and salinity, the kinetic correlation coefficient of Q235 carbon steel is larger when the water content is 30%. In comparison, the kinetic correlation coefficients of pure Al and H62 brass are larger when the water content is 20%.

(3) Through the field corrosion weight loss data and the accelerated soil corrosion test in laboratory weight loss fitting function under high correlation experimental conditions, the laboratory and field corrosion time correlation models of three commonly used earthed metal materials of Q235 carbon steel, pure Al and H62 brass in the typical chloride type saline soil environment in the coastal area of East China are obtained. It is divided into $T_1=18.32T_0$, $T_1=95.56T_0$, and $T_1=15.78T_0$, which provides a theoretical basis for the depreciation strategy of metal equipment in power enterprises.

**Acknowledgments**
The authors thank the State Grid Corporation of China Project (Technology Project: 520800230030).

## References

[1] Wei, W., Wu X.Q., Ke W., et al. (2015) Research progress on corrosion and protection of grounding network materials. Corrosion Science and Protection Technology. 27(3): 273-277.
[2] Sanchez-Rodriguez, J.M., Carcel-Carrasco, J. Pascual-Guillamon, M. (2021) Advance of corrosion analysis in electric protection grounding networks. Dyna. 96(5): 534-539.
[3] Liu, X., Pei, F., Tian, X., et al. (2021) Correlation and evaluation method of indoor and outdoor soil corrosion test of grounding materials. Corrosion & Protection. 42(3): 15-19+27.
[4] Wang, Y., Xu, L., He, H.L., et al. (2018) Evaluation of service life of grounding materials based on soil corrosivity grade. Smart Power. 46(2): 8-12+53.
[5] Zhou, M. (2018) Research on full-life corrosion-resistant grounding device of substation. Electric Power Survey and Design. (S2): 118-124.
[6] Mei, Q.Q., Gong, X.L., Shi, Y.D., et al. (2020) Analysis of salinization degree and its influencing factors of saline soil in Jiangsu coastal area. Journal of Engineering Geology. 28(5): 959-965.
[7] Wu, H.B., Liang, J., Cai, Y.L., et al. (2022) Study on soil quality characteristics of green space in artificial lake coastal zone - A case study of Shanghai Dishui Lake coastal zone. LA Ecology. 38(2): 110-114
[8] Yan, X.X., Zhang, X.L., Wu, X.P. (2023) Progress in development and utilization of salt mud resources. Environmental Protection Science. 49(4): 85-92.
[9] Leng, F.G., Ma, X.X., Ding, W., et al. (2011) Durability analysis of reinforced concrete piles exposed for 17 years in coastal saline soil environment. Building Structures. 41(11): 148-151+144.
[10] Wang, W., Wang, Y., Huang, J.Q., et al. (2021) Accelerated corrosion simulation experiment of steel pipe pile material surface in marine geotechnical environment. Acta Energiae Solaris Sinica. 42(11): 321-328.
[11] Dong, C.F., Li, X.G., Wu, J.W., et al. (2003) Experimental study and data processing of soil corrosion. Corrosion Science and Protection Technology. (3): 154-160.

[12] Nie, X.H., Li, X.G., Li, Y.L., et al. (2012) Accelerated soil corrosion simulation experiments on carbon steel. Materials Engineering. (1): 59-65.
[13] Lin, C.C., Wang, C.X. (2005) Correlation between accelerated corrosion tests and atmospheric corrosion tests on steel. Journal Of Applied Electrochemistry. 35(9): 837-843.
[14] Sun, C., Li, H.X., Zang, S.Q., et al. (1999) Study on the corrosivity of salt soil at Dagang seashore. Environmental Science and Technology. (2): 2-4+35.
[15] Nie, X.H., Li, X.G., Li, Y.L., et al. (2011) Study on the correlation between acceleration ratio and kinetics of soil corrosion acceleration test. Chinese Journal of Corrosion and Protection. 31(3): 208-213.
[16] Tian, M., Liu, S.F., Bu, Z.K. (2008) A research review of grey correlation algorithm model. Statistics and Decision Making. (1): 24-27.

# Study on the influence of thermal aging on the mechanical and magnetic properties of ferritic steel

**Wenqing Jia[1,2], Minyu Fan[1,2], Jing Wan[1,2], Shuang Qi[1,2], Xiangbing Liu[1,2,3], Chaoliang Xu[1,2], Yuanfei Li[1,2], Wangjie Qian[1,2], Qiwei Quan[1,2] and Jian Yin[1,2]**

[1]Suzhou Nuclear Power Research Institute, Suzhou, Jiangsu province 215004 China
[2]National Engineering Research Center for Nuclear Power Plant Safety & Reliability, Suzhou, 215004, China

[3]Corresponding author's e-mail: liuxbing@cgnpc.com.cn.

**Abstract**: Due to the extensive use of ferrite in nuclear power plants, the thermal aging issues of ferritic alloys have attracted significant attention. The present study investigates thermally aged FeCr binary ferritic alloys with different Cr contents using various mechanical and magnetic properties tests. The result indicates that the increase in Cr element content and prolonged thermal aging significantly enhances the strength of alloy materials, along with an important reduction in toughness. Meanwhile, the magnetic characteristic shows a considerable difference in different ferritic alloys with varying Cr contents under the influence of thermal aging. As a potential characterization method, magnetic measurement can be used to determine the thermal aging degree in ferritic alloys and the Cr concentration should be considered.

## 1. Introduction

The ferritic alloy is extensively employed in the nuclear industry due to its advantageous mechanical properties and corrosion resistance features. Similarly, by introducing a small amount of ferrite into the austenitic matrix, the material strength and corrosion resistance can be significantly improved, thus expanding the range of material applications. For example, Z3CN20.09M austenite-ferrite duplex stainless steel, featuring a minor proportion of ferrite, is commonly used as the primary pipe material in nuclear power plants, as it exhibits excellent mechanical characteristics [1]. The FeCrAl ferrite steel, known for its superior strength and outstanding oxidation resistance, is considered a key option for nuclear fuel cladding materials in the future [2]. Furthermore, ferritic-martensitic stainless steel has broad application prospects in fourth-generation fission reactors and nuclear fusion reactors [3].

However, the chromium content is one of the key factors in determining the properties of ferrite, as well as the related ferritic alloy. Although high chromium content enhances strength, it can lead to a considerable decrease in toughness after prolonged exposure to a high-temperature environment [4]. This is primarily attributed to the local enrichment of chromium caused by spinodal decomposition, which adversely affects the performance of steels like CF8 and Z3CN20.09M with a chromium content of 20 wt.% by reducing toughness and increasing the risk of failure [1]. Similarly, the second-generation ATF FeCrAl cladding restricts the chromium content to 13 wt.%, significantly lower than the 22 wt.% in the first generation [2]. Therefore, it is crucial to comprehend the thermal aging effect on the ferrite with various chromium content, which has a significant impact on the optimization of alloy design criteria and the enhancement of the service performance.

The mechanism of thermal aging embrittlement in duplex stainless steel has been widely studied [1]. Generally, decomposition occurs in the ferritic alloy under thermal aging conditions, where the σ' phase appears and the composition fluctuation is greatly influenced by chromium content, aging temperature, and time [4, 5]. As a means to investigate the α' phase formed by spinodal decomposition, the ferritic alloys with varying chromium content have been widely studied [4-6]. Danoix and Auger [5] pointed out that the α' phase formed by spinodal decomposition is significantly higher than that formed through nucleation and growth mechanism. Xiong et al. [4], building upon previous research on FeCr binary ferritic alloy, utilized the phase field simulation method to determine the spinodal line at approximately 20 wt.% Cr and 400°C. However, most of these studies have focused on the spinodal decomposition behavior at the micro-scale [4, 5]. Nevertheless, the properties influenced by thermal aging with varying chromium content need to be explored as well, which can provide a quantitative understanding of the mechanism and new method to probe the spinodal decomposition degree and the performance reduction during thermal aging.

Despite the importance, there is rarely reported literature concerning the impact of thermal aging on the mechanical and magnetic characteristics of ferritic alloy with various Cr content [7]. In the present work, prolonged thermal aging is conducted on different FeCr binary alloys. Then mechanical and magnetic properties were systematically studied. Finally, we discuss the impact of thermal aging and Cr content on the physical properties of the ferritic alloys in the mentioned aspects.

## 2. Experiment procedure

As a primary mechanism of thermal aging damage, previous research on binary ferritic alloy shows that spinodal decomposition occurs under the condition that the Cr content mainly ranges from 10 wt.% to 38 wt.%. While the compositional fluctuation in the alloys is mainly driven by the nucleation-growth mechanism when Cr content is less than 10 wt.%. Specifically, the spinodal line is around 20 wt.%, which may vary in different studies but is always within 40 wt.% [4]. Therefore, four ferritic Fe-Cr binary alloys with different Cr contents of 9.78 wt.%, 18.85 wt.%, 25.45 wt.%, and 37.61 wt.% were prepared by vacuum induction melting method, which is nominated as Fe-10Cr, Fe-19Cr, Fe-25Cr, and Fe-38Cr respectively. More details about the chemical composition are reported in previous studies [8].

Then different quenching processes were implemented to suppress austenite formation and obtain a fully ferritic microstructure in Figure 1. Despite the difference in grain size, the ferrite phase has been verified through X-ray diffraction analysis and FERITSCOPE FMP30 from Helmut Fischer GmbH. More details about the preparation and microstructure of the material in the virgin state have been reported in previous studies [8].

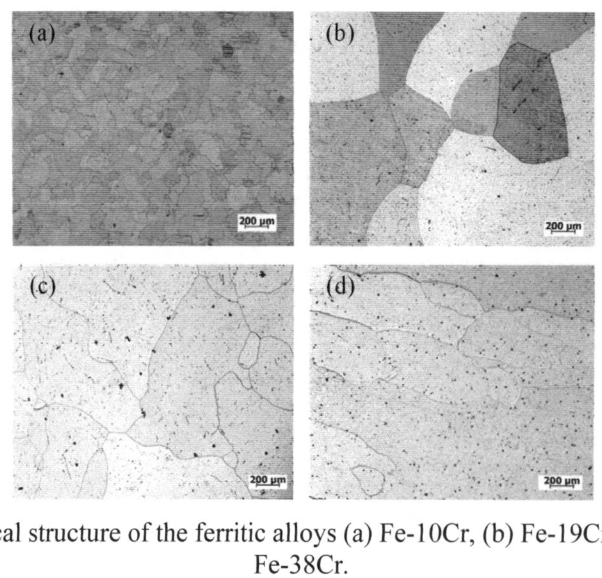


**Figure 1.** Metallurgical structure of the ferritic alloys (a) Fe-10Cr, (b) Fe-19Cr, (c) Fe-25Cr, and (d) Fe-38Cr.

The alloys are exposed to prolonged thermal aging at 400°C for different times. Then the room temperature tensile and V-notch instrumented impact tests were carried out using respectively the SHIMADZU AG-IC 100KN and Zwick RKP 450 instruments at room temperature. It should be noted that the whole process adheres to the ASTM specification.

Besides, studies show that the dislocations interact closely with the magnetic domain walls inside the material, which affects the magnetic properties of materials [7]. Therefore, the investigation of the magnetic parameters could provide a potential nondestructive testing approach to probe the thermal aging of the material. The materials were prepared into dimensions of 1 mm × 1 mm × 3 mm and the magnetic hysteresis loop was obtained using the Lake Shore 7407 vibrating sample magnetometer (VSM). Then the parameters such as the saturation magnetization and coercive force were further analyzed from the measured data.

## 3. Results and analysis

### *3.1 Tensile behavior*

The true stress-strain curves of the various ferritic alloys after prolonged thermal aging are compared in Figure 2. By comparing the tensile properties of ferritic alloys in the virgin state, it can be observed that the yield strength exhibits a significant increase as the Cr content rises, while the ultimate strength shows an upward trend. However, the elongation experiences a sharp decrease due to diminished toughness. The presence of higher concentrations of Cr element exerts a substantial influence on both the strength and toughness, resulting in a more brittle characteristic.

After long-term thermal aging, all the ferritic alloys show a strength increase, which is evident from the rise in both yield strength and tensile strength, with an enhancement effect observed with increasing Cr content. However, due to the substantial reduction in elongation and toughness, high Cr-content ferritic alloys experience brittle fracture before reaching their ultimate tensile strength or even theoretical yield strength after thermal aging. Notably, the decrease in toughness is particularly pronounced in ferritic alloys with Cr contents of 19 wt.% and 25 wt.%.

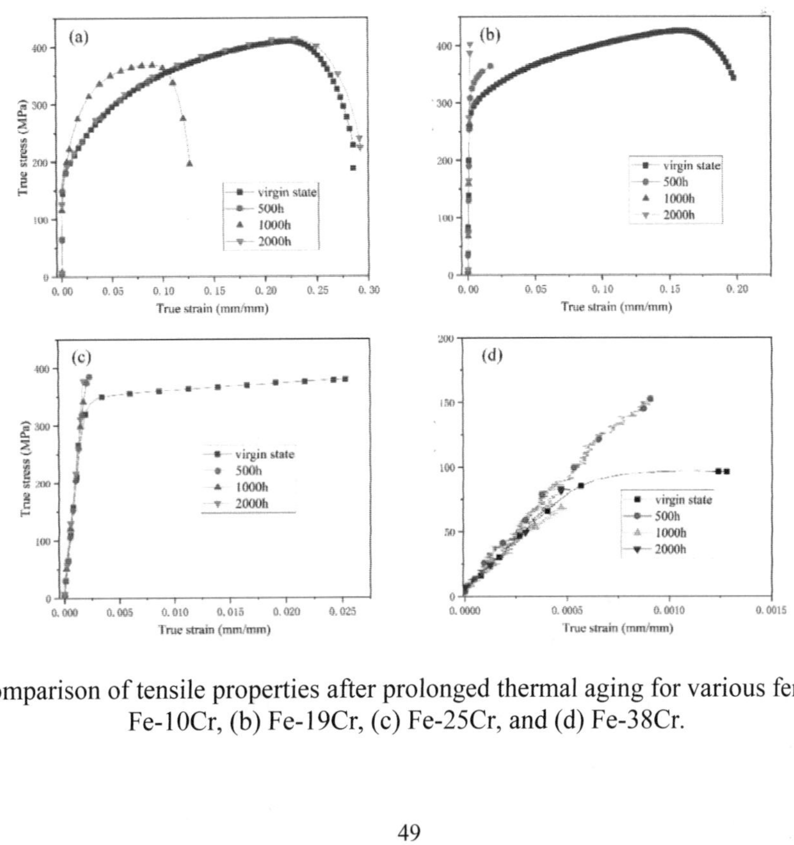


**Figure 2.** Comparison of tensile properties after prolonged thermal aging for various ferritic alloys (a) Fe-10Cr, (b) Fe-19Cr, (c) Fe-25Cr, and (d) Fe-38Cr.

*3.2 Charpy impact test*

The Charpy impact curves of the thermally aged ferritic alloys are compared in Figure 3. By comparing the Oscillo graphic impact properties of the ferritic alloys in the virgin state, it is shown that the Cr content has little impact on elastic deformation, as the curves exhibit similar behavior at the linear elastic stage. Nonetheless, the unloading displacement experiences a substantial decrease as the increasing Cr content. This observation can be attributed to a notable reduction in toughness, which correlates with the decreasing impact energy.

Specifically, both the plastic deformation and final fracture mode during the impact process are affected by the different Cr contents. In both Fe-10Cr and Fe-19Cr ferritic alloys, plastic deformation occurs before reaching the ultimate load, followed by partial stable crack propagation, while the process of stable crack growth is notably more prominent in Fe-10Cr. Finally, the Fe-25Cr and Fe-38Cr ferritic alloys reach the ultimate load, exhibiting a negligible stable crack growth process. Furthermore, the slope of the unloading process is also strongly affected by different Cr content. The unloading angle experiences a significant increase and the Oscillo graphic curve progressively sharpens and steepens during the unloading process, indicating a rapid brittle fracture with instantaneous crack propagation due to reduced toughness.

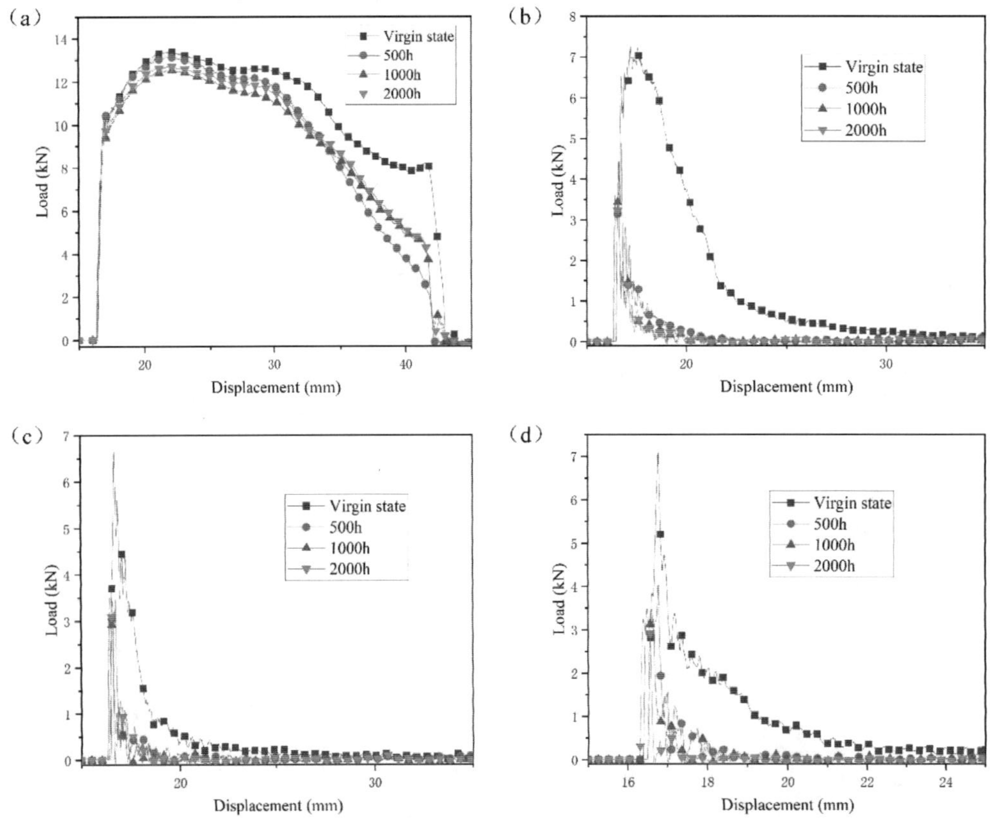


**Figure 3.** Comparison of impact property after prolonged thermal aging for various ferritic alloys (a) Fe-10Cr, (b) Fe-19Cr, (c) Fe-25Cr, and (d) Fe-38Cr.

The impact energy of the ferritic alloys has been further analyzed and the effect of thermal aging is shown in Figure 4. In the virgin state, a global decrease in impact energy along with the rising Cr content is observed, which indicates a substantial increase in material brittleness as the Cr content rises. Compared to the virgin state, the impact toughness of the ferritic alloys significantly decreases with prolonged thermal aging time until the saturation plateau, particularly for the high Cr content alloys. This could be attributed primarily to the increase in material hardness and toughness decrease caused by spinodal decomposition [4-7].

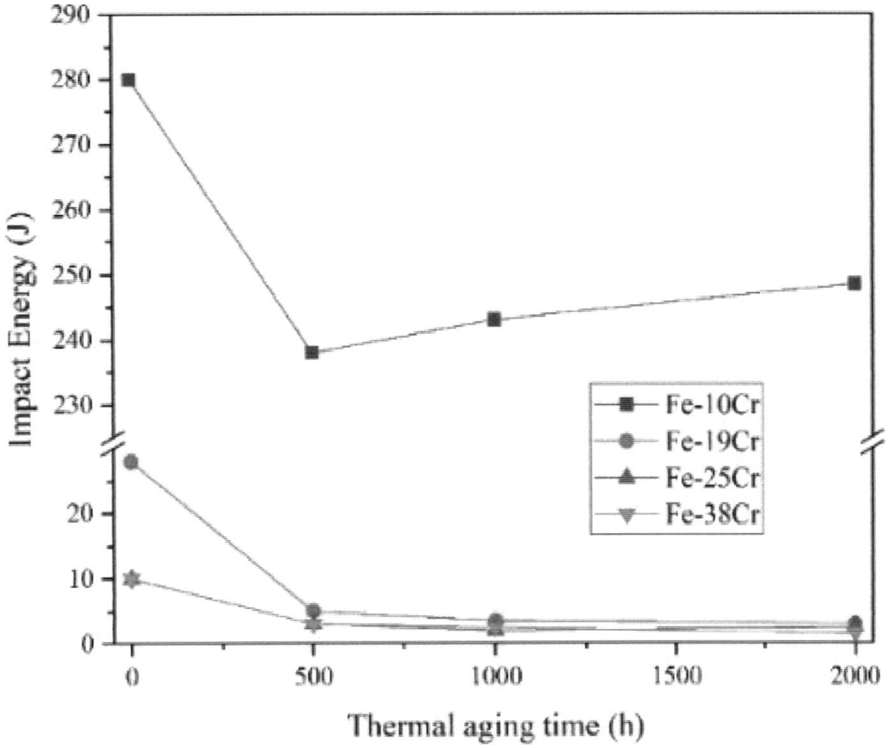


**Figure 4.** Effect of thermal aging on the impact energy for various ferritic alloys.

*3.3 Magnetic properties*
According to microscopic magnetic theory and thermodynamic principles, a stable magnetic state corresponds to a state where the Gibbs free energy within ferromagnetic materials is minimized [9]. In ferromagnetic materials, when a magnetic domain wall is located at the position of a dislocation, the Gibbs free energy decreases and the dislocation acts as a pinning point, hindering the movement of the domain wall and affecting the magnetic characteristics of the material.

The comparison of magnetic hysteresis loops (B-H curve) of the ferritic alloys with varying Cr contents after prolonged thermal aging is shown in Figure 5. Globally, all the materials show a soft magnetic characteristic, featuring narrow hysteresis loops that are symmetric concerning the origin. The materials exhibit high magnetic permeability, low coercive force (Hc), and residual magnetization (Br). However, as can be observed from the enlarged plot near the zero point, there are still differences in the coercive force and residual magnetization of the materials after being subjected to different thermal aging times. By analyzing the hysteresis loop, conventional magnetic characteristic parameters such as coercive force, residual magnetization, and saturation magnetization can be obtained. The following analysis mainly focuses on the variations of the magnetic characteristic parameters of the ferritic alloys during the thermal aging process.

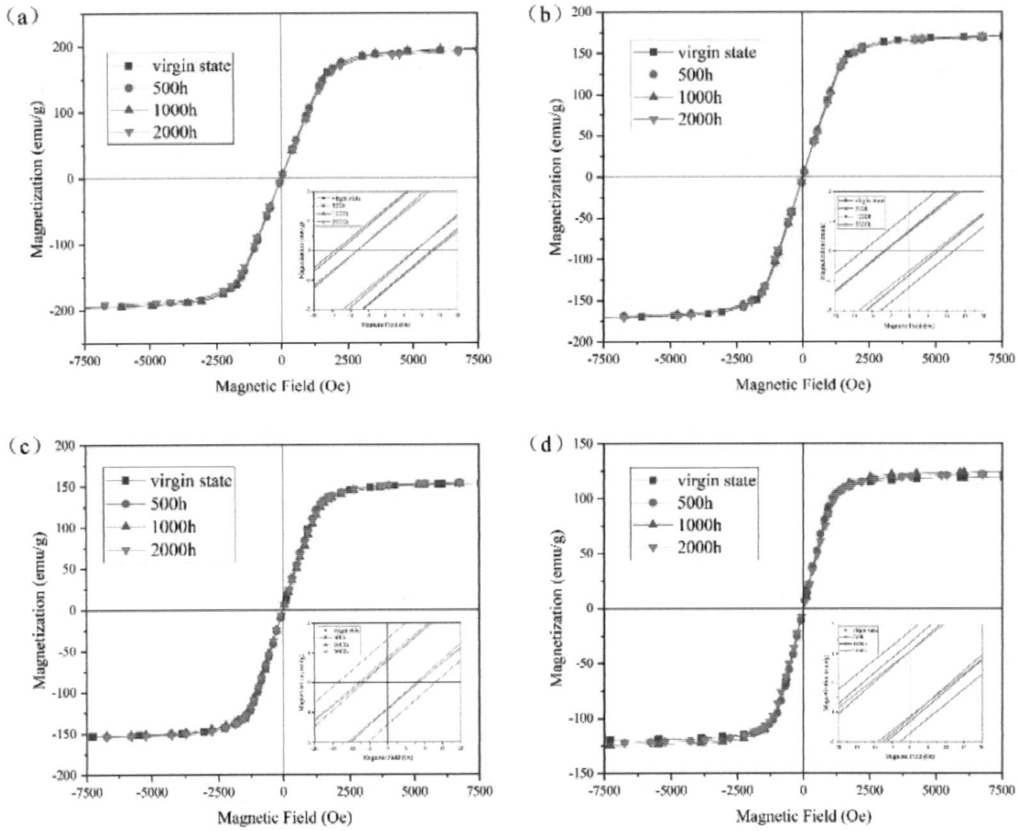


**Figure 5.** Comparison of magnetic hysteresis loop after prolonged thermal aging for various ferritic alloys (a) Fe-10Cr, (b) Fe-19Cr, (c) Fe-25Cr, and (d) Fe-38Cr.

The coercive force represents the magnitude of the external magnetic field required for the demagnetization process to reach a magnetic field intensity of zero. It reflects the resistance encountered by domain walls in overcoming defects during the demagnetization process, which means that a higher coercive force indicates greater resistance encountered during demagnetization. Studies show that the coercive force exhibits complex transformation characteristics due to multiple factors such as defects and lattice distortions, during the damage process such as thermal aging [7]. This poses challenges in the characterization of the damage extent by coercive force, as confirmed by the results of this experiment.

As shown in Figure 6, the variation in coercive force during the thermal aging process differs significantly among the various ferritic alloys. When the Cr element content is within the range of 19-25 wt.%, the coercive force shows regular responsiveness to the thermal aging time. Particularly, the coercive force of Fe-19Cr ferritic alloy exhibits a monotonic decrease with prolonged thermal aging time.

Therefore, the Cr content level should be carefully considered when the extent of thermal aging in ferritic alloys is assessed by coercive force, with an optimal Cr content of around 19 wt.%.

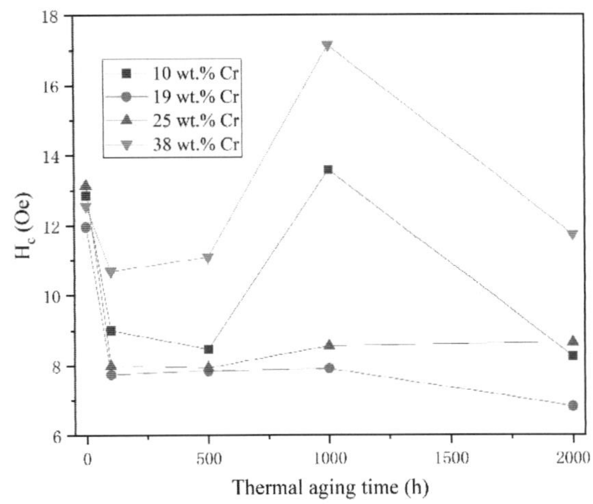


**Figure 6.** Effect of thermal aging on the coercive force for various ferritic alloys.

Previous research indicates that saturation magnetization (Bs) is related to spontaneous magnetization and insensitive to the microstructure of material [7]. However, the residual magnetization is closely associated with the microstructure, defects, and internal stress. From the present experiment results, it can be concluded that the saturation magnetization remains relatively constant with prolonged thermal aging. Therefore, the variation in residual magnetization ratio (Br/Bs) is discussed afterward.

In Figure 7, it can be observed that the variation in the residual magnetization ratio follows a similar trend as the coercive force during thermal aging. For the Fe-19Cr and Fe-25Cr ferritic alloys, the residual magnetization ratio rapidly decreases along with the occurrence of thermal aging damage and then stabilizes. However, ferritic alloys with excessively low or high Cr contents exhibit a significant increase in the residual magnetization ratio after a thermal aging time of 1000 hours. Hence, it is important to consider the reasonable range of compositions when the extent of thermal aging damage is characterized by the residual magnetization ratio.

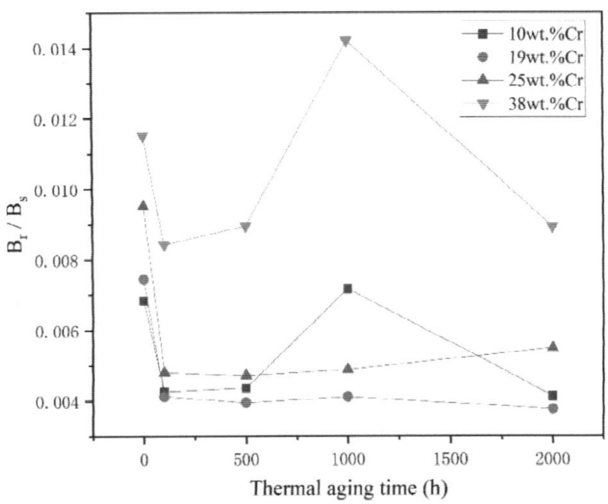


**Figure 7.** Effect of thermal aging on the residual magnetization ratio for various ferritic alloys.

## 4. Conclusion

In summary, to investigate the thermal aging effect on the mechanical and magnetic properties of ferritic alloys, four FeCr ferritic alloys were exposed at 400°C for up to 2000 h. Material mechanical property tests, including tensile and Charpy impact experiments, indicate that an increase in Cr element content and prolonged thermal aging significantly enhance the strength of alloy materials, along with a significant reduction in toughness. Meanwhile, there are considerable differences in the magnetic characteristics of different ferritic alloys with varying Cr contents after prolonged thermal aging. As a potential characterization method, VSM magnetic measurement can be used to determine the thermal aging degree of ferrite and the Cr concentration should be carefully considered.

## Acknowledgments

This work was financially supported by the National Key Research and Development Program of China (2021YFA1600903) and the National Natural Science Foundation of China (12275200, 12075274, 11975171, U23B2072).

## References

[1] Y Fan, T G Liu, L Xin, Y M Han, Y H Lu, and T Shoji (2021) Thermal aging behaviors of duplex stainless steels used in nuclear power plant: A review. J. Nucl. Mater., 544: 152693.

[2] K A Terrani (2018) Accident tolerant fuel cladding development: Promise, status, and challenges. J. Nucl. Mater., 501: 13-30.

[3] C Cabet, F Dalle, E Gaganidze, J Henry, and H Tanigawa (2019) Ferritic-martensitic steels for fission and fusion applications. J. Nucl. Mater., 523: 510-537.

[4] W Xiong, P Hedström, M Selleby, J Odqvist, M Thuvander, and Q Chen (2011) An improved thermodynamic modeling of the Fe–Cr system down to zero kelvin coupled with key experiments. Calphad, 35(3): 355-366.

[5] F Danoix, and P Auger (2000) Atom Probe Studies of the Fe–Cr System and Stainless Steels Aged at Intermediate Temperature: A Review. Mater. Charact., 44(1): 177-201.

[6] A Kuronen, S Granroth, M H Heinonen, R E Perälä, T Kilpi, P Laukkanen, J Lång, J Dahl, M P J Punkkinen, K Kokko, M Ropo, B Johansson, and L Vitos (2015) Segregation, precipitation, and alpha-alpha ' phase separation in Fe-Cr alloys. Phys. Rev. B., 92(21): 214113.

[7] V Vojtech, M Charilaou, A Kovács, A Firlus, S S A Gerstl, R E Dunin-Borkowski, J F Löffler, and R E Schäublin (2022) Macroscopic magnetic hardening due to nanoscale spinodal decomposition in Fe–Cr. Acta Mater., 240: 118265.

[8] W Jia, X Liu, M Fan, C Xu, Y Li, W Qian, Q Quan, and J Yin (2021) A study on mechanical properties of ferritic alloy by physical testing for nuclear power fusion reactors. J. Phys. Conf. Ser., 2083(2): 022088.

[9] S Takahashi, S Kobayashi, H Kikuchi, and Y Kamada (2006) Relationship between mechanical and magnetic properties in cold rolled low carbon steel. J. Appl. Phys., 100: 113908.

# Study on thermal conductivity of micron BN-nano $SiO_2$ combined modified epoxy resin composites

**Yiyi Zhang[1,2,3], Danquan Lan[2], Peng Liu[1], Bo Feng[1], Danyang Xu[2] and Min Wu[1]**

[1]Electric Power Research Institute of Guangxi Power Grid Co., Ltd, Nanning 530000 China

[2]Guangxi Power Transmission and Distribution Network Lightning Protection Engineering Technology Research Center, School of Electrical Engineering, Guangxi University, Nanning, 530004 China

[3]Corresponding author: yiyizhang@gxu.edu.cn

**Abstract:** Epoxy resin has attracted widespread attention in the field of power electronic packaging due to its favorable processing technology, high adhesion, and excellent dielectric properties. The enhancement of thermal conductivity in epoxy resin through the incorporation of diverse particles is a prominent area of interest in the realm of modified insulation materials. In this study, hexagonal boron nitride (h-BN) and silicon dioxide ($SiO_2$) with good thermal conductivity were used as doped fillers to modify epoxy resin (EP). The analysis of experimental results indicated that the thermal conductivity of materials with a total doping amount of 20 wt% and a mass ratio of BN to $SiO_2$ of 70:30 to 60:40 could increase the thermal conductivity by 118.1~126.0%. In contrast to pure EP, the thermal conductivity of composite epoxy resin materials has been significantly enhanced.

## 1. Introduction

With the rapid advancement of microelectronic technologies, there are many kinds of electronic packaging materials, including epoxy resins (EP), silicones, and polyurethanes [1, 2]. Among them, EP is widely used because of its excellent electrical insulation performance, chemical resistance, easy processing, and molding, and is used in the insulation and packaging of electronic components [3-5]. However, as an amorphous polymer, EP has neither free electrons of metal compounds, nor a large number of lattice vibrations, and the overall thermal conductivity is very poor, usually only 0.17 $W \cdot m^{-1} \cdot K^{-1}$. Low thermal conductivity increasingly limits the rapid development of epoxy resins in electrical and electronic applications. Therefore, how to maintain the insulation and dielectric properties of epoxy resin itself, while improving its thermal conductivity, so that it can meet the needs of today and tomorrow, is still a huge challenge.

In recent years, the research on the combined modification of epoxy resin by a variety of particles has set off a boom, and scholars at home and abroad have prepared composite insulation materials with excellent performance by doping epoxy resin with nanoparticles. The selected materials usually include alumina ($Al_2O_3$), boron nitride (BN), silicon dioxide ($SiO_2$), carbon nanotubes, glass fibers, and other materials, and the material scale is micron or nanoscale. Metallic and non-metallic oxide particles usually have high thermal conductivity and excellent insulation and can be mixed uniformly with epoxy resin, which can obtain better electrical properties. BN is considered the most promising candidate, given its excellent mechanical strength, thermal conductivity, and thermal stability [6]. Wang et al. [7] added micron-scale h-BN and carbon nanotubes (CNTs) with a diameter of 112 nm and a length of 50 μm to the epoxy matrix and experimentally found that when 1.06 vol% BN-CNT was added, the epoxy nanocomposite adhesive had high electrical and mechanical properties, such as high Young's modulus, thermal stability, and so on. When the filling amount reaches 3.79 vol%, the thermal conductivity reaches its maximum value of 0.46 $W \cdot m^{-1} \cdot K^{-1}$, which is improved compared with pure epoxy resin or only doped BN. The resistivity of $SiO_2$ is not high, generally at 1015~1016 Ω·cm, but when used as a nanoscale filler, it can show great advantages. Nano-$SiO_2$ can improve the bonding strength inside the material [8]. The molecular structure of nano-$SiO_2$ is characterized by a three-dimensional chain-like arrangement, with various bonding states of hydroxyl groups and a substantial number of unsaturated residual bonds. These can bond with certain groups of the resin to make the overall molecular structure stronger and greatly improve the hardness and strength of the material on a macroscopic scale. Shen et al. [9] found that 2.5 wt% SiC@$SiO_2$ composites can improve the thermal conductivity of polymers, but when particles are modified, epoxy resins with different filling amounts and proportions have a greater impact on their properties. Suitable proportioning schemes need to be studied for multi-particle combined modified epoxy resins to greatly improve their insulation.

In this paper, 15 groups of composite epoxy resins with different filling amounts and different proportions were prepared by doping micron h-BN and nano-$SiO_2$ micro-nanoparticles into the epoxy resin at the same time. By measuring the thermal conductivity of each group, the contribution of the filling amount and the ratio of two particles to the improvement of thermal conductivity and its corresponding microscopic mechanism were analyzed, and how to obtain higher thermal conductivity at lower total filling was discussed.

## 2. Preparation process of epoxy resin composite

### *2.1 Experimental materials*

E-51 bisphenol A type epoxy resin purchased from the United States Dow Chemical Company, methyl tetrahydrophthalic anhydride (MTHPA) purchased from Kunshan Lvxuan Chemical Co., LTD., 2, 4, 6-tri (dimethylamine methyl) phenol (DMP-30) purchased from Kunshan Lvxuan Chemical Co., LTD., silane coupling agent KH550 purchased from Nanjing Trangenesis Chemical Additives Co., LTD., 1 μm h-BN and 30 nm spherical $SiO_2$ purchased from Shanghai Yaogge Alloy Materials Co., LTD.

### *2.2 Material preparation*

The experimental preparation method is consistent with what we described before [10].

*2.2.1 Surface modification.* The particles and coupling agent were added to a sufficient amount of absolute ethanol according to a mass of 20:1, and placed in an ultrasonic dispersion instrument with ultrasonic shaking for 20 min. Then, the suspension after ultrasonic dispersion was heated in a continuous water bath at 50 °C and stirred by an electromagnetic stirrer for 60 minutes to promote the grafting reaction. Then, the reacted suspension is left to stand for some time and washed several times; finally, the slurry obtained after filtration and cleaning is dried, ground, and crushed to obtain particles with surface modification. The surface modification preparation method of micron BN and nano $SiO_2$ is the same.

*2.2.2 Reparation of epoxy resin composites.* The mass ratio of the epoxy resin, the curing agent, and the accelerator is 100:80:0.5, and the mass ratios of BN and $SiO_2$ are set to 100:0, 90:10, 80:20,

**Figure 1.** Composite epoxy finished product material.

**Table 1.** Different material names and corresponding components appear in this paper.

| Number | Name | Proportion (BN: $SiO_2$) | Doping mass fraction |
|---|---|---|---|
| #0 | EP | 0:0 | 0wt% |
| #2 | BN/Si100-10 | 100:0 | 10wt% |
| #3 | BN/Si100-20 | 100:0 | 20wt% |
| #4 | BN/Si90-5 | 90:10 | 5 wt% |
| #5 | BN/Si90-10 | 90:10 | 10wt% |
| #6 | BN/Si90-20 | 90:10 | 20wt% |
| #7 | BN/Si80-5 | 80:20 | 5 wt% |
| #8 | BN/Si80-10 | 80:20 | 10wt% |
| #9 | BN/Si80-20 | 80:20 | 20wt% |
| #10 | BN/Si70-5 | 70:30 | 5 wt% |
| #11 | BN/Si70-10 | 70:30 | 10wt% |
| #12 | BN/Si70-20 | 70:30 | 20wt% |
| #13 | BN/Si60-5 | 60:40 | 5 wt% |
| #14 | BN/Si60-10 | 60:40 | 10wt% |
| #15 | BN/Si60-20 | 60:40 | 20wt% |

70:30, and 60:40, and the total filling amount of mixed particles added to epoxy resin is set to 5 wt%, 10 wt%, and 20 wt%. Composite epoxy resin materials with different filling amounts and ratios were prepared by electromagnetic stirring, water bath heating, ultrasonic dispersion, vacuum deaeration, high-temperature curing, and other processes. The prepared composite epoxy resin material is shown in Figure 1. The naming and corresponding composition are shown in Table 1.

## 3. Thermal conductivity of materials

### *3.1 Thermal performance test principle and experimental system*

When different objects touch each other, or when the temperature inside the object is uneven, heat transfer will occur spontaneously. Heat transfer occurs mainly in three ways: heat radiation, heat convection, and heat conduction. Among them, thermal radiation propagates around the object by electromagnetic waves, does not require any medium, and can be transmitted between substances at the speed of light even in a vacuum. Thermal convection mainly occurs in gaseous and liquid objects, and material exchange occurs directly in parts at different temperatures. Since the volume and density of gases and liquids change significantly at different temperatures, gases or liquids at different temperatures will flow and mix under the action of gravity. The adequate flow of gaseous liquids can quickly reduce the overall temperature gap, which is a very efficient method of heat transfer. Heat conduction is the direct transfer of energy from the movement of adjacent material molecules, and it will occur autonomously from the high-temperature object to the low-temperature object, and from the high-temperature region inside the object to the low-temperature region. And for different substances such as metals, metal compounds, graphene, etc., due to their different microstructure, heat conduction performance differences will also be huge. Owing to the abundance of free electrons inside, these materials such as copper tubes, alloy heat mesh, graphene sheets, etc., usually have better thermal conductivity, and usually can be used as a heat conduction medium in the dissipation system, and polymers such as epoxy resin, cross-linked polyethylene usually have poor thermal conductivity, which can be used as an insulating material wrapped in electronic components. The heat generated by the continuous operation of the components will accumulate in these materials, which brings problems such as heat aging.

In solid objects, heat convection cannot be generated, and the efficiency of thermal radiation is very limited, and the main way of heat conduction is heat conduction. When heat conduction occurs inside the object, it results in the spontaneous transfer of heat from a region of higher temperature to one of lower temperature, creating a temperature gradient with successively decreasing temperatures, and each layer is called the isothermal surface. Layers of isothermal surfaces form a temperature gradient, and when the heat is input, the temperature gradient formed by materials with different thermal conductivity is not the same. Therefore, Fourier's law calculates the thermal conductivity of the object according to the heat passed through the object interface and the corresponding temperature gradient, and the formula is as follows:

$$J_T = -\kappa \frac{dT}{dx} \tag{1}$$

Where $J_T$ represents the heat flux density ($W{\cdot}m^{-2}$), which represents the heat transfer rate in the direction of transmission, that is, in the $x$ direction, on the unit area of the selected cross-section; $\frac{dT}{dx}$ represents the temperature gradient; the proportional constant $\kappa$ of the two represents the heat transport characteristics of the object, that is, the thermal conductivity, also known as the thermal conductivity, in units $W{\cdot}m^{-1}{\cdot}K^{-1}$. The symbol implies the transfer of heat from a region of high temperature to a region of low temperature through conduction.

For the same material of isotropic, uniform medium, the thermal conductivity at the same temperature is constant. When temperature changes, the thermal conductivity will also change accordingly, and some materials will change more obviously, such as pure metals and most liquids other than water. Thermal conductivity will decrease with the increase of temperature, and the thermal conductivity of anisotropic materials in different directions will also be different. In addition to thermal conductivity, there is also the thermal diffusion coefficient as a rate to characterize thermal diffusion:

$$\alpha = \frac{\lambda}{\rho c} \tag{2}$$

Where $\rho$ represents the density of the object in $kg{\cdot}m^{-3}$, $c$ is the specific heat capacity of the object in $J{\cdot}kg^{-1}{\cdot}K^{-1}$, $\lambda$ represents the thermal conductivity, in $W{\cdot}m^{-1}{\cdot}K^{-1}$, and $\alpha$ represents the thermal diffusivity unit $m^2{\cdot}s^{-1}$. Therefore, when the density and specific heat capacity of the material are known, the thermal conductivity and thermal diffusion coefficient are directly proportional.

In this paper, the LFA467 laser thermal conductivity analyzer produced by Netzsch, Germany is adopted. The measuring temperature range of this analyzer can support -100 °C ~ 500 °C, the data acquisition rate is up to 2 MHz, and the measuring range of thermal diffusion coefficient is 0.01 $mm^2{\cdot}s^{-1}$ ~ 2000 $mm^2{\cdot}s^{-1}$. The measurement range of thermal conductivity is 0.1 $W{\cdot}m^{-1}{\cdot}K^{-1}$ ~ 4000 $W{\cdot}m^{-1}{\cdot}K^{-1}$. The device uses the laser pulse of the flash method as the test method, and the principle of this process is illustrated in Figure 2.

The material to be measured needs to be processed to a specified size, placed on a transparent mount (not drawn in the figure), and irradiated with a laser pulse under the material. The laser beam needs to pass through a grating to keep the irradiated area consistent with the material and control the error. Laser pulse irradiation will cause high temperature on the bottom surface of the material, thereby forming a longitudinal temperature gradient, and then the temperature of the top surface can be measured by the infrared probe in a non-contact manner. The temperature gradient of the material can be measured, allowing for the calculation of its thermal conductivity and thermal diffusivity.

Due to the strict requirements of the thermal conduction experiment on the size of the material, the material was processed with fine cutting and grinding processing in this experiment, where the material was unified into rectangular sheets (10 mm × 10 mm), while the thickness and mass were finely measured and the density was calculated. To simulate the long-term operating conditions of the component, the test temperature was selected to be 90 °C. The laser voltage is 250.0 V and the pulse width is 0.60 ms. The standard + pulse correction model was used for calculation. Since repeated measurements of the same material have almost no deviation, each material is flashed three times and calculated.

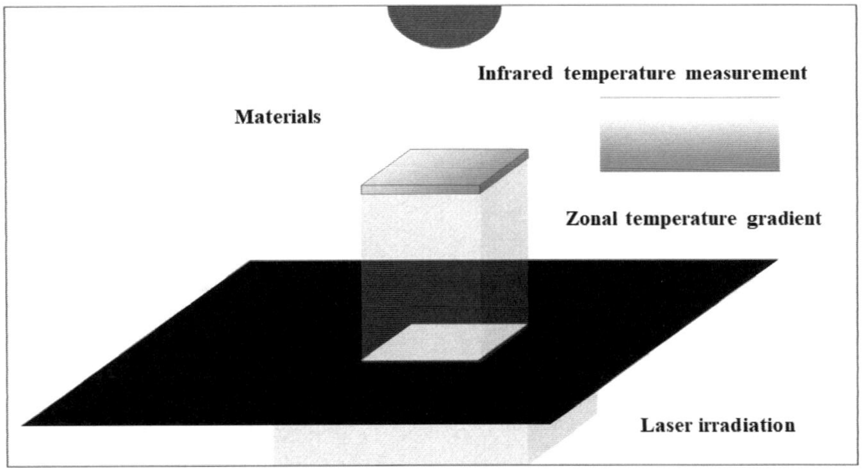


**Figure 2.** Measuring principle of LFA 467 thermal conductivity instrument with flash method.

*3.2 Analysis and discussion*

*3.2.1 Experimental data analysis.* The results of Figure 3 showed that the thermal conductivity of pure epoxy resin is low at 0.177 W·m$^{-1}$·K$^{-1}$. After doping, most materials received some improvement. When the doping and packing amount is 5nwt%, the thermal conductivity of the material is slightly enhanced but not obvious, and the thermal conductivity is from 0.171~0.230 W·m$^{-1}$·K$^{-1}$; when the doping amount is 10 wt%, the results are further improved, and the thermal conductivity distribution is as low as 0.194 W·m$^{-1}$·K$^{-1}$, up to 0.233 W·m$^{-1}$·K$^{-1;}$ when the doping and packing amount was further raised to 20 wt%, the thermal conductivity was qualitatively improved, and the material improvement of BN/Si100 was not obvious, only 0.233 W·m$^{-1}$·K$^{-1}$, but the thermal conductivity of BN/Si70 and BN/Si60 increased to 0.386 W·m$^{-1}$·K$^{-1}$ and 0.400 W·m$^{-1}$·K$^{-1}$, respectively, which is 218.1% and 226.0% of the thermal conductivity of the original pure epoxy resin, which means it has been greatly improved compared with the original material.

From the point of view of the filling amount, the increase of the filling amount on the thermal conductivity is not linear, but there is a threshold effect: a low filling amount, such as 5 wt% and 10 wt% can not play a significant effect, and when the filling amount reaches a certain value, such as 20 wt%, thermal conductivity will get a sudden increase. From the perspective of the ratio, the appropriate introduction of nano-$SiO_2$ can increase the thermal conductivity, and this change can only be seen with a small trend at 5 wt% and 10 wt%, but at 20wt%, gradually increasing the ratio of nano-$SiO_2$ can achieve an approximately linear increase in thermal conductivity. At 30%-40%, the growth gradually slows down and saturation occurs. The mass ratio of 3:1 is better than 2:2 and 4:0 when using micron BN and nano-$Al_2O_3$ for combined modification, which is close to the conclusion of researchers [11]. It shows that micron BN is still needed as the main force to enhance the thermal conductivity of epoxy resin, and the addition of nano $SiO_2$ can promote BN's effect, but it cannot maintain high thermal conductivity when nano $SiO_2$ is the main doping filler.

*3.2.2 Discussion of the microscopic mechanism of thermal conductivity.* According to the measurement results of the broadband dielectric analyzer, the real spectrum of the dielectric constant of 15 groups of BN/$SiO_2$/EP composite materials was plotted according to different proportions, and the relationship between the spectrum of the real part of the dielectric constant and the filling amount was analyzed. Generally, the dielectric constant real part of pure epoxy resin and each composite epoxy resin material decreased with the increase of frequency, gradually decreasing from 4.2~4.4 to 3.8~4.1, by about 10% with frequency. This is because the epoxy resin is a macromolecular structure, and the polarization process is dominated by dipole polarization. This process takes a certain amount of time, and the higher the applied electric field frequency is, the more difficult it is for the dipole steering rate to keep up with changes in the external electric field. Therefore, in the high-frequency region, the internal electric field of the material gradually weakens with increasing frequency, so the overall dielectric constant continues to decrease.

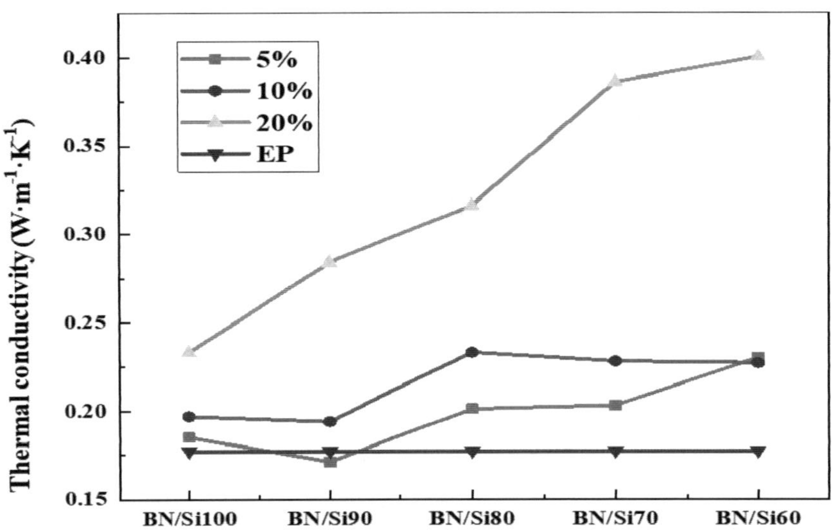


**Figure 3.** The thermal conductivity of composite epoxy resin materials was measured.

In solid materials, the internal heat conduction process is mainly in the form of energy transfer between the thermal movement of molecules or atoms and the surrounding particles. The amplitude of the thermal motion of the molecules in the micro is relative to the macro temperature, and the higher the temperature of the object is, the faster the molecular and atomic vibration rate and the greater the kinetic energy will be. In solid materials, two types of particles are mainly involved in heat transfer, namely, free electrons that can transfer motion within the material and lattice vibration. The energy in lattice vibration is quantized and each part is called phonon. The two can be calculated superimposed:

$$k = k_e + k_p \tag{3}$$

Where $k$ represents the thermal conductivity of the material, $k_e$ represents the thermal conductivity corresponding to the free electron heat conduction, and $k_p$ is the thermal conductivity corresponding to the phonon heat conduction. According to the object characteristics of different materials, the size and proportion of $k_e$ and $k_p$ are greatly different. There are a large number of free electrons in metal materials, the activity space of electrons is large, and the migration rate is higher than that of phonons, so the migration of electrons and the vibration transmission between them play a dominant role, which makes metals have high $k_e$ values and thermal conductivity [12]. In some crystalline compounds, the constituent atoms can be orderly arranged in the lattice, and the atoms vibrate in their respective crystal positions, thus interacting with the surrounding atoms, which is an approximately elastic process. Therefore, the heat vibration can be transferred to the surrounding area with high efficiency in the form of elastic waves. In this way, the lattice vibration is more obvious, the phonons can gradually diffuse from the direction of decreasing concentration gradient, and the material has a large $k_p$, such as AlN and BN [12], and can also have good thermal conductivity. Materials such as graphene and carbon nanoparticles have both a crystal structure and a "large Π bond" composed of many free electrons, both of which can efficiently transfer heat, and the thermal conductivity can usually exceed 2000 $W \cdot m^{-1} \cdot K^{-1}$. The aggregate material is a polymer structure, which is connected by a large number of the same units to form a long chain, and then folded to form a network structure. Therefore, when the heat is transferred in it, a variety of vibration modes will occur, such as stretching vibration, swaying vibration, and torsional vibration. These vibrations are anharmonic coupled vibrations, in which the atoms in the polymer do rotational motion or disordered vibrations near the equilibrium position, and the energy is dispersed to the surrounding groups. This form of vibration travels very fast. Therefore, amorphous polymers such as epoxy resins and polypropylene have very low thermal conductivity. If crystals can be formed inside the polymer, the regular lattice structure can make the polymer have some of the thermal conductivity characteristics of crystal compounds, such as high-density polyethylene, whose thermal conductivity can reach 0.55 $W \cdot m^{-1} \cdot K^{-1}$, which is significantly higher than that of epoxy resin and polypropylene [13].

When one or more substances are doped in the polymer, the path of heat transfer is changed. This mechanism is usually well explained by the theory of thermal conductivity networks [14]. To enhance the thermal conductivity of composite materials, it is necessary to construct a continuous path with low thermal resistance, providing a shortcut for heat transfer. So the heat is mainly propagated along the thermal network, and the ability to build a thermal network plays a crucial role. The content and distribution of the fillers determine the integrity of the thermal conductivity network. When the amount of doping is small, as shown in Figure 4(a), the doped particles will be dispersed in the polymer matrix in isolation, forming a "sea-island" two-phase system. In this way, only the parts with high thermal conductivity in the local area are in a series structure with the epoxy matrix, so an effective thermal conductivity network cannot be formed. This is reflected in the experiment, where the thermal conductivity of 10 wt% doped micron BN composite epoxy resin is only 11.3% higher than that of pure EP. As the filling amount increases, the doping filler can form a through-line thermal conductivity path, and the thermal conductivity network is thus generated. In Figure 4(b), the stacked micron BNS form a thermal conduction channel that is in parallel with the epoxy matrix, so that heat can be transferred quickly between them. This corresponds to the 31.6% increase in the thermal conductivity of the composite epoxy resin at the filling amount of 20 wt% micron BN in the experiment. Generally, continuing to increase the BN content can establish more thermal conductivity channels in the epoxy matrix, form a dense thermal conductivity network, and further improve thermal conductivity [15]. On this basis, the introduction of a small amount of nano-$SiO_2$ can slightly improve the thermal conductivity, as shown in Figure 4(c). The micron BN forms a thermal conduction skeleton, in which dispersed $SiO_2$ can act as bridges and enhance the efficiency of the thermal conductivity network. This corresponds to a 60.5% and 78.5% increase in the thermal conductivity of BN/Si90 and BN/Si80, respectively, at a filling volume of 20 wt%. When the proportion of $SiO_2$ is further increased, as shown in Figure 4(d), more nano-SiO2 will disrupt the stacked structure of the layered micron BN, causing the direction of dispersion of some BN particles to change, thereby improving the longitudinal, that is, the thermal conductivity in the direction of measurement. However, the thermal conductivity network still needs micron BN as the skeleton due to the low thermal conductivity of $SiO_2$ itself, so there is a peak value in the increase of $SiO_2$, corresponding to the experimental thermal conductivity of BN/Si70 and BN/Si60 increased by 118.0% and 126.0%, respectively, and there is an obvious saturation effect.

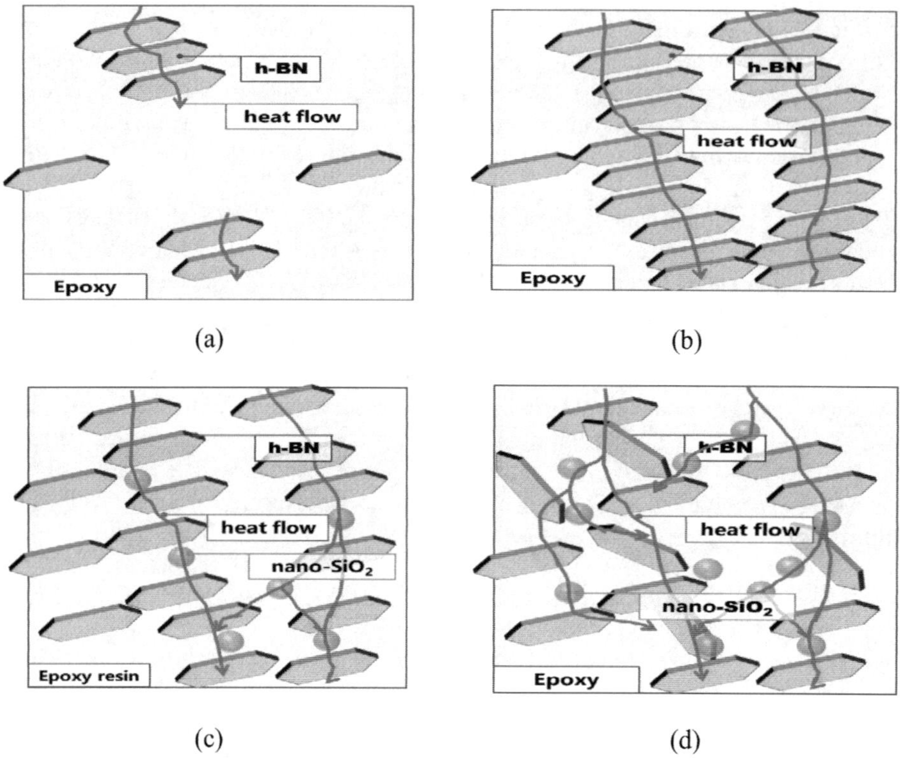


**Figure 4.** The relationship between the spectrum of the real part of the dielectric constant and filling with the ratio at 20 wt% filling: (a) less BN filling, (b) more BN filling, (c) micron BN doped with less nanometer $SiO_2$, (d) micron BN doped with more nanometer $SiO_2$.

In summary, when the filling amount reaches 20 wt%, the addition of $SiO_2$ can further enhance the thermal conductivity of the composite epoxy resin, resulting in a significant improvement in its overall thermal conductivity. Among them, BN/Si70-20 and BN/Si60-20 materials achieved improved thermal conductivity.

## 4. Conclusion

In the micron BN-nano $SiO_2$ composite modified epoxy resin system, micron h-BN acts as the backbone of the thermal conduction network, and its filling amount plays a vital role in improving thermal conductivity. The small filling amount is not enough to bring a significant improvement in thermal conductivity and cannot build a thermal conduction network. When the filling amount is increased to enough (20 wt% and above), the thermal conductivity will be significantly improved; nano-$SiO_2$ acts as a connection to the thermal conduction network, which further enhances the overall thermal conductivity and achieves the effect of a higher filling amount in a single particle. At the same time, nano-$SiO_2$ can disrupt the single arrangement of the original micron BN and promote the longitudinal distribution of the heat conduction network. When the micron h-BN content is sufficient, appropriate filling with $SiO_2$ will further enhance the thermal conductivity. However, as the total filling amount is 20 wt%, and the nano-$SiO_2$ reaches 30% or more in the doped amount, the thermal conductivity is saturated and the increase rate slows down. In the materials in this experiment, the thermal conductivity of BN/Si70-20 and BN/Si60-20 reached 0.386 $W \cdot m^{-1} \cdot K^{-1}$ and 0.4 $W \cdot m^{-1} \cdot K^{-1}$, respectively, compared with pure EP, which is improved by 118.0% and 126.0%, respectively, which effectively improves the thermal conductivity of EP under the condition of lower total fill, and can reduce the occurrence of heat aging.

**Acknowledgment**
This work was supported in part by the National Natural Science Foundation of China (52277138) and in part by the China Postdoctoral Foundation (2022MD713732).

**References**
[1] Y. Wen *et al.*, "Advances on Thermally Conductive Epoxy-Based Composites as Electronic Packaging Underfill Materials-A Review," *Advanced Materials,* vol. 34, no. 52, p. 2201023, 2022/12/01 2022.
[2] Y.-J. Wan, G. Li, Y.-M. Yao, X.-L. Zeng, P.-L. Zhu, and R. Sun, "Recent advances in polymer-based electronic packaging materials," *Composites Communications,* vol. 19, pp. 154-167, 2020/06/01/ 2020.
[3] C. Yu *et al.*, "Enhanced through-plane thermal conductivity of boron nitride/epoxy composites," *Composites Part A: Applied Science and Manufacturing,* vol. 98, pp. 25-31, 2017/07/01/ 2017.
[4] Y. R. Lee *et al.*, "Graphite oxides as effective fire retardants of epoxy resin," *Macromolecular Research,* vol. 19, no. 1, pp. 66-71, 2011/01/01 2011.
[5] C. Xiao, Y. Tang, L. Chen, X. Zhang, K. Zheng, and X. Tian, "Preparation of highly thermally conductive epoxy resin composites via hollow boron nitride microbeads with segregated structure," *Composites Part A: Applied Science and Manufacturing,* vol. 121, pp. 330-340, 2019/06/01/ 2019.
[6] J. Sun *et al.*, "Boron nitride microsphere/epoxy composites with enhanced thermal conductivity," *High Voltage,* vol. 2, no. 3, pp. 147-153, 2017/09/01 2017.
[7] S. Wang *et al.*, "Development of high thermally conductive and electrically insulated epoxy nanocomposites with high mechanical performance," *Polymer Composites,* vol. 42, no. 9, pp. 4217-4226, 2021/09/01 2021.
[8] C. Lu, K. Tieu, and D. Wexler, "Significant enhancement of bond strength in the accumulative roll bonding process using nano-sized SiO2 particles," *Journal of Materials Processing Technology,* vol. 209, no. 10, pp. 4830-4834, 2009/06/01/ 2009.
[9] D. Shen *et al.*, "Enhanced thermal conductivity of epoxy composites with core-shell SiC@SiO2 nanowires," *High Voltage,* vol. 2, no. 3, pp. 154-160, 2017/09/01 2017.
[10] Y. Zhang, D. Lan, P. Liu, D. Xu, and B. Peng, "Preparation and dielectric properties of micron BN-nano SiO2 combined modified epoxy resin composites," *Guangxi Electric Power,* vol. 46, no. 01, pp. 1-7, 2023.
[11] W. Ma, H. Zhao, L. Cui, L. Zhang, G. Du, and X. Han, "Insulation Properties of Epoxy Resin Modified by $Al_2O_3/TiO_2$ Compound," *Polymer Materials Science & Engineering,* vol. 37, no. 02, pp. 61-66, 2021.
[12] Z. Han and A. Fina, "Thermal conductivity of carbon nanotubes and their polymer nanocomposites: A review," *Progress in Polymer Science,* vol. 36, no. 7, pp. 914-944, 2011/07/01/ 2011.
[13] T. Zhang, X. Wu, and T. Luo, "Polymer Nanofibers with Outstanding Thermal Conductivity and Thermal Stability: Fundamental Linkage between Molecular Characteristics and Macroscopic Thermal Properties," *The Journal of Physical Chemistry C,* vol. 118, no. 36, pp. 21148-21159, 2014/09/11 2014.
[14] C.-Y. Hsieh and S.-L. Chung, "High thermal conductivity epoxy molding compound filled with a combustion synthesized AlN powder," *Journal of Applied Polymer Science,* vol. 102, no. 5, pp. 4734-4740, 2006/12/05 2006.
[15] T. Yao, W. Bian, and Y. Yang, "Thermal Conductivity and Insulating Property of Epoxy Composites with Micro-BN and Nano-$Al_20_3$," *High Voltage Engineering* vol. 47, no. 01, pp. 251-259, 2021.

# Comparative study on microstructure and hardness of P92 steel before and after long service

**Yangyang Fu [1, 2], Gaocan Luan [1], Pengfei Zhao [1], Qinzheng Ma [1], Baoyin Zhu [1], Xianxi Xia [1] and Xiao Jin [1, 3]**

[1]Suzhou Nuclear Power Research Institute, Suzhou Jiangsu 215004, China
[2]National Engineering Research Center for Safety and Reliability of Nuclear Power Plants, Suzhou Jiangsu 215004, China

[3]Corresponding author's e-mail: jin_xiao2002@163.com

**Abstract:** Due to its exceptional mechanical properties, the P92 steel is extensively utilized in power plants for high-temperature and high-pressure components, such as main steam and reheated steam pipes. This study aims to characterize the microstructure and Brinell hardness of domestic and imported P92 steel before and after 50 kh of service under identical conditions using SEM (scanning electron microscope), EBSD (electron back scattering diffraction), TEM (transmission electron microscope), and a Brinell hardness tester. The findings reveal that the hardness of domestic P92 steel decreases from 222 HB to 200 HB after 50,000 hours (50 kh) of service, while the imported P92 steel experiences a slight decrease from 205 HB to 203 HB. During prolonged service, the martensitic lath broadening rate in domestic P92 steel surpasses that in imported P92 steel, which primarily accounts for the significant decline in hardness. Additionally, the presence of large-sized Laves phases at matrix lath boundaries, lath bundle boundaries, and grain boundaries in both domestic and imported P92 steels also contributes to the reduction in hardness.

## 1. Introduction

As energy and environmental issues become increasingly prominent, there is an urgent need for power generation methods with higher energy conversion efficiency. In recent years, more and more ultra-supercritical units have been developed, significantly improving power generation efficiency [1, 2]. However, the high-temperature and high-pressure steam poses severe challenges to the service performance of materials. P92 steel is widely used in important components such as main steam and reheat steam pipelines in ultra-supercritical units due to its excellent high-temperature creep resistance and other comprehensive properties [3-5]. Research shows that the heat-resistant martensitic steel P92 performs well due to its complex strengthening mechanisms, including solid solution strengthening, dislocation strengthening, precipitation phase dispersion strengthening, etc. Alloying elements in P92 steel, such as W, Mo, Cr, and Fe, form substitutional solid solutions, causing lattice distortion and strengthening the matrix. Nonmetallic small atoms like C, N, and B can form interstitial solid solutions with Fe. Elements like V and Nb act as strong carbon and nitrogen-forming elements, forming fine dispersed MX compounds that strengthen dispersion [6]. Cr, Mo, Fe, and C can also form $M_{23}C_6$ phases for precipitation strengthening [7]. During prolonged high-temperature service, Laves phases may appear, typically precipitating at grain boundaries, effectively impeding dislocation movement and martensitic lath transformation, thus enhancing the stability of the microstructure through precipitation strengthening. Fine martensite and high-density dislocations prevent grain deformation and dislocation slip during high-temperature use, making them one of the most important strengthening mechanisms for P92 steel [8]. P92 steel strengthens through fine lath-like martensite, grain boundaries, and precipitated phases. Studies on the microstructure evolution and strengthening mechanisms of P92 steel during high-temperature creep conducted by researchers have found that the growth of $M_{23}C_6$ phases can decrease the creep fracture strength of P92 steel. It shows that the initial formation of Laves phases can effectively enhance the creep fracture strength. Still, once they grow to a certain size, they may reduce the creep fracture strength [9]. Lee et al. [10] have found that the formation of creep voids is related to the precipitation of Laves phases at grain boundaries, and the hardness of P92 steel decreases rapidly with the extensive precipitation of Laves phases. Research on domestic P92 steel subjected to 650°C aging treatment revealed that the coarsening rate of lave phases in domestic P92 steel is higher than in imported P92 steel, resulting in lower creep performance [11].

There is a considerable amount of research on the microstructure evolution and high-temperature creep performance of P92 steel after aging [12-14]. However, most studies focus on imported P92 steel, with relatively few studies on domestic P92 steel. Moreover, many studies on P92 steel use long-term high-temperature creep tests to accelerate the simulation of microstructure transformation and performance changes during service. However, the duration of these tests is short, making it difficult to fully reflect the microstructure evolution and performance changes of P92 steel under long-term service conditions [15, 16]. This significantly constrains the effective development of heat-resistant steels for ultra-supercritical units in China. In response to this situation, this paper uses scanning electron microscopy (SEM) and transmission electron microscopy (TEM) to study the microstructure evolution of imported and domestic P92 steel before and after 50 kh of service. Hardness values are the only quantifiable representation of mechanical parameters achievable on-site, serving as an important indicator for evaluating whether materials meet standards and for metal supervision [17]. This paper provides data support for promoting the domestication of P92 steel through intuitive Brinell hardness testing of the mechanical properties of domestic and imported P92 steel before and after service.

## 2. Material and experiment procedure

This experiment focuses on the left-side domestic P92 and right-side imported P92 main steam pipelines of an ultra-supercritical unit boiler in a domestic power plant. Northern Heavy Industries manufactures the domestic P92 steel, and the imported P92 steel is produced by the German company V&M. The study investigates the microstructure and Brinell hardness of the domestic and imported P92 steel after 50 kh of service. A comparison is made with the microstructure and Brinell hardness of the samples not in service (original samples retained during power plant construction). The designations for domestic and imported P92 steel, both before and after 50 kh of service, are as follows: Domestic P92 (0 h), Domestic P92 (50 kh), Imported P92 (0 h), and Imported P92 (50 kh). The actual operating temperature of the main steam pipeline is 605°C, and the actual working pressure is 27.56 MPa. The composition of the material before and after 50 kh of service is shown in Table 1.

**Table 1.** Chemical composition of P92 steel as received and after serviced for 50000 hours (wt %).

| Type | C | Mn | Cr | Mo | V | Ni | Nb | W | B | P | S |
|---|---|---|---|---|---|---|---|---|---|---|---|
| Domestic P9(0 h) | 0.11 | 0.45 | 9.40 | 0.47 | 0.20 | 0.28 | 0.08 | 1.68 | 0.001 | <0.001 | <0.010 |
| Imported P9(0 h) | 0.10 | 0.45 | 9.55 | 0.46 | 0.22 | 0.27 | 0.066 | 1.83 | 0.002 | <0.001 | <0.010 |
| Domestic P92(50 kh) | 0.08 | 0.41 | 8.89 | 0.47 | 0.18 | 0.24 | 0.05 | 1.74 | 0.003 | 0.008 | <0.003 |
| Imported P92(50 kh) | 0.10 | 0.42 | 8.91 | 0.48 | 0.18 | 0.25 | 0.05 | 1.57 | 0.001 | 0.012 | <0.003 |
| ASTM-SA335P92 | 0.07~ 0.13 | 0.30~ 0.60 | 8.50~ 9.50 | 0.30~ 0.60 | 0.15~ 0.25 | ⩽ 0.40 | 0.04~ 0.09 | 1.50~ 2.00 | 0.001~ 0.006 | ⩽ 0.020 | ⩽ 0.010 |

Metallographic and hardness samples were cut from the cross-sections of domestic and imported P92 steel pipes using wire cutting. The samples were ground with sandpaper of different roughness (grit size) in a perpendicular cross direction until a 2500-grit finish was achieved. Subsequently, a polishing machine was used to polish the prepared samples to a mirror-like state. Before and after service, the samples were subjected to Brinell hardness testing using an XHBT-3000ZIII Brinell hardness machine to analyze the trend of hardness changes. For optical microscopy (OM) metallographic samples, ferric chloride hydrochloric acid solution was used for corrosion. Backscattered electron (BSE) samples for electron microscopy and electron backscatter diffraction (EBSD) samples were electrolytically polished with a 10% alcoholic chloric acid solution. The polishing voltage was set at 20 V, and the polishing time ranged from 15 to 20 seconds. The main focus was observing the size, content, and distribution of precipitated phases in the microstructure. Square thin slices with a thickness of 1 mm and a length of 10 mm were cut, mechanically polished to 60 μm, punched into circular slices with a diameter of 3 mm using a punch machine, and then further thinned using a double-jet thinning process to prepare transmission electron microscopy (TEM) samples. Through transmission electron microscopy, observations were made on the morphology and width of lath-like structures and the morphology and size of precipitated phases.

## 3. Result and discussion

### *3.1 Microstructure analysis*

Figure 1 shows the metallographic structures of domestic and imported P92 steel before and after 50 kh of service. Figures 1(a) and 1(c) depict the metallographic structures of domestic and imported P92 steel in the non-service state. It can be observed that the metallographic structures in the non-service state for both domestic and imported P92 steel consist of lath-like tempered martensite. The laths are densely packed, and no δ-ferrite is present. The lath boundaries are clear, and lath bundles are distinctly visible. Figures 1(b) and 1(d) show the metallographic structures of domestic and imported P92 steel after 50 kh of service. Both domestic and imported P92 steel maintain a well-preserved martensitic phase structure with no apparent degradation. Stable lath-like martensite is a primary strengthening mechanism for P92 steel. Martensite is formed by the shear transformation of austenite during rapid cooling. Multiple laths with similar orientations combine to form a lath bundle, and the size of these bundles is closely related to the size of individual laths. Previous research indicates that during the recovery process, precipitated phases, mainly $M_{23}C_6$, form at the original austenite grain boundaries, martensite lath boundaries, and subgrain boundaries, with the appearance of Laves phases and MX phases with increasing service time [13, 18].

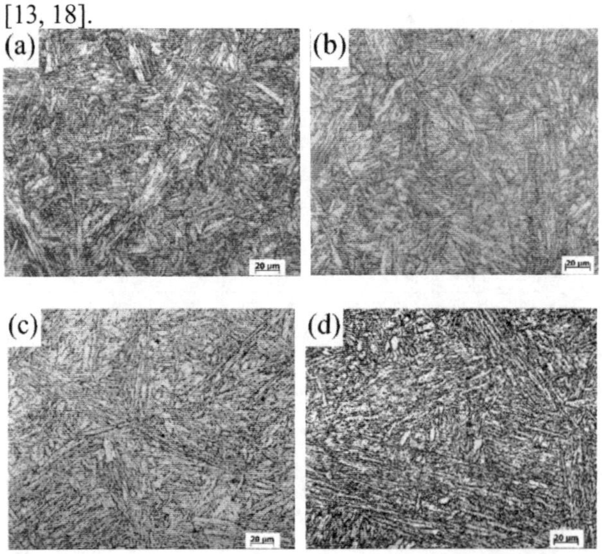


**Figure 1.** Microstructure of domestic and imported p92 steel under different service states: (a) domestic p92 (0 h); (b) domestic p92 (50 kh); (c) imported p92 (0 h); (d) imported P92 (50 kh).

*3.2 Hardness analysis*

Figure 2 presents the Brinell hardness profiles of domestic and imported P92 steel before and after service. Before service, the hardness of domestic and imported P92 steel was approximately 222 HB and 205 HB, respectively. After 50 kh of service, the hardness values decrease to 200 HB for domestic P92 steel and 203 HB for imported P92 steel. The hardness reduction for domestic P92 steel is approximately 10%, while the hardness of imported P92 steel shows no significant decrease.

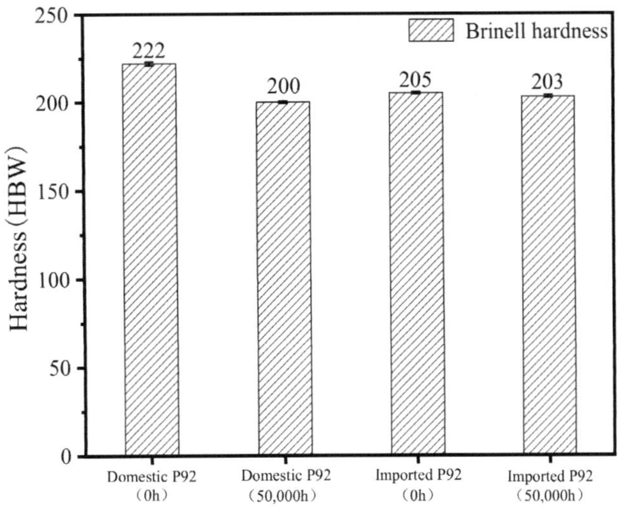


**Figure 2.** Brinell hardness graph of domestic and imported P92 steel under different service states.

*3.3 Backscattered electron analysis*

Figure 3 comprises backscattered electron (BSE) images of domestic and imported P92 steel in different service states. BSE imaging brightness correlates with the atomic number of the sample, where higher atomic numbers result in increased brightness. This allows for qualitative analysis of precipitated phases, effectively distinguishing the locations and distributions of $M_{23}C_6$ and Laves phases in P92 steel [19]. Previous studies indicate that in P92 steel with non-service, the predominant precipitated phases are $M_{23}C_6$ and MX, without Laves phases due to the tempering temperature being higher than the temperature for Laves phase formation (approximately 725°C) [20]. Figures 3(a) and 3(b) show BSE images of domestic and imported P92 steel in the non-service state, revealing $M_{23}C_6$ as the primary precipitated phase at martensite lath boundaries and interiors. The size of $M_{23}C_6$ phases is similar for both domestic and imported P92 steel, approximately 110 μm. The phases are distributed mainly at lath boundaries, lath bundle boundaries, and original austenite grain boundaries. The presence of abundant dislocation structures at these interfaces provides significant energy fluctuations, facilitating the aggregation of alloying elements at boundaries and promoting the generation of precipitated phases. According to the chemical composition of P92 steel, MX phases (V, Nb) (C, N) should form; however, these may be challenging to detect under scanning electron microscopy conditions due to their low content and small size. Figures 3(c) and 3(d) show BSE images of domestic and imported P92 steel after 50 kh of service. Laves phases, characterized by a chemical composition mainly of $Fe_2W$, exhibit higher brightness than $M_{23}C_6$, characterized by a chemical composition mainly of $Cr_{23}C_6$ in BSE images [21]. The growth mechanism of Laves phases at high-temperatures primarily involves grain boundary diffusion [16], leading to a faster growth rate than $M_{23}C_6$. Figure 3 reveals that the size of Laves phases in imported P92 steel is slightly larger than in domestic P92 steel. Compared to the non-service P92 steel, Laves phases have grown significantly, with sizes surpassing 200 nm. This exceeds the beneficial strengthening limit size for Laves phases mentioned by Lee et al. [10] (130 nm critical value). The presence of large-sized Laves phases formed in the matrix after long-term service is identified as one of the contributing factors to the hardness reduction in both domestic and imported P92 steel.

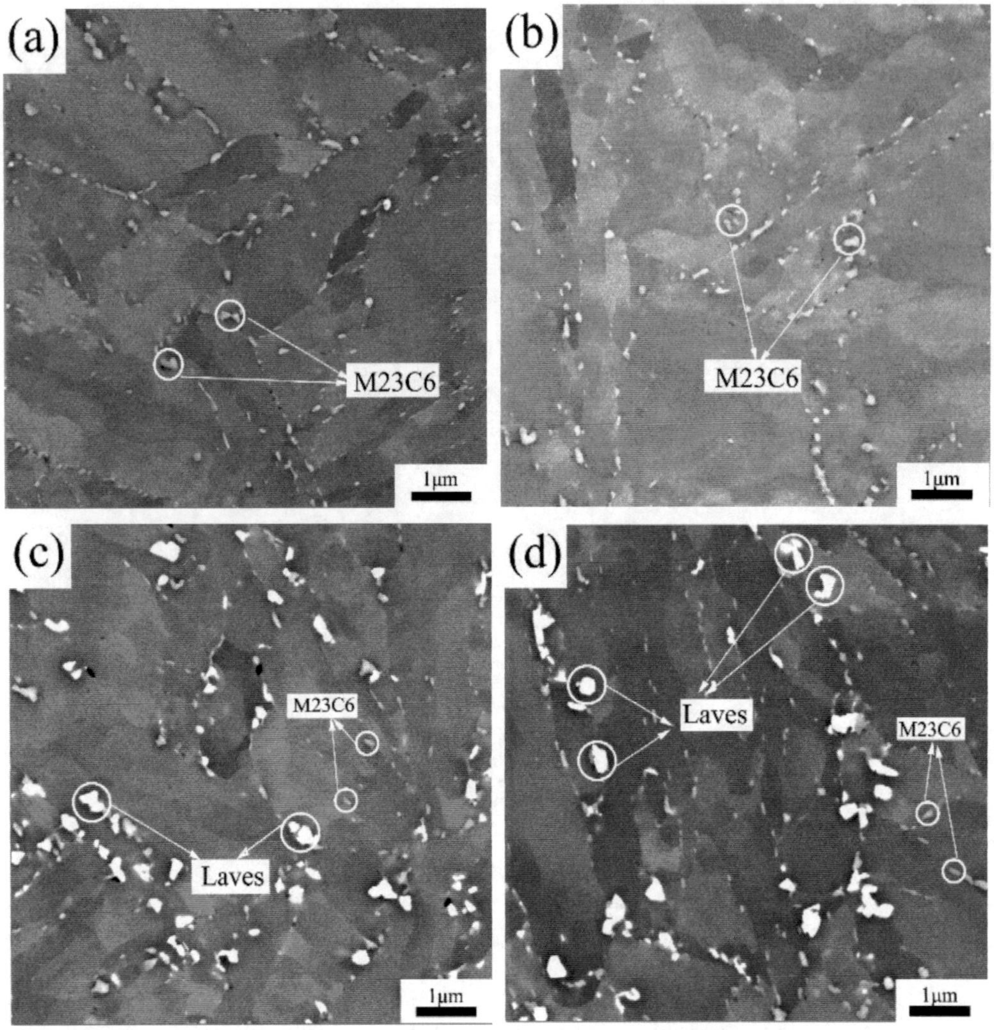


**Figure 3.** BSE graph of domestic and imported P92 steel in different service states.
(a) domestic p92 (0 h); (b) domestic p92 (50 kh); (c) imported p92 (0 h); (d) imported P92 (50 kh).

*3.4 Transmission electron microscopy analysis*

Figure 4 presents transmission electron microscopy (TEM) images of domestic and imported P92 steel in different service states, and the corresponding statistics of lath width variations are shown in Figure 5. Figures 4(a) and 4(c) illustrate that martensite laths of domestic and imported P92 steel are straight in the non-service state, with minimal evidence of precipitated phases. Figures 4(b) and 4(d) show TEM images after 50 kh of service, revealing the presence of precipitated phases at lath boundaries. The quantity and size of precipitated phases in domestic P92 steel are noticeably higher than in imported P92 steel. Image Pro software was used to analyze the width changes of martensite laths statistically, and the results in Figure 5 indicate that the average width of martensite laths for domestic P92 steel increased from approximately 250 nm before service to about 520 nm after 50 kh of service—a widening rate exceeding 100%. The average width of martensite laths for imported P92 steel increased from around 260 nm before service to about 440 nm after 50 kh of service—a widening rate below 70%. These results show that the stability of martensite lath bundles in imported P92 steel is higher than in domestic P92 steel.

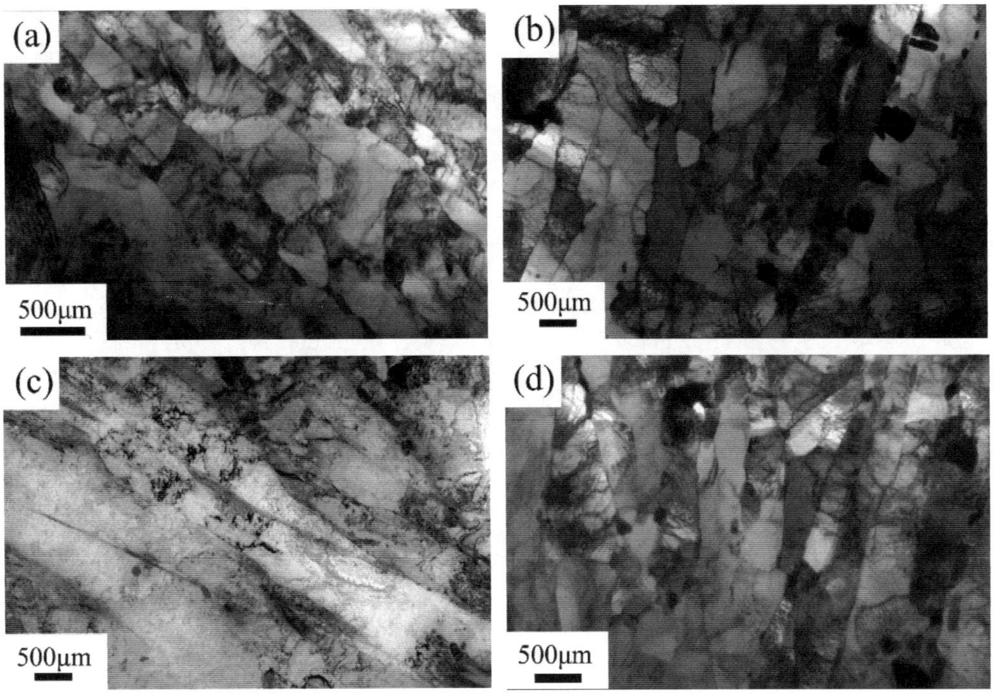


**Figure 4.** TEM graph of domestic and imported P92 steel in different service states. (a) domestic p92 (0 h); (b) domestic p92 (50 kh); (c) imported p92 (0 h); (d) imported P92 (50 kh).

The fine structure of martensitic laths is one of the most crucial strengthening mechanisms for P92 steel. P92 steel undergoes a heat treatment process involving quenching and tempering, forming an over-saturated solid solution martensite lath structure. Combined with abundant lath boundaries, dislocation structures, and alloying elements dissolved within, this structure imparts high strength and hardness to P92 steel. As P92 steel operates at high-temperatures, the martensite lath structure widens. Previous research indicates that martensite lath strengthening accounts for over 60% of the strengthening mechanisms in P92 steel, and the widening of martensite laths is a primary reason for the decrease in strength and hardness of P92 steel [8].

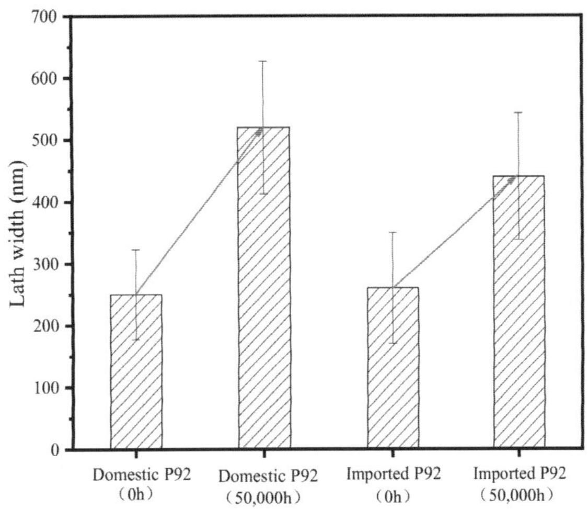


**Figure 5.** The width variation of lath martensite of domestic and imported P92 steel under different service states.

*3.5 Electron back scattered diffraction analysis*

Figure 6 shows the EBSD grain boundary maps of domestically produced and imported P92 steel before and after 50 kh of service. The red lines represent small-angle grain boundaries with orientation differences between 2° and 10°, while the black lines represent large-angle grain boundaries with orientation differences greater than 10°. Figures 6(a) and 6(b) depict the EBSD grain boundary orientation maps for domestic P92 steel before and after service. The black large-angle grain boundaries correspond to the grain boundaries of the original austenite or lath bundle interfaces. In contrast, the red small-angle grain boundaries correspond to lath block interfaces. Small-angle grain boundaries are unstable subgrain boundaries, and their changes over service time are more significant compared to large-angle grain boundaries. Statistical results using Channel 5 software in Figure 6 indicate that the proportion of small-angle grain boundaries in domestic P92 steel increased from approximately 47% before service to about 53% after service, an increase of 6%. In imported P92 steel, the proportion of small-angle grain boundaries increased from 49% before service to 52% after service, an increase of 3%. When the laths are narrow, EBSD is challenging to measure the orientation difference between adjacent laths, yielding information only on the orientation difference of lath blocks. As the laths widen, EBSD gradually measures the orientation difference between adjacent laths, manifested as an increase in the number of small-angle grain boundaries in EBSD. Small-angle grain boundaries are a concentrated manifestation of interfaces such as lath block or lath boundaries in P92 steel. The more severe the widening of the laths is, the greater the number of small-angle grain boundaries is. After long-term high-temperature service, the martensite structure recovers, increasing the measurable subgrain boundaries such as lath bundle or lath boundaries, reflected in the number of small-angle grain boundaries. The changes in small-angle grain boundaries in domestic and imported P92 steel are relatively small. However, regarding numerical changes, the variation in small-angle grain boundaries in imported P92 steel is smaller, indicating a more stable lath structure.

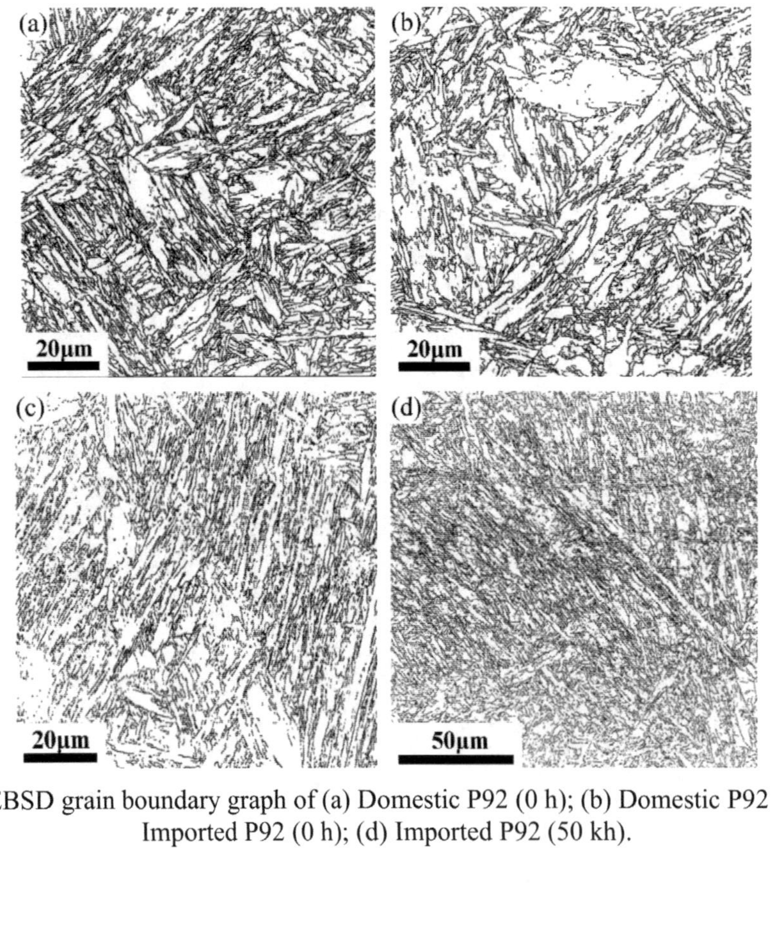


**Figure 6.** EBSD grain boundary graph of (a) Domestic P92 (0 h); (b) Domestic P92 (50 kh); (c) Imported P92 (0 h); (d) Imported P92 (50 kh).

## 4. Conclusions

In this study, through the use of SEM, EBSD, TEM, and Brinell hardness testing, a comparative analysis of the microstructure and Brinell hardness of domestically produced and imported P92 steel before and after 50 kh of service was conducted. The conclusions are as follows:

(1) After 50 kh of service, the hardness of domestically produced P92 steel decreased from 222 HB to 200 HB, a reduction of approximately 10%. In contrast, the hardness of imported P92 steel decreased from 205 HB to 203 HB, with almost no change in hardness.

(2) The key factor leading to a significant decrease in hardness in domestic P92 steel during prolonged service is the higher rate of martensite lath widening compared to imported P92 steel. Additionally, the presence of large-sized Laves phases in domestically produced and imported P92 steel at locations such as matrix lath boundaries, lath bundle boundaries, and grain boundaries during service is one of the contributing factors to the decrease in hardness.

**Acknowledgment**

This work was supported by the Guangdong Major Project of Basic and Applied Basic Research (No. 2019B030302011) and the National Natural Science Foundation of China (U21B2076).

**References**

[1] Hald J, Viswanathan V and Abe F (2006) Energy drivers for materials research and development. Energy Materials Materials Science & Engineering for Energy Systems, 1(1): 1-6.
[2] Ennis P. J, and Czyrska-Filemonowicz A (2003) Recent advances in creep-resistant steels for power plant applications. Sadhana, 28(3-4): 709-730.
[3] Nagode A, Kosec L, Ule B, and Kosec G (2011) Review of creep resistant alloys for power plant applications. Metalurgija, 50(1).
[4] Vaillant J C, Vandenberghe B, Hahn B, Heuser H, and Jochum C (2008) T/P23, 24, 911 and 92: new grades for advanced coal-fired power plants-properties and experience. International Journal of Pressure Vessels & Piping, 85(1-2): 38-46.
[5] Maruyama K, Sawada K and Koike J I (2001) Strengthening mechanisms of creep resistant tempered martensitic steel. Isij International, 41(6): 641-653.
[6] Hawk J A and Jablonski P D (2017) Heat-Resistant Advanced 9% Cr Steel for Fossil Energy Power Generation. Annual TechConnect World Innovation Conference & Expo.
[7] Buzzichelli G, and Anelli E (2002) Present status and perspectives of European research in the field of advanced structural steels. ISIJ International, 42(12): 1354-1363.
[8] Jin X, Zhu B Y, Li Y F, Zhao Y F, and Zhang G D (2020) Effect of the microstructure evolution on the high-temperature strength of P92 heat-resistant steel for different service times. International Journal of Pressure Vessels and Piping, 186(1).
[9] Jiang J, Zhu L H, and Wang Y F (2012) Effect of microstructure evolution on property of P92 steel during high-temperature creep-rupture tests. Journal of Chinese Society of Power Engineering.
[10] Lee J S, Armaki H G, Maruyama K, Muraki T, and Asahi H (2006) Causes of breakdown of creep strength in 9Cr–1.8W–0.5Mo–VNb steel. Mater.sci.eng, 428(1/2): 270-275.
[11] Geng L Y, Guo X F, Gong J M, Jiang Y, and Yang H C (2014) Comparison of microstructure, tensile and creep properties of domestic and imported p92 heat-resistant steel. Materials for Mechanical Engineering, 38(1): 68-73.
[12] Chen J H and Ning B Q (2014) Research status of microstructure evolution and strengthening methods of P92 steel in the process of high-temperature creep. Materials Review.
[13] Ennis, P J, ZielinskaLipiec, Wachter and CzyrskaFilemonowicz (1997) Microstructural stability and creep rupture strength of the martensitic steel p92 for advanced power plant. ACTA MATER, 45(12): 4901-4907.
[14] Obiko J, Chown L H and Whitefield D J (2019) Microstructure characterisation and microhardness of P92 steel heat treated at the transformation temperatures. IOP Conference Series: Materials Science and Engineering, 655: 012-014.

[15] Liu F, Li T, Liang J, Yao B, and Wang C (2011) Effect of thermal aging on microstructure and mechanical properties of p92 steel weld joints. Proceedings of the Csee, 31(14): 121-126.
[16] Saini N, Mulik R S, and Mahapatra M M (2018) Study on the effect of ageing on laves phase evolution and their effect on mechanical properties of P92 steel. Materials Science and Engineering A, 716(14): 179-188.
[17] Li X, Zhang Z, Du B, and Peng X (2012) Microstructure and hardness of P92 steel. Heat Treatment of Metals, 37(5): 38-41.
[18] Cerjak H, Hofer P and Schaffernak B (1999) The influence of microstructural aspects on the service behaviour of advanced power plant steels. Isij International, 39(9): 874-888.
[19] Korcakova L, Hald J and Somers M A J (2001) Quantification of laves phase particle size in 9CrW steel. Materials Characterization, 47(2): 111-117.
[20] Dimmler G, Weinert P, Kozeschnik E and Cerjak H (2003) Quantification of the laves phase in advanced 9-12% Cr steels using a standard SEM. Materials Characterization, 51(5): 341-352.
[21] Hald J. (2008) Microstructure and long-term creep properties of 9-12% Cr steels. International Journal of Pressure Vessels & Piping, 85(1–2): 30-37.

# The effect of dispersion degree of calcium carbonate particles in polyvinyl chloride matrix on the mechanical properties of composites

**Shijun Luo[1], Sining Pan[1,2], Hongbin Chang[1], and Xiaolin Zhou[1]**

[1] School of Artificial Intelligence, Hezhou University, Hezhou, Guangxi, 542899, China

[2] Corresponding author's e-mail: 201807037@hzxy.edu.cn

**Abstract:** By establishing representative volume units of polyvinyl chloride (PVC) matrix composites filled with calcium carbonate ($CaCO_3$) particles with different agglomeration degrees, finite element simulation analysis is conducted to predict the degree of improvement of the mechanical properties of the $CaCO_3$/PVC composites by the dispersion degree of calcium carbonate particles. When the agglomeration coefficient is 0.4, the stress concentration phenomenon of the $CaCO_3$/PVC composites is the weakest, and the stress distribution of the entire $CaCO_3$/PVC composites is relatively uniform. At the same time, the reinforcement effect of the $CaCO_3$/PVC composites's stiffness is also the best, increasing by 13.8%.

## 1. Introduction

Calcium carbonate ($CaCO_3$) is abundant in nature, easy to extract, and inexpensive. Making it into powder and filling it into a plastic matrix can effectively improve the mechanical properties of plastics and reduce the production cost of plastic parts. Many scholars have used various preparation methods to obtain $CaCO_3$ particles with different crystallization modes and studied their filling effects, effectively improving the utilization rate of calcium carbonate and improving the mechanical properties of calcium carbonate-filled polymer matrix composites [1-3]. The surface of calcium carbonate powder is hydrophilic and oil repellent, and when compounded with polymer, it is difficult to evenly disperse in the polymer matrix, forming a serious agglomeration phenomenon, leading to a decrease in the mechanical properties of the composite material, thereby affecting its use effect [4-5]. To achieve a good filling effect, surface modification of calcium carbonate is necessary to improve its dispersibility and compatibility in composite materials [6-7]. Scholars have used various experimental methods to modify the surface of calcium carbonate particles to improve their dispersion effect in polymer matrices [8]. The experimental method makes it difficult to predict the degree to which the dispersion of calcium carbonate particles improves the mechanical properties of composite materials and has certain unpredictability. It is also difficult to explore the laws and mechanisms by which the dispersion of calcium carbonate particles improves the mechanical properties of composite materials.

This article establishes representative volume units (RVE) of calcium carbonate-filled polyvinyl chloride (PVC) composites with different dispersion degrees, and further uses the finite element method to simulate and predict the improvement effect of dispersion degree on the mechanical properties of the composite materials. From a microscopic perspective, the modification mechanism of the modified calcium carbonate dispersion effect on $CaCO_3$/PVC composites is studied.

## 2. Composite material components

This article uses PVC as the matrix of composite materials, filled with calcium carbonate particles to enhance the mechanical properties of PVC materials. The most common shape of calcium carbonate powder particles is a sphere. This article uses spherical calcium carbonate particles with a diameter of 1 micrometer as filling materials. The specific material characteristic parameters are shown in Table 1.

**Table 1.** Performance parameters of each component of $CaCO_3$/PVC composites.

| Material | Density (kg/m$^3$) | Young's modulus (MPa) | Poisson's ratio | Tensile strength (MPa) | Compressive strength (MPa) |
|---|---|---|---|---|---|
| $CaCO_3$ | 2700 | 68800 | 0.31 | 30 | 30 |
| PVC | 1380 | 6000 | 0.47 | 50 | 50 |

PVC as the matrix in composites is the object for us to improve mechanical properties. Due to the high Young's modulus of calcium carbonate, the improvement in the mechanical properties of the PVC matrix is mainly manifested in stiffness. The performance of stiffness enhancement is described by the increase in Young's modulus. The calculation method for Young's modulus E of elastic materials is shown in Equation (1).

$$E = (F/S)/(\Delta L/L) \tag{1}$$

In the equation, "$F$" is the tensile force exerted on the material; "$S$" is the cross-sectional area; '$\Delta L$' is the elongation, and "$L$" is the original length.

## 3. Establishing a simulation model

### *3.1 Establish representative volume units*

The dispersion degree of $CaCO_3$ in the PVC matrix is described by the agglomeration coefficient $K$, which is the ratio of the volume of aggregated $CaCO_3$ to the volume of all filled $CaCO_3$, as shown in Equation (2).

$$K = V_a/V_s \tag{2}$$

In the equation, $V_a$ is the volume of aggregated $CaCO_3$; $V_s$ is the total volume of all $CaCO_3$.

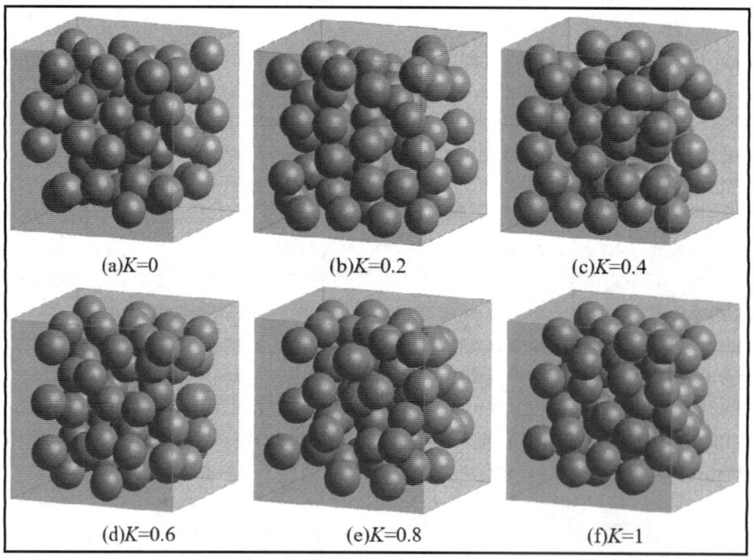


**Figure 1.** RVEs of each agglomeration coefficient.

According to previous research [9], spherical calcium carbonate particles are filled into PVC, and the composite material has the optimal tensile and compressive properties when the proportion of calcium carbonate in the volume of the composite material is 25%. On this basis, this article uses 25% as the filling ratio of calcium carbonate particles. In Digimat software, representative volume units (RVE) with aggregation coefficients $K$ of 0, 0.2, 0.4, 0.6, 0.8, and 1 are established by setting different clustering coefficients. The dimensions of RVEs are all side lengths of 5 μm. The cube of m is shown in Figure 1.

*3.2 Establishing a finite element simulation model*

The representative volume units of the composite materials established above are imported into the Workbench simulation platform of ANSYS 2020 R2 software, the setting of various material parameters (as shown in Table 1) is completed, and automatically the mesh using tetrahedrons is divided. Fixed constraints are applied on one side of the six faces of the RVE, and a tensile load of 40 MPa is applied on the opposite side, with a loading time of 1 s. The final simulation model obtained on the workbench platform is shown in Figure 2.

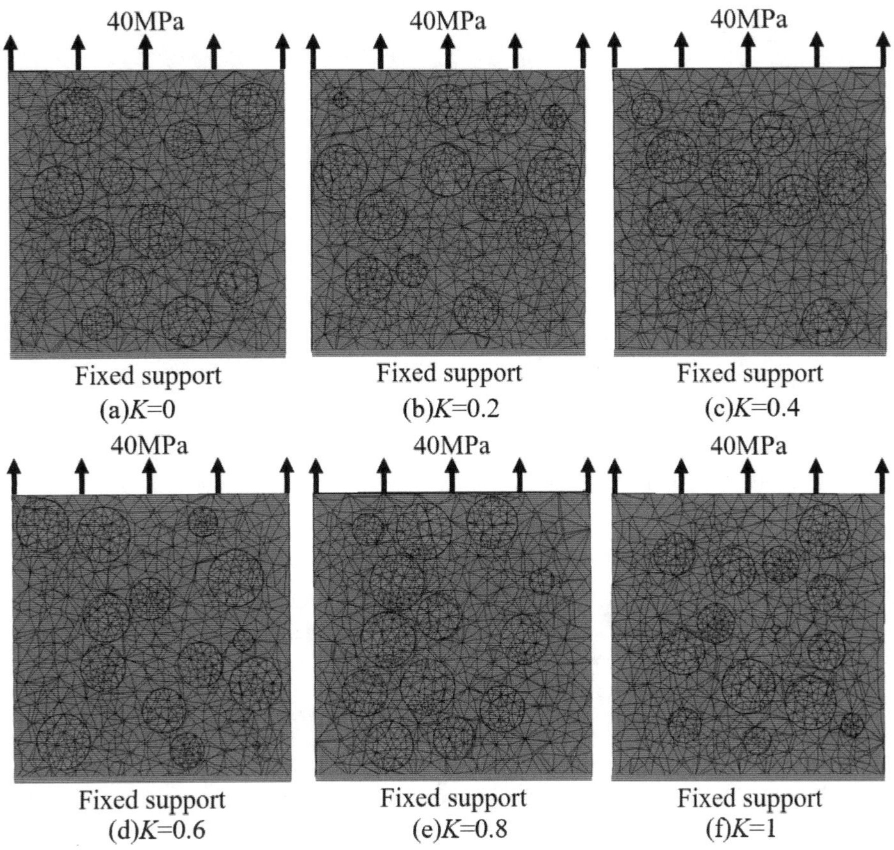


**Figure 2.** Simulation model.

## 4. Results and discussion

*4.1 Stiffness enhancement effect*

Under the tensile load of 40 MPa, the RVE of Calcium carbonate PVC composite material undergoes elastic deformation in the load direction. The deformation results obtained from the simulation of $CaCO_3$/PVC composites' RVEs with various aggregation coefficients in the workbench are displayed in cloud graph form, as shown in Figure 3.

Journal of Physics: Conference Series **2720** (2024) 012011 doi:10.1088/1742-6596/2720/1/012011

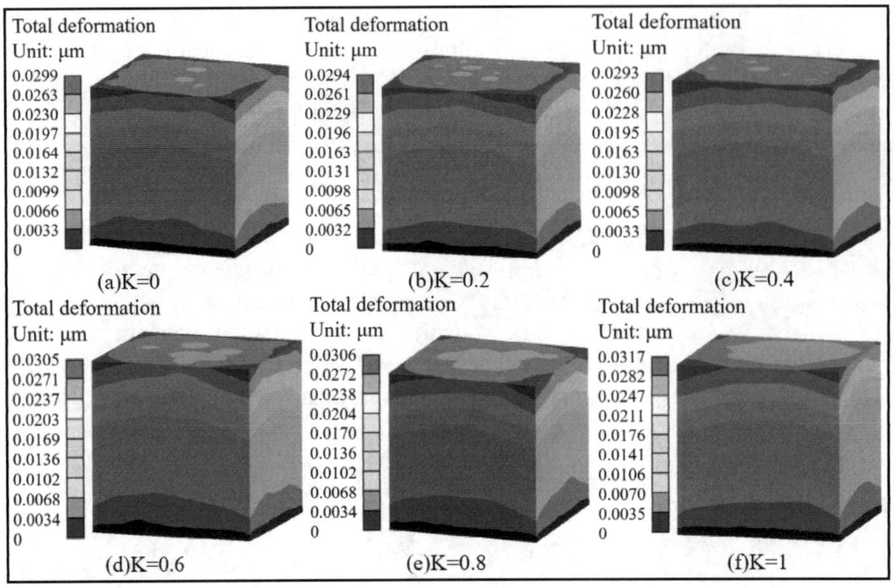


**Figure 3.** Deformation cloud chart results of $CaCO_3$/PVC composites' RVEs.

By using the deformation value of $CaCO_3$/PVC composites' RVEs and combining it with Equation (1), the Young's modulus E of $CaCO_3$/PVC composites' RVEs can be calculated, as shown in Table 2. The reinforcement ratio of Young's modulus of composites compared to pure PVC is shown in Figure 4.

**Table 2.** Deformation values and Young's modulus of $CaCO_3$/PVC composites' RVEs.

| Parameter | Agglomeration coefficient K | | | | | |
|---|---|---|---|---|---|---|
| | 0 | 0.2 | 0.4 | 0.6 | 0.8 | 1 |
| Total deformation (μm) | 0.0299 | 0.0294 | 0.0293 | 0.0305 | 0.0306 | 0.0317 |
| Young's modulus (MPa) | 6688.96 | 6802.72 | 6825.94 | 6557.38 | 6535.95 | 6309.15 |

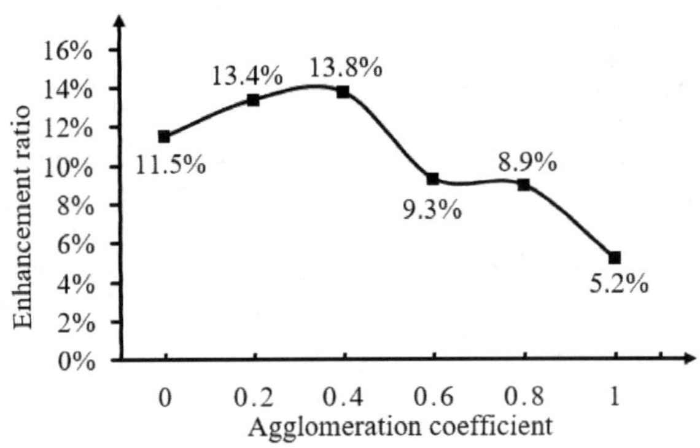


**Figure 4.** Enhancement ratio of Young's modulus.

According to Table 2 and Figure 4, it can be seen that under the same filling ratio of calcium carbonate particles, the stiffness enhancement effect of the composite material decreased. The cohesion coefficient shows an upward trend before 0.4, and a downward trend when it is greater than 0.4. When the aggregation coefficient is between 0.2 and 0.4, the enhancement effect is better; When the aggregation coefficient is 0.4, the reinforcement effect of the composite material's stiffness is the best, with a Young's modulus value of 6825.94 MPa. Compared to pure PVC material, the Young's modulus is increased by 13.8%. The results indicate that the presence of appropriate agglomeration of filled calcium carbonate particles in the PVC matrix helps to improve the stiffness of the $CaCO_3$/PVC composites. When the agglomeration phenomenon is severe, the effect of improving the stiffness of the $CaCO_3$/PVC composites rapidly decreases.

*4.2 Improvement effect of stress concentration*

The stress cloud diagram of the simulation results is shown in Figure 5. The maximum stress values of $CaCO_3$/PVC composites' RVEs are shown in Figure 6. By observing Figure 6, it can be inferred that the stress concentration phenomenon is the most obvious at the connection between calcium carbonate particles and PVC, which is also the location where cracks are most likely to occur. When the agglomeration coefficient is 0.4, the stress concentration phenomenon is the weakest, with a maximum stress value of 187.5 MPa. The stress distribution of the entire composite material is relatively uniform.

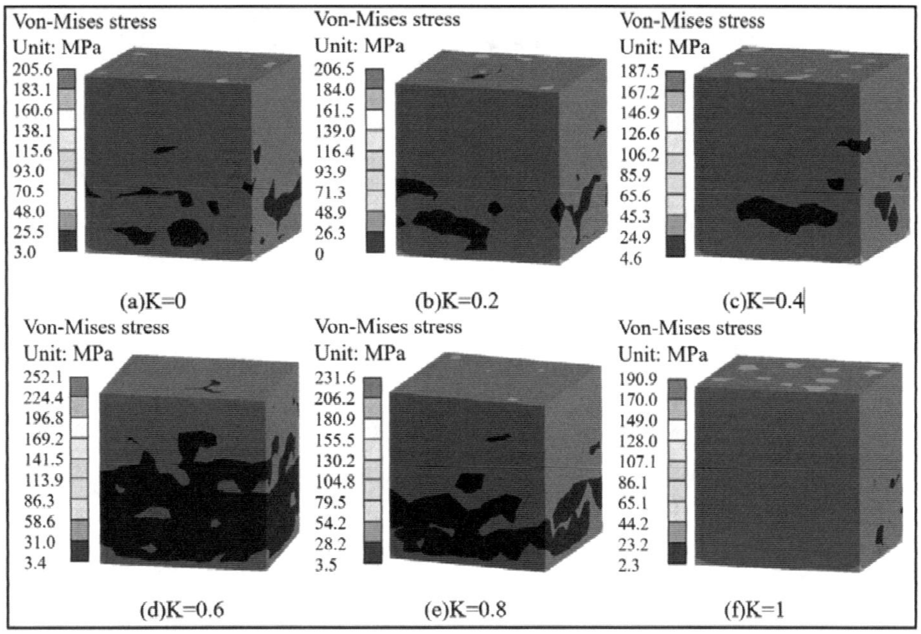


**Figure 5.** Stress cloud diagram of $CaCO_3$/PVC composites' RVEs.

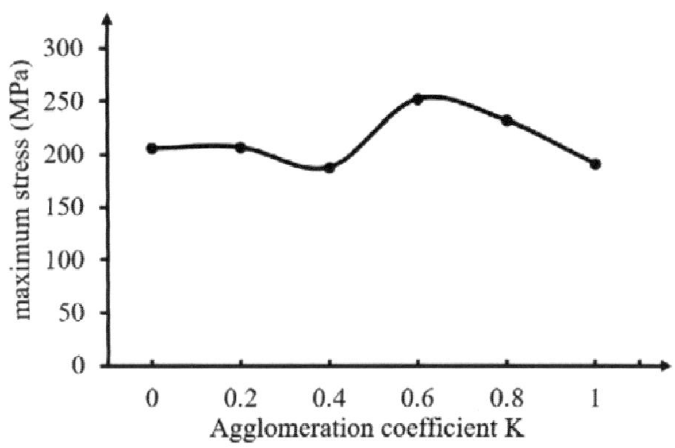


**Figure 6.** The maximum stress values of RVEs.

Based on the above analysis, in PVC composite materials filled with calcium carbonate particles, there is a small amount of agglomeration phenomenon inside the matrix that helps to promote the improvement of composite material stiffness and reduce stress concentration. The optimal agglomeration coefficient is 0.4.

## 5. Conclusion

When the agglomeration coefficient is between 0.2 and 0.4, the stiffness enhancement effect of the $CaCO_3$/PVC composites is better; When the agglomeration coefficient is 0.4, the reinforcement effect of $CaCO_3$/PVC composites stiffness is optimal.

When the agglomeration coefficient is 0.4, the stress concentration phenomenon of the $CaCO_3$/PVC composites is weakest, and the stress distribution of the entire $CaCO_3$/PVC composites is relatively uniform.

## Acknowledgments

This research was supported by National Natural Science Foundation of China (52065018), the Science and Technology Base and Talent Project of Guangxi Province in China (AD20297043), the Natural Science Foundation of Guangxi Province in China (2021JJB160112), the Scientific Research and Technology Development Program Project of Hezhou City (Technology in Hezhou City 202227), and the Research Fund for the Enhancement of Basic Research Competence of Young and Middle-aged Teachers in Universities and Colleges in Guangxi (2023KY0737).

## References

[1] Wu W, Peng H, Zou J and Ye H (2023) Comparative Analysis of Different Calcium Carbonate on PVC Calendering Film Performance. J. SHANGHAI PLASTICS, 51 (04): 41-45.[CrossRef].

[2] Lv J (2023) Application and research progress of nano calcium carbonate in plastics. J. China Powder Industry, 1: 33-36.[CrossRef].

[3] Ren H, Xu J, Ai N, Lvyang G and Deng Z (2023) Synthesis of Calcium Carbonate Powder from Eggshell and Its Process Research. J. Contemporary Chemical Industry, 52 (07): 1581-1584.[CrossRef].

[4] Malak S A (2021) Tensile Stress-Strain Model of Polyvinyl Chloride/Calcium Carbonate (PVC/CaCO3) Nanocomposite Plank. J. Results in Materials, 10: 100193.[CrossRef].

[5] Luo W, Wang Y, and Yang G (2020) Compound Titanate and Its Effect on the Properties of ultrafine Calcium Carbonate/PP Composites. J. Journal of Nanchang Hang Kong University: Social Sciences, 34(3): 53-60.[CrossRef].

[6] Soy U, Findik F, Yetgin S H, Gokkurt T, and Yildinm F (2017) Fabrication and Mechanical Properties of Glass Fiber/Talc/CaCO3 Filled Recycled PP Composites. J. American Journal of Applied Sciences, 14 (9): 878-885.[CrossRef].

[7] Saleh S S M, Suhaimi S M, Akil H M and Mohammad N F (2020) Properties of Carbon Nanotubes-Calcium Carbonate Hybrid Filled Epoxy Composites. J. Materials Science Forum, 1010: 136-141.[CrossRef].

[8] Misra A, Shukla M, Shukla M K, Srivastava D and Nagpal A K (2021) Nano CaCO3 modified multifunctional epoxy nanocomposites: A study on flexural and structural properties. J. Materials Today: Proceedings, 47 (P11): 3295-3300.[CrossRef].

[9] Luo S and Pan S (2021) Numerical study on the effect of CaCO3 ratio on the mechanical properties of CaCO3/PVC composites. J. Journal of Physics: Conference Series, 1820 (1): 012140.[CrossRef].

# Fluidization characteristics of iron ore powder in fluidized bed

**Jianming Pang[1, 2], Yaoxin Song[1, 2], Shiwen Li[1, 2], Chongfeng Yue[1, 2, 3], Zhimin Zhao[1,2] and Lingen Luo[1, 2]**

[1]Resource Application and Alloy Materials Division, China Iron & Steel Research Institute Group, Beijing, 100081, China
[2]Beijing Cisri-nmt Engineering Technology Co., Ltd., Beijing, 100081, China

[3]Corresponding author's e-mail: chongfengyue@163.com

**Abstract.** The fluidization characteristics of iron powder were studied on a self-made cold fluidized bed in the laboratory. The pressure distributed in the axial inside the lifting pipe of a fluidized bed gradually decreases to a constant value from underbody to roof. As the gas flow rate in the fluidized bed increases, the pressure on each axial section first increases and then decreases. When the inflation numbers are fixed, as the operating gas speed increases, the pressure on the same height segment decreases. When the operating gas speed of a fluidized bed is constant, the pressure increases with the increase of inflation numbers. The pressure at different bed heights is affected differently by the operating air speed: at the underbody of the bed, the pressure rapidly decreases with increasing operating airspeed. As the cross-sectional height increases, the influence of operating gas speed on pressure gradually decreases. As the charge increases, the pressure at the underbody of the fluidized bed rises rapidly, while the pressure at the roof changes gently. When iron ore fine powder is fluidized, distribution in axes of particle concentration shows an uneven distribution of upper dilute and lower dense. Under a certain loading number, increasing the operating airspeed will reduce the average concentration of granules in each axial section of the entire bed, and the particle concentration distribution in that section will become uniform. As the number of charges increases, the average concentration of granules in the riser increases, and the difference in particle concentration between the underbody and roof parts of the riser decreases.

## 1. Introduction

As an ironmaking reactor, a fluidized bed has relatively loose requirements for iron ore composition [1]. It can use iron ore powder with fine particle size and wide distribution of partial size and has the advantages of uniform temperature and concentration of the whole bed, high efficiency of heat, and mass transfer. In foreign countries [2], there is much research on ironmaking technology by fluidizing gas reduction, and the development of ironmaking technology by fluidizing gas reduction is fast. In China [3-4], there is relatively little research on the fluidized gas reduction ironmaking technology.

This article systematically analyzes the impact of operating parameters and properties of iron concentrate particles on the pressure and local particle concentration distribution inside the riser through many experiments on the pressure of the gas-solid two-phase flow and local particle concentration in a fluidized bed. This will provide an experimental and theoretical basis for understanding, scaling up design, and optimizing the operation of fluidized bed ironmaking reactors [5].

## 2. Test materials and process

Gas phase: air, at room temperature of 20°C, air density$\rho$=1.205 kg/m$^3$, kinematic viscosity $v$=15×10$^{-6}$ m$^2$/s, kinetic viscosity $\mu$=18.08×10$^{-6}$ Pa•s.

Solid phase: iron concentrate powder provided by Minmetals Yingkou Medium Plate Co., ltd. The test results of iron powder are shown in Table 1.

**Table 1.** Test results of iron powder.

| Sample | Grain size/μm | Specific surface area/m$^2$·g$^{-1}$ | Bulk density/kg·m$^{-3}$ | Angle of repose/° |
|---|---|---|---|---|
| Iron concentrate powder | <75 | 2.5 | 2000 | 40 |

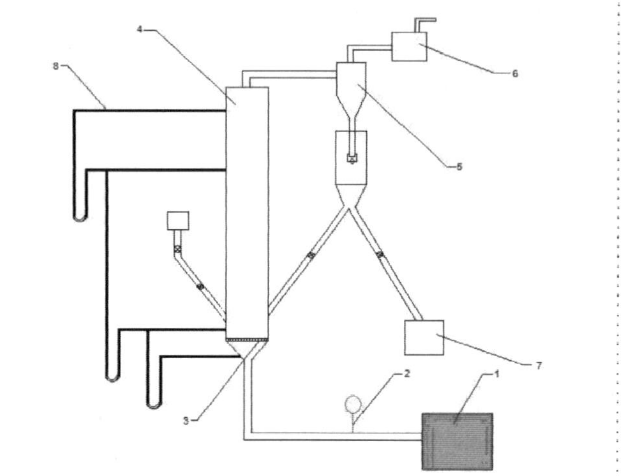

1-Roots blower;
2-Electronic flowmeter;
3-Distribution plate;
4-Circulating bed body;
5-Cyclone separator;
6- Bag dust collector;
7-Discharge tank;
8- U-shaped pressure gauge.

**Figure 1.** Fluidized bed system.

The fluidized system made in the laboratory is shown in Figure 1. The gas inlet uses a conical gas distributor, and the distribution plate is fixed by a flange. Several pressure gauges are opened from the roof to the underbody of the fluidized bed to measure the bed pressure drop of iron ore powder. The inner diameter of the fluidized bed is 140 mm, the height is 1.7 m, and the inlet diameter is 40 mm.

During the experiment, air is transported by Roots blower, passes through a conical gas distributor, enters the fluidized bed riser from the underbody, and is discharged through a cyclone separator and bag filter. Iron ore powder is fed through the silo, and the iron ore powder and gas circulate in a fluidized bed. Fine iron powder particles and gas enter the cyclone through the tangential inlet and undergo complex rotational motion in the cyclone. The particles separated by the tangential rotational motion enter the fluidized bed riser through the reflux valve and continue to circulate for fluidization.

During the experiment, the gas flow rate is measured using an electronic flow meter, and the pressure drop is obtained using a U-shaped pressure gauge.

## 3. Test results and analysis

### *3.1. Pressure distribution in fluidized bed riser*

The pressure in the axis of the riser gradually decreases from underbody to roof to a constant value, which means that the velocity of iron concentrate particles in the axis of the riser increases as the concentration decreases. It can be found that: (1) At the same gas speed ($U_g$), the less the charge is, the more obvious the acceleration of the impact of iron concentrate particles is; (2) Under certain loading conditions, the higher the gas speed ($U_g$) is, the greater the increase in velocity of iron powder particles will be; (3) Iron ore powder always accelerates upwards in axis of the riser to the fully developed part, without any deceleration process. Figure 2 shows the pressure distribution under different operating conditions.

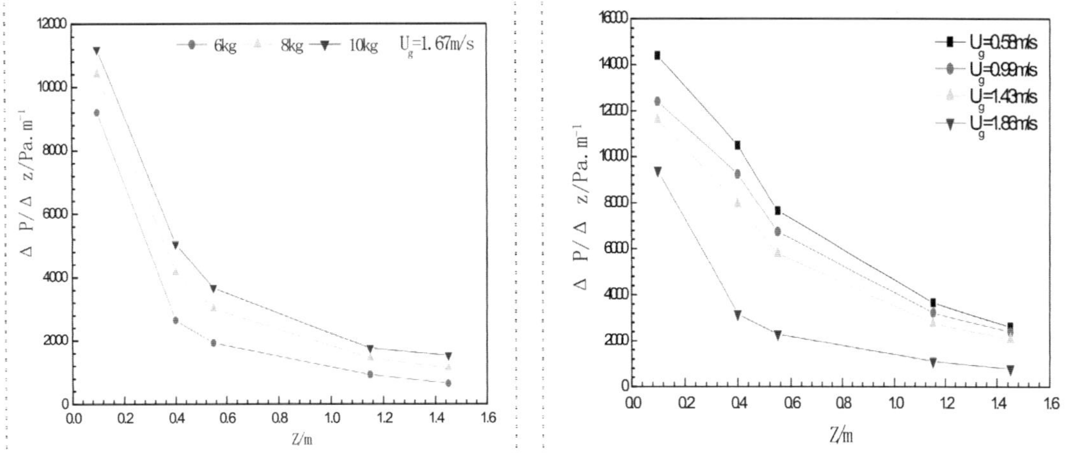


(a) Influence of charge numbers on pressure with constant $U_g$

(b) Influence of $U_g$ on the pressure with constant charge numbers

**Figure 2.** The axial pressure diagram in the riser.

Figure 2 shows the axial pressure diagram in the riser. When the charge remains constant, the pressure drop at the same height decreases with the increase of gas speed ($U_g$). This is because as the gas speed ($U_g$) increases, the speed of iron ore powder increases, and the quantity of iron ore powder in the same height range decreases, the concentration of iron ore powder decreases, and the resistance required to support particles also decreases, resulting in a decrease in pressure. When the gas speed ($U_g$) is constant, the pressure increases with the expansion. The reason for this is that with a certain gas speed and an increase in the number of iron ore powders, the probability of particle fluidization increases, the resistance increases, and therefore, the pressure also increases. Through the analysis of a single curve, the pressure difference change rate in the lower half of the riser is extremely severe. In contrast, the upper half is relatively gentle, which is similar to the phenomenon observed in actual experiments.

### *3.2. The impact of operating parameters on the pressure*

In order to compare the characteristics of the pressure ($\Delta P/\Delta Z$) at different height sections with the operating conditions, five representative sections of the riser are selected for analysis.

#### *3.2.1. The impact t of apparent gas speed on pressure at different height sections.*

Under a certain number of charges, the pressure of each axial section is ($\Delta P/\Delta Z$). Figure 3 shows that as the gas flow speed increases, the pressure on each axial section of the riser first increases and then decreases. The reason is that as the airflow speed increases, the state of the fluidized bed transitions from bubbling fluidization to turbulent fluidization. Therefore, as the number of fluidized iron ore powder particles increases and the weight of particles at the same height increases, the resistance required to maintain the particles will increase, resulting in a corresponding increase in pressure. However, as the gas speed continues to increase, the fluidized bed state changes from a turbulent fluidization state to a rapid fluidization state, the particle velocity will increase, and the particle numbers will correspondingly decrease. The particle concentration will become smaller, and the drag force required to hold the particles will also decrease, so the pressure will also decrease. However, the pressure varies with airspeed at different heights. At the underbody of the bed, the pressure rapidly decreases with the increase of gas speed, and the influence of gas speed on the pressure gradually decreases with the increase of cross-sectional height (Z). In the underbody acceleration section, gas speed has the greatest impact on pressure. At the same operating gas speed, the pressure difference between adjacent sections is very large, indicating that the iron ore powder is in the acceleration stage.

The pressure curve at section z=1.15 m and section z=1.45 m is almost the same, which shows that from the height of section z=1.15 m above, iron ore powder particles are in the fully developed section.

*3.2.2. The impact of charge numbers on pressure at different height sections.*

The pressure at different sections of the riser shows a common trend of increasing with the increase of the charge amount, as shown in Figure 2. However, the change rule of pressure at different height sections is obviously different, and the distribution of iron concentrate particles shows the upper dilute and lower concentrated. As the loading numbers increase, the pressure at the underbody of the bed layer rapidly increases while the pressure at the roof of the bed changes gently. Through the analysis, it is considered that the increase of the charge numbers in the accelerating section at the underbody of the riser will increase the concentration of iron concentrate particles and increase the collision and agglomeration behavior between iron concentrate particles. This will consume more energy, reduce the energy used to accelerate iron concentrate particles, and reduce the effective drag force on iron concentrate particles. Thus, the acceleration effect on particles is weakened, and the pressure at the underbody of the bed rises rapidly. In the dilute phase section of the upper part of the riser, the collision and agglomeration between particles are less, and the energy loss is less. The energy of the gas is mainly used to transport particles. Therefore, the change of particle conveying rate caused by changing the charge numbers is relatively stable, so the pressure and the charge numbers show a relatively stable linear relationship, as shown in the z = 0.4 m and z =1.45 m curves in Figure 4.

Figure 4 also shows that from the height of section z=1.45 m, the relationship between the pressure and the charge numbers tends to be stable; that is, the pressure is not affected by the height of the section, and at this time, the state of particles in the riser is in a fully developed state. This also fully proves the description of the axial development of two-phase circulation in the riser in Figure 2.

*3.3. Study on the concentration of iron ore powder*

There are many methods for measuring the concentration of iron powder particles in a fluidized bed. The simplest method is to obtain the average concentration of granules in a certain area through pressure difference measurement [6]. Due to the complexity of the two-phase circulation in a fluidized bed, many issues need to be analyzed and simplified before conducting actual experimental research. This experimental research also followed this research approach without considering the friction between air and iron ore powder particles and the bed wall. According to static analysis, the pressure drop is completely converted into the resistance of gas to iron powder particles in the bed. The formula is as follows [7]:

- $$\Delta P \cdot A = (A \cdot \Delta Z)[\varepsilon_s \rho_s g + (1 - \varepsilon_s)\rho_g g] \qquad (1)$$

Where $\Delta P$ is bed pressure drop, Pa; $\Delta Z$ is the bed height difference, m; A is the cross-sectional area of fluidized bed, $m^2$; $\varepsilon_s$ is the average concentration of iron ore powder; $\rho_s$ is particle density of iron ore powder, $kg{\bullet}m^{-3}$; $\rho\_g$ is gas density, $kg{\bullet}m^{-3}$; g is the acceleration of gravity, $m{\bullet}s^{-2}$.

The expression of $\varepsilon_s$ can be obtained by reverse calculation:

- $$\varepsilon_s = 1 - \frac{\rho_s - \frac{\Delta P}{\Delta Z}}{\rho_s - \rho_g} = \frac{\frac{\Delta P}{\Delta Z} - \rho_g}{\rho_s - \rho_g} \qquad (2)$$

In the formula, $\Delta P$ is measured by the test, and $\rho_s$, $\rho_g$, and bed height difference $\Delta Z$ are known. Therefore, the average concentration of granules in the fluidized bed can be calculated.

*3.3.1. Axial distribution of iron ore powder concentration.*

Distribution in axes of the particle concentration of iron ore powder in the riser under different operating conditions, as shown in Figure 5. Distribution in axes of the particle concentration presents the uneven distribution of the upper thin and the lower thick when the iron ore powder is fluidized in the fluidized bed. Under the experimental conditions, distribution in axes of the average concentration of particles in the cross-section is basically monotonic exponential distribution.

*3.3.2. The impact of operating gas speed on particle concentration of each section.*
The variation of the iron ore powder particle concentration at five height sections with the operating gas speed is plotted in Figure 6.

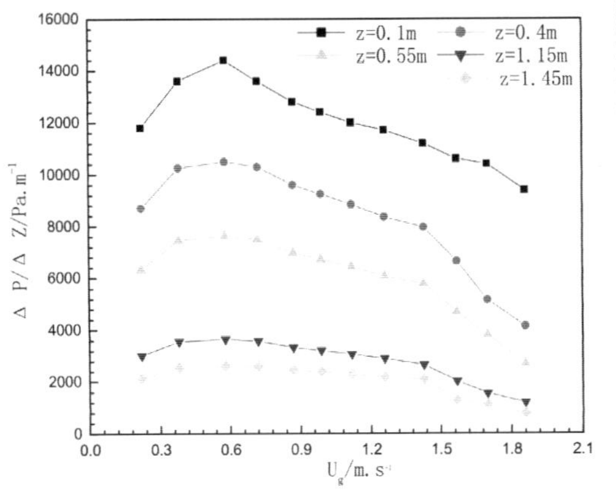


**Figure 3.** Variation of pressure with apparent gas speed at different axial sections.

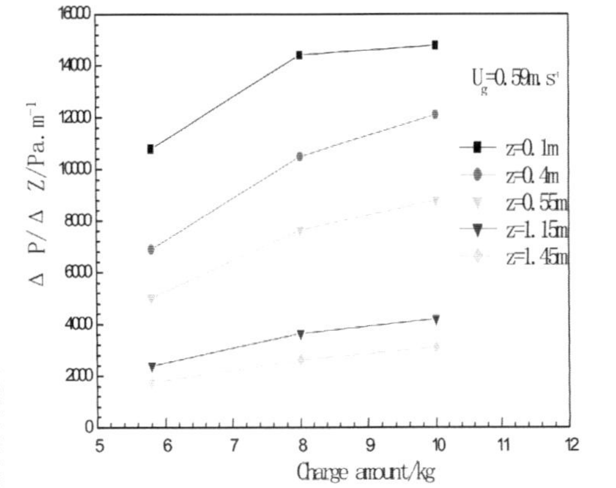


**Figure 4.** Variation of pressure with charge numbers at different axial sections.

When the loading capacity remains constant, the apparent gas speed increases. For iron ore powder with the same distribution of partial size, the concentration of iron ore powder particles at any position in the riser first increases. When the apparent gas speed reaches a certain value, the particle concentration at any position in the riser begins to decrease. The reason for this is that with the increase of gas flow speed, the process of fluidized bed transition from bubbling bed to turbulent bed occurs, and the number of particles increases, resulting in an increase in particle concentration. As the gas speed increases, the particle speed also increases, so particles enter a fast fluidization state, and the particle concentration will sharply decrease. Comparing the particle concentration distribution in various axial cross-sections, it can be found that the higher the axial position is, the smaller the particle concentration is. That is to say, the particle concentration at the underbody of the bed is higher than that at the roof of the bed, indicating that the particle concentration gradually decreases in the axial overall. The higher the axial position is, the more similar and gentler the curve of particle concentration will be, changing with the operating gas speed.

*3.3.3. The impact of charge numbers on particle concentration.*
Figure 7 shows the impact of the charge numbers on the iron ore powder particle concentration at different sections. When the operating gas speed is fixed, the concentration of iron ore powder particles in each section increases with the increase of the charge numbers. According to the analysis, with the increase of the charge numbers, the probability of iron ore powder particles being fluidized increases, so the particle concentration on each section also increases. The fluctuation of particle concentration in each section is different when the charge numbers are increased. In the middle part of the riser, the particle concentration is greatly affected by changes in the number of charges. But in any case, it still conforms to the distribution characteristics of "upper thin and lower thick".

For iron powder with the same distribution of partial size, increasing the operating gas speed can reduce the average concentration of granules in each axial section of the entire bed under certain charge numbers, and the distribution of iron powder particle concentration in this section becomes uniform. As the dosage increases, the average concentration of granules in the riser increases, and the difference between the lower particle concentration and upper parts of the riser decreases. The particle concentration of iron ore powder in the entire bed becomes uniform. The reason for the larger charge is that the larger the fluidized particles are, the higher the average concentration of granules is. Under the same conditions, the larger the particle size of iron ore powder is, the lower the average concentration of granules in the bed will be.

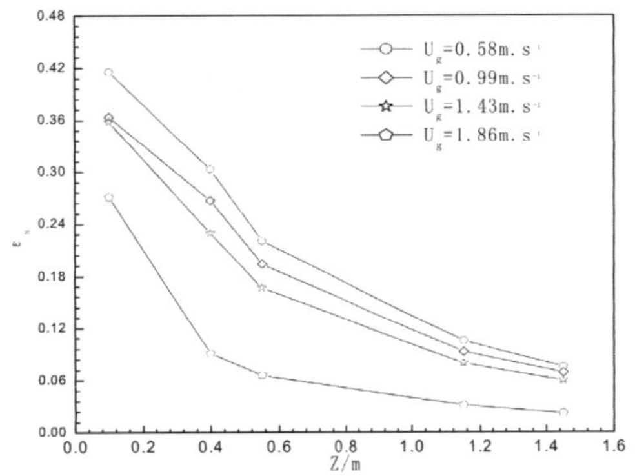


**Figure 5.** Axial distribution of particle concentration under different operation conditions.

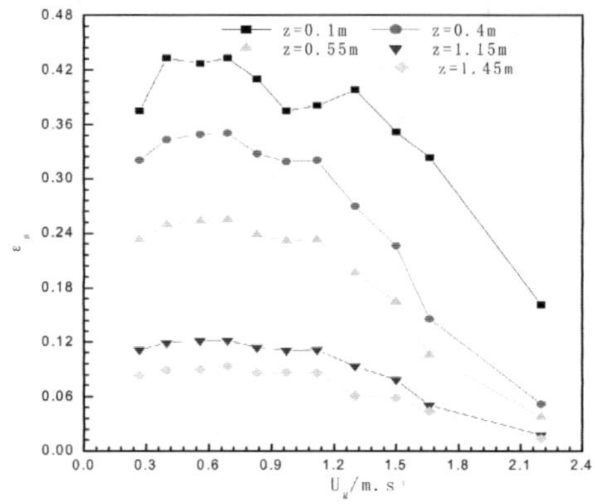


**Figure 6.** The impact of operating gas speed on iron ore powder particle concentration at different axial sections.

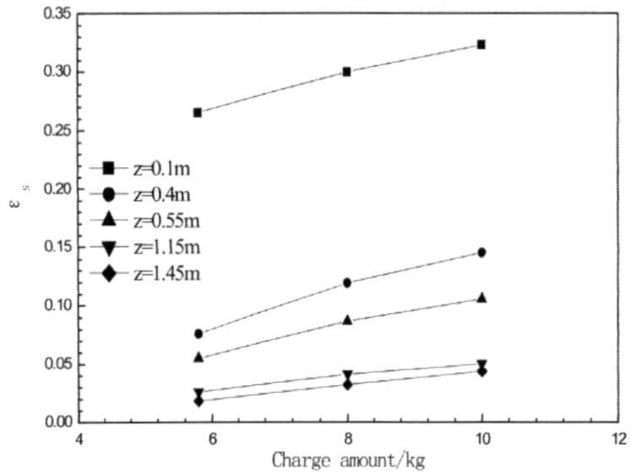


**Figure 7.** The impact of charge numbers on particle concentration at different axial sections.

## 4. Conclusions

In this paper, the distribution of pressure and particle concentration of iron ore powder in axial and the influencing factors in CFB are studied. The main results are as follows:

(1) Distribution in axes of pressure inside the lifting tube gradually decreases from underbody to roof to a constant value. When the charge volume remains unchanged and the gas speed increases, the pressure on the same height segment decreases. When the gas speed remains constant, the pressure increases with the increase of the charge volume.

(2) As the gas speed increases, the pressure on each axial section of the riser first increases and then decreases. At different bed heights, the pressure varies with the change of gas speed. At the underbody of the bed, the pressure rapidly decreases with the increase of gas speed, and the influence of airspeed on the pressure decreases with the higher cross-sectional. As the dosage increases, the pressure at the underbody of the bed changes significantly, while the pressure at the roof of the bed does not change significantly.

(3) When iron ore powder particles are fluidized in a fluidized bed, distribution in axes of particle concentration shows an uneven distribution of upper dilute and lower dense. Under a certain loading

number, increasing the operating gas speed will reduce the average concentration of granules of each axial section of the entire bed. As the number of charges increases, the average concentration of granules in the riser increases, and the difference between the particle concentration in the lower part of the riser. The higher the axial site is, the smaller the particle density will be.

(4) When the operating gas speed is fixed, the particle concentration on each cross-section basically increases with the increase of charge. The fluctuation of particle concentration on each part is different when the number of charges increases. In both ends of the riser, as the number of charges increases, the increase in particle concentration is not significant. In the middle of the riser, the concentration of particles is largely affected by the charge change.

**Acknowledgments**

The work described in this paper was supported by the "Renewable Energy Technology" Key Special Project of the National Key R&D Plan (No. 2023YFB4204002), Chengde Science and Technology Plan Project (No. 202305B021), Youth Fund of China Steel Research Technology Group Co., Ltd (No. KNJT03-JTOM-21013), General Project of Sichuan Vanadium and Titanium Industry Development Research Center (No. 2020VTCY-Y-02).

**References**

[1] Zhu Q S, Li H Z, and Zou Z (2020)Theory and application of fluidized bed particle residence time adjustion [M]. Beijing: Chemical Engineering Press. pp. 158-190.

[2] Feng F, Song G, Xiao J, et al. (2019)Pisupati, Carbon deposition on Ni-based catalyst with TiO2 as additive during the syngas methanation process in a fluidized bed reactor[J]. Fuel, 235: 85-91.

[3] Yang S, Wang S, Wang H. Impact of polydispersity on the flow dynamics in the riser of a circulating fluidized bed [J]. (2021)Powder Technology, 381: 489-502.

[4] Razmi H, Soltani G A, and Mohebbi A (2019) CFD simulation of an industrial hydrocyclone based on multiphase particle in cell (MPPIC) method [J]. Separation and Purification Technology, 209:851-862.

[5] Bhui B and Vairakannu P (2019)Prospects and issues of integration of co-combustion of solid fuels (coal and biomass) in chemical looping technology [J]. Journal of Environmental Management, 231:1241-1256.

[6] Ayeni O O, Wu C L, Nandakumar K, et al. (2016)Development and validation of a new draglaw using mechanical energy balance approach for DEM–CFD simulation of gas-solid fluidized bed [J]. Chemical Engineering Journal, 302:395-405.

[7] Zhang Y, Zhao Y, Lu L, et al. (2017)Assessment of polydisperse drag models for the size segregation in a bubbling fluidized bed using discrete particle method[J]. Chemical Engineering Science, 160:106-112.

# Effect of crosslinking agent types on the tensile properties of polyisoprene rubber

**Junjie Zuo[1], Zhigang Wei[1,2], and Zelin Zhang[1]**

[1]Anhui University of Technology, Maanshan, Anhui Province, 243032 China

[2]Corresponding author's e-mail: zhigwei@163.com

**Abstract**: The impacts of crosslinking bond types on the equilibrium kinetics and uniaxial tensile properties of polyisoprene rubber were investigated by using molecular dynamics (MD) simulations. We analyze the changes in mean square displacement, glass transition temperature, and end-to-end vector autocorrelation function of different crosslinked systems during the equilibrium process. It was found that the mobility of polysulfide crosslinked systems was more limited compared with that of monosulfide and disulfide bonds. In addition, during uniaxial stretching, the polysulfide-bonded crosslinked system could exhibit better tensile strength, and the stress response was significantly improved. The above research results help to understand the mechanical properties and deformation mechanism of crosslinked polyisoprene rubber and provide certain reference values for improving the performance of related materials.

## 1. Introduction

Rubber materials find extensive use in diverse fields, including automobile, electronics, construction, and medical devices, owing to their excellent wear resistance, corrosion resistance, adhesion, and high elasticity [1]. However, in its unprocessed state, it often fails to meet the requirements of practical working conditions due to insufficient strength. Therefore, to enhance the mechanical performance of the material, a common practice involves heating and curing it with a vulcanizing agent to form a cross-linked network structure. Alternatively, fillers like silica are added, or the composite with fibers and metals is employed to improve its properties. Among them, vulcanization treatment has been widely researched and developed, with a certain degree of reliability and controllability. In the traditional vulcanization process, sulfur decomposes to form sulfur molecules with different numbers of sulfur atoms, including mono-sulfur (-S-), disulfide (-S-S-), and poly-sulfur (-Sx-) atoms. Among them, single sulfur atoms may participate in short-chain cross-linking, while double sulfur atoms may participate in forming longer cross-linking chains than single sulfur atoms. Multisulfur atoms typically refer to structures containing three or more sulfur atoms, and these structures may give rise to extremely long cross-linking chains. Therefore, different sulfur molecules, when reacting with the double-bonded carbon atoms on the rubber polymer chains, will form cross-linking networks of varying degrees. To explore the mechanisms and distinctions in the cross-linking networks formed by vulcanized cross-linking agents of different lengths, further investigation at the molecular scale is necessary. However, traditional experimental methods make it difficult to accurately control the formation of cross-linking agent types and are unable to obtain the specific positions and movements of molecules or atoms.

Given the complex internal structural evolution during the stretching deformation of cross-linked systems, the swift progress in computer technology has offered a novel research approach for this purpose. Molecular dynamics simulations combine classical mechanics with statistical mechanics, establishing a relationship between microscopic structure and macroscopic performance by calculating and analyzing the mechanical properties of materials. For instance, Chen et al. [2] simulated the thermal conductivity and mechanical behavior of polyisoprene at different degrees of polymerization. Rottach et al. [3], through molecular dynamics simulations, investigated the influence of cross-linking on the constitutive behavior and internal structure of polymer networks. However, there is a dearth of research utilizing computational simulation methods to investigate how cross-linking agent types affect the mechanical properties of rubber materials.

Therefore, taking polyisoprene rubber as an example, molecular dynamics simulation methods were used to construct network models cross-linked by different types of sulfur molecules (composed of single sulfur, disulfide, and polysulfide atoms). The influence of crosslinking agent types on the dynamic mechanical properties of the IR rubber system during equilibrium and stress response during uniaxial tension was analyzed and studied. From the perspective of molecular scale, the close relationship between the structure of crosslinking agents and the mechanical properties of infrared rubber during stretching has been deeply revealed.

## 2. Simulation model and methods

### *2.1 Modelling*

The model was constructed by using Material Studio (MS) molecular simulation software developed by Accelrys. Through the Visualizer module, the established isoprene molecular monomer was generated into a high degree of polymerization polyisoprene rubber molecular chain. Each established molecular chain was subjected to structure optimization and energy minimization. Then, the Monte Carlo method was used in the Amorphous Cell module to fill the rubber molecular chains into the unit cell lattice and establish a periodic model. The molecular chain length is 100, and the number of chains is 30. The adoption of the PCFF force field could effectively simulate the static and dynamic mechanical properties of polyisoprene rubber. Cross-linking bonds were then introduced into the model via MS Perl scripts, by randomly selecting two double-bonded carbon atoms on two of the rubber molecular chains as reaction sites. If the distance $d$ between them matches $0.25\ \text{nm} < d < 0.75\ \text{nm}$, the sulfur atoms will bond with the C atoms in the C=C bond. The C=C bond is converted to a C-C bond, and the disulfide bond structure is attached to the rubber molecular chain via a carbon-sulfur bond (C-S). Finally, the Forcite geometry optimization was performed by using the Smart algorithm to eliminate local irrational structures in the model and to adjust the distance of the C-S bond and angle in the model. In this study, a crosslinking agent molecule was composed of 1, 2, or 4 sulfur atoms (Ncs=1, 2, and 4) to simulate the monosulfide, disulfide, and polysulfide crosslinking agents. The modeling process is shown in Figure 1.

### *2.2 Molecular dynamics simulation methods*

Structural optimization of different crosslinked IR rubber models and computational analysis of kinetic properties were carried out by using LAMMPS software. Before conducting the dynamic equilibrium simulation of the cross-linking system, it was necessary to employ periodic boundary conditions to eliminate the finite size and edge effects in the simulation [4]. The NVT and NPT ensembles were employed to bring the cross-linking system to an equilibrium state. Initially, the model underwent a stepwise dynamic simulation within the NVT ensemble. We use the Nose Hooper method to control the temperature at 500 K to achieve short-term equilibrium of the system and reduce internal stress. Subsequently, the NPT ensemble was employed for a $1\times10^{5}$ step dynamic simulation, maintaining the temperature at 500 K to obtain the optimal structure of the model. Finally, the temperature was lowered to 300 K for a 1 ns NPT equilibrium simulation to bring the system to a stable state. After reaching equilibrium, different cross-linking models were subjected to tensile simulations along the x-axis by using an engineering strain rate of $1e^{-5} fs^{-1}$, with a maximum tensile strain of 3.0 and a stretching time of 300 ps. To eliminate the random deviation of the data, 100 steps were used as a reference to take 1 sample every 10 steps forward, a total of 5 samples were taken, and the stress values of these 5 samples were averaged. The molecular visualization was conducted by using OVITO software.

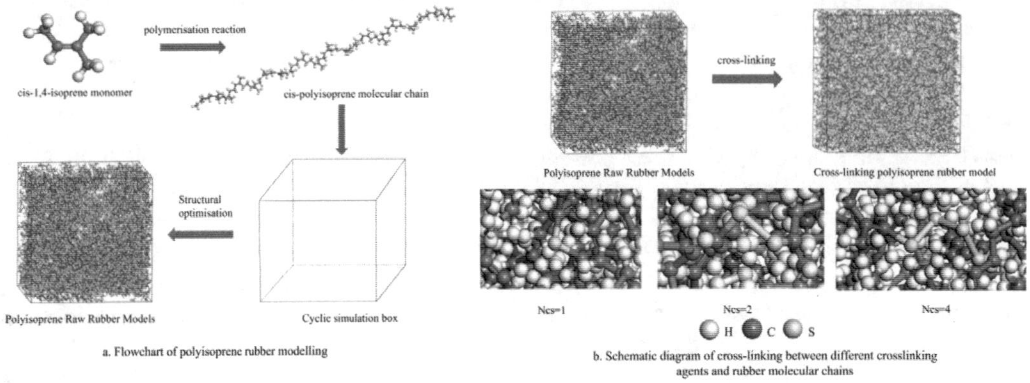


**Figure 1**. Modeling process of vulcanized polyisoprene rubber model.

## 3. Results and discussion

*3.1 Dynamic equilibrium performance analysis*

This section calculates and analyzes the changes in mean-square displacements, end-to-end vector autocorrelation functions, and glass transition temperatures of the IR rubber model. The goal is to elucidate and predict the impact of crosslinking agent types on their kinetic properties during the equilibrium state process.

The influence of crosslinker type on the ability of polymer segments to move can be further characterized by calculating the system's glass transition temperature. Staged cooling was used to reduce the temperature of the system from 500 K to 100 K, with each stage cooling by 50 K. The statistically collected data were plotted against the density of the simulated box versus temperature. The glass transition temperature of the system was obtained by linearly fitting the temperature at the intersection point. Figure 2 gives the variation rule of the glass transition temperature of IR rubber with the parameters of the type of cross-linking bond in the system. With the increase of Ncs, the glass transition temperature of IR rubber is gradually increased. This means that higher Ncs can more effectively inhibit the movement of molecular chains within the crosslinking system, which makes the molecular movement of the material near the glass transition temperature more strongly restricted. Therefore, high Ncs cross-linking systems require higher temperatures to begin movement, and this result is consistent with the experimental law [5].

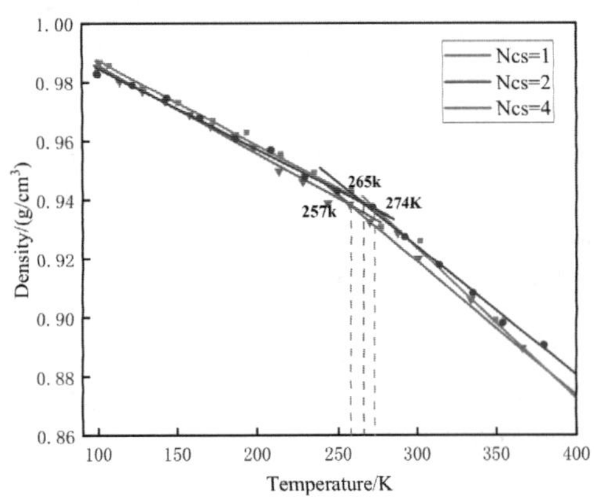


**Figure 2**. Density versus temperature for different crosslinked IR rubber systems.

Mean Squared Displacement (MSD) serves as a metric to characterize the average distance traveled by particles in space over time and is extensively employed to delineate the translational dynamics of the system [6]. Figure 3 (a) illustrates the time evolution of the MSD curves for molecular chain segments within various cross-linked IR rubber systems at a temperature of 300 K. Looking at the whole picture, with the increase of relaxation time, the diffusion rate of crosslinked IR rubber system gradually tends to level off, and the system tends to be stable. With the increasing Ncs, the change of molecular chain means square displacement in the cross-linked IR rubber system decreases. This indicates that the increasing Ncs can more effectively limit the diffusion ability of the molecular chain in the system.

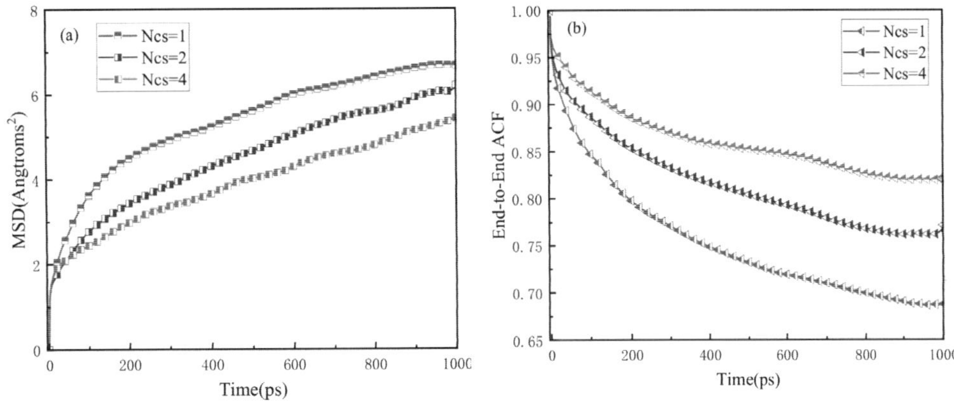


**Figure 3**. (a) Mean square displacement; (b) End-to-end vector autocorrelation function over time for different crosslinked IR rubber systems.

The end-to-end vector autocorrelation function $P(t)$ described the correlation between terminal units of polymer chains and was commonly employed to characterize the rotational dynamics of polymer chains or segments [7], as given by:

$$P(t)=\langle e(t)\cdot e(0)\rangle \quad (1)$$

Where $e(t)$ and $e(0)$ denote the end-to-end unit vectors for each molecular chain at time $t$ and the initial time, respectively; < > denotes the average of the system by using different time origins. The end-vector autocorrelation function of the molecular chain in the crosslinked IR rubber system shows a decreasing trend with the increase in relaxation time. This trend slows down with the increase of Ncs in the system, which reflects that the rotational ability of the molecular chain segments gradually weakens with the increase of Ncs in the system. The primary reason for this phenomenon is the constraint of crosslinking on the molecular chains, leading to the inhibition of rotational motion of the chain segments. This constraint effect significantly intensifies with the increase of Ncs in the system.

### *3.2 Uniaxial tensile properties*

Uniaxial tensile simulation experiments were carried out to investigate the effect of cross-linking bond type on the mechanical properties of IR systems. A uniaxial stretching simulation was performed along the x-axis direction at a constant strain rate of $1e^{-5}fs^{-1}$, and the total volume of the simulated box was maintained constant during stretching. Figure 4 presents the variation rule of the stress-strain curve with the Ncs parameter for the IR rubber model during stretching. From Figure 4 (a), when the strain is less than 0.15, the stress increases linearly with the strain, which satisfies Hooke's law and exhibits quite good elasticity. The system reaches its maximum yield limit when the strain is 0.15. This is because as the strain increases, the molecular chains gradually become more ordered. However, within a certain strain range, the interaction between chain segments is insufficient to resist the increase of external strain. When a certain strain is reached, the interactions between the chain segments begin to affect the overall behavior of the material, leading to a reduction in stress. This critical value is often referred to as the "yield point" or "stress maximum" of the material. After reaching the yield point, the stress slowly decreases with the increase of strain, which is a strain-softening phenomenon. A turnaround occurs after the stress decreases to a local minimum. When the strain is greater than 1, the value of stress starts to enhance continuously with the increase in strain, and after the strain is greater than 2.5, the stress increases exponentially. The stress values of the system at the same strain are positively correlated with the magnitude of Ncs. For instance, at a strain of 3, the stress response of the system with Ncs=4 is significantly higher than that of the system with Ncs=1, which is approximately four times higher.

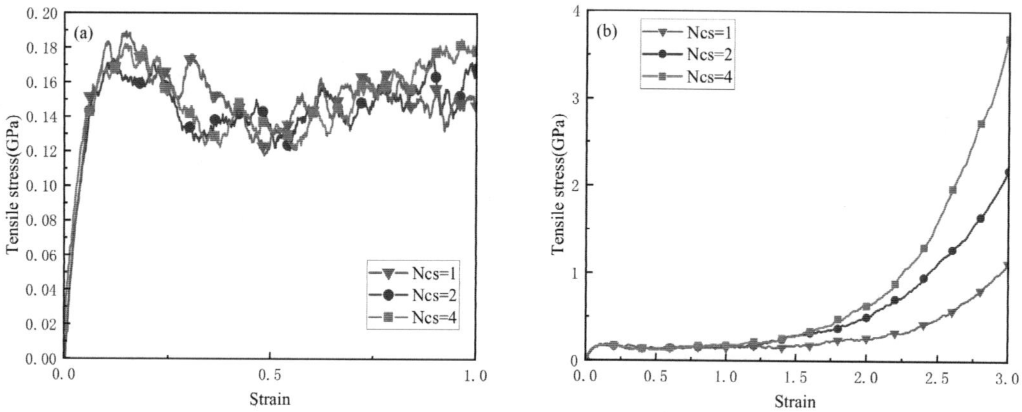


(a) Elongation of 100%; (b) Elongation of 300%.

**Figure 4**. Variation of stress-strain curves for different Ncs systems.

## 4. Conclusions

The effect of the cross-linking bond type Ncs on the structure and properties of polyisoprene rubber materials was systematically investigated through molecular dynamics simulation (MD) methods. Analyses showed that an increase in Ncs within the system during equilibrium diminished the motility of the molecular chain segments. This was specifically manifested as an increase in the glass transition temperature, which represents the chain segment mobility of the entire IR system. Simultaneously, there was a decrease in the mean square displacement of the molecular chain, indicating a reduction in the translational ability of the entire chain. Additionally, there was a decrease in the terminal vector autocorrelation function, reflecting a decline in the rotational ability of the chain segment. During uniaxial stretching, the IR system exhibits a stronger stress response with the increase of Ncs. The findings from this research contribute to an improved understanding of how cross-linking structures affect the mechanical properties of polyisoprene rubber. This provides valuable insights for enhancing the performance of related materials.

## Acknowledgments

We gratefully acknowledge financial support from the Natural Science Foundation of Anhui Province (2008085MA15) and the Key Projects of Natural Science Research in Higher Educational Institutions of Anhui Provincial Department of Education (023AH051122).

## References

[1] Martínez J. D., Puy N., Murillo R., García T., Navarro M. V., and Mastral A. M. (2013). Waste tyre pyrolysis–A review. *Renewable and Sustainable Energy Reviews*, 23, 179-213.

[2] Chen Z., Tu Q., Fang Z., Shen X., Yin Q., Zhang X., and Pan M. (2022). Molecular Dynamics Studies of the Mechanical Behaviors and Thermal Conductivity of Polyisoprene with Different Degrees of Polymerization. *Polymers*, 14 (22), 4, 950.

[3] Rottach D. R., Curro J. G., Budzien J., Grest G. S., Svaneborg C., and Everaers R. (2006). Permanent set of cross-linking networks: Comparison of theory with molecular dynamics simulations. *Macromolecules*, 39 (16), 5, 521-5, 530.

[4] Heinz H., Lin T. J., Kishore Mishra R., and Emami F. S. (2013). Thermodynamically consistent force fields for the assembly of inorganic, organic, and biological nanostructures: the INTERFACE force field. *Langmuir*, 29 (6), 1, 754-1, 765.

[5] Mansilla M. A., Marzocca A. J., Macchi C., and Somoza A. (2017). Natural rubber/styrene-butadiene rubber blends prepared by solution mixing: Influence of vulcanization temperature by using a semi-EV sulfur curing system on the microstructural properties. *Polymer Testing*, 63,150-157.

[6] Khan M., Wen J., Shakoori M. A., and Zhou A. (2023). Thermophysical properties and condensation of R514A through molecular dynamics simulation. *International Journal of Thermofluids*, 20, 100, 436.

[7] Olowookere F. V., Al Alshaikh A., Bara J. E., and Turner C. H. (2023). Effects of chain length on the structure and dynamics of polyvinyl chloride during atomistic molecular dynamics simulations. *Molecular Simulation*, 49 (15), 1, 401-1, 412.

# Preparation and application research of transition metal phosphides

**Deming Li[1#], Xing Fang[1#], Bo Zhang[1], Songlin Chen[1], Guojun Yuan[1,2]**

[1] Anhui Vocational and Technical College, Hefei, 230011, China

[2]Corresponding author's e-mail: 165378341@qq.com
#Contributed equally to this work

**Abstract.** Due to the serious environmental pollution in the world, the shortcut to harmonious development is to improve traditional disposable energy and identify new energy sources that can be utilized. The current problem that human society must solve is to discover and use efficient and inexhaustible new energy sources, among which electrolytic water technology is an effective way to develop clean energy. In recent years, researchers have focused on producing hydrogen by electrolyzed water, but the intermediate hydrogen evolution reaction and the oxygen evolution reaction are relatively slow kinetic processes, and they cannot give the factory a large range of electrolytic water hydrogen production work and push down the commercialization process, so improving the rate of electrolytic water process must be an efficient catalyst. Transition metal phosphide is one of the non-precious metal catalysts in the electrochemical hydrogen evolution reaction and oxygen evolution reaction. Transition metal phosphide can promote rapid change of chemical reaction rate, and it has the characteristics of being cheap and easy to operate so it has a great prospect in the electrolysis of water. In this experiment, the oxygen evolution performance was optimized based on the crystal surface regulation of cobalt-doped transition metal nickel phosphide. The optimal reduction time and a small overpotential (60 mV) were obtained by using sodium borohydride to reduce the sample.

## 1. Introduction

The global economy's fast-paced growth has resulted in a surge in demand for fossil fuels, coal being one of the primary sources. With industrialization and population growth accelerating in recent years, the world's energy needs have been on a constant rise. Consequently, environmental pollution and resource scarcity are becoming increasingly complex and posing serious threats to human society [1]. The issue of energy has long been a concern for humanity and remains a pressing problem in today's society, including issues such as energy resource depletion and soil pollution. As a result, researchers must explore a series of resources that not only meet societal needs but also reduce environmental pollution. Currently, the enormous consumption of fossil fuels results in significant energy expenditure, and the reserves of these fuels in nature are extremely limited. Therefore, the combustion of these primary energy sources will have a tremendous impact on the environment, making it urgent to develop new energy sources that are environmentally friendly, affordable, and reusable.

Solar energy and wind energy show promising prospects in the field of new energy [2]. Utilizing them for power generation can help solve the energy problem. Electrolyzing water, oxygen evolution reactions, and hydrogen evolution reactions are important directions in current energy research. Electrolyzing water will be one of the key points of future water-splitting kinetics research, as it can produce carbon-neutral fuel. Researchers are showing growing interest in electrolysis catalysts because of their exceptional catalytic properties. Transition metal phosphides have a promising future in the market economy as a new material. Following the discovery of transition metal carbides and nitrides, transition metal phosphides have gradually emerged in the public eye. They are new materials composed of transition metal elements and phosphorus elements, exhibiting multiple properties. These compounds have high efficiency, excellent conductivity, and good stability, thus attracting significant attention from researchers.

The mechanism of electrolyzing water involves the application of an electric current that induces the decomposition of water molecules into hydrogen and oxygen ions. This process forms hydroxide ions by combining the oxygen atoms in water molecules with negative ions, while the hydrogen atoms in water molecules combine with positive ions to form hydrogen ions. This process is known as ionization. In the process of water electrolysis, water molecules undergo dissociation into two distinctions: positively charged hydrogen ions and negatively charged hydroxide ions. By separating these two different ions and guiding them to different locations, they can be employed in the synthesis of hydrogen gas and oxygen gas for production purposes. The apparatus for the electrolysis of water typically consists of a power source, an anode, a cathode, and an electrolyte solution.

Under standard gas conditions, the water electrolysis reaction has a standard Gibbs free energy change of -237.2 kJ $mol^{-1}$. According to the relationship between the water electrolysis potential and the Gibbs free energy equation, it can be inferred that the theoretical voltage required for the decomposition of water is 1.23 V. However, in actual conditions, factors such as electrochemical polarization, concentration polarization, and solution resistance can cause the electrode potential to be higher than the theoretical equilibrium potential. Many of the costs of industrial alkaline hydrogen production through water electrolysis come from electricity expenses. It is estimated that approximately 4.5 to 5.5 kWh of electric energy is needed for about 1 $m^3$ $H_2$. Therefore, reducing the overpotential of the electrolysis water reaction requires the addition of efficient electrocatalysts to improve the mutual conversion of energy and reduce energy consumption to lower the cost of hydrogen production. Up to now, precious metals such as Pt, $RuO_2$, and $IrO_2$ have exhibited the best catalytic activity and high current density for water electrolysis [3-4]. However, due to their low content and high cost, their large-scale use in various industries is limited. Therefore, it is necessary to promote the development of non-precious metal catalysts to replace the position of precious metals.

Researchers discovered the excellent performance of transition metal phosphides in the late 20th century. In the early 1970s, Levy R. B. and Boudart M. [5] found that tungsten exhibited good catalytic properties, gradually replacing other metal materials in catalytic applications. Oyama S. T. [6] identified the significant potential of molybdenum phosphide in catalyst applications. In the early 21st century, Stephanie J. Sawhill et al. synthesized $Ni_xP_y/SiO_2$ and $NixPy/Al_2O_3$ catalysts through temperature-programmed reduction for hydrodesulfurization reactions [7]. Sun Xuping's research group successfully prepared a series of self-supported hydrogen evolution catalysts using a low-temperature phosphidation strategy, such as carbon cloth-supported $Ni_2P$ nanosheets [8], foam copper-supported $Cu_3P$ nanowires [9], and titanium sheet-supported CoP nanosheets [10], all exhibiting excellent hydrogen evolution catalytic performance. In 2005, José Rodriguez et al. demonstrated through density functional theory (DFT) calculations that the Ni and P atoms on the $Ni_2P$ (001) crystal surface act as hydride and proton acceptors, respectively, with the P atom weakening the interaction between Ni and H, indicating that the (001) crystal surface of $Ni_2P$ should exhibit hydrogen evolution catalytic behavior similar to NiFe hydrogenases [11]. Additionally, Professor Schaak's research group at the University of Pennsylvania in the United States used an organic phosphorus source method to synthesize cobalt phosphides, including CoP and $Co_2P$. Subsequently, Professor Schaak designed and prepared CoP and $Co_2P$ catalysts with similar sizes and morphologies to investigate the differences in their catalytic activity [12]. Transition metal phosphides play important roles in various fields, particularly in the field of catalysis, where they have gained significant attention from researchers. In the field of photocatalysis, transition metal phosphides can be used for organic pollutant degradation and hydrogen production from water. The research field of transition metal phosphide is also expanding [13-14].

Based on the research progress mentioned above, transition metal phosphides have achieved significant results as catalysts for the OER. As the scientific level of research continues to advance, the focus has shifted from characterizing the basic properties of catalysts to studying catalytic mechanisms. Simultaneously, it has been discovered that their use as catalysts can accelerate catalysis.

## 2. Materials and methods

### *2.1 Materials*

All the chemicals were Cobaltous nitrate hexahydrate ($Co(NO_3)_2 \cdot 6H_2O$, AR) and Sodium borohydride ($NaBH_4$, AR) were purchased from Adamas-beta L-Aspartic acid ($C_4H_7NO_4$, AR) was purchased from Sinopharm Chemical Reagent Co., Ltd.

### *2.2 Catalyst Preparation*

Add 2.619 g of cobalt (II) nitrate hexahydrate and 1.197 g of L-aspartic acid into a 500 mL beaker. Next, combine 45 mL of deionized water with 54 mL of ethylene glycol to create a mixture. Place the beaker on a hotplate with a magnetic stirrer and stir vigorously. Wait for complete dissolution and then slowly add 9 mL of 2 M NaOH solution dropwise. Continue stirring for 0.5 hours until the solution turns pink. Next, transfer three 36 mL portions of the resulting mixture into three 50 mL stainless steel autoclaves using graduated cylinders. Place one piece of dried nickel foam into each autoclave. React the autoclaves at 180°C in a high-temperature drying oven for 5 hours. After the reaction is complete, pour out the solution. There will be pink precipitates on the nickel foam. Wash the foam with deionized water and anhydrous ethanol three times. Transfer the cleaned foam into an evaporating dish and dry it in the high-temperature drying oven at 60°C for three hours. Let it cool and keep it for further use. Place the obtained samples in a double-temperature zone tube furnace. Raise the temperature from ambient to 300°C and heat for 3 hours. Then allow it to cool back to room temperature. Finally, collect the resulting black phosphide product. Weigh 5.7 g of sodium borohydride using an electronic balance and dissolve it in 50 mL of water in a beaker. At this point, the solution will bubble, which will disappear after a certain time. Transfer the black product obtained above into three separate 25 mL beakers labeled 1, 2, and 3. Then pour the dissolved sodium borohydride solution into each of the three beakers. Let the first reaction proceed for 1 hour, the second reaction for 2 hours, and the third reaction for 3 hours. After the respective time intervals, rinse each set with deionized water and anhydrous ethanol. Once cleaned, transfer them to an evaporating dish. Vacuum-dry them at 60°C in a high-temperature drying oven for 5 hours to collect the black product. The samples are designated as Co1-NixPy, Co2-NixPy, and Co3-NixPy, respectively.

### *2.3 Characterizations*

The powder X-ray diffraction (XRD) measurement was conducted using Cu Kα radiation with a wavelength of 0.1541 nm on the Ultima IV X-ray diffractometer with a current of 40 mA and a respective voltage of 40 kV. The 2θ scans covered the range of 25°-65° with a speed of 10° $min^{-1}$.

### *2.4 Catalytic Activity Testing*

Electrochemical testing is conducted using an electrochemical analyzer instrument (CHI660E) in a standard three-electrode system. The sodium hydroxide solution prepared as described earlier is used as the electrolyte. In the three-electrode system used, the reference electrode is a graphite carbon rod, and the working electrode is a platinum electrode clip. The testing is carried out under these conditions.

## 3. Results and discussion

### *3.1. X-ray diffraction analysis (XRD)*

XRD characterization was performed on the three sets of samples, and the resulting XRD patterns are shown in the following three figures. Phase analysis was conducted using XRD. From the three graphs, it is evident that all measured samples exhibit distinct diffraction peaks, indicating the excellent crystalline structure of the catalyst prepared in the previous steps. In Figure 1a, the peak positions of Co1-NixPy are observed at 40.8°, 44.6°, 47.3°, and 54.2°, corresponding to the (111), (201), (210), and (300) crystal planes of Co1-NixPy, respectively (PDF card number: 03-0953). In Figure 1a, the peak positions of Co2-NixPy are observed at 41.6°, 46.9°, and 48.9°, corresponding to the (321), (420), and (312) crystal planes of Co2-NixPy, respectively (PDF card number: 22-1190). Last, Figure 1b shows a similar pattern to Figure 1c, with comparable peak positions and crystal planes. In conclusion, it can be observed that after reduction by sodium borohydride, all samples have completely transformed into nickel phosphide, indicating pure nickel phosphide. The treatment with sodium borohydride leads to structural damage, resulting in defects. These defects may potentially have an impact on the subsequent oxygen evolution reaction. The increase in treated defect sites is primarily due to the disruption of the crystal structure of NixPy, and it is not significantly related to the doping of the Co element. This ultimately results in significant changes in peak position in the image of Figure 1c compared to Figures 1a and 1b, leading to a lower hydrolysis potential due to structural modifications.

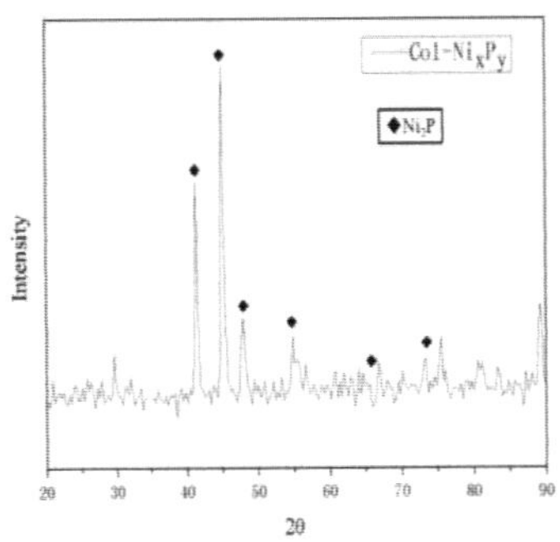


(a)

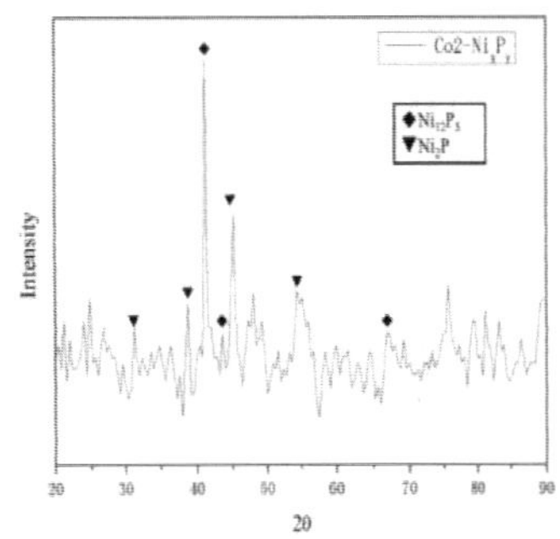


(b)

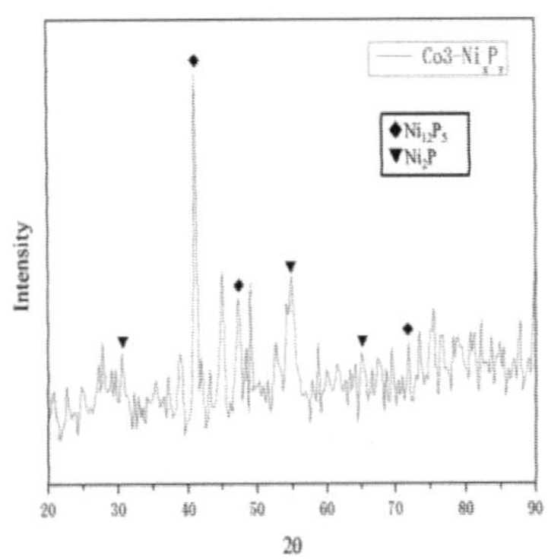


(c)

**Figure 1.** XRD patterns of a) Co1-NixPy, b) Co2-NixPy, and c) Co3-NixPy.

### *3.2. Cyclic voltammetry (CV)*

This electrochemical test allows for the assessment of the reversibility of electrode reactions. Next, three sets of samples were subjected to scanning tests at different scan rates for observation and comparison. From these three graphs, it can be observed that as the scan rate increases, the difference between the two ends of the curve becomes larger, indicating excellent stability of the catalyst material. This portion refers to the pre-activation of the LSV test before conducting the actual test. The CV images obtained at different scan rates in Figure 2c exhibit good separation and significant current changes, indicating that Co3-NixPy has better electronic transport properties. This assures a lower hydrolysis potential at the microscopic level.

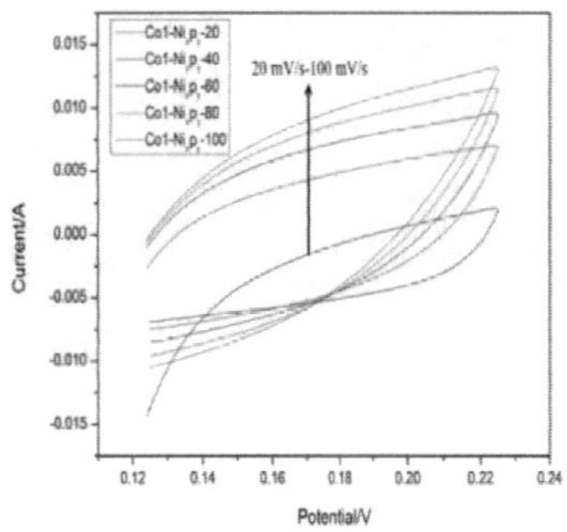


(a)

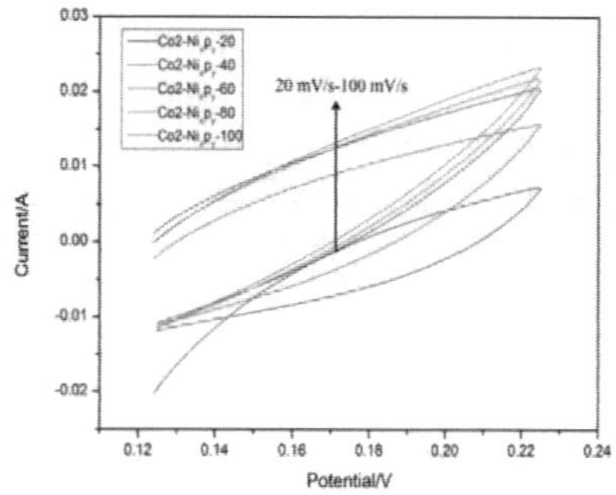


(b)

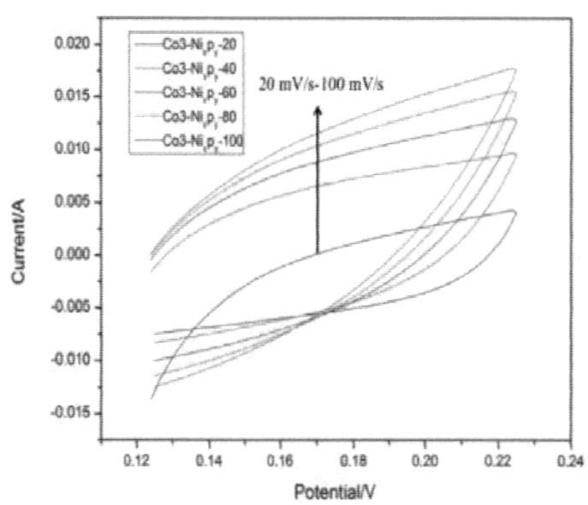


(c)

**Figure 2.** CV plots of NixPy samples at different scan rates: a) Co1-NixPy, b) Co2-NixPy, and c) Co3-NixPy.

### *3.3. Linear sweep voltammetry (LSV).*

This is the primary method in electrochemical testing used for quantitatively analyzing the OER performance. Next, three sets of samples were tested and compared. From Figure 3 , it can be observed that at 10 mA/cm$^2$, the overpotentials for the three samples Co1-NixPy, Co2-NixPy, and Co3-NixPy are 310 mV, 225 mV, and 60 mV, respectively. The experimental results indicate that the data for Co3-NixPy is the smallest, thus exhibiting the best OER performance.

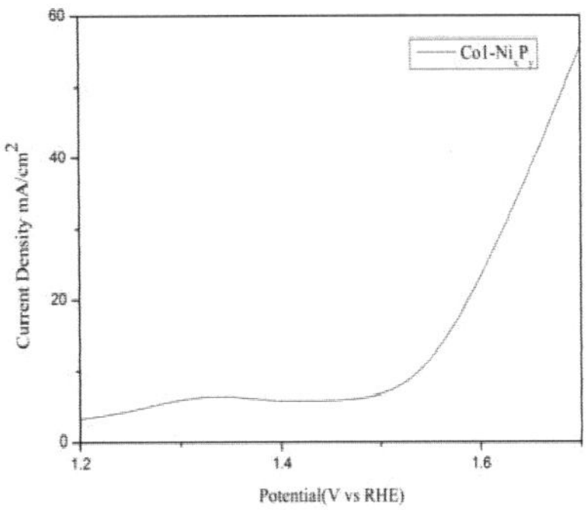


(a)

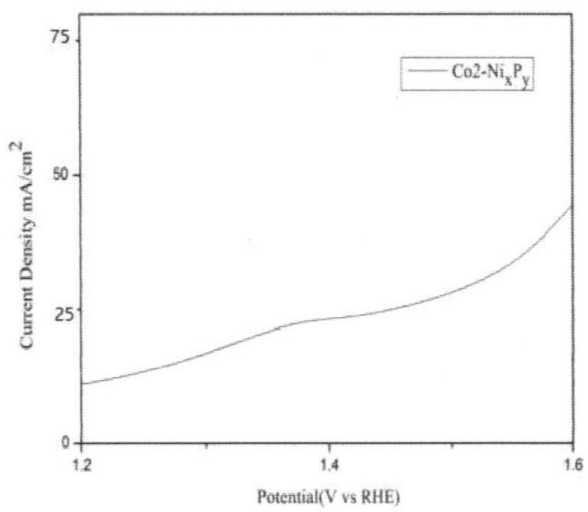


(b)

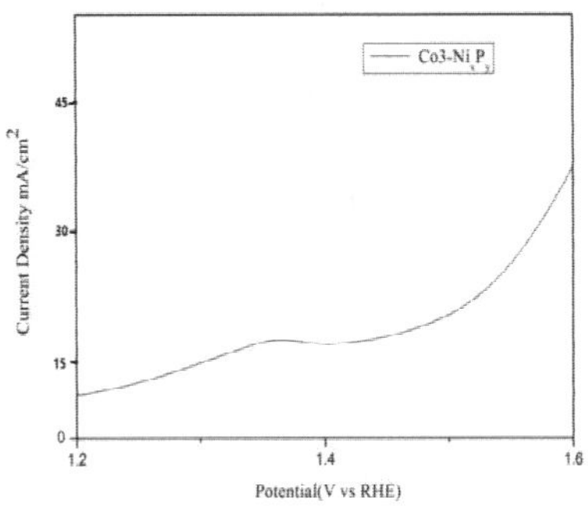


(c)

**Figure 3.** LSV test plots of NixPy samples: a) Co1-NixPy, b) Co2-NixPy, and c) Co3-NixPy.

## 4. Conclusion

In this study, transition metal phosphide nickel underwent sodium borohydride treatment, which had a certain effect on the crystal structure of nickel phosphide. Synthesis of nickel phosphide was achieved through the solvothermal method, and the above images show that cobalt compounds were loaded onto nickel phosphide. The results indicate that different crystal structures of nickel phosphide have varying effects on OER performance, with Co1-NixPy > Co2-NixPy > Co3-NixPy, demonstrating that Co3-NixPy has the best catalytic performance.

## Acknowledgments

This work was funded by Scientific research planning project in Anhui Province [KJ2020A1037,2023AH051431, 2022AH040283, 2022AH052075]; Anhui Vocational and Technical College School-level scientific research [2023xjkytd4, 2023xjyyq1].

## References

[1] Lewis N S, Nocer D G (2006). *Proc Natl Acad Sci,* 103(43): 15729-15735. DOI: 10.1073/pnas.0603395103.

[2] Peng J. C., Shi Y.D., Xiao Z.S. (2023). *Materials Research Bulletin*, 166: 112341. DOI: /10.1016/j.materresbull.2023.112341.

[3] Sheng W, Myint M, Chen G J, et al (2013). *Energy & Environmental Science*, 6(5): 1509. DOI: 10.1039/C3EE00045A.

[4] Seh W Z, Kibsgaard J, Dickens F C, et al. (2017) *Science*, 355: eaad4998. DOI: 10.1126/science. and 4998.
[5] Levy R B, Boudart M (1973). *Science*, 181(4099): 547-549. DOI: 10.1126/science.181.4099.547.
[6] Oyama S T (2003). *Journal of Catalysis,* 216(1/2): 343-352. DOI: 10.1016/S0021-9517(02)00069-6.
[7] Sawhill J S, Layman A K, Wyk V R D, Engelhard H M, Bussell E M (2005). *Journal of Catalysis*, 231(2): 300-313. DOI: 10.1016/j.jcat.2005.01.020.
[8] Ping J, Qian L, Sun X P (2014). *Nanoscale*, 6(22): 13440-13445. DOI: 10.1039/C4NR04866K.
[9] Tian J, Liu Q, Cheng N, Asiri M A, Sun X P (2014). *Angewandte Chemie International Edition,* 53(36): 9577-9581. DOI: 10.1002/anie.201403842.
[10] Li Y.Q., Hao Y G., Ali U., et al (2023). *Inorg. Chem. Front.,* 10: 5719. DOI: 10.1039/D3QI00971H.
[11] Liu P, Rodriguez A J (2005). *Journal of the American Chemical Society*, 127(42): 14871-14878. DOI: 10.1021/ja0540019.
[12] Popczun E, Read C, Roske C, et al (2014). *Angew. Chem. Int. Ed.*, 53(21): 5427–5430. DOI: 10.1002/anie.201402646.
[13] Song W., Xu G., Kun L., Guiting W., et al (2023). *Journal of Alloys and Compounds*. 968(15): 171939. DOI: 10.1016/j.jallcom.2023.171939.
[14] Mao, S.H., Su, W.X., Liu, C.Y., Zhao, F., et al (2023). *Integrated Ferroelectrics*. 237(1): 161-169. DOI: 10.1080/10584587.2022.2143187.

# Preparation and application of superhydrophobic polytetrafluoroethylene materials and micro/nanoindentation

**Jianfeng Zhou[1], Xiangru Zhou[2], Jianjun Zhou[3] and Zengzhen Mi[4, 5]**

[1]Chongqing Municipal Housing and Urban-Rural Development Administrative Approval Service Center, Chongqing, 400010, China
[2]College of Electromechanical Engineering, Beijing University of Chemical Technology, Beijing, 102202, China
[3]Guangzhou Haige Communication Group Co., Guangzhou, Guangdong, 510663, China
[4]College of Mechanical Engineering, Chongqing University of Technology, Chongqing, 400054, China

[5]Corresponding author's e-mail: lomoer@cqut.edu.cn

**Abstract.** A method for preparing superhydrophobic polytetrafluoroethylene materials by micro- and nanoimprinting is discussed. Surfaces with superhydrophobic properties were prepared by designing and imprinting micro- and nano-structures on polytetrafluoroethylene materials. The experiments on weather resistance and durability revealed that the microstructure of screens of different mesh sizes was processed onto the surface of PTFE material by micro-nano thermal imprinting to make it hydrophobic and oleophobic, and retained the original excellent properties of corrosion resistance and low surface attachment, etc. The material processed by the new method has a wide range of application prospects in various fields.

## 1. Introduction

Polytetrafluoroethylene (PTFE), with a melting point of 327°C and a white waxy color, is a thermoplastic. At temperatures above its melting point, it will appear transparent. PTFE has a wide range of applications in many fields such as defense, energy, medical, construction, household goods, etc., and is of great research value. Due to its strong internal C-F atomic bonds, PTFE has a very high melt viscosity and is able to have microstructures processed on its surface by micro-nano thermal embossing.

Ordinary PTFE has poor weathering resistance, however, processing PTFE materials by the micro-nano thermal embossing process can make them hydrophobic while maintaining their original excellent properties. After this treatment, the material can still keep its original excellent performance unchanged.

Superhydrophobic materials can effectively reduce the corrosion efficiency of metals and prolong the service life of materials by virtue of their excellent anti-corrosion properties, thus reducing costs. Liu et al. prepared a superhydrophobic magnesium alloy surface by electrochemical process, which successfully protected the internal metal from corrosion. This superhydrophobicity also means that the material is somewhat oleophobic, and the rolling down of the droplets can take away some of the impurities, making the material extremely self-cleaning. Ma et al. prepared clothes made of copolymer fibers by means of electrospinning, which is not only comfortable and soft but also has a strong self-cleaning ability. Jia et al. prepared a superhydrophobic coating by spraying carbon black nanoparticles, PDMS, which was also used to protect the internal metal from corrosion, PFDTES mixture, using epoxy resin as binder. The coating achieved a product contact angle of 161° and a rolling angle of 1.4°, which extended the water freezing time on the surface of the material from 30 to 120 seconds. However, most of the current methods for preparing superhydrophobic surfaces use chemical methods, which are costly and have low production efficiency. Meanwhile, the weather resistance of the products when facing complex environments has yet to be strengthened. Therefore, how to prepare superhydrophobic materials with high weathering resistance in large quantities, quickly and at low cost is still a worthy research problem [1-4].

## 2. Superhydrophobic properties analysis

Superhydrophobic surfaces refer to material surfaces with a static contact angle $\theta_e$ greater than 150° and a rolling angle $\alpha$ lower than 10°. The static contact angle $\theta_e$ and rolling angle $\alpha$ are common parameters to describe the surface infiltration of materials.

Surface infiltration refers to the nature of a liquid replacing an otherwise incompatible gas or liquid on a solid surface to form a new and stable solid-liquid mixing interface. The contact Angle is the Angle formed by the tangent line of the gas-gas interface and the solid-liquid interface at the junction of the solid-liquid-gas interface. When the contact angle $\theta_e$ is not less than 90°, it means that the liquid cannot infiltrate the solid surface, and when the contact angle $\theta_e$ is less than 90°, the liquid can infiltrate the solid surface [5].

When the solid surface is an absolutely ideal surface, the relationship between the solid-liquid-gas three-phase interface tension and the contact angle $\theta_e$ can be expressed by Young Equation, as shown in Formula (1):

$$cos\theta_e = (\gamma_{SV} - \gamma_{S1}) / \gamma_{1V} \quad (1)$$

Where $\gamma_{sv}$, $\gamma_{sl}$, and $\gamma_{lv}$ are the interfacial tension between solid-gas, solid-liquid, and liquid-gas, respectively.

As can be seen from the above formula, the lower the surface energy of the solid material is, the greater the contact angle of the droplet is. However, the material has a surface roughness, so the actual contact area is larger than the ideal contact area. The roughness factor can be introduced in the formula [6], as shown in Formula (2):

$$cos\theta_w = rcos\theta_e \quad (2)$$

Where $\theta_w$ is the contact angle of the droplet on the roughness surface; r is the roughness factor, the ratio of the actual surface area to the geometric projected area.

As can be seen from Formula (2), when the material is hydrophobic, $\theta_w$ will increase as r increases, that is, when the hydrophilic surface increases its roughness, the hydrophobic capacity of the surface will be enhanced.

When the surface roughness of the material is large, the air forms an air cushion in the groove. When the Wenzel model fails, the surface contact state is converted to the Cassie model [7]. In this model, the residual gas in the solid surface groove forms a new solid-liquid-gas composite interface with the droplet and the solid surface, as described in Formula (3):

$$\cos\theta = R_f \cos\theta_0 - f_{LA}(R_f \cos\theta_0 + 1) \quad (3)$$

When the static contact angle is greater than 150°, the rolling angle is higher than 10°; if the static contact angle is greater than 150°, the 10° is lower. Because the surface tension of oil droplets is generally smaller than water, the superhydrophobic materials are generally oil-hydrophobic. In the face of the polluted environment, the self-cleaning ability is strong. Because of its excellent nature, it plays an important role in preventing corrosion, ice, pollution, resistance, and other aspects. PTFE materials prepared with micro and nanoscale structures by direct replication method and plasma treatment method have stronger anti-ice attachment properties than general hydrophobic materials [8].

At present, the main ways to prepare superhydrophobic surfaces are sol-gel, phase separation, anodic oxidation, chemical etching, self-assembly, and so on. These methods are basically chemical methods, which are difficult to use in industrial production due to the complex process and low production efficiency.

## 3. Experiments and analysis

### *3.1 Experimental materials and sample preparation*
The materials used in the experiments are mainly 304 stainless steel sieves of 300 mesh, 600 mesh, 800 mesh, 1200 mesh, and 1500 mesh; 0.2 mm, 0.3 mm, and 0.5 mm PTFE membranes; anhydrous ethanol. The correspondence between sieve mesh and aperture size is shown in Table 1:

**Table 1.** Correlation between mesh number and pore size

| Mesh number (mesh) | Pore size (diameter) / um | Mesh number (mesh) | Pore size (diameter) / um |
|---|---|---|---|
| 200 | 75 | 1000 | 13 |
| 300 | 48 | 1200 | 11 |
| 400 | 38 | 1500 | 8 |
| 500 | 25 | 2000 | 6.5 |
| 600 | 23 | 5000 | 2.6 |

Through the analysis, it was found that on the 0.3 mm processed PTFE membrane, the water beads were full of ellipsoidal, slightly tilted and then rolled down, with better hydrophobic properties, and the material had a rough, uniform surface. Therefore, the thickness of 0.3 mm PTFE membrane was finally chosen as the experimental material.

### *3.2 Analysis of the results*

#### *3.2.1 Embodiment of hydrophobic properties and the effect of screen mesh size on contact angle*
The higher the mesh number of the sieve mesh is, the smaller its pore size is, and the larger the surface roughness of the processed material is, which enhances the hydrophobic properties of the material. At the same time, changing the mesh size of the sieve mesh does not affect the diameter of the wire used to weave the sieve mesh, so the processed microgroove structure remains almost unchanged. However, as the aperture of the screen mesh decreases, the denseness of the grooves increases significantly, allowing more gas phase to be mixed into the solid-liquid phase, which in turn enhances the hydrophobic properties of the material. In summary, in order to obtain the best superhydrophobic PTFE material, a 1500 mesh screen should be selected for processing.

#### *3.2.2 Influence of the temperature of heat preservation on the surface microstructure and hydrophobicity of the material*
The contact angle increases and then decreases slowly with the increase of heating temperature. From the contact angle theory, the larger the contact angle is, the better the hydrophobicity of the measured material surface is. At the heating temperature of 180°C, the contact angle for the maximum value of 148°, when the hydrophobicity of the processed material is the best. The heating temperature and holding temperature should be selected as 180°.

#### *3.2.3 Effect of holding pressure on the surface microstructure and hydrophobicity of the material*
The experimentally selected heating temperature is the heat deformation temperature of PTFE. When the material has softened to the point where microstructures can be machined, appropriate pressure is required to fully and uniformly replicate the microstructures on the mold surface to the surface of the material to be processed. The maximum pressure that can be exerted by the laboratory machine is about

5 MPa, so 1 MPa, 2 MPa, 4 MPa, and 5 MPa were selected. A 1500 mesh screen was chosen, the heating temperature was 180°, and the holding time was 120 s [9, 10].

*3.2.4 Weatherability enhancement experiments*

Before hot embossing, anhydrous ethanol mixed with silica microspheres is uniformly applied to the screen mesh as a mold, and after all the anhydrous ethanol has evaporated, the mold is placed on the PTFE sample to be processed for hot embossing. While the screen microstructure was processed on the surface of the material, the silica microspheres were transferred to the surface of the PTFE film with the aid of the screen. The temperature of the hot embossing was selected to be 180°C, the holding time was 120 s, the applied pressure was 5 MPa, and the screen mesh was selected to be 1500 mesh, 1200 mesh, and 300 mesh. After the hot embossing was completed, the mold was removed while it was still hot. A drop of water was applied to the surface of the product with a rubber-tipped burette and its morphology was observed. Then the contact angle was measured and the results are shown in Figure 1:

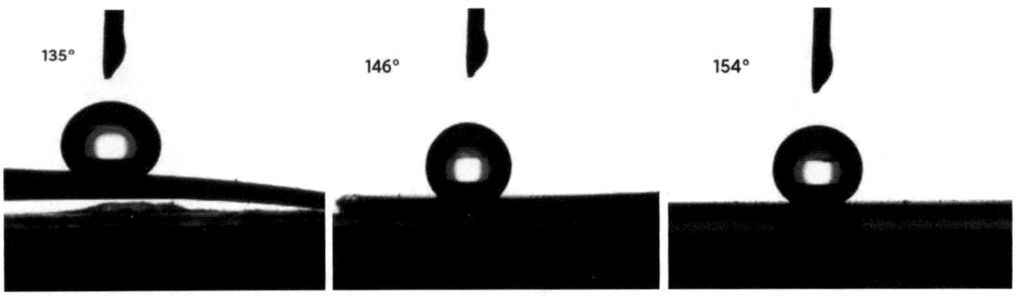


(a) 300 mesh (b) 1200 mesh (c) 1500 mesh

**Figure 1.** Contact angle of the sample after adding silica microspheres processing.

According to the results of the contact angle measurement, it can be seen that after adding silica microspheres, the hydrophobicity of the samples processed through different screen mesh is improved: the contact angle of the samples processed after 300 mesh is 135°; the contact angle of the samples processed after 1200 mesh is 146°; the contact angle of the samples processed after 1500 mesh is 154°, which can satisfy the requirements of superhydrophobicity. Next, the 1500 mesh samples were selected for friction, folding, and static tests, and then their contact angles were measured. The results are shown in Figure 2:

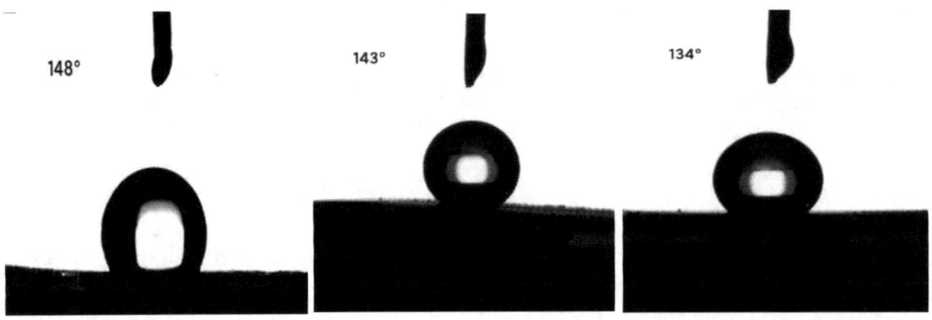


(a) Contact angle after friction; (b) Contact angle after folding; (C) Contact angle after resting.

**Figure 2.** Contact angle measurement after durability and permanence test.

As shown in Figure 2, the surface of the material has a groove-like microstructure processed with a screen as a mold. Attached to the surface of the groove structure are silica microspheres attached to the screen before processing, which are transferred to the surface of the material by thermal imprinting, and they form a new secondary microstructure with the screen indentation. The silica balls in the process of embossing will inevitably be processed on the surface of the material with more tiny indentations. These new secondary microstructures on the material hydrophobicity are critical to the improvement of the hydrophobicity of the material. At the same time after a long period of time after the static, the surface of the groove structure of the material is destroyed, but the secondary microstructures will not be altered or affected by the less. These microstructures contribute to the maintenance of the material's hydrophobicity, as shown in Figure 3.

**Figure 3.** The surface of the screen with $SiO_2$ microspheres attached and PTFE film under electron microscope.

In the direct hot embossed product, after durability and long-lasting tests, the contact angle decreased from 148° to 115°, 109°, and 95°, respectively; and after adding silica microspheres and then after testing the contact angle, it decreased from 154° to 148°, 143°, and 134°, respectively. According to the contact angle measurements of the samples before and after rubbing, folding, and resting, compared with the contact angle measurements of the samples thermally embossed after the addition of silica microspheres before and after the durability and permanence tests, it can be found that the reduction of the contact angle of the samples decreased substantially. The hydrophobicity angle of the samples processed by adding silica microspheres is significantly larger than that of the samples processed by adding silica microspheres, indicating that the samples processed by adding silica microspheres are still able to maintain a better hydrophobicity after friction, folding, and resting, and the durability and endurance are significantly stronger than that of the products made by direct hot embossing.

## 4. Results and discussion

After weathering and durability tests, the hydrophobicity decreases dramatically. It was found by scanning electron microscopy that the microstructure of the groove surface of the sample after friction and folding appeared to be smoothed and filled, which caused the microstructure of the material surface to be destroyed, the roughness to be reduced, the proportion of meteorology at the solid-liquid interface to be reduced, and the hydrophobicity to be reduced. Therefore, on the basis of previous hot embossing, using anhydrous ethanol mixed with silica microspheres evenly coated on the screen. After drying this as a mold, under the conditions of a heating temperature of 180°C, holding pressure of 5 MPa and holding time of 120 s, 300 mesh, 1200 mesh, and 1500 mesh were used for hot embossing again. The contact angle measurements of the improved experimental results showed that the contact angle of all the samples was greatly improved, with the contact angle of the samples made through 1500 mesh reaching 154°, which meets the requirements for superhydrophobicity. The decrease in contact angle of these samples after durability and permanence tests was significantly lower than that of the samples processed without the addition of silica microspheres, indicating that the addition of silica microspheres greatly enhanced the weatherability and permanence of the processed materials.

**Acknowledgments**

This work was financially supported by the National Natural Science Foundation of China (No. 61901068), Natural Science Foundation of Chongqing, China (No. CSTB2023NSCQ-MSX0911).

**References**

[1] Lu X, Chen Y, Yan W, et al. Amphiphobic polytetrafluoroethylene membrane with a ring-on-string-like micro/nano structure for air purification [J]. Journal of Membrane Science, 2022, 652:120476-. DOI:10.1016/j.memsci.2022.120476.
[2] Isakova T A, Petrova P N. New Materials Based on Polytetrafluoroethylene and Modified Filling Materials in Mining Industry [J]. IOP Conference Series: Earth and Environmental Science, 2021, 773(1):012068 (5pp). DOI:10.1088/1755-1315/773/1/012068.
[3] Isakova T A, Petrova P N. New Materials Based on Polytetrafluoroethylene and Modified Filling Materials in Mining Industry [J]. IOP Conference Series: Earth and Environmental Science, 2021, 773(1):012068 (5pp). DOI:10.1088/1755-1315/773/1/012068.
[4] Masalehdan T, Eskandarzade M, Tutunchi A, et al. Two-dimensional clay nanosheet-reinforced polytetrafluoroethylene composites and their mechanical/tribological studies [J]. Materials Today Communications, 2021, 26. DOI:10.1016/j.mtcomm.2021.102026.
[5] Zhang Y, Yang F. Impression deformation of carbon-microsphere and polytetrafluoroethylene composite [J]. Materials Science and Engineering B, 2021, 270:115199. DOI:10.1016/j.mseb.2021.115199.
[6] Rasheed L M, Mansoor R. Synthesis and Characterisation of Polytetrafluoroethylene Composite for Electrical Insulation and Dielectrics [J]. IOP Conference Series: Materials Science and Engineering, 2021, 1090(1):012104 (6pp). DOI:10.1088/1757-899X/1090/1/012104.
[7] Ambroziak A, Klosowski P. Influence of cyclic frozen and defrost on mechanical properties of polytetrafluoroethylene (PTFE)-coated woven fabrics [J]. Archives of civil engineering, 2023.
[8] Dorigo W , Fargion A , Bassoli G ,et al.Autologous saphenous vein and heparin-bonded expanded polytetrafluoroethylene as graft materials for below-the-knee femoro-popliteal bypass in patients with critical limb ischemia: A propensity score-matched analysis[J].The surgeon : journal of the Royal Colleges of Surgeons of Edinburgh and Ireland, 2022, 20(2):85-93.DOI:10.1016/j.surge.2021.02.001.
[9] Zhao J, Wang G, Zhu W, et al. Lightweight and strong polypropylene/talc/polytetrafluoroethylene foams with enhanced flame-retardant performance fabricated by microcellular foam injection foaming [J]. Materials & Design, 2022, 215:110539-. DOI:10.1016/j.matdes.2022.110539.
[10] Kisic M D D. The Influence of Alumino-Silicate Matrix Composition on Surface Hydrophobic Properties [J]. Science of sintering, 2019, 51(2).

# Influences of temperature on the property of ITO/Teflon/Ag in atomic oxygen environment

**Liang Guo [1,2], Tiecheng Qiu [1], Qian Zhang [1], Zhanping Guo [1], and Ming Shan [1]**

[1] Beijing Institute of Spacecraft Environment Engineering, Beijing 100094, China

[2]Corresponding author's e-mail: guoliang_1113@163.com

**Abstract:** This paper uses ground simulation equipment for an atomic oxygen environment to conduct experimental research on ITO/Teflon/Ag, which is used in spacecraft using an atomic oxygen environment. The atomic oxygen integrated flux selected in the study is $9.1\times 10^{19}$ atoms/cm$^2$, with a vacuum degree of $10^{-2}$ Pa, and sample temperatures of 20°C, 70°C, and 120°C, respectively, were selected to investigate the possible impact of sample temperature on atomic oxygen environment simulation experiments. After the test, a high-precision microelectronic balance was used to test the mass loss of the sample and calculate the reaction rate of the material. The thermal emissivity and solar absorption ratio of the sample were tested using TEMP 2000A and LPSR 300, respectively. Through experiments and analysis, it was found that as the sample temperature increased, the ITO/Teflon/Ag reaction rate with atomic oxygen gradually increased but remained at $10^{-25}$. Atomic oxygen and temperature have a significant impact on the thermal physical properties of the material.

## 1. Introduction

Atomic oxygen (from now on referred to as AO) mainly exists in the low Earth orbit atmospheric environment. When the spacecraft is in orbit, AO will collide with the outer surface material of the spacecraft at a relative speed of 7-8 km/s and undergo chemical reactions with it, causing mass loss of the outer surface material of the spacecraft, thereby reducing its various performance and ultimately affecting the on-orbit life of the spacecraft [1, 2].

Chemical reactions are influenced by various factors, and environmental temperature is one of the important factors affecting the speed of chemical reactions. In foreign in-orbit experiments, polyimide films were selected as the research object to investigate the various performance changes of polyimide films in AO environments under different temperature conditions [3, 4, 5]. Some universities and scientific research institutions in China have also conducted extensive research on the performance changes of surface materials used in spacecraft under an AO environment and formed a series of research results, providing data support for the selection of thermal control materials used in spacecraft, especially in the field of manned spaceflight. However, there is relatively little research on the material performance changes in the AO environment under different temperature conditions in China [6].

ITO/Teflon/Ag, a commonly used thermal control material for spacecraft, is widely used in the outer surface coating of aerospace. This article will take ITO/Teflon/Ag as the research object, select AO environments under different temperature conditions, and study its performance changes.

## 2. Research methods

### *2.1. Ground simulation equipment*

This research-oriented experiment was conducted on the ground AO simulation equipment, consisting mainly of a vacuum container, a vacuum pumping system, a three-dimensional moving mechanism, a beam emission device, a microwave generation system, and an overall control and data acquisition system [5]. Figure 1 shows the structural schematic diagram of the equipment.

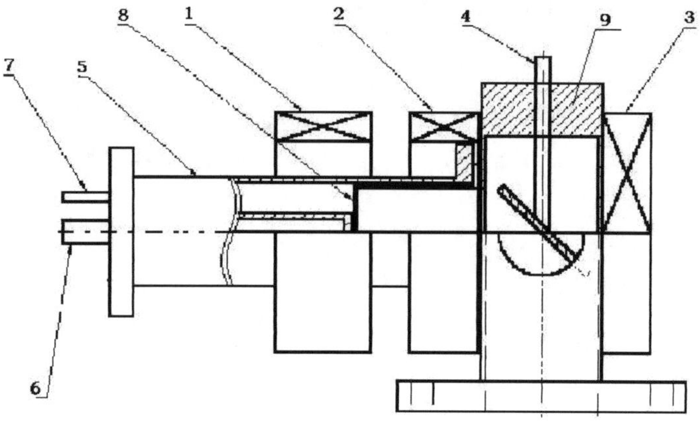


1, 2, 3-Magnetic block; 4-Platform for target practice; 5-Resonance field; 6-Adjustive conductor; 7-Wire; 8-Silica cup; 9-Reaction chamber.

**Figure 1.** The schematic structure of the AO environmental simulation equipment.

### *2.2. Test device*

The experimental device used in this experiment is shown in Figure 2. The sample target platform and sample temperature control device are installed on the existing three-dimensional moving mechanism, and the installation diagram of the entire device is shown in Figure 3. The sample target has good contact with the cooling water tank to control the temperature of the sample. The sample temperature control mainly consists of cooling and thermal cycles. In addition, effective connection methods are adopted to disassemble the sample target rapidly.

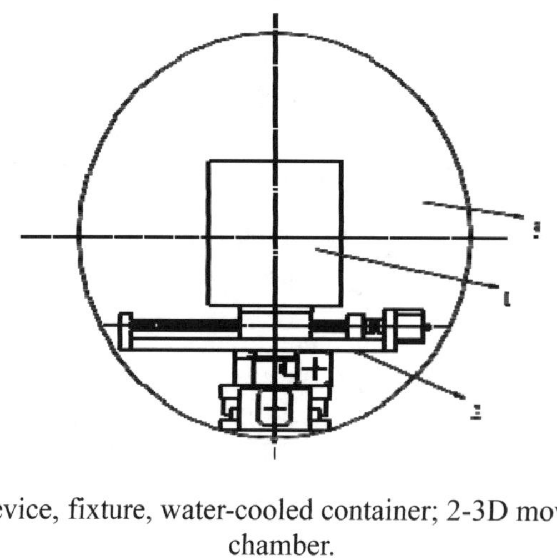

1-Sample installation device, fixture, water-cooled container; 2-3D moving machine; 3-Vacuum chamber.

**Figure 2.** The schematic structure of the experimental device.

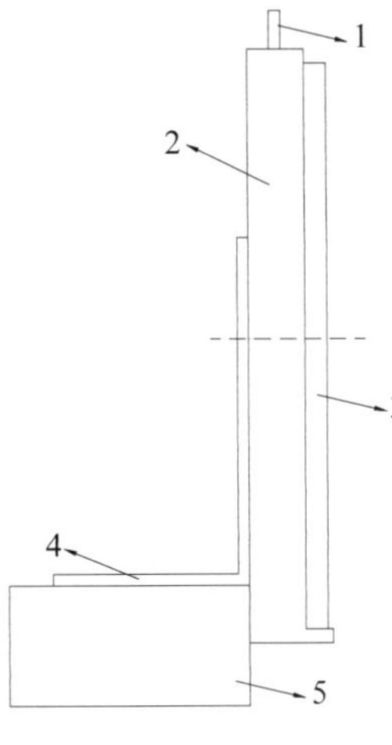


1-Water pipe; 2-Water-cooled container; 3-Sample installation device; 4-Fixture; 5-3D moving machine.

**Figure 3.** The schematic installation of the experimental device.

### *2.3. Test parameters*

Before the experiment, the AO beam was calibrated at different distances from the axial center of the AO source to the neutralization plate. The selected material was Kapton H thin film. The AO beam density at different distances from the neutralization plate was calculated based on the mass changes of the Kapton H thin film before and after the experiment. The beam density of AO calculated by this formula is $2.1\times10^{15}$ atoms/cm$^2$•s.

The AO integrated flux was selected as $9.1\times10^{19}$ atoms/cm$^2$.

During the test, the vacuum degree inside the container is better than $2\times10^{-2}$ Pa.

The sample temperatures were selected as 20°C, 70°C, and 120°C, with a temperature deviation of ± 2°C.

Before the experiment, a cesium telluride detector was used for VUV testing, and the results indicate that there is no accompanying VUV near the sample installation device.

### *2.4. Testing instrument*

Before and after the experiment, the surface morphology of the sample was observed using an OLYMPUS GX51 optical microscope. A ME215S high-precision microelectronic balance measured the quality change of the sample. The thermal emissivity and solar absorption ratio of the sample were measured by TEMP 2000A and LPSR 300, respectively.

## 3. Experiment results and analysis

### *3.1. Appearance changes*

The surface morphology of ITO/Teflon/Ag samples was shown in Figure 4, observed using an OLYMPUS GX51 optical microscope before and after the experiment. It can be seen that after the AO exposure test with the fluence of $9.1\times10^{19}$ atoms/cm$^2$, the surface morphology of ITO/Teflon/Ag showed significant changes compared to before the test. Figure 3 shows that as the sample temperature increases, the surface defect density of ITO/Teflon/Ag gradually increases.

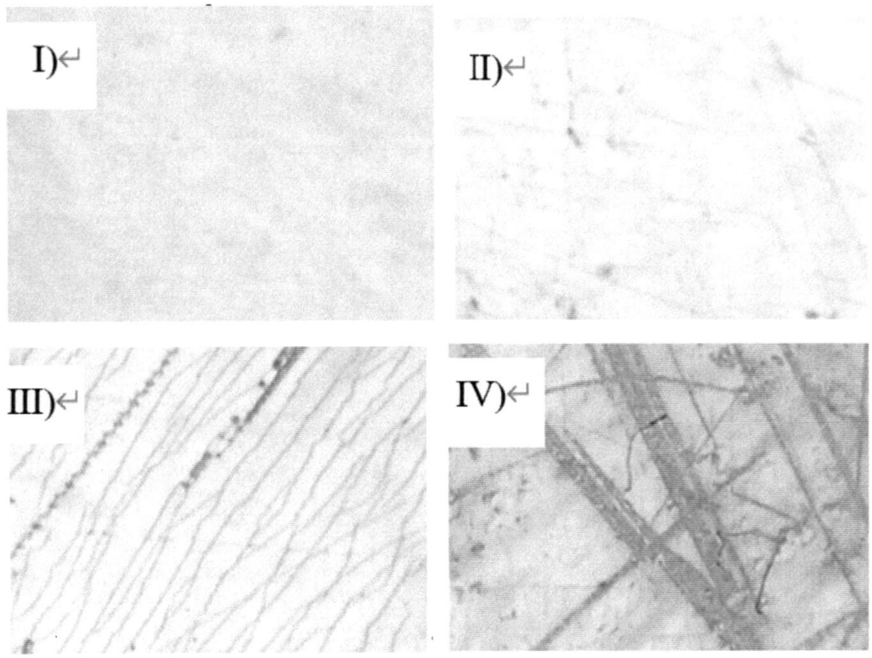


(I) Original sample; (II) after 20°C test; (III) after 70°C test; (IV) after 120°C test.

**Figure 4.** Surface morphology of ITO/Teflon/Ag before and after the experiment (600×).

### *3.2. Reaction rate of materials*

Before and after the experiment, the mass of ITO/Teflon/Ag samples was tested using a ME215S high-precision microelectronic balance (with a measurement accuracy of $10^{-5}$ g) produced by the German company Sedoris. The reaction rate between the sample and AO was calculated based on the mass change of the sample, as shown in Table 1. From Table 1, it can be seen that sample temperatures of 20°C, 70°C, and 120°C, after the AO exposure test with the fluence of $9.1\times10^{19}$ atoms/cm$^2$, the mass loss of ITO/Teflon/Ag samples gradually increased, and the reaction rate of the material also increased with the increase of sample temperature.

**Table 1.** Quality changes and ITO/Teflon/Ag reaction rate in AO and temperature composite effect test.

| Temperature (°C) | Quality before test (g) | Quality after test (g) | Mass loss (g) | Reaction rate (cm$^3$/atom) |
|---|---|---|---|---|
| 20 | 0.02740 | 0.02737 | 0.00003 | $1.95\times10^{-25}$ |
| 70 | 0.02513 | 0.02509 | 0.00004 | $2.60\times10^{-25}$ |
| 120 | 0.02644 | 0.02635 | 0.00009 | $5.86\times10^{-25}$ |

### *3.3. Performance variation in thermophysical properties*

The test measured the thermal emissivity and solar absorption ratio of ITO/Teflon/Ag using TEMP 2000A and LPSR 300. For effective comparison, a new sample is selected at each target temperature, and only the sample is heated to the target temperature without AO irradiation. The other test environments are the same, and the results are shown in Table 2.

From Table 2, it can be seen that as the sample temperature increases after the AO exposure test with the fluence of $9.1\times10^{19}$ atoms/cm$^2$, the change trend of the thermal emissivity of ITO/Teflon/Ag samples is consistent with the joint effect of AO and temperature and the single temperature effect. Based on the data, it can be determined that the change in the thermal emissivity of ITO/Teflon/Ag is mainly caused by temperature.

From Table 2, it can also be seen that the solar absorption ratio of ITO/Teflon/Ag exhibits a complex trend with temperature changes. Temperature mainly causes a decrease in solar absorption ratio, while AO mainly causes an increase in solar absorption ratio. At 70°C, the effect of temperature is balanced with that of AO, and the influence of AO is greater than that of temperature at other experimental temperatures.

**Table 2.** Comparison of ITO/Teflon/Ag thermophysical performance parameters before and after AO and temperature composite effect test.

| Temperature | ε (before) | ε (after) | $\alpha_s$ (before) | $\alpha_s$ (after) |
|---|---|---|---|---|
| 20°C in AO | 0.725 | 0.746 | 0.111 | 0.131 |
| 20°C | 0.702 | 0.711 | 0.108 | 0.105 |
| 70°C in AO | 0.710 | 0.705 | 0.101 | 0.097 |
| 70°C | 0.71 | 0.698 | 0.096 | 0.090 |
| 120°C in AO | 0.719 | 0.708 | 0.116 | 0.148 |
| 120°C | 0.722 | 0.713 | 0.126 | 0.098 |

## 4. Conclusion

Through this experiment, the following conclusions can be drawn:

1) As the temperature increases, the AO and ITO/Teflon/Ag reaction rate gradually increases but remains at $10^{-25}$. The ITO film provides a good protective effect on the substrate material.

2) When AO and temperature interact with ITO/Teflon/Ag, the change in thermal emissivity of the sample is mainly caused by temperature. In contrast, the change in the solar absorption ratio of the sample is more complex. It shows that the change in solar absorption ratio of the sample caused by a single AO and single temperature is opposite. The effect of temperature and AO on the solar absorption ratio of ITO/Teflon/Ag reaches equilibrium at 70°C. Both AO and temperature within the experimental dose range cause changes in the thermophysical properties of the sample.

This experiment shows that changes in temperature impact the reaction rate between materials and AO, as well as the changes in the thermal physical properties of materials. However, the mechanism is not yet clear, and further research is needed to provide data support for material selection on the outer surface of spacecraft.

## References

[1] Jiao Z L, Jiang H F, Jiang L X, et al. The test conditions in the study of synergistic effects of atomic oxygen and far ultraviolet irradiation in low Earth orbit [J]. Spacecraft Environment Engineering, 2022, 39(3): 230-236.
[2] Cui Z Y, Wang L S, Qian Y H, et al. Atomic Oxygen Erosion-Corrosion to Polyimide in Low Earth Orbit and Its Protective Measures [J]. Corrosion & Protection, 2021, 42(11): 12-19.
[3] CARR J A, JOHNSON L, BOYD D, et al. LISA-T part three: the design and space environments testing of a thin-film power generation and communication array [J]. Acta Astronautica, 2023, 205: 267-280.
[4] VERKER R, BOLKER A, CARMIEL Y, et al. Ground testing of an on-orbit atomic oxygen flux and ionizing radiation dose sensor based on material degradation by the space environment [J]. Acta Astronautica, 2020, 173: 333-343.
[5] MORRISSEY L S, RAHNAMOUN A, NAKHLA A. The effect of atomic oxygen flux and impact energy on the damage of spacecraft metals[J]. Advances in Space Research, 2020, 66(6): 1495-1506.
[6] Guo L, Qiu T C, Guo Z P, Shan M, Zheng L D, Song X H, 2022.Research of the Property Changes of ITO/Kapton/Al Based on Different Temperature in Atomic Oxygen Environment (AMCE 2022. Maoming, Virtual, China).

# Research on the performance of pavement crack repair materials based on foaming polyurethane

**You Zhou[1,2], Fei Chen[1] and Zhoucong Xu[1]**

[1]China Merchants Chongqing Communications Technology Research & Design Institute Co., Ltd. & National United Engineering Laboratory of Road Engineering in Mountainous Areas and Disaster Prevention and Reduction Technology, Chongqing 400067, China

[2]Corresponding author's e-mail: 81289587@qq.com

**Abstract.** In this paper, Pucs, a new two-component foaming polyurethane material, is selected to carry out a feasibility study on its application in pavement crack repair. Based on the two common crack repair materials in current projects, tests of pavement performance of repair materials are carried out, such as bonding performance, water damage resistance performance, high-temperature performance, and aging resistance performance. The comparative analysis of the results reveals that Pucs have better pavement performance than conventional crack repair materials. This indicates that Pucs is a new kind of material with excellent performance in pavement crack repair and can be widely used in the field of crack repair.

## 1. Introduction

Crack is one of the most common diseases of asphalt pavement. If not treated as soon as possible, it will further extend after repeated application of traffic loads and rainwater, and lead to mud-pumping, void, breakup, potholes, and other damages [1-3], which can seriously affect driving comfort and the service life of the road [4-5]. The SHRP points out that if pavement cracks can be repaired in time, the life of asphalt pavement can be extended by 10 to 20 years, and the maintenance costs can be saved by more than 50%.

At present, crack maintenance mainly focuses on crack pouring and surface filling processes for repair. Major repair materials are ordinary asphalt, modified asphalt, fiber asphalt, emulsified asphalt, rubber asphalt sealant, silicone jointing sealing compound, sand-gravel hot mix asphalt, etc, with common shortcomings of insufficient bonding strength with the original pavement cracks. Perhaps due to their poor elasticity and toughness as well as inconsistent deformation of the original pavement, they are prone to make the repaired pavement re-cracking and fail [6]. Thus, traditional crack repair materials cannot fundamentally solve the existing problems of pavement cracks. In recent years, polymer materials represented by polymer grouting technology as a new type of crack repair materials are gradually attracting attention, such as epoxy resin adhesives, modified polyurethane materials, polyurethane/epoxy resin interpenetrating polymer networks, and so on [7-9]. Studies have shown that these materials are effective in pavement crack repair. In this paper, Pucs [10], a two-component foaming polyurethane material, is selected as a crack repair material. And the pavement performance of Pucs is evaluated through a comparative study of repair materials for conventional asphalt pavement cracks, such as bonding, water damage resistance, high temperature, and aging resistance, to study the application prospects of Pucs as a crack repair material.

## 2. Raw materials of pucs

In this paper, a solvent-free two-component polyurethane is used as a crack repair material, which is a kind of chemically reactive foaming polyurethane material. Its indexes before and after the reaction are shown in Table 1 and Table 2. Component A is a mixture of polyether polyols, which is a light yellow transparent liquid; component B is a mixture of isocyanates, which is a brown liquid and needs to be preserved in a lucifugal and dry environment. The material is environmentally friendly, non-toxic and harmless, and easy to operate. When in use, component A and component B are mixed in a volume ratio of 1:1, and 2% water is added as a foam initiator. Then, after being fully mixed, the material can be cured and formed naturally. The whole low-carbon and energy-saving process is carried out at ambient temperature without heating.

**Table 1.** Indexes of physical parameters of two-component polyurethane cementitious materials.

| Order number | Items | Component A | Component B |
|---|---|---|---|
| 1 | Solid content/% | >99 | >99 |
| 2 | Density (25°C)/g·cm3 | 1.01 | 1.23 |
| 3 | Viscosity (25°C)/mPa·s | 720±50 | 350±50 |
| 4 | Hydroxyl value/mgKOH·g-1 | 330±15 | -- |
| 5 | NCO content/wt% | -- | 21.7±0.5% |
| 6 | Moisture content/% | < 0.2 | 0 |
| 7 | Flashpoint/°C | >200 | >200 |

**Table 2.** Chemical reaction parameters and main performance indexes of foaming polyurethane.

| Polyurethane gelling agent | Viscosity after 25 seconds of mixing/mPa·s | Initial foaming time/sec | Cream time /sec | Rise time/sec | Curing time/h |
|---|---|---|---|---|---|
| Index parameters | <600 | 40±10 | 90±25 | 160±25 | 2-6 |
| Polyurethane gelling agent | Expansion coefficient | Hardness (Shore D) | Tensile strength /MPa | Flexural strength/ MPa | Breaking elongation/% |
| Index parameters | 3~4 | 68 | ≥12.5 | ≥4 | ≥150 |

Note: All indexes in the table are tested at an ambient temperature of 25°C.

## 3. Experimental methods

### *3.1 Experimental preparation*

In this paper, the performance of Pucs is studied to repair asphalt pavement cracks. Therefore, all specimens are asphalt mixtures and an AC-13 gradation is adopted. The wheel track plate is cut into 5 cm×5 cm×5 cm cube specimens and 5 cm×5 cm×10 cm cuboid specimens for the pull-out test. The bonding area S is calculated based on the smaller cube specimen ($S = 25\ cm^2$). The bonded pull-out test specimen is shown in Figure 1. In addition, the surface of the Marshall specimen is cut into a crack with a depth of about 15 mm and a width of about 5 mm to stimulate pavement cracks. The cut specimen is shown in Figure 2. After filling various repair materials in the crack, relevant performance tests can be carried out.

**Figure 1.** Pull-out test specimen.

**Figure 2.** Pavement cracks simulation (left: vertical view; right: side view).

*3.2 Performance of materials*

At present, polymer-modified asphalt is a frequently used crack repair material in the market, and matrix asphalt is also used in low-temperature areas. In this paper, SBS-modified asphalt and matrix asphalt No. 70 are selected as the control group to verify the performance of Pucs. Also, the following types of tests are carried out: (1) pull-out test is used to examine the performance of Pucs as crack repair materials; (2) freeze-thaw splitting test is used to study its water damage resistance performance; (3) pull-out test at high temperature is used to study its high-temperature stability; (4) aging test is used to study its aging resistance. In these tests, 3 parallel data are tested for each group and the results are analyzed by taking the average value.

*3.2.1 Bonding performance.* In this paper, the test method for the bonding strength, which is described in Appendix E of the Technical Guide for Design and Construction of Highway Steel Box Girder Bridge Deck Pavement issued by the Ministry of Communications, is imitated to test the bonding performance of Pucs. The test device and method for bonding strength are shown in Figure 3.

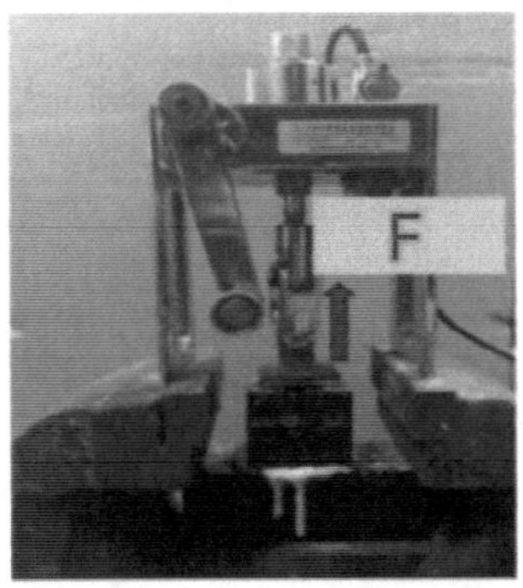


**Figure 3.** Test device and method for bonding strength.

*3.2.2 Water damage resistance performance.* According to the T0729-2000 freeze-thaw splitting test methods for asphalt mixtures in JTG E20-2011 Standard Test Methods of Bitumen and Bituminous Mixtures for Highway Engineering, the frost and thaw condition is changed in the test to keep the specimens in -18°C freezer for 12 h, and then keep them in 60°C constant temperature water baths for 12 h. In this process, each specimen is freeze-thawed twice, and other operations refer to T0729-2000. The water damage resistance performance of the materials is evaluated by the tensile strength ratio (TSR) of the specimens.

*3.2.3 High-temperature performance.* In this paper, the high-temperature performance of the materials is evaluated by testing the bonding performance of the specimens at high temperatures, so the test method is almost the same as that of the bonding strength in 2.2.1, except that the test temperature is raised to 60°C.

*3.2.4 Aging resistance performance.* The UV aging test chamber is used to simulate the aging process of the materials under natural light irradiation. The aging condition is that UVA340 lamps continuously illuminate for 100 h with a radiation intensity of 200 w/m$^2$. Then, according to the

T0709-2011 Marshall stability test method of asphalt mixtures in JTG E20-2011 Standard Test Methods of Bitumen and Bituminous Mixtures for Highway Engineering, the Marshall stability of the specimens before and after aging is tested respectively. And the aging resistance of the materials is evaluated by analyzing the percentage decrease in stability before and after aging.

## 4. Results and analysis of the experiment

### *4.1 Bonding performance*

The primary task of examining the availability of materials as crack repair materials is to examine their firm bonding to the original crack interfaces, otherwise, it is prone to cause interracial peeling and then repair failure. In this paper, Pucs is applied evenly on the bottom surface of the 5 cm×5 cm×5 cm cube specimen which then is bonded to the middle part of the 5 cm×5 cm×10 cm cuboid specimen. After curing at room temperature for 8 h, performance tests are carried out. The asphalt specimens of the control group are also formed with this method, and the binder dosage of all specimens is 4 g. The results of the bonding strength of each group of specimens at 25°C are shown in Table 3.

**Table 3.** Test results of bonding strength at ambient temperatures.

| Bonding materials | Bonding strength (25°C)/MPa | | | |
|---|---|---|---|---|
| | Test results | | | Average value |
| Pucs | 1.01 | 1.03 | 1.01 | 1.02 |
| SBS modified asphalt | 0.52 | 0.51 | 0.54 | 0.52 |
| Matrix asphalt No. 70 | 0.43 | 0.39 | 0.38 | 0.40 |

According to the test results of the bonding strength in Table 3, it can be seen that among the three crack repair materials, Pucs has the highest bonding strength which is more than two times higher than that of SBS-modified asphalt and matrix asphalt No. 70, two conventional crack repair materials. This fully shows the excellent bonding performance of Pucs with asphalt concrete interfaces and also proves that it can be used as a crack repair material for asphalt pavement.

### *4.2 Water damage resistance performance*

Water damage is the most common and the most serious disease that causes early damage to asphalt pavement, so the water damage resistance of a certain material can directly affect the application prospect of the material. In this paper, freeze-thaw splitting tests are used to evaluate the water damage resistance performance of Pucs. The crack simulation specimen in Figure 2 is filled with different repair materials after being tested, the splitting tensile strength of the repair specimens before and after freeze-thaw cycles is tested respectively, and then the TSR is calculated. The test results are shown in Table 4 and Figure 4.

**Table 4.** Test results of freeze-thaw splitting test.

| Repair materials | Groups | Splitting tensile strength/MPa | | | | TRS/% |
|---|---|---|---|---|---|---|
| | | Test results | | | Average value | |
| Pucs | Before freeze-thaw test | 1.18 | 1.22 | 1.22 | 1.21 | 99 |
| | After freeze-thaw test | 1.18 | 1.19 | 1.22 | 1.20 | |
| SBS modified asphalt | Before freeze-thaw test | 1.00 | 1.01 | 1.03 | 1.01 | 95 |
| | After freeze-thaw test | 0.96 | 0.98 | 0.95 | 0.96 | |
| Matrix asphalt No. 70 | Before freeze-thaw test | 0.87 | 0.94 | 0.92 | 0.91 | 90 |
| | After freeze-thaw test | 0.81 | 0.82 | 0.84 | 0.82 | |

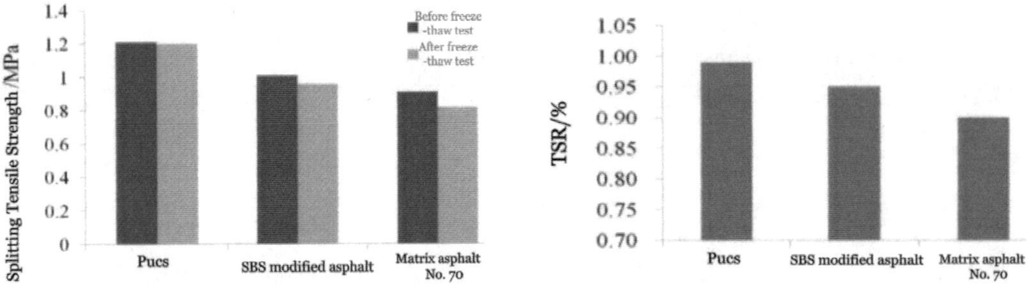


(a) Splitting tensile strength (b) TSR

**Figure 4.** Test results of the freeze-thaw splitting test.

According to the test results of freeze-thaw splitting tests in Table 4 and Figure 4, it can be seen that the order of the splitting tensile strength of the three crack repair specimens before and after freeze-thaw cycles is Pucs>SBS modified asphalt>matrix asphalt No. 70, which indicates that Pucs shows better structural strength before and after water damage. Based on TRS, it can be seen that after 2 freeze-thaw cycles, the splitting tensile strength of Pucs is almost unchanged (TRS=99%), the splitting tensile strength of SBS-modified asphalt decreases slightly (TRS=95%), and the splitting tensile strength of matrix asphalt No. 70 shows an obvious reduction (TRS=90%). Thus, Pucs has good water damage resistance which is better than conventional asphalt crack repair materials.

*4.3 High-temperature performance*

Asphalt is an amorphous polymer. Its fluidity increases with temperature, so the bonding performance of conventional asphalt crack repair materials are susceptible to decline due to softening at high temperature, which will cause repair failure. In this paper, pull-out tests at high temperatures are designed to examine the high-temperature performance of Pucs. The fabrication and testing devices of the specimens are the same as the pull-out tests at ambient temperature, and the testing temperature is raised to 60°C. The high-temperature performance of the materials is evaluated by the percentage decrease in their bonding strength at high temperatures. The test results are shown in Table 5.

**Table 5.** Test results of the bonding strength of different repair materials at high temperatures.

| Bonding materials | Bonding strength (60°C)/MPa | | | |
|---|---|---|---|---|
| | Test results | | | Average value |
| Pucs | 0.90 | 0.89 | 0.93 | 0.91 |
| SBS modified asphalt | 0.40 | 0.42 | 0.41 | 0.41 |
| Matrix asphalt No. 70 | 0.25 | 0.28 | 0.26 | 0.26 |

According to the test results of the bonding strength in Table 5, it shows that the bonding strength of Pucs remains the highest among the three crack repair materials at a high temperature of 60°C. Based on the test results of the bonding strength at ambient temperature in Table 3, it can be seen that the bonding strength of Pucs at 60°C is still higher than that of SBS-modified asphalt at ambient temperature, showing good bonding performance and high-temperature performance. Through further comparative analysis of the data in Table 3 and Table 5, it can be seen that the bonding strength of the three crack repair materials is in a downward trend with increasing temperature. The bonding strength of Pucs, SBS-modified asphalt, and matrix asphalt No. 70 decreases by 11%, 21%, and 35%, respectively, among which the decent degree of Pucs is the smallest, which means that the high-temperature performance of Pucs is better than that of the conventional asphalt crack repair materials.

*4.4 Aging resistance performance*

Sunlight exposure is one of the important factors causing a decline in the durability of asphalt pavement, with ultraviolet rays having the greatest influence on the durability of the materials. So in this paper, the ultraviolet aging process is stimulated through UV aging chamber tests to evaluate the aging resistance of the materials. Marshall specimens repaired with different materials are subjected to ultraviolet aging under the set conditions, and then their Marshall stability is tested. The test results are shown in Table 6 and Figure 5.

**Table 6.** Test results of aging resistance performance.

| Repair materials | Groups | Marshall stability/kN | | | | Percentage decrease of stability/% |
|---|---|---|---|---|---|---|
| | | Test results | | | Average value | |
| Pucs | Before aging | 12.1 | 12.2 | 12.0 | 12.1 | 95 |
| | After aging | 11.4 | 11.7 | 11.3 | 11.5 | |
| SBS modified asphalt | Before aging | 10.2 | 10.4 | 10.4 | 10.3 | 92 |
| | After aging | 9.7 | 9.3 | 9.6 | 9.5 | |
| Matrix asphalt No. 70 | Before aging | 9.1 | 8.9 | 8.9 | 9.0 | 89 |
| | After aging | 7.9 | 8.2 | 8.0 | 8.0 | |

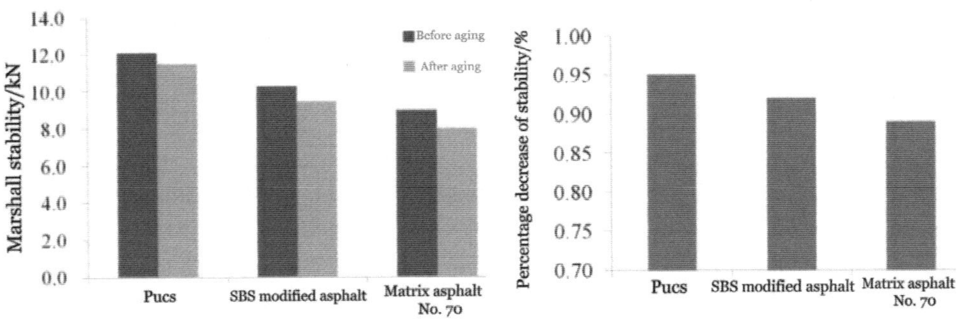


(a) Marshall stability before and after aging (b) Percentage decrease in stability

**Figure 5.** Test results of aging resistance performance.

According to the test results of aging resistance performance in Table 6, the Marshall stability of the specimens is decreased after ultraviolet aging, and they are all subjected to ultraviolet aging to a certain extent. Among them, the Marshall stability of Pucs is the highest, and even the Pucs specimens after ultraviolet aging have higher stability than asphalt repair specimens before aging. According to the comparison of the Marshall stability before and after aging, it can be seen that the stability of the three materials decreases in the following order: Pucs$<$SBS modified asphalt$<$matrix asphalt No. 70. The results show that Pucs is a better crack repair material than asphalt and that Pucs has better aging-resistance performance. Pucs material is a kind of synthetic material. The addition of a UV absorber to component A raw materials of Pucs can effectively absorb some UV light, thereby delaying aging and exhibiting a better aging resistance. This also shows the variability and controllability of synthetic materials in terms of performance.

## 5. Conclusion

In this paper, Pucs, a new crack repair material based on foaming polyurethane is introduced, which is formed by on-spot mixing and reaction of polyether polyol components and isocyanate components at room temperature. Based on SBS-modified asphalt and matrix asphalt No.70, the two common crack repair materials, the pavement performance of Pucs is evaluated in terms of bonding strength, water damage resistance, high-temperature stability, and aging resistance. The conclusions are as follows:

(1) With simple preparation and easy operation, Pucs material is safe, environmentally friendly, low-carbon, energy-saving, non-toxic, and harmless.

(2) Pucs material is cured and formed through a chemical reaction. With a fast curing time, it can be used as a rapid crack repair material. According to different temperatures, traffic can be opened within 2-6 hours, greatly shortening the road occupying time for repair.

(3) Under the same test conditions set in this paper, Pucs material shows better bonding performance than conventional crack repair materials.

(4) Under the same test conditions set in this paper, Pucs material demonstrates better water damage resistance and high-temperature stability than conventional crack repair materials.

(5) Under the same test conditions set in this paper, Pucs material exhibits better-aging resistance than conventional crack repair materials and can extend the service life of the road.

**Acknowledgments**

Chongqing technological innovation and application development key special projects (CSTB2022TIAD-KPX0102), Chongqing communications science and technology project (CQJT-CZKJ2023-05), and Chongqing communications science and technology project (CQJT-CZKJ2023-08).

**References**

[1]Yang D. Y. (2015). Application Research of High Polymer Grouting in Rapid Repair of Internal Defects of Expressway Asphalt Pavement (Master Dissertation, Wuhan Institute of Technology).

[2] N. Hoang ,Q. A. Nguyen. (2019) .novel method for asphalt pavement crack classification based on image processing and machine learning.Engineering with Computers,35(2),487-498.

[3] M. Jared ,U. A. Maheswar, B. Manik .(2021) .Decision Trees for Selecting Asphalt Pavement Crack Sealing Method . Transportation Research Record,2675(1),172-183.

[4] Xiao X. H. (2021). The Research of Polyurethane Repair Materials for Asphalt Pavement Crack Maintenance (Master Dissertation, Changsha University of Science & Technology).

[5] Zhao H. L., Liu Q. F. & Ju Y. Z. (2009). Research on the Genesis and Integrated Treatment of Asphalt Pavement Cracks. Journal of Shandong Agricultural University (Natural Science Edition), (4),4.

[6] K. L. Smith, A. R. Romine. (2001). Materials and Procedures for Sealing and Filling Cracks in Asphalt-surfaced Pavements — Manual of Practice. National Academy of Sciences (America).

[7] Deng J. T. (2010). Application Study of Modified Polyurethane for Crack Repair of Asphalt Pavement on Highway. Concrete and Cement Products, (3),23-25.

[8] Zhou X. Y. (2012). The Research of Grafted Polyurethane/Epoxy Resin Interpenetrating Network Polymer Grouting Materials (Master Dissertation, Changsha: Hunan University).

[9] Gao L. (2014). Research on the Performance and Technology of Crack Grouting Materials for Asphalt Concrete Pavement (Master Dissertation, Central South University).

[10] Wang H. M.,Yang J. G. & Li R. K. (2017). Study on the Performance of Millimeter Scale Ultrathin Color Anti-slip Surface Material Puas. Technology of Highway and Transport,(2),5-10.

# Study on the effect of double mineral admixtures on the properties of high water materials

**Yuying Zheng[1], Qiufeng Wang[1,2] and Yuanyuan Zhao[1]**

[1]Xuhai College, China University of Mining and Technology, Xuzhou, Jiangsu, 221000, China

[2]Corresponding author's e-mail: zhengyuying@zgkydxxhxy.wecom.work

**Abstract:** In this paper, different contents of fly ash, silica fume, slag, and Nano-$SiO_2$ were added to the high water material. Then, fly ash-silica fume, slag-silica fume, and Nano $SiO_2$-silica fume are mixed into the high water material. Through the strength test of two groups of specimens, the influence of different types of mineral admixtures and blending ratios on the physical and mechanical properties of high-water materials was studied. The mechanical parameters of high water doped materials were measured by ultrasonic method, and the internal pore structure of high water doped materials was analyzed. In the case of double mixing, the optimal mixing amount is 4% fly ash and 6% silica fume, and the 7 d compressive strength of this group of specimens is increased by 15.85% compared with the original high water material. Through ultrasonic detection, the first wave is smooth and stable, the longitudinal wave amplitude curve is flat, the peak difference is slight, and the periodic amplitude is uniform, indicating that the structure of this group of specimens is dense. The shear modulus is increased by 20.31%, Young's modulus is 1.3 times, and the volume modulus is 8.9 times that of the original high water material. The research findings suggest that high water materials can be modified by double mixing.

## 1. Introduction

Based on the two-carbon target, applying industrial waste to building materials effectively reduces carbon and sequesters carbon [1]. High water material has no chemical pollution and good pumping ability. It can quickly fill and treat large areas of goaf and has been used in many fields such as backwall grouting [2].

The new high water material used in this paper is divided into A and B. A and B materials can be pumped separately for a long time without condensation and will condense a certain strength shortly after mixing[3]. However, it is prone to condense rapidly at high reaction temperatures, uneven material slurry, and water layering and secretion phenomena[4]. Sun et al. [5] found that the reaction temperature rise of fly ash high-water material with 60% content decreased by 22.9%, and the cost decreased by 54.6%. Feng et al. [6] believed that adding 15% fly ash into high-water materials was ideal, which reduced the peak strength by 25%, increased the residual strength by 50%, and reduced the elastic and deformation modulus. Zheng et al. [7] proposed that the optimal dosage of blast furnace slag (BFS) in cement-based materials is 30%. Shi et al. [8] found that the optimal content of fly ash was 20%, and the optimal content of carbide slag in fly ash was 5%. Some experimental results show that nano-silica (NS) can significantly improve the physical and mechanical properties of cement-based materials [9].

The research is mainly focused on the properties of high-water materials doped with a single material, while the research on high-water materials doped with two different materials is less [10].

## 2. Raw materials and test methods

### 2.1. Composition of raw materials

To break through the limitation of applying high water materials and improve the cement substitution effect of fly ash in high water materials, fly ash, slag, silica fume, and nano silica were used to replace part of cement in high water materials by single or double mixing. The chemical composition and basic properties of raw materials were tested to analyze the effect of mineral admixtures on high water content materials. It can be seen that the chemical composition of calcium sulfoaluminate cement (CSA), slag (S), fly ash (FA), and silica fume (SF) used in the test are shown in Table 1.

**Table 1.** Chemical composition of raw materials wt%.

| Name | CaO | $SiO_2$ | $Al_2O_3$ | MgO | $K_2O$ | $Fe_2O_3$ | $Na_2O$ | $TiO_2$ |
|---|---|---|---|---|---|---|---|---|
| CSA | 40.76 | 7.41 | 33.66 | 0.92 | 0.29 | 2.43 | 0.16 | 1.26 |
| S | 39.69 | 33.75 | 15.14 | 7.47 | 0.5 | 0.44 | - | 0.61 |
| FA | 7.55 | 51.28 | 18.07 | 0.82 | 1.75 | 6.14 | 2.19 | 0.73 |
| SF | 0.31 | 94.9 | 1.22 | 0.92 | 0.40 | 0.13 | - | - |

To improve the testing effect, the experiment adopted ultrafint-slag powder with a density of 2.86 g/cm$^3$, a specific surface area of 0.944 m$^2$/g, and a median particle size D50 of 3.815 $\mu m$. The density of ultra-fine fly ash is 2.90 g/cm$^3$, the specific surface area is 0.424 m$^2$/g, and the median particle size D50 is 1.729 $\mu m$. The specific surface area of silica fume used is 21.709 m$^2$/g, most particle size distribution is below 1.0 $\mu m$, and D50 is about 200 nm. The three materials are very different in chemical composition, and the activity is not completely the same. Nano-$SiO_2$ (NS) with a particle size of 10 nm and a specific surface area of 250±30 m$^2$/g were used in the experiment. Because the particle size is very small and the surface adsorption capacity of particles is strong, it is difficult to achieve ideal results. Therefore, nano was first prepared as nano dispersion liquid with a nanometer content of 25% and a PH value 7 as a neutral liquid.

### 2.2. Composition of raw materials

The experiment adopted a high water-cement ratio of 1.5:1. It should be noted that when adding nano admixtures because the nano dispersion was used, the water content in the dispersion was also considered when calculating the water consumption. The method of adding mineral admixtures was to replace calcium sulfoaluminate cement in material A with its equal mass, add additive A, and mix evenly with water three times to obtain A slurry. The B material was mixed with additive B and then mixed with water. Finally, the A slurry and B slurry were mixed.

High water material without admixture Jz-0 was used as the reference group. The amount of fly ash, slag, silica fume, and Nano-$SiO_2$ is determined according to the current research results of domestic and foreign scholars. The results proposed by Zheng et al. [11] show that the effect is that doped with Nano-$SiO_2$ is the first, doped with silica fume is the second, the strength of high-water materials doped with slag is slightly increased, and the strength of high-water materials doped with fly ash decreases with the increase of fly ash content.

When conducting a double mixing test, we replace the cement content with 5% fly ash in the high water content material and then use slag to replace 1% and 3% of cement in equal amounts, numbering the specimens as FS-1 and FS-2. If 1% or 3% silica fume is used instead of cement, the specimen numbers are FSF-1 and FSF-2. Specimens with 1% and 2% nano substitution are numbered FNS-1 and FNS-2, respectively. When the replacement amount of fly ash is 10%, then we use slag to replace 1% and 3% of cement in equal amounts, numbering the specimens as FS-3 and FS-4. If 1% or 3%

silica fume is used instead of cement, the specimen numbers are FSF-3 and FSF-4. Specimens with 1% and 2% nano substitution are numbered FNS-3, FNS-4, and FNS-5.

A full-automatic pressure testing machine was used for the strength test of the specimen. The test acceleration was 2.4±0.2 kN/s, as shown in Figure 1. The size of the specimen was 40 mm×40 mm×160 mm, as shown in Figure 2. The number of specimens in each group produced and maintained at the same age and under the same conditions is three. After curing to the age of 1 d, 3 d, and 7 d, the flexural test was carried out first, and the average flexural strength of the three specimens was taken. After the specimens were broken, the axial compression test was carried out on the 6 specimens, and the average compressive strength was taken.

The HKB-type intelligent ultrasonic automatic testing instrument was used for ultrasonic testing of test specimens, as shown in Figure 3. According to the testing results, the elastic modulus, shear modulus, and other material parameters of the specimens were calculated, and the material structure was analyzed.

**Figure 1.** Figure with pressure testing machine.

**Figure 2.** Figure with the specimen.

**Figure 3.** Figure with HKN-B type intelligent ultrasonic crosswave automatic tester.

## 3. Test results and analysis of mechanical properties

### *3.1. Compression test results and analysis of double-doped high water materials*

#### *3.1.1. Test results of fly ash replacement 5%*

- According to the research in Section 3.1, adding fly ash could improve the working performance of high-water materials. Still, the early activity of fly ash was low, which reduced the early mechanical properties of materials. Adding slag, silica fume, and Nano-$SiO_2$ could improve the strength of high water materials to different degrees. To maintain the early physical and mechanical properties of high water materials, fly ash was used to replace cement in high water materials to obtain greater economic benefits. In this study, fly ash and silica fume, fly ash, slag, fly ash, and Nano-$SiO_2$ were mixed into high-water materials in different proportions, and the early mechanical properties of the specimens were tested.
- As for the double-doped high-water material with a 5% replacement amount of A material, it could be seen from Figure 4(a) that the double-doped effect of the FSF group was the best among the three groups of the FSF group (fly ash-silica fume), FS group (fly ash-slag) and FNS group (fly ash-nano). Due to the low pozzolash activity of fly ash in the early stage, fly ash replaces part of cement in the same amount, reducing hydrated calcium silicate gel. Therefore, the lower the fly ash content in several groups of double-doped specimens is, the higher the compressive strength is.
- Figure 4(b) shows the 1 d~7 d compressive strength growth rate of each experimental group. It could be seen from Figure 2(b) that the experimental group effect of the two materials mixed with high-water materials was superior to that of single fly ash. Although the 7-d compressive strength of the double-admixture test group was close to that of the Jz-0 group, only the FSF-1 group had a higher 7-d compressive strength than the Jz-0 group.

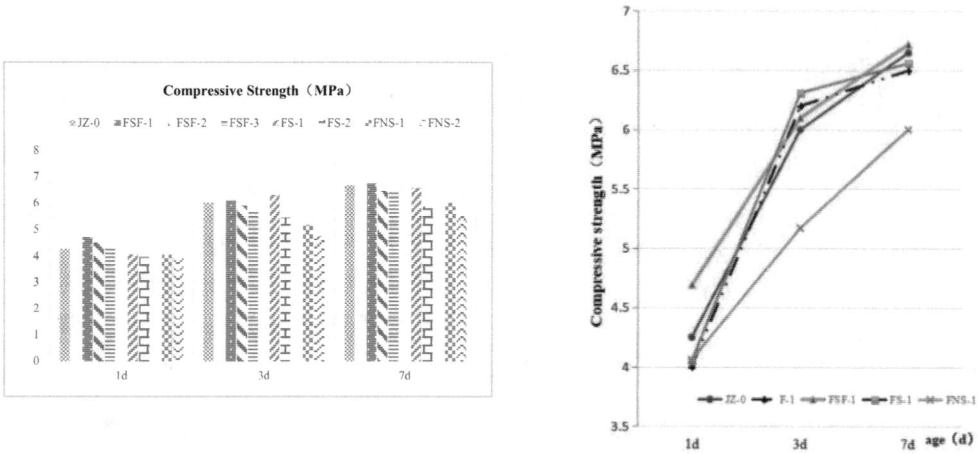


(a) Compressive strength (b) Growth rate of compressive strength at each age

**Figure 4.** Figure with the influence of replacing 5% cement with double mixed mineral admixture on compressive strength of high water material.

- The early strength of specimens in group FSF 1 increased faster from 1 d to 3 d, especially the 1 d compressive strength increased by 10% compared with Jz-0 and 17% compared with group F-1. Although the increase rate of the strength of the FSF 1 group in 3 d~7 d was significantly reduced, the compressive strength of the FSF 1 group at 7 d age is 1.05% higher than that of Jz-0 and 3.38% higher than that of the F-1 group.
- The compressive strength of group FS-1 at 7 days is 0.92% higher than that of group F-1 but still lower than that of Jz-0.
- However, the mixed use of fly ash and nano admixtures significantly reduced the early compressive strength of the high water material, even lower than that of the high water material with fly ash alone. This indicates that Nano-$SiO_2$ and fly ash with small content do not synergize when the water-cement ratio is large.
- Therefore, the optimal ratio of mineral additives was 1% fly ash content and 4% silica fume content in these tests.

*3.1.2. Test results of fly ash replacement 10%*

- With the increase in the proportion of fly ash, the compressive strength of the specimen decreased continuously. It could be seen from the results of a single-doped test that when fly ash replaced 10% of A material in high-water material, the strength of material 7 d decreased seriously, which was much lower than that of group Jz-0.
- When the substitution amount of double admixture was 10%, it could be seen from Figure 5 that the compressive strength of specimens in the three groups of FSF (fly ash-silica), FS (fly ash-slag), and FNS (fly ash-nano) was significantly higher than that of specimens in the F-2 group (10% fly ash alone). The effect of double admixture in the FSF group was the best.
- When fly ash was mixed with silica fume, the addition of silica fume made up for the deficiency of the early strength of fly ash, and the compressive strength was even higher than that of the Jz-0 specimens. The compressive strength of specimens in group FSF-4 at 1 d, 3 d, and 7 d ages was increased by 50%, 67%, and 40% compared with group F-2 with the same cement replacement amount. The compressive strength of FSF-4 specimens at 1 d, 3 d, and 7 d age increased by 20%, 26.7%, and 15.8% compared with Jz-0 specimens, respectively.
- The composition of Nano-$SiO_2$ is the same as that of silica fume, and the specific surface area is larger, and the particle size is smaller. From the test results of additive alone, we know that the effect of single doping Nano-$SiO_2$ was very obvious, and the improvement of compressive strength of

high-water materials was better than that of silica fume. However, the double mixing test of nano and fly ash did not achieve the expected effect. Moreover, the more Nano-$SiO_2$ was mixed, the worse the test effect was. The compressive strength of the FNS-3 group mixed with Nano-$SiO_2$ and fly ash was 7.27% higher than that in the F-2 group but 16.69% lower than that in the Jz-0 group.

- The effect of fly ash and slag mixed samples in the FS group was between the FSF and FNS groups. The compressive strength of the FS-3 group was 14.55% higher than that of the F-2 group but 5.26% lower than that of the Jz-0 group.

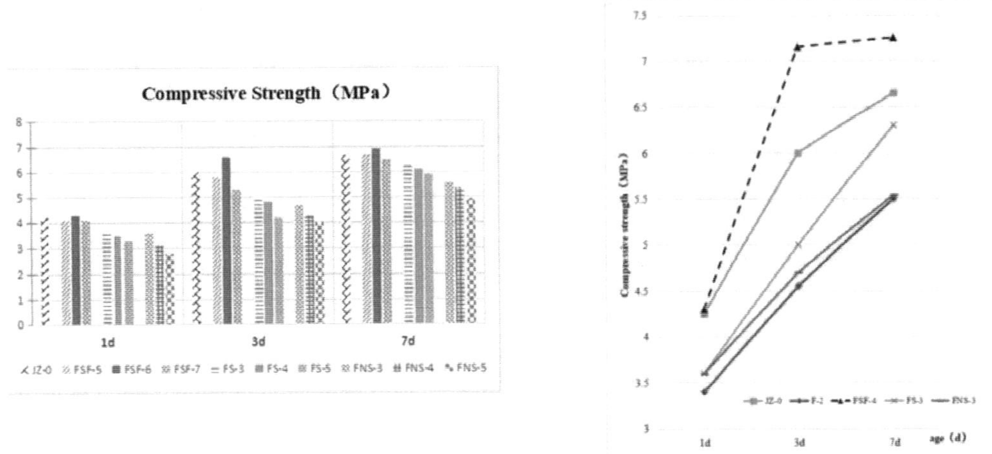


(a) Compressive strength (b) Growth rate of compressive strength at each age

**Figure 5.** Figure with the influence of replacing 10% cement with double mixed mineral admixture on compressive strength of high water material.

*3.2. Analysis of test results*

The reasons for the test results are analyzed as follows:

- The strength of high water material could be improved by adding fly ash and silica fume.

Fly ash and silica fume have great differences in activity and physical properties. They can promote each other. The silica fume particles used in the experiment were small and reacted quickly with the initial hydration products, and the hydration product particles adhered to the silica fume particles. Silica fume plays a "nucleation role", which promotes the crystallization, growth, aggregation, and condensation of the hydration reactants. On the other hand, the content of $SiO_2$ in silica fume is 94.9%, as shown in Table 1. C-S-H and other gel materials produced by $SiO_2$ and $Ca(OH)_2$ filled the structural void. The high water material structure was denser. Therefore, the strength effect of high-water materials mixed with fly ash and silica fume was the best, especially the hydration speed was the fastest at 1-3 d age.

- The compressive strength of the high-water material with double fly ash and nano decreased with the increase of nano content.

The composition of Nano-$SiO_2$ is the same as that of silica fume; the surface area is larger, and the particle size is smaller. Therefore, the effect of nano-nanoparticles alone in high-water materials was very obvious, and the effect of improving the compressive strength of high-water materials was better than that of adding silica fume. However, the double mixing nano and fly ash did not achieve the expected effect. Nano silica can promote hydration reactions and can react quickly with calcium hydroxide. First, the cement in grade A was replaced by fly ash in equal amounts, which reduced the calcium silicate hydrate gel generated during hydration. Secondly, the absorption of $Ca^{2-}$ by the rapid hydration reaction of Nano-$SiO_2$ led to the low concentration of $Ca^{2-}$ in the solution. The formation and growth of hydration products in high water materials were delayed. The low concentration of $Ca(OH)_2$ could not stimulate the pozzolanic activity of fly ash. Therefore, the compressive strength of high water materials with fly ash and nano materials decreased seriously. The higher the amount of Nano-$SiO_2$ is, the lower the strength of the high-water material is.

- Although the strength of the high-water material with slag and fly ash was higher than that with single fly ash, it was still lower than the reference group.

Blast furnace slag (BFS) has a fast hydration reaction due to its amorphous properties and high volcanic ash activity. To obtain better activity, the ultrafine slag used in this experiment had a smaller particle size and could quickly consume a large amount of $OH^-$. In addition, slag contains more CaO than fly ash. With high alkalinity, more $Ca(OH)_2$ was generated after hydration, and calcium silicate hydrate with low alkalinity was generated with $SiO_2$. This gel product had high strength. Slag could replace cement for hydration reaction and provide strength, but the strength increase could not make up for the early strength decline caused by fly ash incorporation.

## 4. Ultrasonic testing of high water materials with double mineral admixtures

The Jz-0 basic group and FS-4, FSF-4, and FNS-3 test groups in 10% fly ash double-doped groups were selected for ultrasonic detection using HKB intelligent ultrasonic automatic testing instrument. The vertical and horizontal wave shapes of several groups of specimens were compared and analyzed.

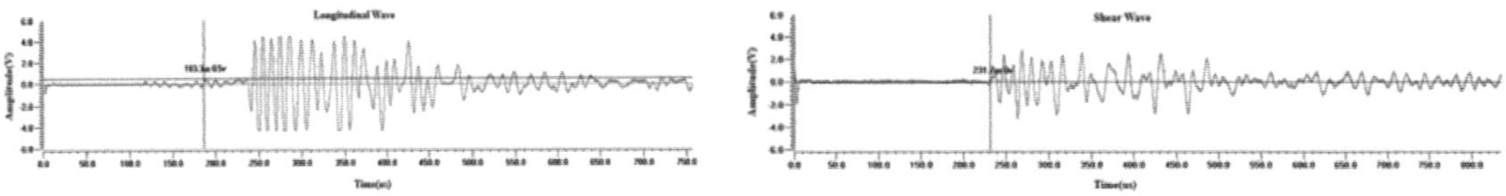

(a) The longitudinal wave of the Jz-0 specimen (b) The shear wave of the Jz-0 specimen

**Figure 6.** The ultrasonic wave of the JZ-0 specimen.

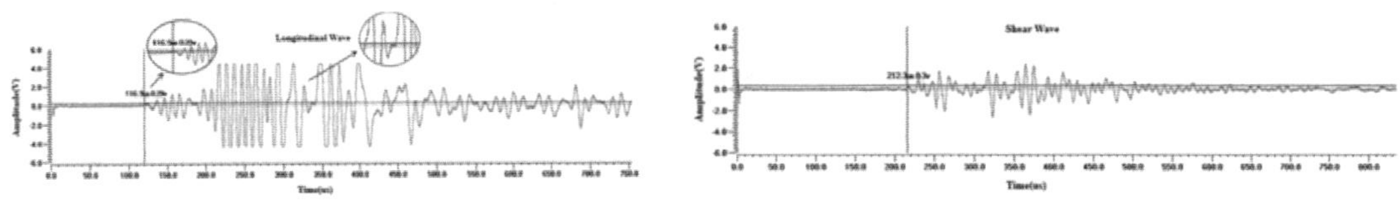

(a) The longitudinal wave of the FSF-4 specimen (b) The shear wave of the FSF-4 specimen

**Figure 7.** An ultrasonic wave of the FSF-4 specimen.

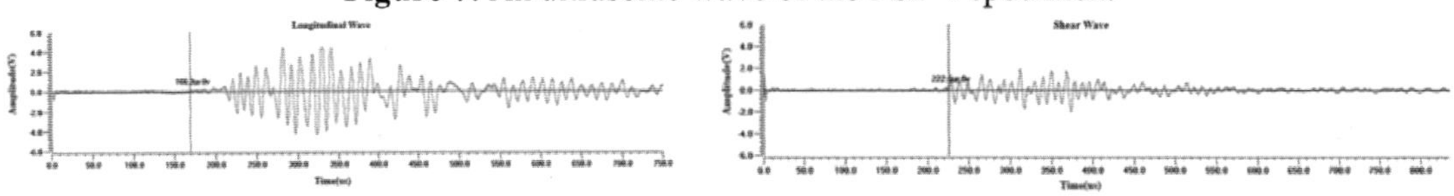

a The longitudinal wave of the FNS-3 specimen b The shear wave of the FNS-3 specimen

**Figure 8.** An ultrasonic wave of the FNS-3 specimen.

Generally speaking, with the increase of the porosity of the material, the wave velocity of the longitudinal and transverse waves will decrease, which also indicates the decrease of the elastic modulus of the material. However, the shape, aspect ratio, distribution, and other factors of the pore of the material will also affect the elastic modulus and Poisson's ratio of the material.

### *4.1. Ultrasonic waveform analysis*

In the uniform cement mortar medium model, the ultrasonic wave is stable, and the ultrasonic law is attenuated. The ultrasonic wave propagates in materials, and the acoustic properties of materials and their defects will affect the propagation of the ultrasonic wave. Ultrasonic wave produces reflection, refraction, and wave pattern conversion at different interfaces, and the uniformity of material structure can be understood non-destructively through ultrasonic detection.

- Head-wave analysis. As shown in Figure 6~8, the ultrasonic head wave movement of each group was greater, which was caused by the porosity defects of high water materials. Compared with other groups, the head wave of the FSF 4 group was smooth and stable, the amplitude curve of the p-wave was flat, the peak difference was small, and the periodic amplitude was uniform.
- Waveform and crack. However, at the time of 320 s, the longitudinal wave of the FSF-4 specimen showed the loss of a lower peak, and the shear wave was also attenuated (Figure 8). This was because the wave impedance between cement and cavity or crack was very different, and the sound wave would have a total reflection on the interface of the crack. The reflection wave formed by the P-wave and the first arrival wave of the S-wave would interfere with each other, and the S-wave could not propagate in the crack, resulting in the envelope surface of the S-wave being more chaotic. The energy of this part of the sound wave was reduced, and after the P-wave reflection, due to the attenuation of the wave, part of the waveform would also be missing. If a large crack was encountered in the propagation process, the wave had a wide range of total reflection at the interface, and the sound wave propagating outside the crack was less, forming a trapezoid area with weak energy.
- Waveform analysis. The ultrasonic waveform of FS-4 was similar to that of FSF-4, but the period and amplitude were relatively small, indicating that the two groups of high-water materials were of uniform material. The amplitude of the longitudinal wave of the FNS-3 specimen varied greatly. Still, no significant waveform loss indicated more holes in the high-water material after adding nano and fly ash. Still, the distribution was relatively uniform without large voids.

**Table 2.** Test mix ratio of the two materials added

| Specimen number | Bulk density | P-wave velocity | S-wave velocity | Poisson's ratio | Young's modulus | Shear modulus | Lamet constant | Bulk modulus |
|---|---|---|---|---|---|---|---|---|
| JZ-0 | 1.35 | 962.7 | 718.78 | 0.28 | 1.21 | 0.7 | -0.26 | 0.32 |
| FSF-4 | 1.38 | 1368.69 | 753.6 | 0.34 | 1.97 | 0.77 | 1 | 1.51 |
| FS-4 | 1.36 | 1098.15 | 705.47 | 0.15 | 1.53 | 0.67 | 0.28 | 0.73 |
| FNS-3 | 1.31 | 872.89 | 690.55 | 0.13 | 0.85 | 0.64 | -0.14 | 0.17 |

### *4.2. Basic physical property parameters of materials*

After processing, the data of each group are shown in Table 2. It could be seen that the amplitude and wave velocity of group FSF-4 were larger than those of the other groups, and the elastic modulus was also the highest. The compressive strength test also showed that group FSF has the highest compressive strength. Compared with the JZ-0 group, Young's modulus, bulk modulus, and shear modulus of the doped high-water material experimental group were improved. The Young's modulus of the FSF-4 group was 2.3 times that of the original high water material, the bulk modulus was 8.9 times that of the raw material, and the shear modulus was increased by 20.31%. Young's modulus, volume modulus, and shear modulus increased with the decrease of porosity, indicating that the high-water material mixed with fly ash and silica fume (FSF-4) was more compact in its internal structure.

## 5. Conclusion

Based on the new high-water material, this paper studies the influence of fly ash, slag, silica fume, and nano mixture on high-water material through compressive test and ultrasonic detection, and the following conclusions can be drawn:

- Fly ash, silica fume, and slag are common industrial wastes. If they can be used to replace part of the cement in high-water materials, the cost can be reduced, the reaction temperature can be controlled, and the high-water materials can be prevented from setting too fast.
- When the fly ash content was 5%, the 7-d compressive strength of flyash-silica fume (FSF) was 1.05% higher than that of JZ-0 and 3.38% higher than that of the F-1 group. The 7-d compressive strength of the high-water material with fly ash-slag was 0.92% higher than that of F-1 but still lower

than that of JZ-0. However, the strength of the fly ash-nano silica high water material decreased obviously with the increase of the nano content.

- When the fly ash content was 10%, the compressive strength of the FSF-4 specimens of high water material mixed with coal ash and silica fume at the 1 d, 3 d and 7 d ages was 50%, 67% and 40% higher than that in the F-2 group of single fly ash with the same replacement amount. The compressive strength of FSF-4 specimens at the 1 d, 3 d, and 7 d ages increased by 20%, 26.7%, and 15.8% compared with that of JZ-0 specimens, respectively. The compressive strength of the FS-3 group of high water material mixed with coal ash and slag at the 7 d age was 14.55% higher than that of the F-2 group but 5.26% lower than that of the JZ-0 group. The compressive strength of the NFS-3 group doped with Nano-$SiO_2$ and fly ash was 7.27% higher than that of the F-2 group doped with 10% fly ash but 16.69% lower than that of the JZ-0 group.
- Through ultrasonic detection, it was obtained that Young's modulus of the FSF-4 group was 2.3 times that of the original high-water material, the bulk modulus was 8.9 times that of the raw material, and the shear modulus was increased by 20.31%. The high-water material (FSF-4) mixed with fly ash and silica fume was known to be dense in the internal structure. The fly ash and slag material mixed with high water was more uniform but less dense than the FSF-4 specimen. There were more holes in nano and fly ash double-doped high water materials, but the distribution of holes was more uniform, and there was no large cavity.

At present, the high water material has high water content, a small Poisson ratio, low hardness, and low strength, which limits the popularization and application range of the material. Adding mineral waste to high-water materials not only saves environmental protection and costs but also improves the strength of high-water materials, which provides a reliable idea for further research of the material and can broaden the application of materials in the future.

**References**

[1] HUANG Feng. (2021) Research on Optimization and Application of Ultra-high Water Material Filing System. Shandong Coal Science and Technology, 02: 71-72.

[2] Helena Monteiro, Bruna Moura, Nelson Soares. (2022) Advancements in nano-enabled cement and concrete: Innovative properties and environmental implications, 56: 104736.

[3] Ahmed S. Kadhim, Alaa A. Atiyah, Shakir A.Salih. (2020) Properties of self-compacting mortar containing nano cement kiln dust, 20: 499-504.

[4] SAAFI M, ANDREW K, TANG P L, et al. (2013) Multifunctional properties of carbon nanotube fly ash geopolymeric nanocomposites, 49: 46-55.

[5] SUN Dao-sheng, HU Mei-Mei, WANG Ai-Guo. (2016) Research and Preparation of High Water Filing Material with High Dosage of Fly Ash. Bulletin of the Chinese Ceramic Society, 35(4): 1074.

[6] FENG Bo, LIU Chang-Wu, XIE Hui. (2018) Experimental Study and Analysis of the Mechanical Properties of High-water-content Materials Modified with Fly Ash. Chinese Journal of Engineering, 40(10): 1188-1195.

[7] ZHENG Ke-Ren, SUN Wei, JIA Yan-Tao. (2005) Effects of Slag Dosage on Hydration Products and Pore Structure of Cement Paste at High Water to Binder Ratio. Journal of the Chinese Ceramic Society, 33(04): 520-524.

[8] SHI Song, LIU Chang-Wu, WU Hai-Kuan, CHEN Kang-Liang. (2021) Study on Physical and Mechanical Properties of Modified High Water Filing Material with Fly Ash and Calcium Carbide Slag. Materials Reports, 35(7): 07027-07032.

[9] P. Z. Wang, R. Trettin, V. Rudert. (2005) Effect of fineness and particle size distribution of granulated blast-fumace slag on the hydraulic reactivity in cement systems. Adv.Cem.Res, 17(4): 161-166.

[10] TAO Rui, HU Xiao-Xia, DU Jia-Heng, WANG Pan, HOU Dong-Shuai. (2023) The Effect of CaO in Fly Ash on the Hydration Mechanism and Properties of Magnesium Phosphate Cement. Construction and Building Materials, 406: 133328.

[11] ZHENG Yuying, XIE Wei, WANG Qiu-Fen, et al. (2020) Effect of silica fume and nano $SiO_2$ on properties of high water materials. IOP Conference Series: Earth and Environmental Science (EES), 571: 012107.

# Crystallization, thermal properties, rheological properties, mechanical properties, and morphology of thermoplastic polyurethanes/calcium sulfate whiskers composites

**Daohai Zhang [1, 2, 3,4], Xiao Zhan[3], Guilan Liu[1], Yanyan Tan[3], Lianjun Shi[1], Qianlin Chen[2], and Tian Xie[1]**

[1]Postdoctoral Research Station, Wengfu (Group) Co., Ltd., Guiyang, Guizhou, 550000China
[2]Postdoctoral Station of Mechanical Engineering, Guizhou University, Guiyang, Guizhou, 550025 China
[3]School of Chemical Engineering, Guizhou Minzu University, Guiyang, Guizhou, 550025 China

[4]Correspondence author's e-mail: zhangdaohai6235@163.com

**Abstract**: The research shows that thermoplastic polyurethane (TPU) which is easy to process has great application potential in practical application. In this study, a calcium sulfate whisker (CSW) with high strength and stability was used as a reinforcing filler to improve the processability of flexible TPU. In this paper, TPU/CSW composites were prepared by the melt blending method, and the effects of modified CSW on the thermal, crystalline, and melt properties, as well as the rheological and mechanical properties of TPU composites were investigated. The results showed that when the mass content of modified CSW was 10%, the maximum values of $\Delta H_c$ and $\Delta H_m$ were 222.81°C and 9.24 J/g, respectively, and the thermal properties of the composites were the most stable. When the content of modified CSW is 5%, the maximum impact strength, bending strength, and tensile strength are 50.5 KJ/m$^2$, 38.8 MPa, and 927.7 MPa, respectively, and the mechanical properties of the composites are the best. The excellent stability and high strength of TPU/CSW composites show great promise for various applications in transportation and building materials.

## 1. Introduction

Thermoplastic polyurethane (TPU) is a synthetic organic polymer material composed of oligomeric polyols and diisocyanate hard segment, which possesses excellent properties such as high strength, strong tensile resistance, and easy processability, making it widely utilized in aerospace, aviation, and mechanical parts industries [1-3]. However, the mechanical strength and thermal stability of TPU are poor, which limits its application and development in various fields[4-5]. In recent years, domestic and foreign research on TPU blending modification technology has been more active through the addition of modifiers to improve the thermal stability and processability of TPU, because of maintaining the original excellent properties of TPU [6].

Calcium sulfate whisker (CSW) is a new inorganic nano-filler with a nearly perfect crystal structure, which is not only cost-effective but also has special properties such as excellent thermal stability, insulation, high tensile strength, and elastic modulus, witnessing wide applications as an excellent filler to enhance the machinability and thermal stability of polymers [7-8]. However, as the modification of TPU with inorganic materials becomes more common, it is not possible to improve the mechanical, crystalline, and processability properties of TPU at the same time. The reason is that CSW whiskers have a high aspect ratio, and CSW is often agglomerated within TPU when preparing TPU/CSW composites [9]. Improving the overall performance of CSW in TPUs, including dispersion, mechanical properties, and processability, poses a significant challenge in current research [10].

Previous studies have demonstrated that silane coupling agents are commonly employed to enhance the dispersion of inorganic fillers and further promote the nucleation and crystallization of composites, acting as a “bridge” between inorganic and organic materials [11]. He et al. [12] prepared isotactic polypropylene (IPP)/CSW composites by surface modification of CSW using γ-aminopropyl triethoxysilane (KH550). The results showed it can improve the interfacial compatibility between the CSW filler and the IPP matrix, and then improve the mechanical properties, thermal stability, and crystalline properties of the IPP/CSW composite system. Jayachitra et al. [13] performed three types of silane modifications on lignocellulosic raw fibers extracted from coconut inflorescences, namely, KH550, KH560, and KH570 before hybridization. The results showed that KH570 silane-modified inflorescence fibers blended with glass fibers and reinforced epoxy composites were found to exhibit maximum tensile and flexural strengths of 102.6 MPa and 166.89 MPa, respectively.

To address these issues, the TPU/CSW composite was subjected to surface modification by using KH570 and subsequently prepared through a melt blending and extrusion pelletizing process. The melt was blended with TPU to enhance the interfacial compatibility between CSW and TPU while reducing the agglomeration of CSW in the TPU matrix and improving the mechanical processability of the composites. The resulting TPU/CSW composites were then analyzed for their thermomechanical properties, crystallization, melting behaviors, rheological properties, and mechanical properties to investigate the effects of different levels of KH550-modified CSW on the thermal stability, crystallization, and rheological and mechanical properties of the TPU/CSW composite system.

## 2. Experiment

### *2.1 Materials*

TPU (903TU) was purchased from Shenzhen Plastic Sea Rubber & Plastic Technology Co., Ltd. Calcium sulfate (CSW with a length of 30~90 μm, diameter of 2~3 μm, and effective mass fraction of 99%) was obtained from Jinan Qingyuyuan New Material Co., Ltd.

### *2.2 Preparation of CSW modifier and CSW surface modification*

We measured 17 mL of anhydrous ethanol, 2 mL of distilled water, and 1 mL of KH570. Then we transfered the mixture to a beaker and stir magnetically for 4 h at room temperature to obtain the formulated modifier.

The CSW was placed in a blast oven at 100°C for 12 h. 250 g of CSW was weighed and 20% of the solid volume of the modifier was slowly added drop by drop, and the reaction was carried out with magnetic stirring at 60°C for 3 h. Afterward, the precipitate was washed and dried at 100°C for 8 h to obtain the modified CSW.

### *2.3 Preparation of TPU/CSW composites*

The TPU was placed in a blast drying oven at 100°C for 12 h. The additional amounts of KH570 modified CSW were set to 0%, 5%, 10%, 15%, and 20% of the mass of TPU. Using the melt co-mixing method, the dried TPU and modified CSW were put into a high-speed mixer and mixed well and then put into a twin-screw extruder (medium control temperature set at 185 ~ 230°C) for extrusion, and finally, the samples were pelletized with a cutting machine. The pellets were dried in an oven at 100°C for 12 h and then put into a plastic injection moulding machine for injection moulding to obtain the required sample strips.

### *2.4 Characterization*

The TG-DTG curve was obtained by using a thermogravimetric analyzer (TGA-55, TA Instruments, USA). Enthalpic properties of samples were conducted by using NETZSCH DSC214. The rheological properties were tested by using a Thermo Fisher Scientific MARSIII rotational rheometer. The morphologies of samples were characterized by scanning electron microscopy (FEI Quanta 250). The samples were evaluated by tensile test according to GB/T1040-79, using a universal material testing machine (SNAS, MTS Industrial Systems Co., Ltd., China)

## 3. Results and discussion

### *3.1 Thermal properties of TPU/CSW composites*

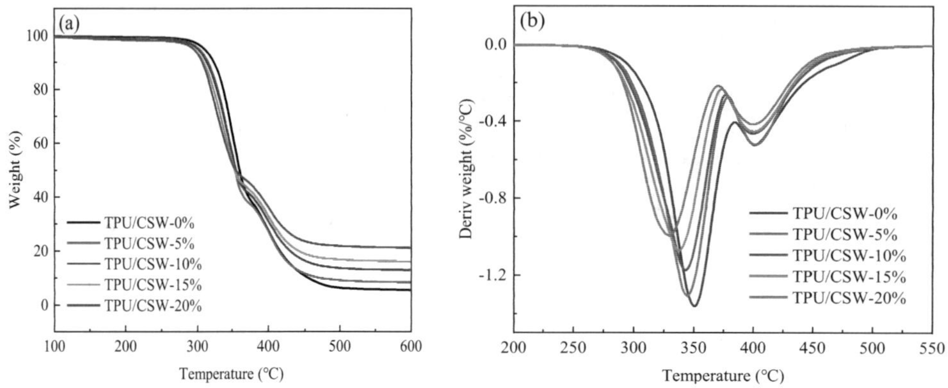


(a) TG (b) DTG

**Figure 1.** Curves of TPU/CSW composites with different additions.

**Table 1.** TG and DTG data of TPU/CSW composites with different additions.

| Samples | $T_{2\%}$ (°C) | $T_{5\%}$ (°C) | $T_{10\%}$ (°C) | $T_{\mathrm{DTG\ peak1}}$ (°C) | $T_{\mathrm{DTG\ peak2}}$ (°C) | Vestigital (%) |
|---|---|---|---|---|---|---|
| TPU/CSW-0% | 291.4 | 310.6 | 324 | 352.3 | 400.6 | 5.53% |
| TPU/CSW-5% | 281.1 | 302.5 | 315.7 | 346.8 | 402.5 | 8.38% |
| TPU/CSW-10% | 278.5 | 300.4 | 313.6 | 350.7 | 401.2 | 12.88% |
| TPU/CSW-15% | 258.1 | 294.6 | 308.8 | 346.9 | 401.9 | 16.03% |
| TPU/CSW-20% | 258.4 | 294.2 | 306.7 | 326.1 | 397.6 | 21.17% |

The TG and DTG curves and data of TPU/CSW composites with different mass fractions of CSW content are shown in Figure 1 and Table 1, which overall show that the thermal decomposition process of TPU/CSW composites with different CSW fractions mainly occurs from 300°C to 450°C. When the temperature reaches 352.3°C and 400.6°C, two weight losses occur, mainly due to the decomposition of TPU. The $T_{\mathrm{DTG\ peak1}}$ appearing at 320°C ~ 360°C is the maximum weight loss peak of the composite. As the CSW content gradually increased, the DTG peak temperature shifted to the left, and the peak height gradually decreased, i.e., the temperature at which the TPU/CSW composites started to decompose shifted towards lower temperatures. As shown in Table 1, the $T_{2\%}$, $T_{5\%}$, and $T_{10\%}$ of TPU/CSW composites gradually decreased with increasing CSW content, because CSW as a TPU filler enhanced the intermolecular forces between matrix and filler, produced a rapid heat transfer effect, and acted as a thermal catalyst for degradation, resulting in lower $T_{2\%}$, $T_{5\%}$, and $T_{10\%}$. However, the residual carbon rate of TPU/CSW composites gradually increased with the increase of CSW content, with the maximum residual carbon rate of 21.17% for TPU/CSW-20%, indicating that the filling of CSW reduced the mass loss of TPU composites and facilitated the formation of residual carbon, and the low thermal transfer efficiency could not degrade HDPE/CSW composites effectively.

*3.2 Crystallization and melting behavior of TPU/CSW composites*

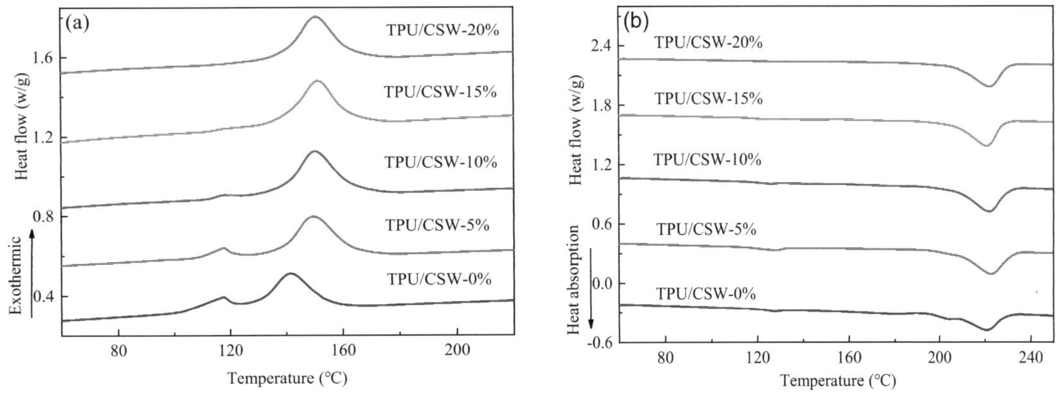


(a) Ramp-up curves; (b) Ramp-down curves

**Figure 2.** DSC curves of TPU/CSW composites with different additions.

**Table 2.** DSC melting and crystallization data of pure TPU and TPU/CSW composites.

| Samples | $T_c$ (°C) | $\Delta H_c$ (J/g) | $T_m$ (°C) | $\Delta H_m$ (J/g) |
|---|---|---|---|---|
| TPU/CSW-0% | 141.4 | 12.74 | 220.9 | 11.36 |
| TPU/CSW-5% | 149.9 | 19.23 | 222.8 | 19.24 |
| TPU/CSW-10% | 150.3 | 22.12 | 221.7 | 19.09 |
| TPU/CSW-15% | 151.6 | 21.74 | 221.2 | 17.91 |
| TPU/CSW-20% | 150.6 | 21.48 | 222.0 | 18.19 |

(Note: $T_c$: crystallization temperature; $T_m$: melting temperature; $\Delta H_c$: enthalpy of melting; $\Delta H_m$: enthalpy of crystallization.)

The DSC curves and the melting and crystallization data of TPU/CSW composites with different CSW contents are shown in Figure 2 and Table 2, respectively. The TPU/CSW composites exhibit a single melting or crystallization peak, but with an increase in CSW content, the peak gradually increases and shifts towards higher temperatures. From the data presented in Table 2, it can be observed that as the content of CSW increases, there is a gradual increase in the crystallization and melting temperatures of TPU/CSW composites. Specifically, there is an increase of 9.2°C and 8.74°C in the melting and crystallization temperatures, respectively. In addition, the values of both melt enthalpy ($\Delta H_c$) and crystallization enthalpy ($\Delta H_m$) exhibited an initial increase followed by a decrease upon increasing CSW content, but both were greater than those of pure TPU, indicating that the thermal stability of the TPU/CSW composites increased with the addition of CSW. Notably, the largest values of $\Delta H_c$ and $\Delta H_m$ for TPU/CSW-10% were 222.81°C and 9.24 J/g, respectively, indicating that the TPU/CSW composites had the greatest thermal stability at 10% CSW content.

### *3.3 Rheological properties of TPU/CSW composites*

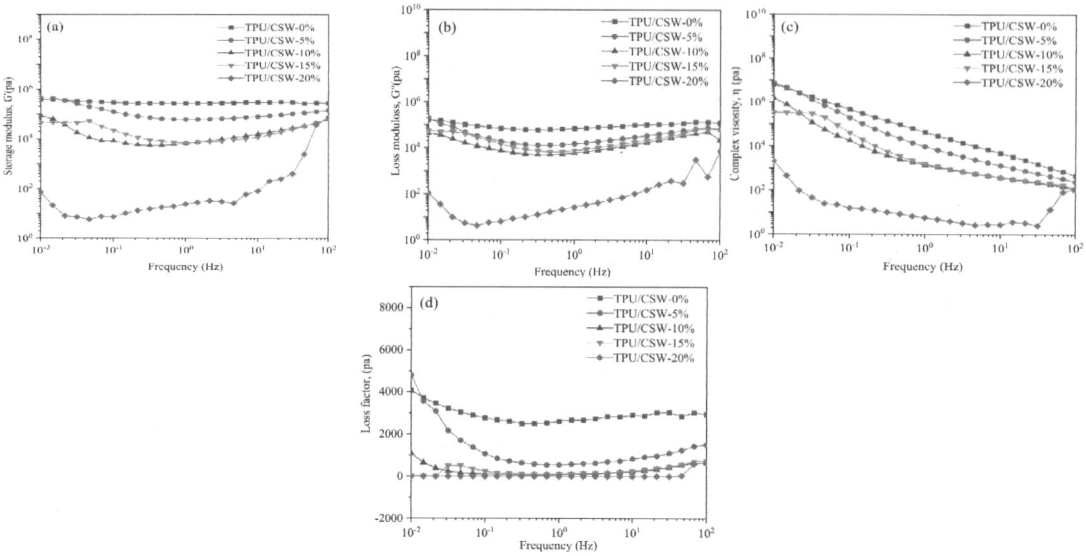


**Figure 3.** (a) Energy storage modulus G′; (b) Loss modulus G″; (c) Complex viscosity η*; (d) Loss factor τ for TPU/CSW composites with different additions.

Polymer rheology measurements are a valuable technique for evaluating the flow and deformation characteristics of polymeric materials, specifically, which serve as an effective method for gaining insights into microstructural information of filled particles within a polymer melt. The energy storage modulus of the melt G′ represents the amount of recoverable energy stored in the melt and the loss modulus G″ represents the amount of non-recoverable energy released from the melt. Figure 3 displays the energy storage modulus G′, loss modulus G″, complex viscosity η*, and loss factor τ of TPU/CSW composites with different CSW contents. The results reveal a non-Newtonian characteristic of the TPU/CSW blended system, where shear viscosity decreases with increasing shear rate. The G′, G″, η*, and τ of the pure TPU are all lower than those of the TPU/CSW composites. The value of G′ and G″ for TPU/CSW composites exhibits a decline as the frequency increases, while the value of η* primarily exhibits a decrease with frequency. This is due to the reduced duration of shear force action with increasing frequency, leading to a rise in energy loss and an increase in resistance to the movement of molecular chains within the material. As a result of the latter, the deformation of macromolecular chains is unable to keep pace with the changes in shear external force. The system then changes to an elastic behavior, with G' and G" decreasing initially and subsequently increasing as shear frequency rises. In the low-frequency region, the TPU macromolecular chains have enough time to deform. The viscoelasticity decreases, and G′, G″, and $\eta^*$ also decrease.

### *3.4 Mechanical properties of TPU/CSW composites*

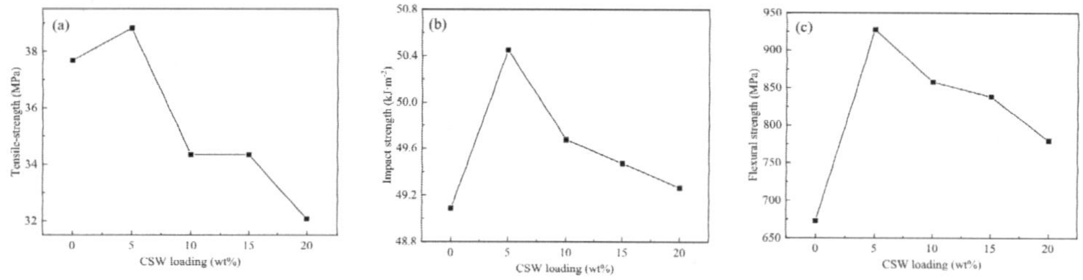


(a) Tensile strength; (b) Impact strength; (c) Bending strength.

**Figure 4.** Plots of TPU/CSW composites with different additions.

As shown in Figure 4, with the addition of CSW content, the impact and bending strengths of TPU composites increased compared to pure TPU, with the highest impact, bending, and tensile strengths of 50.5 KJ/m$^2$, 38.8 MPa, and 927.7 MPa, respectively, when the CSW content was added at 5%. This phenomenon can be explained by the special structure of TPU and the increasing content of CSW whiskers, which increases the concentration in the plastic phase of TPU, resulting in complete micro-phase separation and improving the mechanical properties of TPU/CSW composites to some extent. When the mass fraction of CSW content is greater than 5%, CSW will appear as agglomerates and have an uneven dispersion in the composite process. When the composite is subjected to external forces, a large internal stress will be generated within it, resulting in the concentration of internal forces. Dry cracking will occur when the force has been sustained throughout the process, which will lead to a reduction in the tensile strength, bending strength, and impact strength of TPU/CSW composites. The mechanical properties are then reduced.

### *3.5 Morphological characterization of TPU/CSW composites*
Figure 5 (a) shows the SEM image of pure TPU, which exhibits strong continuity and flat cross-section. Figures 5 (b)-(e) show that the KH570-modified CSW particles are uniform in size, CSW is well dispersed in the TPU matrix, and there is no gap at the interface between CSW and the matrix, indicating good compatibility between the two. Figures 5 (b)-(c) show that the SCSW exhibits a uniform insertion of the sharp end into the TPU matrix in a radioactive manner, and more CSW whiskers can be observed with the increase of the filling amount. A more uneven fracture surface near the CSW is found in Figure 5 (c), because more CSW appear to be agglomerated in the TPU matrix, hindering the dispersion of the CSW, which explains the decrease in mechanical properties mentioned above.

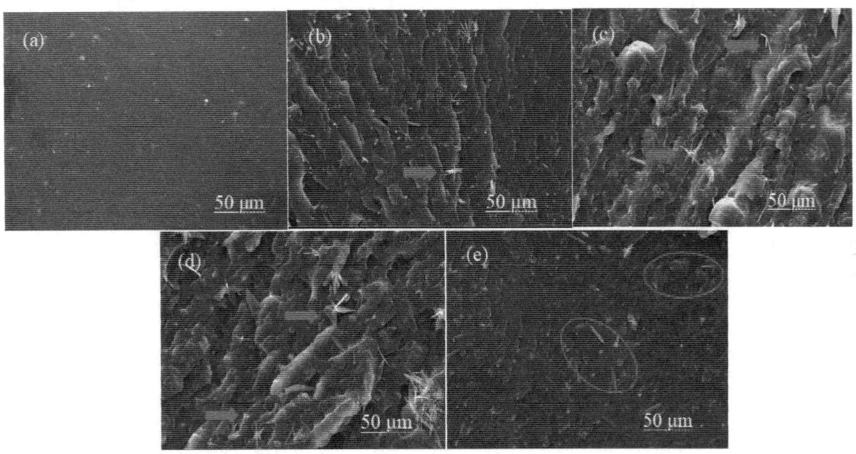


(a) TPU/CSW-0%; (b) TPU/CSW-5%; (c) TPU/CSW-10%; (d) TPU/CSW-15%; (e) TPU/CSW-20%.

**Figure 5.** Cross-sectional morphology of TPU/CSW composites with different additions.

## 4. Conclusions
In this work, TPU/CSW composites were prepared by using the melt mixing method. The incorporation of modified CSW of a silane coupling agent (KH570) at a content of 20% resulted in a maximum residual carbon rate of 21.17% in the TPU/CSW composites. The thermal stability of the TPU/CSW composites initially increased and then decreased with increasing modified CSW content. The maximum values of $\Delta H_c$ and $\Delta H_m$ at 222.81°C and 9.24 J/g were achieved when the modified CSW content was 10%. The fluidity of the composites exhibited a gradual decrease as the modified CSW content increased, primarily due to agglomeration. The mechanical properties of the TPU/CSW composites were best when the modified CSW content was 5%, with maximum impact, flexural strength, and tensile strength of 50.5 KJ/m$^2$, 38.8 Mpa, and 927.7 MPa, respectively. These results contribute to the advancement of high-strength and easily processed polymers in various fields.

**Acknowledgments**
This work was supported by the Postdoctoral Fund of Postdoctoral Research Workstation of Wengfu (Group) Co., Ltd. (WF-001-2021-QT-00120).

**References**

[1] Ran Z., Mengqi L., Yajun C., et al. (2023). Fire behavior and mechanical properties of phosphorus containing silica gel coated ammonium polyphosphate as fire retardants in thermoplastic polyurethane (TPU) [J]. Journal of Polymer Research, 30 (10): 1-10.

[2] Jeongwon M., Sunhee L., Bian Y., et al. (2023). Dyeability of 3D printed thermoplastic polyurethane (TPU) material using turmeric and lac natural dyes [J]. Fibers and Polymers, 24 (5): 1, 727-1, 739.

[3] Zhang Daohai, Pei Meng, Xie Jinhui, et al. (2023). Effect of thermal oxygen aging time on the properties of HDPE/$Na_2SO_4$ composites [J]. Journal of Guizhou Normal University (Natural Sciences), 41 (2): 1-14, 128.

[4] Shaokang L., Haihua W., Bin C., et al. (2023). Study on FDM preparation and properties of RGO/Ni/PLA/TPU composite materials [J]. Journal of Alloys and Compounds, 968: 172, 119.

[5] Wang X., Liu J., Zheng Y., et al. (2022). Biocompatible liquid metal coated stretchable electrospinning film for strain sensors monitoring system [J]. Science China Materials, 65 (8): 2, 235-2, 243.

[6] Tie G., HanChi X., XinChao W., et al. (2023). The study on the morphology and compression properties of microcellular TPU/Nanoclay tissue scaffolds for potential tissue engineering applications [J]. Polymers, 15 (17): 36-47.

[7] Handawy M. K., Snegirev A. Y., Stepanov V., et al. (2020). Kinetic modeling of pyrolysis and combustion of high-density polyethylene-based polymer mixtures [J]. Journal of Physics: Conference Series, 1, 565 (1): 012, 097.

[8] Dou Q. and Duan J. (2016). Melting and crystallization behaviors, morphology, and mechanical properties of [beta]-polypropylene/polypropylene-graft-maleic anhydride/calcium sulfate whisker composites [J]. Polymer Composites, 37 (7): 2, 121-2, 132.

[9] Muhammad H. U. Z., Jian W., Zonglin P., et al. (2023). Mechanical properties and thermal stability of modified calcium sulfate whisker-reinforced polydimethylsiloxane/ethylene–vinyl acetate copolymer composites [J]. Iranian Polymer Journal, 32 (10): 1, 335-1, 346.

[10] Li Qingbiao, Liu Hao, Nie Chenchen, Xie Guiming, Che Zhaomei, Zhu Dehui, Guo Lei, Xiang Yuan, and Shi Wei (2022). PMMA-grafted calcium sulfate whiskers for applications as fillers in PVC [J]. Polymers, 14 (19): 4, 199-4, 204.

[11] Yuan W., Cui J., and Xu S. (2016). Mechanical properties and interfacial interaction of modified calcium sulfate whisker/Poly (Vinyl Chloride) composites [J]. Journal of Materials Science & Technology, 32 (12): 1, 352-1, 360.

[12] He B., Lin X. F., and Zhang Y. F. (2018). Effect of a novel compound nucleating agent calcium sulfate whisker/β-nucleating agent dicyclohexyl-terephthalamide on crystallization and melting behavior of isotactic polypropylene [J]. Journal of Thermal Analysis and Calorimetry, 132 (2): 1, 145-1, 152.

[13] R. J., S. S. R. M., Kumar B. A., et al. Physical and mechanical behavior of silane-modified coconut inflorecence/glass fibril-fortified hybrid epoxy composites [J]. Advances in Materials Science and Engineering, 2022, 2022: 1-6.

# Influence of modified phosphogypsum content on the properties of HDPE/phosphogypsum composites

**Daohai Zhang [1, 2, 3,4], Jingyu Du [3], Guilan Liu [1], Meng Pei [3], Lianjun Shi [1], Qianlin Chen [2], and Tian Xie [1]**

[1]Postdoctoral Research Station, Wengfu (Group) Co., Ltd., Guiyang, Guizhou, 550000 China
[2]Postdoctoral Station of Mechanical Engineering, Guizhou University, Guiyang, Guizhou, 550025 China
[3]School of Chemical Engineering, Guizhou Minzu University, Guiyang, Guizhou, 550025 China

[4] Correspondence author's e-mail: zhangdaohai6235@163.com

**Abstract**: Phosphogypsum has received increasing attention for solid waste with high production and environmental pollution. This study treated phosphogypsum with the silane coupling agent 3-(methacryloyloxy) propyltrimethoxysilane (KH570) to create modified HDPE/phosphogypsum composites. The modified phosphogypsum was then melted and blended with high-density polyethylene (HDPE). Thermal weight loss (TG), differential scanning calorimetry (DSC), and rheological and mechanical property testing were used to study the impact of the changed phosphogypsum content on the properties of the composites. The outcomes demonstrate that even at low filling levels, the modified phosphogypsum may be evenly distributed throughout the matrix. Furthermore, the addition of increased phosphogypsum filling enhances the mechanical characteristics of the HDPE matrix.

## 1. Introduction

High-density polyethylene (HDPE) is widely used as a plastic material due to its cost-effectiveness, ease of processing, and minimal requirements [1]. However, HDPE has drawbacks, including easy deformation, poor mechanical qualities, and low surface hardness [2]. To address these drawbacks, modifying the filling of HDPE can lower costs while enhancing HDPE performance [3]. Calcium sulfate dihydrate ($CaSO_4 \cdot 2H_2O$), a by-product of the manufacturing of phosphoric acid [4], is the major component of phosphogypsum. Phosphogypsum is often dumped in piles due to factors such as high production costs, waste of resources and pollutants (such as F, S, and heavy metals) polluting the environment, and low recovery rate. (e.g. F, S, and heavy metals). Researchers have done a lot of research on the use of phosphogypsum as a filler. The use of phosphogypsum as HDPE filler not only enhances the comprehensive properties of the material but also realizes the solid waste utilization of phosphogypsum, which is of great significance for broadening the application field of phosphogypsum. However, phosphogypsum powder is easy to agglomerate and has poor dispersion [5], and the surface modification of phosphogypsum is required to enhance the interface compatibility with polymers.

Recently, Luo et al. [6] have tried to use phosphogypsum as a filler powder to prepare composite materials. Li et al. [7] prepared Nano-$CaSO_4$ by using phosphogypsum and found that Nano-$CaSO_4$ had a significant reinforcing effect on linear HDPE. Sun et al. [8] found that the modified phosphogypsum whiskers had a good reinforcing effect on HDPE. Zhang et al. [9] modified phosphogypsum by using KH550, KH560, and KH570 and prepared HDPE/phosphogypsum composites, and found that the modified phosphogypsum could significantly improve the mechanical properties of the composites and KH560 and KH570 had the best modification and strengthening effect. Despite the positive outcomes of the research above, the process of converting phosphogypsum into whiskers for utilization proved to be labor-intensive. Based on the aforementioned findings, the effect of phosphogypsum filling quantity on composite properties is examined in this study to offer a theoretical foundation for the use of phosphogypsum in polymers.

## 2. Experiment

### *2.1 Main raw materials and reagents*

High-density polyethylene (HDPE): an industrial material (Maoming Branch, China Petroleum & Chemical Corporation); Silane coupling agent: 90%, KH570 (Kangjin New Material Technology Co., Ltd.); Anhydrous ethanol: ≥99.7% (Chongqing Wansheng Chuandong Chemical Co., Ltd.); Phosphogypsum: Urnford (Group) Co., Ltd.

### *2.2 Main equipment and instruments*

Electrothermal constant temperature blast drying oven: DHJ-101 (Zhengzhou Yuhua Instrument Manufacturing Co., Ltd.); Magnetic stirrer: DF-101S (Zhengzhou Yuhua Instrument Manufacturing Co., Ltd.); Twin-screw mixing and extruding machine: AK36 (Nanjing Keya Chemical Equipment Co., Ltd.); Plastic injection molding machine: YCT132-B (Taizhou Huatai Speed Control Motor Factory); Electronic Universal Testing Machine: 3365B1979 (Yingst Company, USA); Cantilever beam impact tester: 3365B1979 (Yingst Company, USA); Differential Scanning Calorimeter (DSC): DSC214 (NETZSCH, Germany); Thermal weight loss analyzer (TG): TG209F3 (NETZSCH, Germany); Rotational rheometer: (MARSIII, Thermo Fisher Scientific).

### *2.3 Sample preparation*

Firstly, a certain amount of HDPE was weighed and dried at a constant temperature of 100°C for 12 h, and 17 mL of absolute ethanol, 2 mL of distilled water, and 1 mL of silane coupling agent KH570 were taken and placed in a magnetic mixer for a constant reaction for 4 h. Secondly, HDPE composites were prepared with modified phosphogypsum as filler. We weigh the dry phosphogypsum, add a modifier with a volume of 20% of the solid volume, slowly add it drop by drop, stir magnetically at 60°C for 3 h to modify, set the centrifuge speed of 3, 000, centrifuge for 3 minutes, dry the precipitate, and obtain the modified phosphogypsum. Then, the dried HDPE and phosphogypsum were put into the high-speed mixer for mixing evenly according to Table 1 and then put into the twin-screw extruder to control the temperature of 170~200°C. After being chopped into pellets by a cutting machine, the samples were then dried for 12 h at 100°C before being inserted into an injection molding machine to create sample strips.

**Table 1.** HDPE/Phosphogypsum formulation.

| Name | Modified phosphogypsum mass (g) | HDPE mass (g) |
|---|---|---|
| HDPE/phosphogypsum-10% | 80 | 720 |
| HDPE/phosphogypsum-20% | 160 | 640 |
| HDPE/phosphogypsum-30% | 240 | 560 |
| HDPE/phosphogypsum-40% | 320 | 480 |

### *2.4 Performance testing and structural characterization*

We weigh 5-10 mg of the sample strip in an aluminum crucible and heat the sample from 50°C to 600°C at a temperature increase rate of 10 K/min under the atmosphere of $N_2$ to obtain the thermal degradation curve. Then, we weighed 5-10 mg of the sample strip in the aluminum crucible, and under the atmosphere of $N_2$, the sample was heated from 50°C to 250°C with a temperature rise and fall rate of 10

K/min and then decreased to 50°C after a certain time of constant temperature, and increased to the end of 250°C to observe the crystallization and melting curve of the sample. Rheological performance test: We cut 1 cm by 1 cm samples on the testing platform, separated by 0.5 mm, and frequency testing is conducted at the temperature of 230°C. Impact performance: Under GB/T 1843-2008 "Determination of Impact Strength of Plastic Cantilever Beams", the impact performance test of the specimen is carried out. Tensile properties: Referring to the tensile standard GB/T1040-79, the tensile standard specimen is dumbbell-shaped, and the corresponding dimensions are 180×10×4 mm. Bending performance: Referring to the bending standard GB/T1042-79, the bending standard specimen is a rectangular strip with a size of 80×10×4 mm.

## 3. Results and discussion

### *3.1 Mechanical properties of HDPE/phosphogypsum composites*

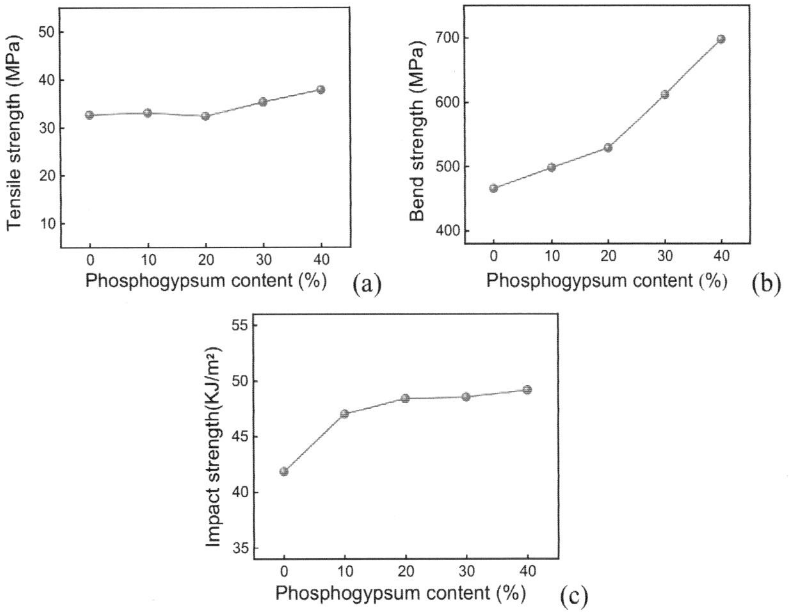


a) Tensile strength; b) Bend strength; c) Impact strength.

**Figure 1.** Mechanical properties of HDPE/phosphogypsum composites.

The impact of phosphogypsum on the mechanical characteristics of composites made of HDPE and phosphogypsum is depicted in Figure 1. The tensile strength of pure HDPE is 32.69 MPa, as shown in Figure 1 (a). When the mass fraction of phosphogypsum is 40%, the tensile strength of the composite increases to 37.89 MPa, and the modified phosphogypsum has high adhesion to the surface, improving the stiffness of the composite. The composites will have a reduction in fracture expansion under stress. The HDPE will gradually become stronger as the modified phosphogypsum is added, and the tensile strength of the composite will gradually increase. It can be seen from Figures 1 (b) and (c) that the bending strength and impact strength of HDPE composites are significantly higher than those of pure HDPE. As the mass fraction of phosphogypsum content increases, the bending strength of HDPE/phosphogypsum composites increases by 6.96%, 13.50%, 31.34%, and 49.79%, respectively, while the impact strength increases by 12.33%, 15.66%, 15.99%, and 17.52%. The enhancement of the bending properties is because the modified phosphogypsum has better interfacial compatibility with the matrix, while the phosphogypsum powder dispersed in the matrix increases the stress point [10]. The enhancement of impact strength is because the presence of inorganic particles produces a stress concentration effect on the composites, which can easily cause microcracks in the surrounding resin and absorb certain deformation work, thus improving the notched impact strength of the composites.

*3.2 Thermal properties of HDPE/phosphogypsum composites*

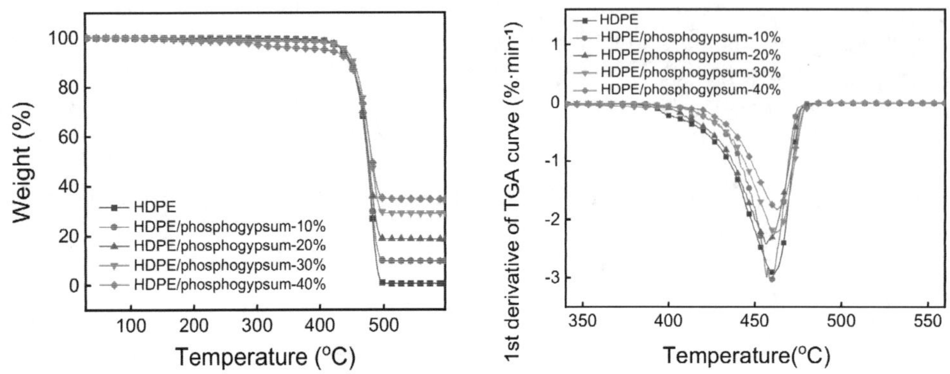


a) TG b) DTG

**Figure 2.** Curves of HDPE/phosphogypsum with different mass fractions of phosphogypsum

**Table 2.** TG and DTG data for HDPE/phosphogypsum with different mass fractions of phosphogypsum.

| Samples | $T_{2\%}$ (°C) | $T_{5\%}$ (°C) | $T_{10\%}$ (°C) | $T_{\mathrm{DTG\ peak}}$ (°C) | Vestigital (%) |
|---|---|---|---|---|---|
| HDPE | 417.4 | 435.5 | 448.2 | 481.9 | 0.62 |
| HDPE/phosphogypsum-10% | 418.7 | 437.9 | 449.8 | 479.9 | 9.62 |
| HDPE/phosphogypsum-20% | 414.3 | 436.2 | 449.0 | 447.8 | 18.52 |
| HDPE/phosphogypsum-30% | 404.0 | 440.6 | 453.3 | 479.8 | 28.97 |
| HDPE/phosphogypsum-40% | 398.6 | 409.1 | 446.7 | 478.8 | 34.62 |

Figure 2 shows the TG and DTG curves of HDPE/phosphogypsum with different phosphogypsum mass fractions, and the corresponding thermal decomposition temperatures are listed in Table 2. As shown in Figure 2, the weight loss curves of the composite HDPE/phosphogypsum are essentially the same as those of HDPE, with both exhibiting a single weight loss step, and the residual rate of HDPE/phosphogypsum-10% is close to 10%, with residues filled with modified phosphogypsum powder. When comparing the $T_{2\%}$, $T_{5\%}$, $T_{10\%}$, and $T_{DTG\ peaks}$ of the materials, phosphogypsum has no significant effect on the thermal stability of HDPE/phosphogypsum, with HDPE/phosphogypsum-30% composite having the best thermal performance [11].

*3.3 Crystallization and melting behavior of HDPE/phosphogypsum composites*

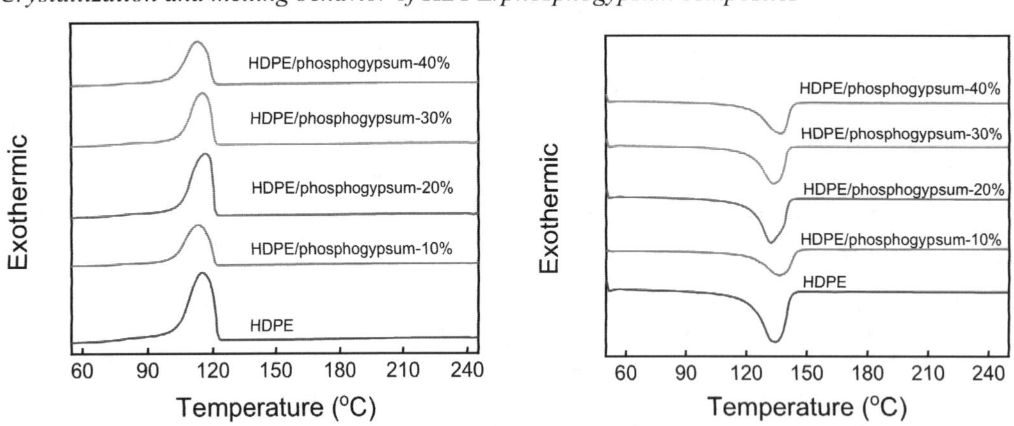


**Figure 3.** DSC curves of HDPE/phosphogypsum with different mass fractions of phosphogypsum content.

**Table 3.** DSC melting and crystallization data of pure phosphogypsum and HDPE/phosphogypsum (10%-40%) composites.

| Samples | $T_c$ (°C) | $\Delta H_c$ (J/g) | $T_m$ (°C) | $\Delta H_m$ (J/g) | $\chi_c$ (%) |
|---|---|---|---|---|---|
| HDPE | 117.97 | 198.73 | 132.60 | 199.98 | 69.19 |
| HDPE/phosphogypsum-10% | 117.24 | 172.08 | 132.67 | 174.37 | 67.03 |
| HDPE/phosphogypsum-20% | 116.94 | 171.87 | 132.75 | 172.86 | 74.76 |
| HDPE/phosphogypsum-30% | 116.19 | 123.66 | 133.68 | 126.36 | 62.46 |
| HDPE/phosphogypsum-40% | 117.04 | 108.93 | 132.76 | 109.67 | 63.24 |

(Note: $T_c$: peak temperature of crystallization; $\Delta H_m$: enthalpy of crystallization; $\chi_c$: degree of crystallization.)

As shown in Figure 3 and Table 3, the peak crystallization temperature of the modified HDPE/phosphogypsum composite ranges from 117.97°C to 116°C. At 94°C, the melting point decreases and the crystallization ability deteriorates. As can be seen from Table 3, the crystallinity of the modified HDPE composites decreases to 62.46%, the flexibility of the material chain increases, the intermolecular force decreases, the movement of the molecular chain segments is easier, and the impact strength of the material becomes significantly stronger. The toughness increases after the impact load is applied to the material, which is consistent with the law of influence in Figure 1.

*3.4 Rheological properties of HDPE/phosphogypsum composites*

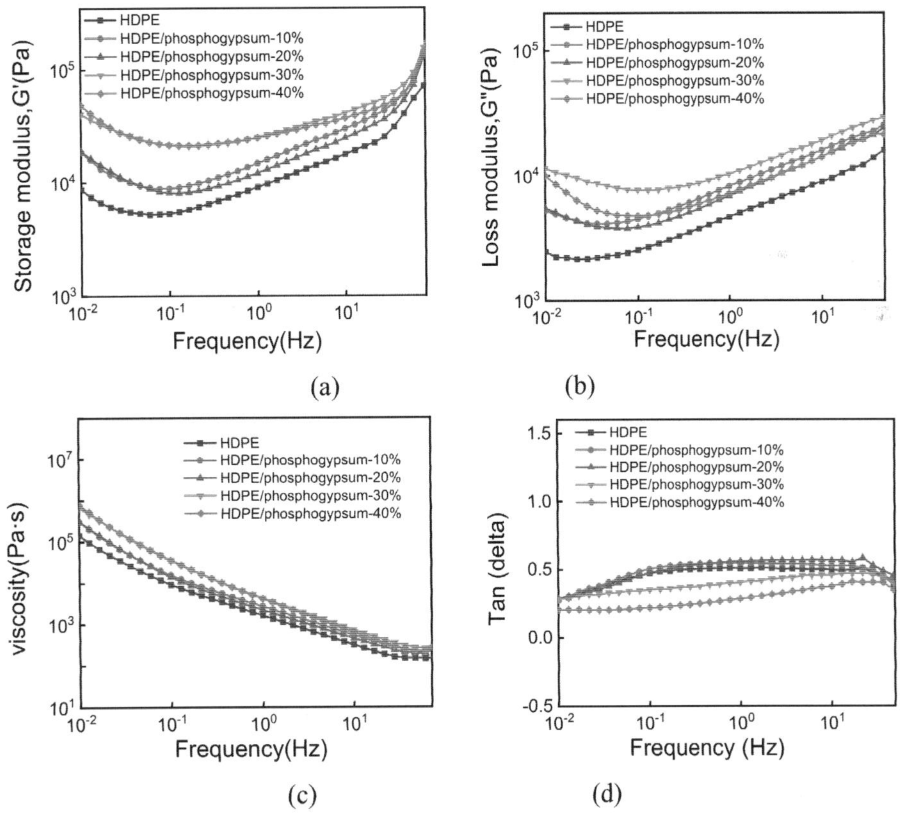


**Figure 4.** Rheology curves of HDPE/phosphogypsum with different mass fractions of phosphogypsum content.

As shown in Figure 4, the addition of 10% phosphogypsum resulted in a more significant decrease in the energy storage modulus, loss modulus, and viscosity of the composites, which was due to the small amount of phosphogypsum acting as a lubricant in the system. However, the viscosity of

HDPE/phosphogypsum composites was increased by the phosphogypsum content, which led to a decrease in the rheological rate of the composites and a decrease in flowability. The addition of phosphogypsum filler hinders the movement of HDPE molecular chains, and more energy is required to make the molecular chains move. When phosphogypsum increases, the loss modulus of the composite material increases. The loss factor of pure HDPE is low, and the addition of phosphogypsum increases the loss factor and the composite viscosity. The storage modulus also increased with the increase of the added amount, indicating that the stiffness of the material increased with the increase of the filling amount.

*3.5 Microstructure of HDPE/phosphogypsum composites*
Figure 5 (a) shows the SEM image of pure HDPE, which has strong continuity and flat sections. Phosphogypsum modified by KH570 is well dispersed in the matrix, and there is no gap at the interface between phosphogypsum and matrix, indicating that the modification has improved the interfacial compatibility between phosphogypsum and HDPE. Phosphogypsum particles are larger after KH570 modification, and the fracture surface is flat. As shown in Figures 5 (b) and (c), phosphogypsum was uniformly distributed in the HDPE matrix, and as the filling amount increased, more powder could be observed in the fracture out.

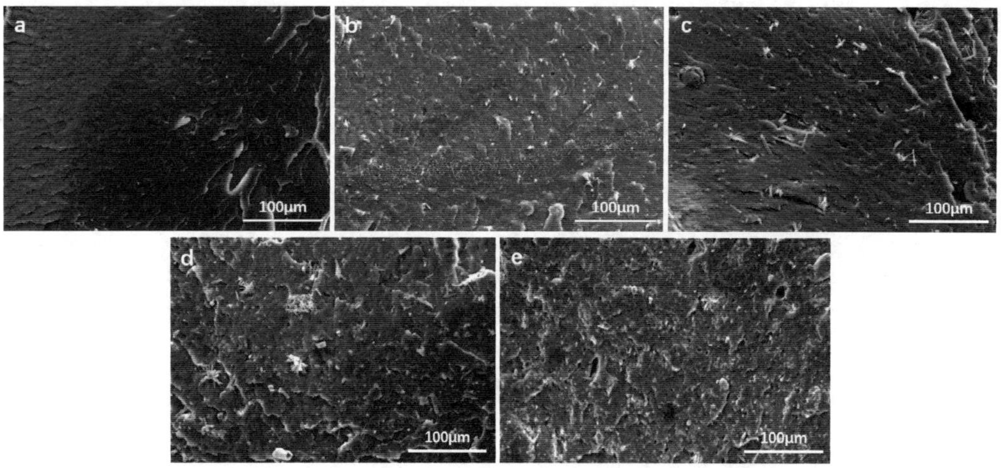


**Figure 5.** SEM images of HDPE/phosphogypsum with different mass fractions of phosphogypsum content: (a) Pure HDPE; (b) HDPE/phosphogypsum-10%; (c) HDPE/phosphogypsum-20%; (d) HDPE/phosphogypsum-30%; (e) HDPE/phosphogypsum-40%.

## 4. Conclusion

The incorporation of modified phosphogypsum yields measurable improvements in the properties of HDPE. As the filling quantity increased, the mechanical properties of HDPE/phosphogypsum composites improved and reached the optimum at 40% mass fraction of phosphogypsum, which was also confirmed by scanning electron microscopy. Impacts on thermal properties remain relatively subtle, and a discernible upward trend in energy storage modulus indicates increased material stiffness. This research remains nascent with areas requiring refinement including addressing deficiencies and enhancing precision through subsequent experiments focused on optimizing the phosphogypsum mass fraction gradient.

**Acknowledgments**
This work was supported by the Postdoctoral Fund of Postdoctoral Research Workstation Wengfu (Group) Co., Ltd. (WF-001-2021-QT-00120).

**References**

[1] Park S. and Fu K. K. (2021). Polymer-based filament feedstock for additive manufacturing. *Composites Science and Technology*, 108, 876.

[2] Zhang M., Biesold G. M., Choi W., et al. (2022). Recent advances in polymers and polymer composites for food packaging. *Materials Today*, 53.

[3] Li S., Dong C., Yuan C., Liu S., and Bai X. (2021). Effects of $CeO_2$ nano-particles on the anti-aging performance of HDPE polymer during friction. *Wear*, 203, 832.

[4] Zhang Daohai, Pei Meng, Xie Jinhui, et al. (2023). Effect of thermal oxygen aging time on the properties of HDPE/$Na_2SO_4$ composites [J]. *Journal of Guizhou Normal University (Natural Sciences)*, 41 (2): 1-14, 128.

[5] Gong S., Li X., Song F., Lu D., and Chen Q. (2020). Preparation and application in HDPE of Nano-CaSO4 from phosphogypsum. *ACS Sustainable Chemistry & Engineering*, 8 (11): 4, 511-4, 520.

[6] Luo R., Xiao F., Zhou Y. Q., et al. (2018). Mechanical properties of PE-HD/phosphogypsum whisker composites [J]. *Engineering Plastics Applications*, 46 (6): 25-29+40.

[7] Li H., Liu X. Y., Fu H., et al. (2019). Application of three silane coupling agents modified phosphogypsum in HDPE [J]. *Plastics Industry*, 47 (6): 137-140+144.

[8] Sun M., Sun Q., Zhang J., et al. (2022). Surface modification of phosphogypsum and application in polyolefin composites [J]. *Environmental Science and Pollution Research*, 29 (44): 66, 177-66, 190.

[9] Zhang Daohai, Pei Meng, Xie Jinhui, et al. (2023). Effect of thermal oxygen aging time on the properties of HDPE/$Na_2SO_4$ composites [J]. *Journal of Guizhou Normal University (Natural Sciences)*, 41 (2): 1-14, 128.

[10] Li Q., Liu H., Nie C., et al. (2022). PMMA-Grafted Calcium Sulfate Whiskers for Applications as Fillers in PVC. *Polymers*, 14 (19): 4, 199.

[11] Lei Yang, Cheng Shuya, Wu Liangxi, et al. (2022). Effect of phosphogypsum and red mud on the mechanical properties of HDPE composites [J]. *Plastics Industry*, 50 (9): 72-76+191.

Journal of Physics: Conference Series **2720** (2024) 012021 doi:10.1088/1742-6596/2720/1/012021

# Synthesis of Nano-aluminum Nitride powder at low temperature by Sol-gel method

**Shuwei Zhang [1,2,3], Weiwei Men[1, 2], Xiao Wu[1, 2], Siyu Zhu[1, 2], and Hui Xu[1, 2]**

[1] AVIC Research Institute for Special Structures of Aeronautical Composite, Jinan 250023, China
[2] Aviation Key Lab of Science and Technology on High-performance Electromagnetic Windows, Jinan 250023, China

[3]Corresponding author's e-mail: 1129393033@qq.com

**Abstract**: In this paper, molecular-level mixed aluminum nitride has been prepared from aluminum sec-butyl alcohol, glucose, and urea by sol-gel processing. Then, the nano-aluminum nitride powder with uniform particle size distribution was obtained by carbothermal reduction nitridation reaction. The effects of different feed ratios, reaction temperatures, and holding times on the phase of AlN powder were systematically studied. The prepared nano-aluminum nitride powder has a uniform particle size and a near-spherical shape, and its grain size is approximately 20~30 nm. The preparation temperature in this method was significantly reduced from the same preparation method of aluminum nitride powder.

## 1. Introduction

Aluminum nitride (AlN) ceramic materials have high thermal conductivity and are matched with the thermal expansion coefficient of silicon to make substrates and packaging materials for large-scale integrated circuits. Transparent AlN ceramics have good transmittance and high-temperature thermal stability for visible light and infrared light. At the same time, these ceramics can maintain high hardness and mechanical strength before decomposition temperature (about 2, 300°C). Therefore, AlN ceramic materials also have high application value in transparent protective armors and high temperature-resistant windows [1-2].

In general, the preparation of high-purity dense AlN ceramics is greatly affected by the purity, particle size, and particle uniformity of AlN ceramic powders. Therefore, the relatively mature preparation methods of AlN powder mainly include the direct nitriding method of aluminum powder, the thermal nitridation reduction method of alumina carbon, and the chemical vapor deposition method. Among them, the aluminum powder prepared by the direct nitriding method has a relatively high oxygen content and a large particle size, resulting in difficulty in preparing high-density AlN ceramics. In addition, high-purity AlN powder can be produced by the chemical vapor deposition method, but it is difficult to achieve mass production due to the constraints of its process conditions. In comparison, the thermal nitriding reduction method of carbon is a cost-effective way even though the reaction temperature is generally above 1, 500°C. We aim to reduce the reaction temperature and prepare uniform nano-sized AlN powder with good moldability and high sintering activity. In this paper, the precursor of AlN was prepared by sol-gel method using aluminum sec-butoxide and glucose as raw materials, and then the precursor was prepared by carbothermal reduction at a temperature lower than 1, 500°C [3-8].

## 2. Experimental procedure

First, $Al_2O_3$ sol (ethanol solution of sec-butanol aluminum) with 59 mL and 0.04 mol, glucose ($C_6H_{12}O_6$), and urea were fully mixed evenly according to the ratio of the amount of the substance as [Al]/[C]/[U] = 4:1:4, 4:2:4, 4:2.5:4, and 4:3:4 at 60°C. Further, the mixed gel was obtained, which was continuously dried to obtain a precursor sample of pale yellow AlN. Then, the precursor was heat-treated under a nitrogen atmosphere, which reached a certain temperature at a heating rate of 8°C/ min and kept for a certain period. After that, it cooled with the furnace to obtain AlN powder. The obtained powder samples were subjected to carbon removal in an air atmosphere at 700°C for 1 h and then characterized by XRD and SEM.

## 3. Experimental results and discussion

### *3.1. Reactions and processes in the experiment*

In the first step, the aluminum sec-butoxide solution reacted with water to produce Al $(OH)_3$ gel [4]. The Al $(OH)_3$ gel encapsulates the urea molecules and the glucose molecules to achieve molecular level mixing. Then, the Al $(OH)_3$ gel was heat-treated at a certain temperature to prepare γ-$Al_2O_3$. The processes are shown in Equation (1).

$$2Al(C_4H_9O_3) \xrightarrow{\text{gel}} 2Al(OH)_3 \xrightarrow{\text{heated}} \gamma - Al_2O_3 \tag{1}$$

The reaction of Equation (2) occurred in the second step. Glucose was pyrolyzed to carbon black under anaerobic conditions, allowing carbon black and $Al_2O_3$ to achieve molecular level horizontal mixing. During this process, the carbon atoms in glucose could not be completely cleaved into carbon black, and too much glucose may cause waste of raw materials and difficulty in removing carbon in subsequent experiments. Therefore, a suitable amount of glucose needs to be explored.

$$C_6H_{12}O_6 \xrightarrow{heated} 6C \tag{2}$$

In the third step, γ-$Al_2O_3$ was subjected to a thermal decomposition reaction of carbon in a nitrogen atmosphere, and γ-$Al_2O_3$ was reduced to AlN by carbon black in this process, as shown in Equation (3).

$$Al_2O_3 + 3C + N_2 \xrightarrow{heated} 2AlN + 3CO \tag{3}$$

The fourth step is the carbon removal of the AlN sample in the air atmosphere [9-11].

$$C + O_2 \xrightarrow{heated} CO_2 \tag{4}$$

### *3.2 Effects of carbon content on phase composition of products*

Figure 1 represents the XRD patterns of the powder samples prepared by four precursors with different [Al]/[C] ratios which thermally reacted at 1, 350°C for 2 h. It can be seen that when [Al]/[C] = 4:1, the phase composition of the product was γ-$Al_2O_3$ and AlN. As the amount of glucose increased, the diffraction peak of γ-$Al_2O_3$ gradually decreased, and a single-phase AlN powder was obtained when [Al]/[C] = 4:2.5. By analyzing the line of [Al]/[C] = 4:1 in Figure 1, it can be known that the AlN powder is directly reacted by γ-$Al_2O_3$ near its phase transition temperature point without the phase transition of γ-$Al_2O_3$→α-$Al_2O_3$. The lattice loosens and internal defects significantly increased due to the increase of the particle activity near the phase transition temperature, so γ-$Al_2O_3$ has higher reactivity than α-$Al_2O_3$, which greatly reduced the preparation reaction temperature of AlN powder.

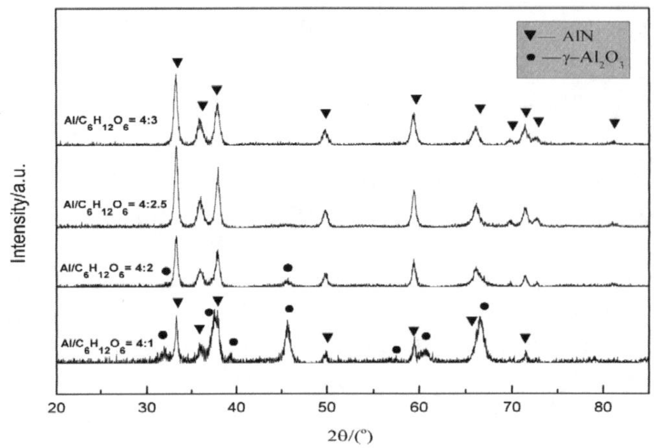


**Figure 1**. XRD patterns of the samples with different [Al]/[C] ratio precursors thermally reacted at 1, 350°C for 2 h.

*3.3 Effects of thermal reaction temperature on phase composition and micro morphology of products*

Figure 2 shows the XRD patterns of the samples obtained from the precursors with [Al]/[C]=4:2.5 which thermally reacted at different temperatures. It can be seen from Figure 2 that the products consist of two phases of γ-$Al_2O_3$ and AlN when the precursors were reacted at 1, 250°C for 2 h. This indicates that γ-$Al_2O_3$ had undergone a carbothermal reduction reaction at 1, 250°C, which further verified that the reaction occurred near the γ-$Al_2O_3$ phase transition temperature (the phase transition temperature of γ-$Al_2O_3$→α-$Al_2O_3$ is about 1, 200°C). The diffraction peaks of the sample at 1, 250°C have weak intensity and wide peak shape, indicating that the sample prepared under this reaction conditions has poor crystallinity and small grain size. Single-phase AlN with a six-direction nanocrystal was obtained when the reaction temperature reached 1, 350°C, and its unit cell parameters are a=b=3.099 and c=4.997 [12-13]. As the reaction temperature increased, the diffraction intensity of the AlN sample was continuously enhanced and the peak shape was narrowed, indicating that the grains of AlN were continuously growing.

Figure 3 indicates that the AlN powder has spherical particles of uniform size when the reaction temperature was 1, 350°C and 1, 450°C. Among them, the particle size of AlN powder is about 20-30 nm at 1, 350°C. The sample still has a uniform particle size when the temperature reaches 1, 500°C, but the morphology has been significantly deformed to increase the particle size to 100 nm.

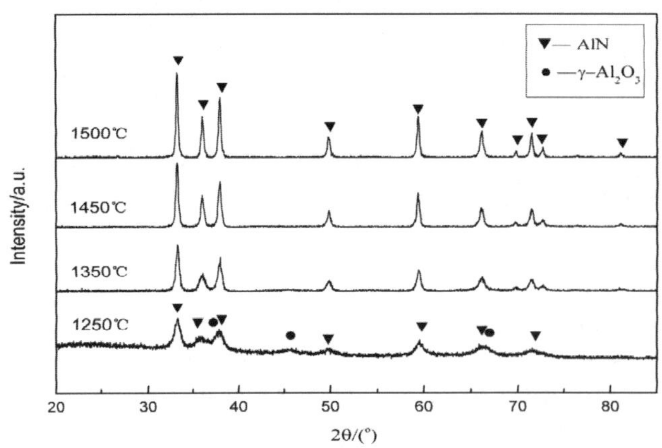


**Figure 2**. XRD patterns of the precursors after thermal reaction at different temperatures for 2 h.

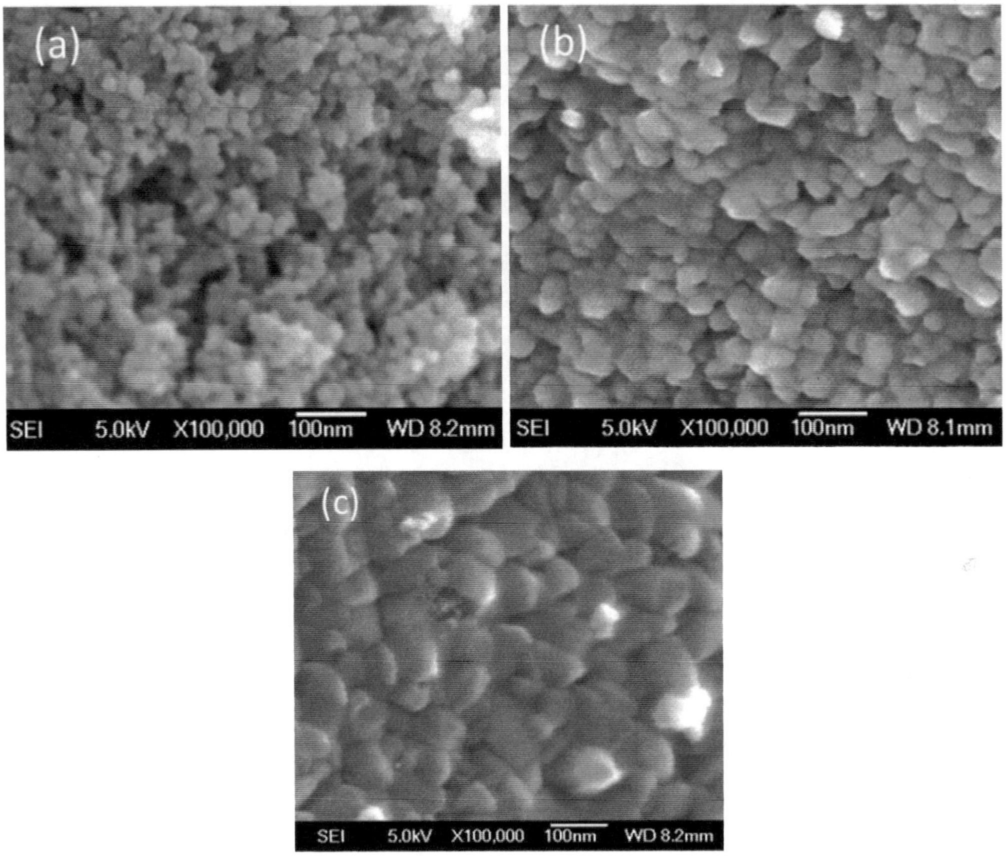


**Figure 3**. SEM micrographs of precursors after thermal reaction at 1, 350°C (a), 1, 450°C (b), and 1, 500°C (c) for 2 h.

*3.4 Effects of thermal reaction time on phase composition of products*

Figure 4 represents the XRD patterns of the samples with [Al]/[C]=4:2.5 reacted at 1, 350°C for 0.5-2 h in a nitrogen atmosphere. Figure 4 reveals that a bit $\gamma$-$Al_2O_3$ did not react completely at 1, 350°C for 0.5 h, but pure phase AlN was obtained as the reaction time extended to 1 h. When the reaction time was 1 h, the broad spectral peaks of the XRD line indicated that the AlN powder had poor crystallinity [14].

According to the Sherrer Equation (where K is a constant, taken as K=0.89; B is the half-width of the diffraction peak), the XRD particle size analysis of the powder samples with different holding times is shown in Table 1. It can be seen from Table 1 that all four powder samples had nanoscale grain sizes, and the grain size was increasing as the nitriding heat reaction time increased.

**Table 1**. The grain size of samples under different process conditions.

| Sample number | Experimental parameters | Grain size (nm) |
|---|---|---|
| 1 | 1, 350°C, 0.5 h, $N_2$ | 14 |
| 2 | 1, 350°C, 1 h, $N_2$ | 18 |
| 3 | 1, 350°C, 1.5 h, $N_2$ | 24 |
| 4 | 1, 350°C, 2 h, $N_2$ | 29 |

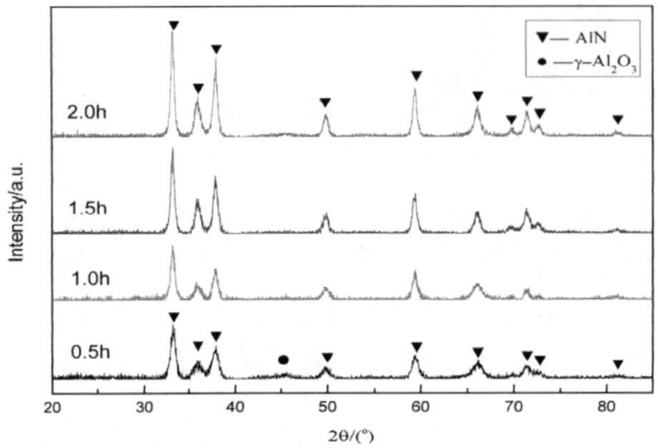


**Figure 4**. XRD patterns of the samples reacted at 1, 350°C for 0.5~2 h.

*3.5 Carbon removal of aluminum nitride powders*

Figure 5 shows the DSC curve of the sample with [Al]/[C]=4:2.5 prepared by reacting at 1, 350°C for 1 h in a nitrogen atmosphere. The DSC curve in Figure 5 indicates that the sample has a significant exothermic peak at about 630°C, which indicates that the residual carbon in the sample was oxidized to carbon dioxide gas. Further, the AlN powder began to be oxidized at 800°C, so the carbon removal temperature of the aluminum nitride powder was selected to be 700°C. Therefore, an off-white AlN powder sample was obtained by removing carbon for 1 h at 700°C in the air atmosphere.

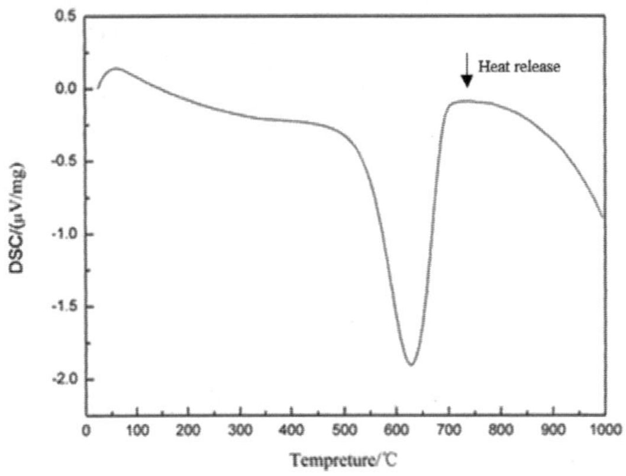


**Figure 5**. DSC curve of the sample with [Al]/[C]=4:2.5 reacted at 1, 350°C for 1 h in nitrogen atmosphere.

## 4. Conclusion

The glucose in the precursor had carbon loss during the cracking process, and the raw material ratio was $n$Al/$n$$C_6H_{12}O_6$=4:2.5. The temperature of this synthesis method was about 200°C lower than that of the mixed nitriding reduction method of alumina and carbon powder. The grain size of the AlN powder increased as the nitriding thermal reaction temperature or the thermal reaction time duration increased. The grain size of the AlN sample was approximately 20 nm with uniform spherical morphology.

## References

[1] Qin M. L., Qu X. H., Lin J. L., Xiao P. A., and Zhu B. J. (2002). Synthesis of Ultrafine Aluminium Nitride Powder by a Low-temperature Carbothermal Reduction Process. J. Inorg. Mater., 17: 1, 054-1, 058.

[2] Tong J. F., Zhou Y., and Chen D. (2002). Synthesis Methods and Mechanism of the AlN Powder. Bull. Chin. Ceram. Soc., 5: 12-16.
[3] Li Z. X., Hao L. C., Zhang J. F., et al. Preparation of aluminum nitride nanomaterials by precursor method. J. China Ceram Soc., 2020, 48 (6): 787-793.
[4] He Q., Qin M. L., Huang M., et al. Synthesis of highly sinterable AlN nanopowders through sol-gel route by reduction-nitridation in ammonia. Ceram. Int., 2019, 45 (12): 14, 568-14, 575.
[5] Chen Guangle, Zhao Chunxia, Liu Jie, et al. The X-Ray Diffraction Analysis of AlN Powder by Precursor Preparation and Nitrified Carbothermal Reduction Process [J]. Vacuum Electronics, 2009, 2: 45-48.
[6] Yang Qinghua, Wang Huanping, Zhang Wenjuan, et al. Effect of Ca-compound on the low-temperature synthesis of AlN ultrafine powders by carbothermal reduction nitride route from sol-gel precursor [J]. Ind. Eng. Chem., 2012, 6, (63): 1, 907-1, 912.
[7] Li L., Ni G. H., Zhao Y. J., et al. Synthesis of nano-AlN powders from Al wire by arc plasma at atmospheric pressure. Ceram Int., 2018, 44 (17): 21, 810-21, 815.
[8] Sun S. Y., Ge Y. Y., Wang Q., et al. Combustion synthesis of fine aluminum nitride powders under micropositive N2 pressure with water additive. J Am Ceram Soc., 2019, 102 (3): 914-918.
[9] Xie X., Sui Y., Huang X. Y., et al. Synthesis of AlN by direct combustion of Mg-Al alloy powder [J]. Inorg. Mater, 2019, 34 (4): 439-443.
[10] Lei M., Yang H., Li P. G., et al. Synthesis and characterization of straight and stacked-sheet AlN nanowires with high purity [J]. Alloys Compd., 2008, 459: 338-342.
[11] Cheng Y. L., Huang X., Xi X., et al. The effect of the urea content on the properties of nano-size AlN powders synthesized by a wet chemical method. Ceram Int., 2018, 44 (5): 5, 774-5, 779.
[12] Pavel A. N., Artem E. K., Nikolay E. O., et al. (2019). Plasma Chemical Synthesis of Aluminum Nitride Nanopowder, Key Engineering Materials, 822, 628-633.
[13] Fattahi M., Vaferi K., Vajdi M., et al. (2020). Aluminum Nitride as an Alternative Ceramic for Fabrication of Microchannel Heat Exchangers: A Numerical Study. Ceramics International, 46, 11, 647-11, 657.
[14] Huang D., Liu Z., Jonathan H., et al. (2019). High Thermal Conductive AlN Substrate for Heat Dissipation in High-Power LEDs. Ceramics International, 45, 1, 412-1, 415.

# Structural safety of compressed air energy storage sealing materials

**Jing Liu [1, 2], Shengjie Di [1], Xi Lu [1] and Yang Ji [1]**

[1] Northwest Engineering Corporation Limited, Power China, Xi'an, Shaanxi, 710065, China

[2]Corresponding author's e-mail: liujing@nwh.cn

**Abstract:** The structural safety of sealing materials is one of the important technical problems of compressed air energy storage. FLAC3D is used to analyze and compare various combination schemes using steel lining and flexible material as sealing materials and evaluate the safety of each scheme, which provides a reference for the design of high-pressure gas storage. The contact surface between the concrete and the steel lining is set up to enhance the relative sliding between them, which is conducive to the coordination of steel plate deformation and the reduction of steel plate stress. The stress of steel plate can be significantly reduced by placing some flexible materials in steel plate and lining concrete, and the lower the modulus of flexible materials is, the better the stress of steel plate structure is.

## 1. Introduction

Energy storage is a key supporting technology to achieve the dual-carbon goal and the energy revolution, and compressed air energy storage is one of the large-scale power energy storage technologies with the most development potential [1~2]. Compressed air energy storage power generation technology is the use of equipment to compress air into high-pressure and high-density gas storage, until the need for power generation to release the high-pressure gas, and then through the equipment to complete the conversion of gas to electricity. Compressed air energy storage has high safety, and has the advantages of large energy storage scale, long discharge time, and long service life, and air can be recycled [3~4]. Relevant universities, research institutes, and enterprises at home and abroad attach great importance to the research, development, and application of compressed air energy storage technology.

So far, the structure which is composed of the surrounding rock, concrete lining, and seal layer proposed by Salter et al. [5] and Zhuang et al. [6] is the most used way to store high-pressure air. Among them, the surrounding rock bears the pressure of the cavern, the lining transmits the pressure, and the sealing layer seals the gas stored in the cavern. How to ensure the structural safety of sealing materials is an important technical problems of compressed air energy storage [7]. There are many reports on the mechanical stability of underground gas storage, but few studies on the structure of the sealing layer [8~11]. Therefore, based on an underground gas storage project, this paper establishes a numerical model to analyze and study the sealing structure of the gas storage under stable high-pressure action, evaluates the tightness and safety of the gas storage, and provides a reference for the design of the high-pressure gas storage.

## 2. Scheme and model

### *2.1. Scheme*

Relying on the site selection of the project, the buried depth is 260 m, the lining thickness is 50 cm, the inner diameter is 10 m, and the operating pressure is 10 MPa. Because of its high strength, reliable performance, and easy processing, steel is one of the preferred materials [12] for compressed air. Relying on the project, steel plates are used as sealing materials, and 15 drums are evenly arranged on the steel plates to enhance the structural safety of the steel plates. Given the possible sealing scheme, its structural safety during operation is calculated and analyzed, and a variety of combination schemes using steel lining and flexible materials as sealing materials are proposed. According to the different sealing structures, sealing materials, and combination types, the following schemes are proposed.

Scheme 1: Rock + concrete + 6 mm thick steel plate, fixed design between steel plate and concrete.

Scheme 2: Rock + concrete + 6 mm thick steel plate, sliding design between steel plate and concrete.

Scheme 3: Rock + concrete + 1 cm thick flexible material 1+6 mm thick steel plate, fixed design between steel plate and concrete.

Scheme 4: Rock + concrete + 1 cm thick flexible material 1+6 mm thick steel plate, sliding design between steel plate and concrete.

Scheme 5: Rock + concrete + 1 cm thick flexible material 2+6 mm thick steel plate, fixed design between steel plate and concrete.

### *2.2. Model*

In this paper, FLAC3D is used to establish the overall model, and the calculation model is shown in Figure 1. The calculation parameters are shown in Table 1, in which the linear elastic model is used for the steel liner sealing layer and the ideal elastic-plastic constitutive model based on the Mohr-Coulomb yield criterion is used for the rest materials.

Coordinate system: the center point of the hole is the center of XYZ coordinate, XY is horizontal, the axis of the hole is horizontal X direction, along the height is vertical Z direction, and vertical upward is positive. It conforms to the Cartesian right-hand coordinate system rule.

**Table 1.** Calculation parameter.

| Materials | Density /kg·m$^{-3}$ | Modulus /MPa | Poisson's ratio | Cohesive /MPa | Angle of internal friction/° | Tensile strength /MPa |
|---|---|---|---|---|---|---|
| Rock | 2300 | 5000 | 0.29 | 1.65 | 33.0 | 1.25 |
| Concrete | 2500 | 32500 | 0.167 | 3.08 | 58 | 1.71 |
| Steel plate | 7850 | 210000 | 0.3 | / | / | 290 |
| Flexible material 1 | 220 | 100 | 0.45 | 1.5 | 20 | 1 |
| Flexible material 2 | 220 | 30 | 0.48 | 1 | 20 | 1 |

(a) Steel plate

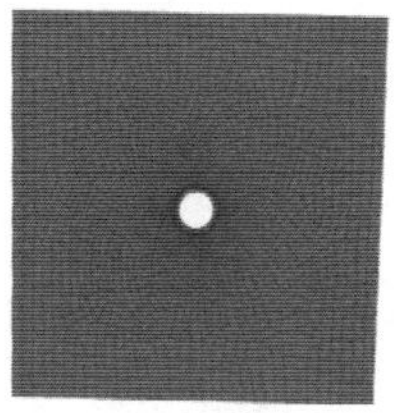
(b) Global model

**Figure 1.** Calculation model.

## 3. Research results

### *3.1. Stress analysis of the sealing layer*

The main stress statistics of steel plates under each scheme are shown in Table 2.

① The principal tensile stress of the steel plate under calculation Scheme 2 (194 MPa) is much less than that under calculation Scheme 1 (723 MPa), and the principal tensile stress under calculation Scheme 4 (57 MPa) is much less than that under calculation Scheme 3 (357 MPa), indicating that the sliding between steel plate and concrete is allowed, and the deformation of steel plate can be coordinated to reduce the stress.

Although steel plate and concrete belong to different materials, the contact surface itself has a certain slippability, but because the internal pressure of compressed air is often relatively large, during the operation period, the steel plate and the concrete surface under the action of internal pressure will be in close contact. Therefore, the design needs special treatment to increase the slippability of the contact surface.

② The main tensile stress of steel plate under calculation Scheme 3 (357 MPa) is much less than that under calculation Scheme 1 (723 MPa), and the main tensile stress under calculation Scheme 4 (57 MPa) is much less than that under calculation Scheme 2 (194 MPa), indicating that certain flexible materials arranged in steel plate and lining concrete can significantly reduce the stress of steel plate.

③ The main tensile stress of steel plate under calculation Scheme 5 (280 MPa) is smaller than that under calculation Scheme 3 (357 MPa), indicating that reducing the modulus of flexible materials can reduce the stress of steel plate.

For support engineering, the steel plate design allows a stress of 290 MPa, and it is recommended to arrange a sealed structure in options 2 and 5. The stress nephogram of the sealing layer of Scheme 2 and Scheme 5 is shown in Figure 2.

**Table 2.** Stress of the sealing layer in each scheme (unit: MPa).

| Calculation scheme | | 1 | 2 | 3 | 4 | 5 |
|---|---|---|---|---|---|---|
| Maximum Principal | MAX | 723.0 | 194.2 | 357.1 | 57.1 | 280.8 |
| Stress (S1) | MIN | 265.9 | 10.2 | 223.7 | 3.81 | 177.2 |
| Minimum Principal | MAX | -38.4 | -12.1 | -4.99 | -6.65 | -8.1 |
| Stress (S3) | MIN | -0.04 | 0 | 0 | 0 | 0 |

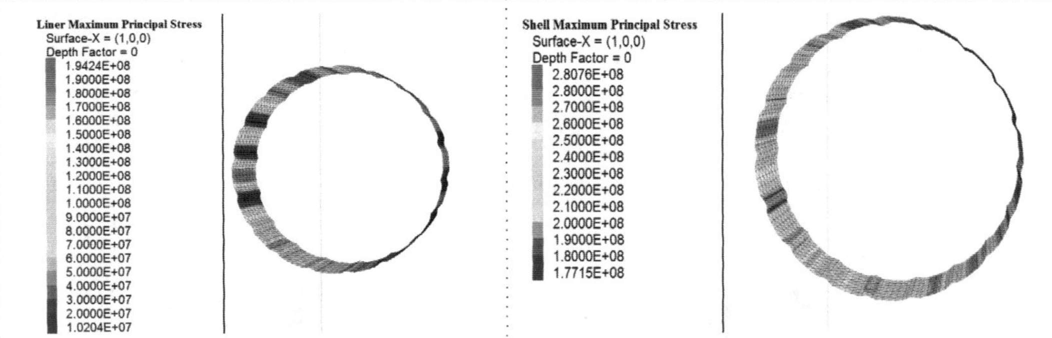


(a) Scheme 2 (b) Scheme 5

**Figure 2.** The maximum principal stress of the sealing laye.

### *3.2. Stress analysis of flexible materials*

In Schemes 3, 4, and 5, some flexible materials are arranged between steel plate and concrete. According to the analysis in Section 3.1, flexible materials can obviously adjust the stress state of steel plates. The main stress results of flexible structures under each scheme are shown in Table 3. Flexible materials are mainly manifested as compressive stress, basically no tensile stress, and no shear-tensile damage. The maximum compressive stress of recommended solution 5 is -14.96 MPa, which meets the stress requirements of flexible materials. As shown in Figure 3, the plastic zone does not appear in the flexible material of Scheme 5.

**Table 3.** Stress of flexible materials in each scheme (unit: MPa).

| Calculation scheme | | 3 | 4 | 5 |
|---|---|---|---|---|
| Maximum Principal Stress (S1) | MAX | 0.65 | 0.51 | -1.92 |
| | MIN | -12.7 | -13.49 | -13.67 |
| Minimum Principal Stress (S3) | MAX | -15.78 | -16.64 | -14.96 |
| | MIN | -2.96 | -1.0 | -4.31 |

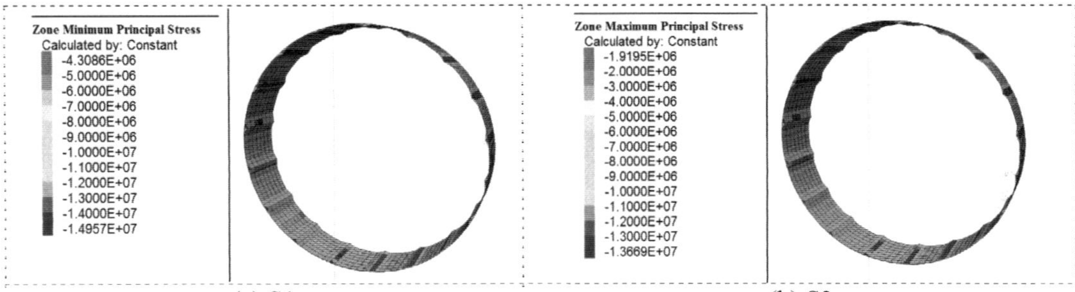


(a) S1 (b) S3

**Figure 3.** Scheme 5 Stress nephogram of flexible material.

*3.3. Displacement analysis*

Under the action of 10 MPa pressure, the displacement is outward expansion. The maximum displacement and vertical displacement of the sealing layer appear at the top and bottom of the free surface. The maximum Y-displacement occurs on the left and right sides, and the Y-deformation is basically symmetrical along the Z-axis. The plate displacement statistics under each scheme are shown in Table 4.

**Table 4.** Displacement of the sealing layer in each scheme (unit: cm).

| Calculation scheme | | | 1 | 2 | 3 | 4 | 5 |
|---|---|---|---|---|---|---|---|
| Displacement | Y | Left | 1.58 | 1.62 | 1.75 | 1.99 | 2.06 |
| | | Right | -1.58 | -1.61 | -1.76 | -1.99 | -2.06 |
| | Z | Top | 1.92 | 2.02 | 2.05 | 2.16 | 2.31 |
| | | Bottom | -1.44 | -1.50 | -1.57 | -1.78 | -1.93 |
| | Magnitude | | 1.93 | 2.03 | 2.09 | 2.18 | 2.31 |

① The transverse displacement of steel plate under calculation Scheme 2 is larger than that under calculation Scheme 1, and the transverse displacement under calculation Scheme 4 is larger than that under calculation Scheme 3, indicating that the tighter the connection surface between sealing layer and lining concrete is, the smaller the displacement of sealing layer is, and the more unfavorable the self-coordination of steel plate is.

② The transverse displacement of steel plate under calculation Scheme 3 is larger than that under calculation Scheme 1, and the transverse displacement under calculation Scheme 4 is larger than that under calculation Scheme 2, indicating that the arrangement of certain flexible materials between sealing layer and lining concrete increases the circumferential displacement of sealing layer, which is conducive to the coordination of steel plate deformation and the reduction of steel plate stress.

③ The transverse displacement of the steel plate under calculation Scheme 5 is larger than that under calculation Scheme 3, indicating that the softer the steel plate is, the more flexible the material arranged in the lining concrete is, and the larger the displacement of the steel plate is, and the more favorable the deformation coordination of steel plate is.

Figure 4 the anisotropic displacement cloud image of Scheme 5.

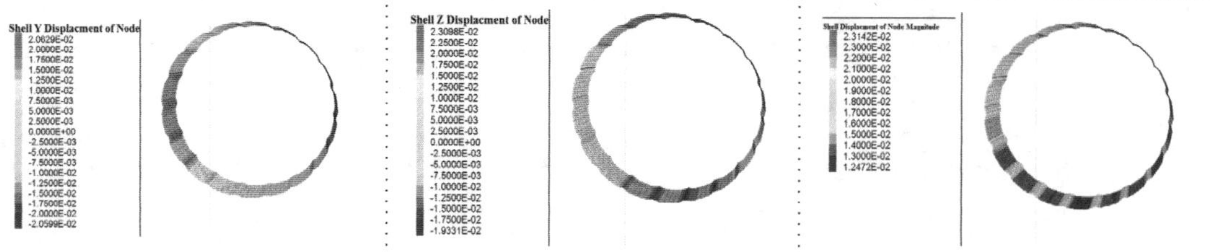


(a) Y-direction (b) Magnitude (c) Z-direction

**Figure 4.** Displacement cloud image of steel plate in Scheme 5.

## 4. Conclusions

This paper analyzes and compares various combination schemes using steel lining and flexible materials as sealing materials. According to the conventional design, steel lining is only placed on the lining concrete surface as sealing material, which is not conducive to the structural stress of the steel plate. If a certain contact surface is set between the lining concrete and the steel plate, the relative sliding between the two is enhanced, which is conducive to the deformation coordination of the sealing layer and the stress reduction. The arrangement of flexible materials in steel plate and lining concrete can also significantly reduce the stress of the sealing layer, and the lower the modulus of flexible materials, the better the stress of the sealing structure. For the design of the compressed air seal layer, the design of the connecting surface between the steel plate and the lining concrete is very important.

## References

[1] Zhang WL,Gu H,Zhang C, et al. (2023). Technical and economic characteristic and development trend of CAES [J]. Science and Technology of energy storage,12(4): 1295-1301.

[2] Ramadan O, Omer S, Jradi M, et al. (2016) Compressed air energy storage analysis for large-scale energy in Suez, Egypt [J]. International Journal of Low-Carbon Technologies, 11: 476-488.

[3] Du JS, Chen J, Jiang DY, et al. (2014) Potential and preliminary feasibility study of CAES in artificial caves [J]. Engineering Science and technology , 2023,55(1):253-264.

[4] Kushnir R, Dayan A, Ullmann A, et al. (2012) Pressure and Temperature variations of compressed air energy storage caverns [J]. International Journal of Heat and Mass Transfer. 55(21): 5616

[5] Salter MG, Macfarlane IM, Willett DC, et al. (1985) The design of underground Compressed air energy storage system [J]. Design of Underground Excavation: 37-44.

[6] Zhuang XY, Huang RQ, Liang C, et al. (2014) A coupled thermo-hydro-mechanical model of jointed hard rock for CASE [J]. Mathematical Problems in Engineering, 2014: 1-11

[7] Zhou Y, XIA CC, ZHAO HB et al. (2017) Stimation method of air leakage rate and mechanical response of the rock in pressurized gas storage lining cavity [J]. Journal of Rock Mechanics and Engineering,2017,36(2):297-309.

[8] Luo X, Wang J, Dooner M, et al. (2015) Development status and application potential of electric energy storage technology [J]. Applied Energy, 2015,137:501-509.

[9] Jiang ZM, Guo J, Tang D. (2021) Thermodynamic model of compressed wet air in compressed air energy storage underground gas storage [J]. Science and Technology of energy storage, 2021,10(2):638-646.

[10] Jiang ZM, Qin SZ, Tang D. (2020) Numerical study on cumulative damage characteristics of surounding rock of compresed gas storage underground gas storage [J]. The geotechnical engineering,42(2):230-238.

[11] Han ZH, GUO SH. (2020) Study on the influence of working fluid and operation mode on compressed air energy storage system [J]. Journal of solar energy. 41(11):11-16.

[12] Johansson, J. (2003) High pressure storage of gas in lined rock caverns [M.S. Thesis] [D]. Sweden: Royal Institute of Technology.

# Research and application of key technology of energy saving and capacity increasing for overhead conductors with carbon fiber composite core reinforced

**Wei Zeng[1,2] and Peng Dang[1]**

[1]Shanghai Electric Cable Research Institute Co., Ltd. State Key Laboratory of Special Cable Technology Shanghai, 200093 China

[2]Corresponding author's e-mail:zengwei@secri.com

**Abstract:** Aluminum conductor composite core (ACCC) has many excellent comprehensive properties and is gradually popularized and applied in overhead transmission lines. As an energy-saving and environmentally friendly capacitance-increasing conductor, ACCC gives full play to the advantages of carbon fiber materials. This article analyzes the test results of ACCC and compares them with traditional aluminum conductor steel reinforced (ACSR) in terms of mechanical and electrical properties and proposes the development trend of ACCC.

## 1. Introduction

With the rapid growth of my country's economy, the electricity demand is rising rapidly. However, the construction of transmission projects is restricted by factors such as long construction periods, difficult land acquisition, and environmental protection. Especially in areas with tight land use, it is extremely difficult to build new transshipments for transmission network construction, and their manufacturing technology and industrial development are closely related to the country's transmission engineering construction and national economic development. On the premise of not building new lines or changing the existing network structure and ensuring the safe and reliable operation of the power grid, improving the conductor transmission capacity and easing the tight power supply through technological upgrading and transformation have become hot and difficult issues in the current transmission network construction industry.

To increase the transmission capacity of existing lines, the most economical and applicable way is to replace the conductors with capacity-increasing conductors without changing the existing conditions as much as possible. This can increase the transmission capacity of the original lines, which depends on the capacity of the capacity-increasing conductors: type and new set long-term operating temperature. For the reconstruction of lines, it is of considerable economic benefit to use the existing corridor conditions, only replace the conductors, and meet the design sag requirements to increase the transmission capacity. The ACCC emerged under these factors. As an energy-saving and environmentally friendly capacitance-increasing conductor[1], it has significant characteristics such as lightweight, high strength, large capacity, low line loss, low sag, high temperature, and corrosion resistance [2]. This article proposes the development trend of ACCC, analyzes its test results, and compares it with ACSR in terms of mechanical and electrical properties.

## 2. Characteristics of ACCC

Compared with traditional wires, ACCC has the advantages of high strength, low line loss, small sag, high-temperature resistance, large carrying capacity, lightweight, corrosion resistance, etc., realizing energy saving, environmental protection, and power transmission safety. Its specific characteristics are as follows.

*2.1Reinforced core with high strength*

The tensile strength of general steel wire is 1240 MPa, high-strength steel wire is 1410 MPa and extra-high-strength steel wire is 1770 MPa. The tensile strength of the carbon fiber composite core can reach 2600 MPa, which is 2.1 times, 1.8 times, and 1.5 times that of the previous steel wire[3]. ACCC and ACSR is shown in Figure 1.

**Figure 1.** ACCC and ACSR.

*2.2 Low line loss and high conductivity*

Since overhead conductors with ACCC do not have the magnetic loss heat effect caused by steel wire materials and have lower operating temperatures under the same load transmission conditions, they can reduce power transmission losses by about 6% compared with ACSR[4]. In addition, the outer surface formed by the trapezoidal cross-section of overhead conductors with ACCC reinforced is much smoother than the surface of the ACSR, thereby increasing the surface roughness coefficient of the wire, which is beneficial to increasing the corona starting voltage of the wire, reducing corona losses and reducing corona noise and radio interference levels. [5]

*2.3 Thermal expansion and elongation are small*

The linear expansion coefficient of the carbon fiber composite core is low ($1.6\times10^{-6}/°C$), which is about 1/8 of the ordinary steel core [6]. Therefore, ACCC has significantly lower thermal expansion characteristics compared with conventional wires. [7]

*2.4 Lightweight*

The density of carbon fiber composite core material is small (1.9 g/cm$^3$), about 1/4 of that of ordinary steel core (7.78 g/cm$^3$) [8]. The mass per unit length of ACCC is about 60% to 80% of that of conventional conductors.

*2.5 Its device characteristics are good when operating at the maximum allowable operating temperature of the conductor for a long time*

## 3. Sag characteristics of ACCC

Taking the common ACSR (JL/G1A-400/50) and ACCC (JLRX1/F2B-400/45) commonly used in overhead conductor lines as representatives (comparison of specific conductor parameters is shown in Table 1), according to the comparison with ordinary ACSR based on the principle of equal cross-sections of stranded conductor, a true type test of 200 m span is carried out under the following test conditions:

wind speed: 1.5~3.1 m/s, air temperature: 20°C, and humidity: 77%. Test site and test line segment structure is shown in Figure 2.

**Figure 2.** Test site and test line segment structure.

### *3.1 Sag characteristics of ACSR (JL/G1A-400/50)*

Figure 3 shows the temperature-sag curve and current-sag curve of ACSR.

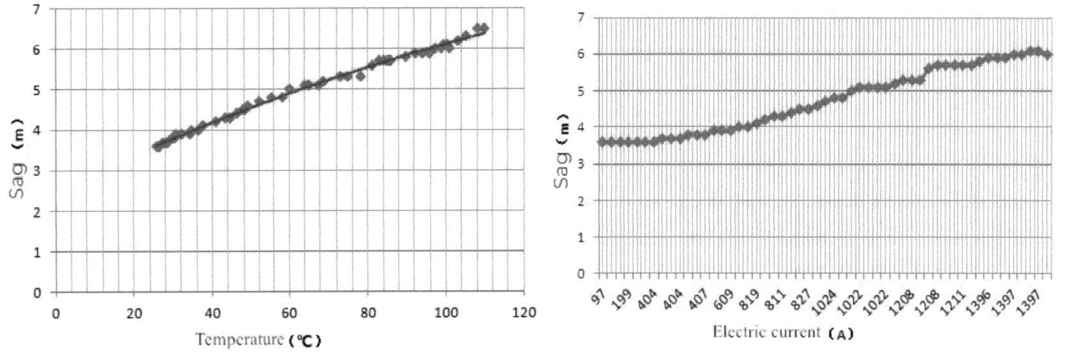


a) Temperature-sag curve b) Current-sag curve

**Figure 3.** The temperature/current-sag curve of ACSR

It can be seen from Figure 3 that the sag of the ACSR increases with the increase in temperature, which is a linear relationship, and there is no temperature migration point.

**Table 1.** Comparison of parameters between ACSR (JL/G1A-400/50) and ACCC (JLRX1/F2B-400/45).

| Conductor Parameter | | Unit | ACSR JL/G1A-400/50 | ACCC JLRX1/F2B-400/45 |
|---|---|---|---|---|
| Conductor structure | Aluminum | Radical / Diameter (Number of floors) | 54/3.07 | 22/2 |
| | Reinforcing core | Radical / Diameter | 7/3.07 | 1/8.0 |
| Cross area | Aluminum | $mm^2$ | 400 | 402.3 |
| | Reinforcing core | $mm^2$ | 51.8 | 50.4 |
| | Total cross-sectional area | $mm^2$ | 451.8 | 452.7 |
| Diameter | | mm | 27.6 | 24.8 |
| Mass per unit length | | kg/km | 1510.5 | |
| Calculate the tensile force | | kN | 122.95 | 143.8 |
| Dc resistance at 20°C | | Ω/km | 0.0724 | 0.0694 |

### *3.2 Sag characteristics of ACCC (JLRX1/F2B-400/45)*

The temperature-sag curve and current-sag curve of ACCC is shown in Figure 4.

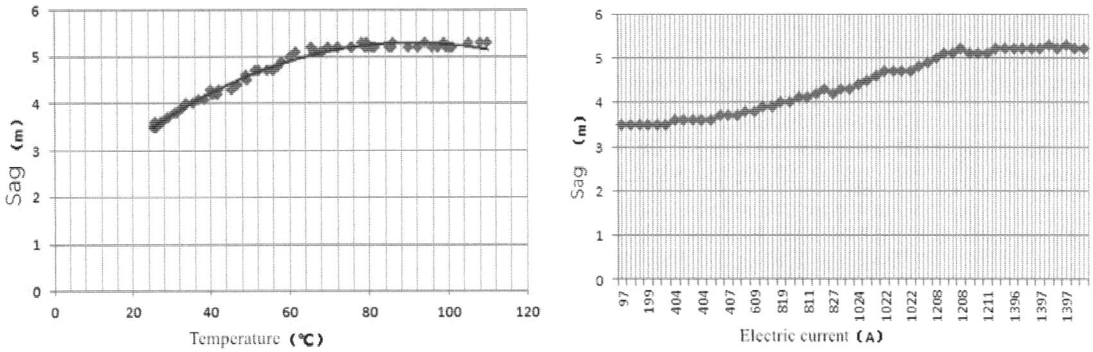


a) Temperature-sag curve b) Current-sag curve

**Figure 4.** The temperature/ current -sag curve of ACCC.

In Figure 4, the sag of the ACCC increases with the increase in temperature, and there is an obvious migration point temperature, which is around 70°C. It can also be seen from the current-sag curve in Figure 4 that as the current increases, the sag also increases, but after reaching a certain current, the sag no longer changes significantly.

### *3.3 Comparison of sag characteristics of ACSR / ACCC*

In Figure 5, the ACSR has no temperature migration point within the test temperature range, while the ACCC has an obvious temperature migration point. ACCC has significantly lower sag characteristics than ACSR. Under the same test conditions, when the temperature rises from 25.6°C to 110°C, the sag of the ACSR increases from 3.6 m to 6.5 m, an increase of 0.8 times, while the sag of the ACCC only increases from 3.5 m to 5.3 m, an increase of 0.5 times; the change in the sag of the ACCC is 62% of that of ACSR. As the operating temperature increases, the sag of the ACSR will continue to increase. The migration point temperature of the ACCC is much lower than that of the ACSR, and above this temperature, the sag increases with the temperature. Significantly lower than ACSR, it can effectively reduce the insulation space in overhead line corridors and improve the safety and reliability of conductor operation.

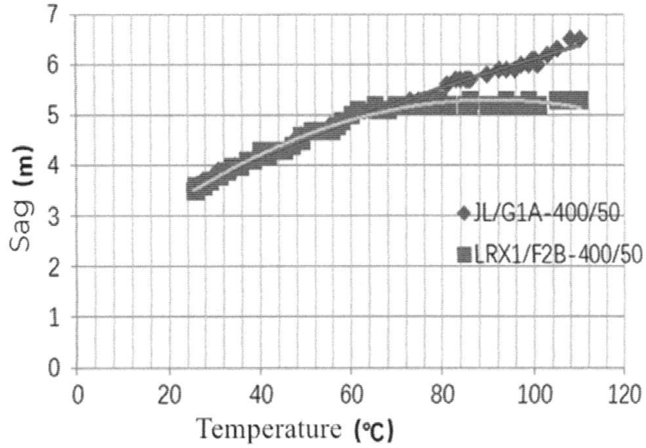


**Figure 5.** Temperature-sag contrast curve of ACSR and ACCC.

## 4. Electrical properties of ACCC

Taking ASCR (JL/G1A-300/40) and ACCC (JLRX1/F2B-300/40) as representatives (see Table 2 for comparison of specific wire parameters), according to the wire parameters in Table 2, commonly used domestic calculation parameters are used to calculate the current carrying capacity at 70°C, 100°C, and 120°C, and the results are shown in Table 3.

**Table 2.** Comparison of parameters between ACSR(JL/G1A-300/40) and ACCC (JLRX1/F2B-300/40).

| Conductor Parameter | | Unit | ACSR JL/G1A-400/50 | ACCC JLRX1/F2B-400/45 |
|---|---|---|---|---|
| Conductor structure | Aluminum | Radical / Diameter (Number of floors) | 24/3.99 | 16/2 |
| | Reinforcing core | Radical / Diameter | 7/2.66 | 1/7.0 |
| Cross area | Aluminum | mm$^2$ | 300 | 300 |
| | Reinforcing core | mm$^2$ | 38.9 | 38.5 |
| | Total cross-sectional area | mm$^2$ | 338.9 | 338.5 |
| Diameter | | mm | 23.9 | 21.5 |
| Mass per unit length | | kg/km | 1132.0 | 891.6 |
| Calculate the tensile force | | kN | 92.36 | 109.6 |
| Dc resistance at 20°C | | Ω/km | 0.0961 | 0.0691 |

The calculation uses the following parameters:

| | |
|---|---|
| Wind speed | 0.5 m/s |
| Sunshine intensity | 1000 W/m$^2$ |
| Conductor surface absorption coefficient | 0.9 |
| Conductor radiation coefficient | 0.9 |
| Ambient temperature | 20-40°C |
| Conductor operating temperature | 70-120°C |

**Table 3.** Comparison of current carrying capacity test data.

| Ambient temperature /°C | Operating temperature /°C | Carrying capacity /A | |
|---|---|---|---|
| | | ACSR JL/G1A-300/40 | ACCC JLRX1/F2B-300/40 |
| 20 | 70 | 750 | 757 |
| | 100 | 968 | 978 |
| | 120 | - | 1091 |
| 25 | 70 | 699 | 705 |
| | 100 | 932 | 943 |
| | 120 | - | 1061 |
| 30 | 70 | 643 | 649 |
| | 100 | 896 | 905 |
| | 120 | - | 1030 |
| 35 | 70 | 582 | 588 |
| | 100 | 857 | 867 |
| | 120 | - | 998 |
| 40 | 70 | 514 | 519 |
| | 100 | 817 | 826 |
| | 120 | | 965 |

In Table 3, the current carrying capacity of ACCC of equal cross-section is equivalent to that of ACSR when the operating temperature is 70°C. However, when the operating temperature of the conductor increases to 120°C, the current carrying capacity of the ACCC is more than 1.45 times that of ACSR. As the ambient temperature increases, at 40°C, the carrying capacity of ACCC is more than 1.88 times that of ACSR.

## 5. Problems with ACCC

ACCC has obvious advantages over traditional conductors in terms of mechanical and electrical properties and has become an ideal new technological product in overhead transmission lines worldwide. At present, ACCC is used both at home and abroad. There are more than 1, 000 ACCCs in operation in the domestic network, and the number of line kilometers is about 20, 000 km. Rod-type composite cores have been used in China for many years and have accumulated rich experience. However, it has also been discovered that this reinforced core has difficult-to-overcome weaknesses, such as poor flexibility, easy to break, difficulty in construction and wiring, and difficulty connecting hardware structures[9]. Many breakage accidents occurred during the application process of rod-type ACCC (Figure 6). Later studies attributed them to damage to the composite core caused by violent construction and fretting damage during operation.

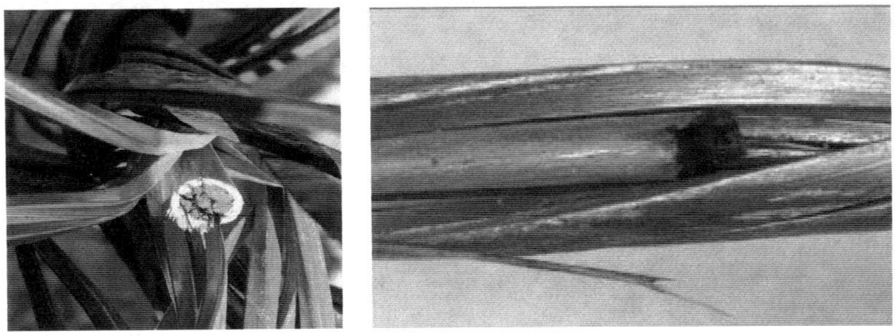

**Figure 6.** Core break of typical accidents.

When the rod-type composite core is impacted by an external force (see Figure 7), the outer glass fiber and epoxy resin composite are shattered and peeled off from the matrix[10]. The cracks in the inner layer of carbon fiber and epoxy resin composite severely damaged the overall mechanical properties of the material, causing the rod-type composite cores to break under a load far lower than the calculated tensile breaking force, causing the conductor to fail.

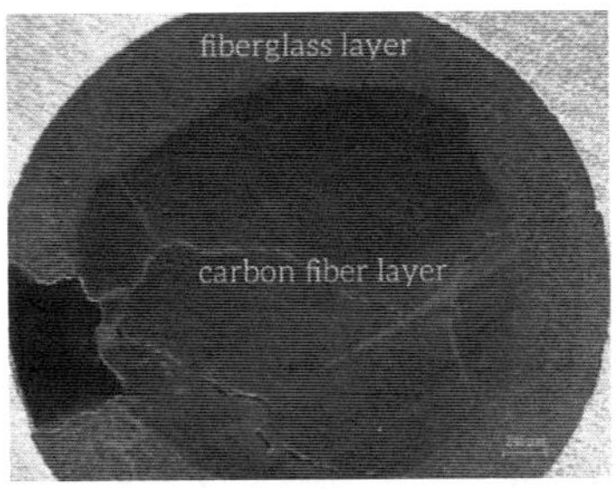


**Figure 7.** Rod-type composite core failure state.

A stranded composite core has been developed in recent years to improve these weaknesses of the rod-type composite core (see Figure 8). Compared with the rod-type structure, it has the advantages of large bending radius, good compression resistance, and simple supporting hardware[11]. At present, overhead electrical strand conductors with reinforced stranded carbon fiber composite cores are used in Hainan, Japan, Southeast Asia, the European Union, and other places and have certain advantages in improving the safety and reliability of the power grid.

Since the overhead electrical strand conductors with stranded carbon fiber composite core reinforced is a new type of conductor that has appeared in recent years, current research on its performance at home and abroad mainly focuses on mechanical properties (high-temperature tensile strength test, tensile holding force test, and stress strain test), material properties (vitrification Transformation temperature, ultraviolet light, acid, alkali, and salt environment), etc. However, there has yet to be in-depth theoretical research on the changes in the mechanical properties of the carbon fiber composite core under different aging environments or multi-factor composite conditions, and the material and structure itself cannot explain it. The change in mechanical properties is because a single artificial accelerated aging test cannot fully simulate the conditions of carbon fiber composite cores under production, transportation, wiring, and operating conditions. If overhead electrical strand conductors with stranded carbon fiber composite core reinforced are to be used in overhead conductors in the future, basic theoretical research and testing must be strengthened, especially their online detection technology in production, online monitoring technology during wiring and operation, and re-establish user's confidence in its use[12].

a) Stranded composite core and stranded composite core conductors

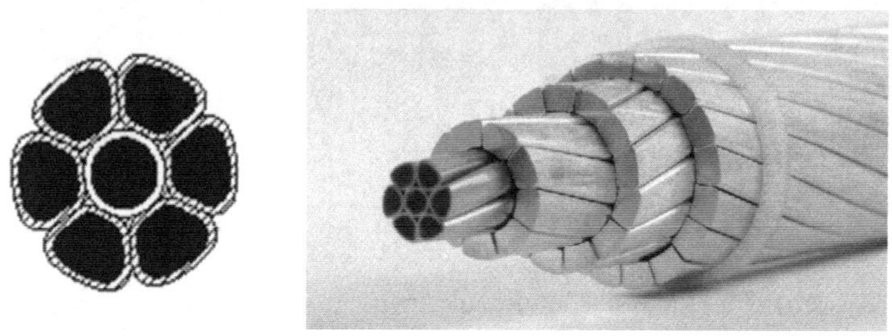

b) Metal-clad stranded composite core and metal-clad stranded composite core conductors

**Figure 8.** Overhead electrical strand conductors with stranded carbon fiber composite core reinforced of different structural types.

## 6. Conclusions

1) ACCC gives full play to the advantages of carbon fiber materials. They have corrosion resistance characteristics, high tensile strength, high elastic modulus, low sag, small thermal expansion coefficient, and lightweight. They are an excellent capacity-increasing conductor.

2) ACCC has good sag characteristics below and above the inflection point temperature. The sag is much smaller than ACCC at the same operating temperature. When the operating temperature exceeds the inflection point temperature, the sag has No changes.

3) ACCC of equal cross-section and ACSR can greatly increase the current carrying capacity under the same temperature conditions. In reconstructing old lines, not only can the capacity be doubled, but the weight of the conductors is also light. The existing line corridors can be used without replacing the poles and towers to increase the transmission capacity of the unit corridor.

4) A stranded composite core was developed based on the shortcomings of the rod-type carbon fiber composite core. However, in-depth basic theoretical research and online detection technology are needed for new stranded carbon fiber composite core conductors, which will help promote the application of carbon fiber composite core conductors and have important practical significance for power grid capacity expansion and reconstruction projects and new large-capacity lines.

## References

[1] MA Minzheng, PAN Chunping", Application research on aluminium conductor multi-strand carbon-fiber core cable in new power transmission lines", Southern Energy Construction, vol. 4, 2017, pp.82-87.

[2] LIANG Xuming, YU Jun, YOU Chuanyong", A survey of aluminium conductor with reinforced core of composite material", Power System Technology", 2006, pp.1-6.
[3] ZHOU Chao, ZHAO Shijie, ZHI Peng, et al.", Stress analysis of ACCC conductor under tension and bending loads", Hubei Electric Power, vol.43, 2019, pp.24-33.
[4] JIANG Wendong, ZHANG Yong", Application of ACCC to capacity increasing projects for lines", East China Electric Power, vol.37, 2009, pp.418-421.
[5] ZHANG Chunlei, HU Ping, HE Fengsheng", Analysis on the structure, texture and properties of carbon fiber composite core for an overhead conductor. Southern Power System Technology", vol.6, 2012, pp.104-107.
[6] WAN Jian, LIU chen, JIANG Ming, YANG Lei, HAO Yujing", Radial pressure test and simulation on carbon fiber composite core for overhead conducto. Southern Power System Technology", vol. 12, 2018, pp.21-26.
[7] ASTM D790. Standard Test Methods for Flexural Properties of Unreinforced and Reinforced Plastics and Electrical Insulating Materials.
[8] ASTM B976 -21 Standard Specification for Fiber Reinforced Aluminum Matrix Composite (AMC) Core Wire for Aluminum Conductors Aluminum Matrix Composite.
[9] Melro AR, Camanho PP, Pires FMA, Pinho ST", Micromechanical analysis of polymer composites reinforced by unidirectional fibres: Part I-Constitutive modelling", Solids Struct. vol.50, 2013, pp.1897-1905.
[10] Melro AR, Camanho PP, Pires FMA, Pinho ST", Micromechanical analysis of polymer composites reinforced by unidirectional fibres: Part II-Micromechanical analyses", Solids Struct, vol.50, 2013, pp.1906-1915.
[11] Bosze EJ, Tsai YI, Barjasteh E, et al.", Long-term durability of the composite core conductor ACCC/TW", In; International conference on overhead lines. 2008, pp.516-525.
[12] B.N. Fedulov, A.A. Safonov, M.M.Kantor, S.V. Lomov", Modelling of thermoplastic polymer failure in fiber reinforced composites", Composite Structures, vol. 163, 2017, pp.293-301.

# Study on sinusoidal lapping micro-structured surface by abrasive stone with staggered abrasive particles

**Qi Liu[1], Xingshan Li[1,2], Yushan Lyu[1], Shenrong Hu[1] and Zhaotong Shang[1]**

[1]School of Mechanical Engineering, Shenyang Ligong University, Shenyang, Liaoning, 110159 China

[2]Corresponding author's e-mail: lxs7241@ sina. com

**Abstract.** In order to manufacture structured surfaces, firstly, the abrasive stone model with staggered abrasive particles is established with mathematical methods. Secondly, the motion of abrasive particles on the abrasive stone is designed. Finally, the surface morphology of the workpiece is simulated. The simulated workpiece morphology under different machining parameters is compared, and the influence of different machining parameters on the workpiece surface morphology is obtained. The results indicate that the microstructured surface can be lapped by using the abrasive stone of staggered abrasive particles. As the amplitude and frequency of the abrasive stone increase, the number of protrusions increases, and the area of individual protrusions decreases.

## 1. Introduction

In recent years, biomimetics has discovered that the characteristic structures on biological surfaces can achieve specific functions. For example, the micron-scale convex hull on the surface of lotus leaves can increase the contact angle with water [1-2], causing water droplets on the lotus leaf surface to quickly roll off [3]. The grooves on the surface of the shell can reduce resistance to water [4]. The structure of the abdomen of a pangolin causes the soil attached to its surface to quickly detach [5]. Due to the superior performance of structured surfaces, people machined it with chemical etching, electrochemical etching, and other [6-7] methods, and conducted research on structured surfaces. The drag reduction effect of three different shapes of grooves was compared by Li et al. [8]. It was found that the drag reduction effect of V-shaped grooves is better than that of U-shaped and L-shaped grooves. Yousfi et al. [9] machined different surface morphologies on the surface of the engine cylinder with honing, making it more effective in reducing drag. The effects of production time and contact pressure between lapping stones and workpieces on the surface morphology of workpieces during superfinishing were studied by Chang et al. [10].

The above research provides many ways to manufacture micro-structured surfaces, but there are few studies on using lapping methods for structured surfaces. This article introduces the design of abrasive stones with staggered grain patterns, the workpiece morphology is simulated, and the impact of workpiece surface morphology with different machining parameters is analyzed.

## 2. Arrangement of abrasive particles

Due to the fact that the machined part is the outer circular surface of the workpiece, the abrasive stone also needs to have a curved surface corresponding to it, so that all the abrasive particles on the abrasive stone participate in lapping, ensuring machining efficiency. The abrasive stone with a staggered arrangement of abrasive particles is shown in Figure 1.

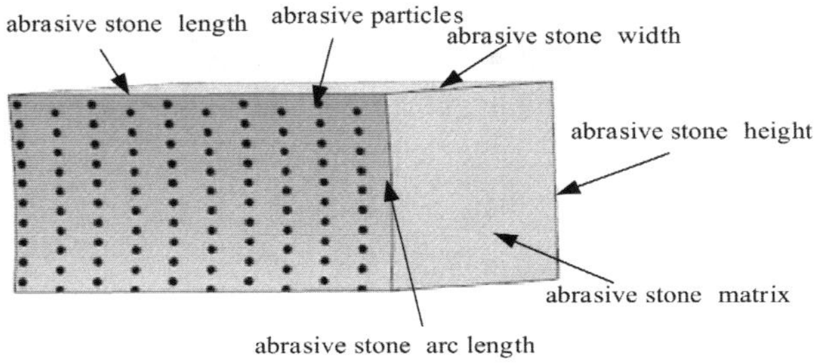


**Figure 1.** Abrasive stone with a staggered arrangement of abrasive particles.

When machining the outer circular surface with abrasive stone, the abrasive particles move around the axis of the workpiece. Considering that not all abrasive grains are involved in lapping during abrasive lapping, during the simulation, the abrasive particles are considered hemispherical and evenly divided into equal parts for lapping. In the workpiece coordinate system, the coordinates of the abrasive particle position points for each equal portion obtained from this are as follows:

$$\begin{cases} \theta = j \times \Delta\theta \\ R_A = \frac{D_d}{2} + r_k[1 - \sin(k \times \psi)] \\ Z_{Aij} = a \times \frac{i}{I} - r_k \cos(k \times \psi)(j = odd \quad number) \\ Z_{Aij} = a \times \frac{i}{I} - r_k \cos(k\psi) + 0.5 \times \frac{a}{I}(j = even \quad number) \end{cases} \tag{1}$$

Where $D_d$ is the diameter of the workpiece being processed; $r_k$ is the radius of the abrasive particle; $\Psi$ is the bisection angle of the abrasive hemisphere; $I$ and $J$ are the number of rows and columns of abrasive particles; $a$ is the arc length of the abrasive stone; $k$ is the average fraction of abrasive particles.

## 3. Motion of abrasive particles

During superfinishing[11], the abrasive stone oscillates back and forth along the axis, and the workpiece rotates clockwise along its axis, which can be converted into a counterclockwise rotation of the abrasive stone. The angle of rotation after time $t$ is $\omega_S t$, and $\omega_S$ is the angular velocity of the abrasive stone rotation. The abrasive stone axial harmonic feed is $Z=A \times Sin(\omega_S t)$, where A is the amplitude of the abrasive stone. The motion diagram is shown in Figure 2.

$$\begin{cases} x' = x\,cos(\omega_d t) + y\,sin(\omega_d t) \\ y' = -x\,cos(\omega_d t) + y\,sin(\omega_d t) \end{cases} \tag{2}$$

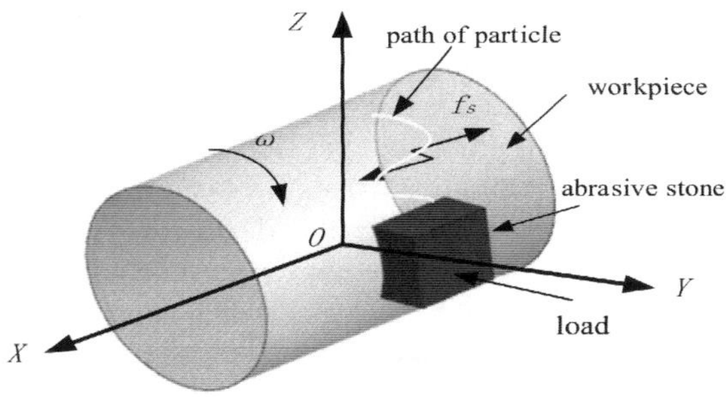


**Figure 2.** Sketch of superfinishing lapping motion.

## 4. Simulation results and analysis

*4.1 Simulation strategy*

The simulation strategy adopted in this article is as follows. Firstly, the shape of the workpiece is constructed. Secondly, machining parameters are input to simulate abrasive motion. Then, the surface of the workpiece is meshed and replaced with the trajectory points of the abrasive particles. Finally, the new workpiece morphology is output.

*4.2 Analysis of simulation results*

(1) The Effect of abrasive stone amplitude on the morphology

In morphology simulation, the workpiece speed is set to 25 r/min, the oscillation frequency of the abrasive stone is 405 times/min, the speed ratio is 16.2, and the amplitudes of the abrasive stone are 300, 350, and 400 μm, respectively. The simulation diagram of the workpiece morphology is shown in Figure 3. As shown in Figure 3, as the amplitude of the abrasive stone increases, the area of a single triangular protrusion decreases. According to calculations, the amplitudes are 300, 350, and 400 μm. The average area of a single protrusion of the workpiece at m is 1.35, 1.27, and 1.19 mm$^2$, respectively. The influence of different machining parameters on the workpiece morphology is shown in Table 1.

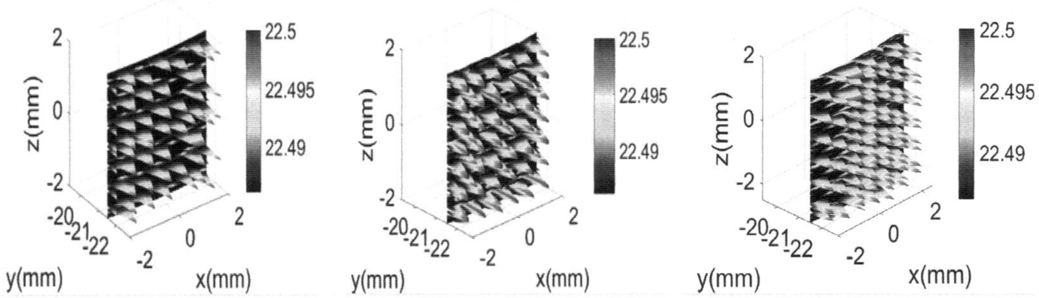


(a) A=300 μm (b) A=350 μm (c) A=400 μm

**Figure 3.** Workpiece morphology under different abrasive stone amplitudes.

**Table 1.** Effect of different abrasive stone amplitudes on the morphology of workpieces.

| Workpiece Rotate Speed (r/min) | Amplitude of Abrasive Stone (μm) | Frequency of Abrasive Stone (times/min) | Rotational Speed Ratio | The Average Area of a Single Protrusion on the Surface of the Workpiece (mm$^2$) |
|---|---|---|---|---|
| | 300 | | | 1.35 |
| 25 | 350 | 405 | 16..2 | 1.27 |
| | 400 | | | 1.19 |

(2) Effect of speed ratio on workpiece morphology

In morphology simulation, the workpiece speed is set to 25 r/min, and the amplitude of the abrasive stone is selected to be 400 μm. The oscillation frequency of the abrasive stone is 395, 400, and 405 times/min. The ratio of the number of abrasive stone oscillations to the workpiece speed is 15.8, 16, and 16.2. The effect of different speed ratios on the morphology of the workpiece is studied. The simulation diagram of the workpiece morphology is shown in Figure 4. From Figure 4, it can be seen that as the speed ratio increases, the area of a single triangular boss decreases. From the calculation, it can be seen that when the speed ratio is 15.8, 16.0, and 16.2, the area of a single boss on the surface of the workpiece is 1.37, 1.26, and 1.19 mm$^2$, respectively. As the speed ratio increases, the surface area of a single platform on the workpiece decreases. The influence of different machining parameters on the workpiece morphology is shown in Table 2.

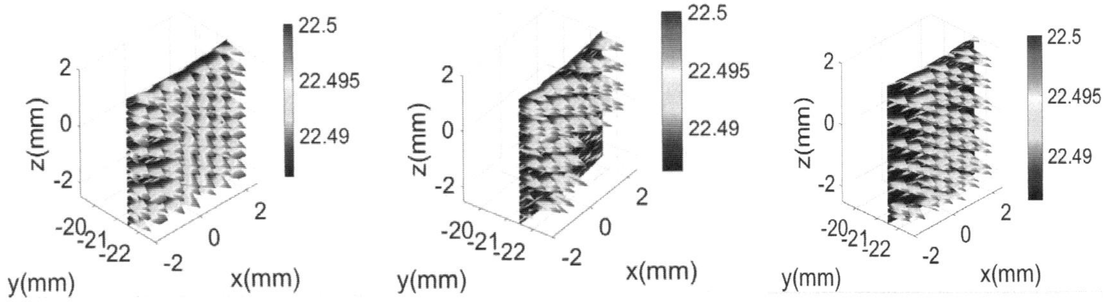


(a) The speed ratio is 15.8. (b) The speed ratio is 16. (c) The speed ratio is 16.2.

**Figure 4.** Workpiece morphology under different speed ratios.

**Table 2.** Effect of different abrasive stone frequencies on the morphology of workpieces.

| Workpiece Rotate Speed (r/min) | Amplitude of Abrasive Stone (μm) | Frequency of Abrasive Stone (times/min) | Rotational Speed Ratio | The Average Area of a Single Protrusion on the Surface of the Workpiece ($mm^2$) |
|---|---|---|---|---|
| 25 | 400 | 385 | 15.8 | 1.37 |
| | | 400 | 16 | 1.26 |
| | | 405 | 16.2 | 1.19 |

## 5. Conclusion

A micro-structured surface can be machined on the surface of the workpiece by using the abrasive stone with staggered abrasive particles and superfinishing lapping. As the amplitude and oscillation frequency of the abrasive stone increase, the number of protrusions on the surface of the workpiece increases, and the area of individual protrusions decreases.

## Acknowledgments

The authors gratefully acknowledge the support for this work from the Chinese National Natural Science Foundation (No.51875368).

## Reference

[1] Jianyong Zheng, Mingqiang Zhong, Jie Feng. Research progress and industrialization status of self-cleaning surfaces based on superhydrophobic principles [J]. Chemical Progress, 2010,29(02): 281-284.

[2] Rajeev G, Rajeev V, Saurabh K, et al. A critical review on recent progress,open challenges, and applications of corrosion-resistant superhydrophobic coating[J]. Materials Today Communications, 2023,34: 105-201.

[3] Yonghua W, Zhongbin Z, Jinkai X, et al. One-Step method using laser for large-scale preparation of bionic superhydrophobic & drag-reducing fish-scale surface[J].Surface and Coatings Technology, 2021.409: 126-801.

[4] Yifeng Fu, Xiuqin Bai, Chengqing Yuan, et al. The synergistic effect and progress of ship anti-fouling and drag reduction based on biomimetics [J]. Ship Science and Technology, 2014, 36(9): 7-12.

[5] Luquan Ren. Research and application of biological desorption wear resistance and biomimetic [J]. Technology Jilin Province, Jilin University, 2006,01.

[6] M Sun, L Luo, L Xu, et al. Artistic lotus leaf by casting[J]. Langmuir, 2005,21(19): 8978-8981.

[7] B Wu, M Zhou, J Li, et al. Super hydrodynamic surfaces fabricated by microstructures of stainless steel using a FEM to second laser [J]. Applied Surface Science, 2009,256 (1): 61-66.

Journal of Physics: Conference Series **2720** (2024) 012024 doi:10.1088/1742-6596/2720/1/012024

[8] Yongcheng Li, Hua Zhang. Numerical simulation study on dragging reduction performance of flat biomimetic groove surfaces [J]. Ship Science and Technology, 2023,45(09): 28-31.
[9] Yousfi M, Mezghani S, Demirci I, et al. Tribological performance of elliptic and circular texture patterns produced by innovative honing process [J]. Tribology International, 2016,100.
[10] Shih-Hsiang Chang, Thomas N. Experimental analysis on the evolution of super finished surface texture [J]. Journal of Materials Processing Technology, 2008,203 (1/3): 365-371.
[11] Tadeusz M, Tomasz F, Adam K, et al. Computer aided system for superfinishing process control[J]. Procedia Technology, 2016,22: 48-54.

# Additive manufacturing and performance research of aluminum-based ceramic cores

**Haifeng Song[1,3], Jinqiu Tang[1] and Hongchang Zhang[2]**

[1]Jianglu Machinery Electronic Group Co., Ltd Xiangtan, 411100 China
[2]Beijing North Hengli Technology Development Co., Ltd Beijing, 100089 China

[3]Corresponding author's e-mail: 953103847@qq.com

**Abstract:** As a ceramic core for forming the cooling structure of the intrinsic cavern of turbine blades, the quality of tubular turbine blades is directly affected by their accuracy and performance. In this paper, based on the DLP photocuring technology, the 50% by volume alumina porcelain suspensions with low viscosity were successfully prepared. Secondly, the effects of printing parameters on the curing characteristics were explored to obtain the best printing parameters. The alumina porcelain properties were analyzed for the heating treatment process effect. As the temperature increases, its mechanical properties increase. In addition, the porosity of the patulous orifice reached 48%. The flexural strength of the alumina porcelain component was 26 MPa, which met the requirements of a high-temperature compound metal blade for the porcelain core.

## 1. Introduction

The aero-engine is the core component of the spacecraft, and the turbine blade that determines the actual performance of the engine is one of the most critical parts of the aero-engine [1-3]. Manufacturing a complex cooling structure inside the blade to reduce the wall temperature of the turbine blade and improve its performance has become a critical link in advanced engine manufacturing. The shape of the internal cavity of the blade determines the cooling method of the blade, and the complexity of the inner cavity depends on the complex structure that the ceramic core can manufacture. Therefore, the ceramic cores become one of the bottlenecks in making blades. The conventional fabricating process of porcelain core mainly includes heated pressing injection molding, transfer molding core, grouting molding, and gel injection molding. The ceramic core produced by traditional methods has the merit of eximious fabrication exactness, uniform size, and mass production. However, they also have a long development cycle and the process is complex and challenging to control [4]. It is difficult to manufacture hollow turbine blades with complex cavities. It needs to design and manufacture expensive molds. Therefore, it is urgent to explore a new ceramic core manufacturing method to make up for the defects of the traditional preparation process. The emergence of ceramic additive manufacturing technology (3D printing) provides an excellent solution to these problems in time.

Stereolithography additive manufacturing technology has the characteristics of high manufacturing precision, considerable design freedom, fast forming speed of complex structures, and fast iteration of products [5-7]. It can compensate for the shortcomings of ceramic core manufacturing complex cavity structures, complex processes, long cycles, and high costs. Light curing technology reduces the process of core pressing, mold preparation, body trimming, and molding filler preparation, thereby reducing the time cost of mold manufacturing and core processing, which provides excellent convenience for rapid verification and iteration of turbine blades. Therefore, the application of photocuring technology to manufacture ceramic cores has unique technical advantages and potential, and it has also become one of the research hotspots of ceramic core manufacturing.

Many universities and research institutions have used photocuring ceramic additive manufacturing technology to prepare ceramic cores. Halloran's team at the University of Michigan [8] successfully prepared an integrated ceramic core-shell by the SLA process using fused silica as a printing material. The in-layer plane size of the printed blank is about 0.7% smaller than the model size, while the size perpendicular to the printing layer is 0.3% larger than the model size. After sintering, the sintering shrinkage in these two directions is about 10.7 ± 0.2%. Bae et al. [9] studied the cristobalite conversion rate and strength of the integrated core at elevated temperatures. The results show that due to the shrinkage of β-type to α-type during the transformation process, many micro-cracks are formed. Regarding aluminum-based ceramics, Li et al. [10] explored the affection of temperature on the properties of alumina ceramics manufactured by photocuring 3D printing technology. When the sintering temperature was 1280°C, the flexural strength of the alumina porcelain component was 24 MPa. The porosity of the patulous orifice was 37.6%, while the shrinkage rates in the long, wide, and high directions were 2.1%, 2.3%, and 3.8%, respectively. The research on the aluminum-based core is relatively few and not systematic. Therefore, detailed research is necessary.

Based on the DLP photocuring technology, this paper focuses on the key issues affecting the manufacturing accuracy in the additive fabricating process of aluminum-based ceramic cores and explores the slurry properties, sintering process control, and molding accuracy control of alumina porcelain cores.

## 2. Experiments

### *2.1 Slurry preparation*

Alumina porcelain suspension contains alumina powder ($D_{50}$ = 500 nm), light-sensitive colophony, and affix. The alumina suspension can be up to 50% by volume. Figure 1 (a) exhibits the alumina porcelain particle size distribution. Alumina powder has a lognormal granum allocation. Figure 1 (b) shows the alumina porcelain scanning electron microscope (SEM) micrograph. Based on the previous work [11], photosensitive resins were prepared. Acidic copolymers are used as dispersants. The addition of dispersant is 2 wt.% of the total weight of the ceramic powder. Irgacure 819 photoinitiator and photoinhibitor purchased from Aladdin were used. The amount of photoinitiator and photoinhibitor was 0.5 wt.% and 0.02 wt.% of light-sensitive colophony, respectively.

**Figure 1.** (a) Granum allocation and (b) SEM micrograph.

### *2.2 3D printing preparation process*

$Al_2O_3$ particles, photosensitive resin, photoinitiator, and the dispersing agent were blended in a planetary mill for 5 hours. Alumina porcelain printing suspension was then defoamed in a vacuum for

10 minutes. Alumina ceramic samples were fabricated by the DLP fabrication machine. The machine has a 405 nm ray origin and a 50 μm pixel DLP chip.

*2.3 Degreasing*

Based on thermogravimetric analysis, the photosensitive resin thermolysis is localized at 450°C. At 600°C, the photosensitive resin thermolysis is essentially finished. So we set the temperature holding point at 450°C and the temperature holding point at 600°C. First of all, the alumina porcelain raised the temperature to 450°C, and the warming velocity was 0.5 °C/min. The alumina porcelain was kept warm for 120 minutes. Second, the alumina porcelain was enhanced to 600 °C at a warming velocity of 1 °C/min. The alumina porcelain was kept warm for 120 minutes. Finally, the alumina porcelain was then enhanced to 1000°C for 120 minutes at a warming velocity of 2 °C/min. The sintering alumina porcelain was set to decrease to 600°C with a cooling velocity of 2 °C/min. The samples were then cooled naturally to room temperature.

*2.4 Characterization*

The laser particle size analyzer was employed to analyze the alumina porcelain particle size and distribution (Winner 2000ZDE). The TGA equipment was used to investigate the weight loss behavior of the suspension when heated to 1000°C (NETZCH STA 449F5). By using the Archimedes method, the porosity and density of alumina ceramics were tested. SEM (Hitachi S-8020) was employed to examine the printed alumina porcelain fracture. The Instron 5965 machine was utilized for the bending test. An average of ten samples is taken for each flexural test result.

## 3 Results and discussion

*3.1 Performance analysis of slurry*

Before printing alumina-based ceramic cores by the DLP process, preparing photocuring slurry with good curing performance, suitable viscosity, and high solid content is necessary. Well-dispersed photocuring slurry is the key to obtaining high-density and uniform microstructure of ceramic products. It is also an essential factor in improving the macroscopic properties of materials. The first step was to conFigure the 50% by-volume alumina porcelain suspension. The viscosity of the slurry was tested as shown in Figure 2. When no dispersant is added, The alumina porcelain suspension exhibits high viscosity, with a static viscosity reaching 27,500 cp. Viscosity decreases as the shear rate increases. When 2wt.% BYK111 was added, the rheology of the slurry was improved. To further control the thixotropic properties of the slurry, thixotropic agents were added to the slurry.

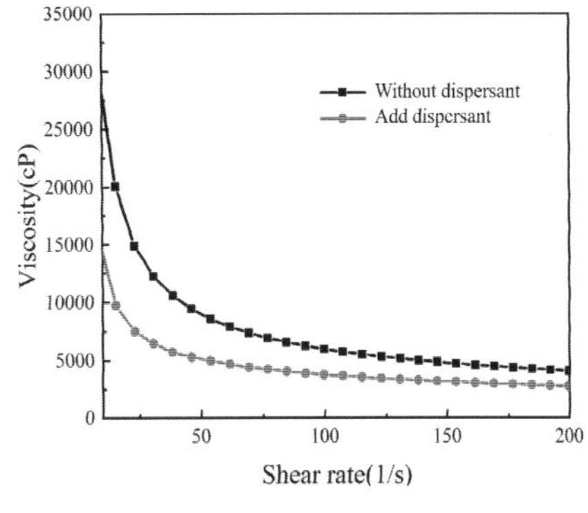


**Figure 2.** The viscosity of alumina suspension.

Before the printing process, it is essential to assess the curing depth of the alumina suspension, with the corresponding outcomes presented in Figure 3 and Table 1. After a curing time of 0.01 s, the alumina slurry can achieve a curing depth of 100 μm. As the illumination time increases, the curing depth is further improved. When the illumination time is 0.52 s, the single-layer printed sample warps.

The single-layer printed sample cracks when the illumination time increases to 0.56 s. When the illumination time increases to more than 0.6 s, the single-layer printed sample appears overexposed. Finally, the printing parameters were selected as 0.4 s and the printing layer thickness was 50 μm, which can ensure that each layer can be fully bonded.

**Table 1.** The curing depth of alumina slurry under different illumination time.

| Samples | 1 | 2 | 3 | 4 | 5 |
|---|---|---|---|---|---|
| Illumination time (s) | 0.01 | 0.1 | 0.3 | 0.4 | 0.52 |
| Curing depth (μm) | 100 | 120 | 150 | 160 | 180 |
| Samples | 6 | 7 | 8 | 9 | 10 |
| Illumination time (s) | 0.56 | 0.6 | 0.7 | 1 | 1.5 |
| Curing depth (μm) | 190 | 200 | 210 | 310 | 430 |

**Figure 3.** Single-layer cured sample photos of alumina slurry under different illumination time.

Thermogravimetric analysis should be carried out before degreasing the printed body. The appropriate degreasing process should be selected according to the thermal degradation curve so that the organic matter can be slowly degraded and discharged from the body to reduce the phenomenon of degreasing cracking. Thermogravimetric analysis of the suspension is shown in Figure 4. The decomposition of most photosensitive resins typically occurs at 450°C, with the process essentially completed by 600°C. To allow the resin to decompose adequately during the heating process, the heating rate should be set relatively slowly before 450°C. After 600°C, the resin in the printing body is completely volatilized.

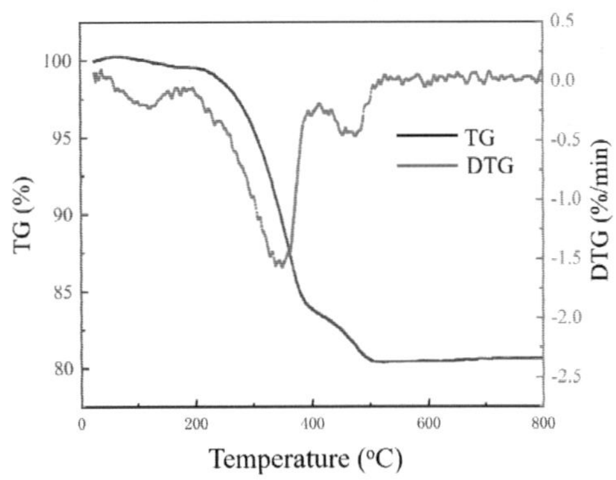


**Figure 4.** TG-DTG curves of the alumina porcelain suspension.

### 3.2 *Performance analysis of sintered samples*

The experiment involved examining the shrinkage, density, and porosity of alumina porcelain under various sintering temperatures, as detailed in Table 2. The alumina porcelain shrinkage rate is also enhanced by raising the sintering temperature. The abnormality at 1400°C is attributed to the sintering expansion phenomenon observed in the alumina porcelain. Furthermore, the shrinkage of alumina porcelain is consistent across all three dimensions, which can ensure that the accuracy of the spline is within the predictable range after sintering. Concurrently, with the elevation of sintering temperature, there is a corresponding increase in the alumina porcelain density. The trend is in line with the rate of contraction. (Each data point in Table 2 is derived from 10 samples and averaged.)

**Table 2.** The shrinkage, density, and porosity of alumina porcelain.

| Sintering temperature(°C) | 1300 | 1400 | 1500 |
|---|---|---|---|
| Shrinkage ratio(length) | 1.79% | 0.46% | 5.90% |
| Shrinkage ratio(width) | 1.92% | 0.77% | 6.14% |
| Shrinkage ratio(height) | 1.52% | 0.2% | 6.35% |
| Density($g/cm^3$) | 2.10 | 2.02 | 2.41 |
| Porosity | 45% | 48% | 32% |

Figure 5 shows the microstructure of the alumina body after degreasing and sintering at different temperatures. There is no apparent sintering polymerization phenomenon of alumina ceramic particles after degreasing. There are some voids between alumina ceramic particles. The connection between large particles depends on the adhesion of some small alumina ceramic particles, resulting in a loose distribution of ceramic particles. At 1300°C, the alumina ceramic particles began to neck further, and the temperature also reached the diffusion activation energy of the atoms and began to diffuse and rearrange. The alumina ceramic particles sintered at 1300°C and 1400°C changed from spherical to non-spherical. The alumina porcelain particles sintered at 1500°C are further tightly bonded and the ceramic material is dense. The alumina porcelain density is enhanced by increasing the alumina ceramic sintering temperature. The alumina porcelain properties are improved by the close arrangement of the ceramic grains.

**Figure 5.** Microstructure of alumina green body after degreasing (a) and sintered alumina porcelain at different temperatures (b) 1300°C, (c) 1400°C and (d) 1500°C.

The bending strength of alumina after different heat treatment temperatures is shown in Figure 6. The ceramic bending strength increases as the temperature increases. The flexural strength is 24 MPa at 1300°C. The scanning electron microscope also shows that many pores are initially sintered, but not dense sintered. At the same time, it is observed from the picture that the particle shape of alumina begins to change from spherical to non-spherical. When the temperature rises to 1400°C, many spherical particles are transformed into non-spherical particles, which expands the sample volume, and the bending strength increases slightly, only 26 MPa. The alumina porcelain particles are tightly bonded at 1500°C. The ceramic grains are closely arranged. The flexural strength upgraded to 105 MPa. Combined with the performance of the core, the appropriate bending strength is between 20 and 30 MPa. Considering its dissolution performance, its porosity needs to be greater than 35%. Therefore, the 1300°C and 1400°C samples can meet the requirements of the core. The alumina porcelain core was heated to 1400°C and the shrinkage rate was degraded. This temperature is the best heat treatment temperature for aluminum-based cores.

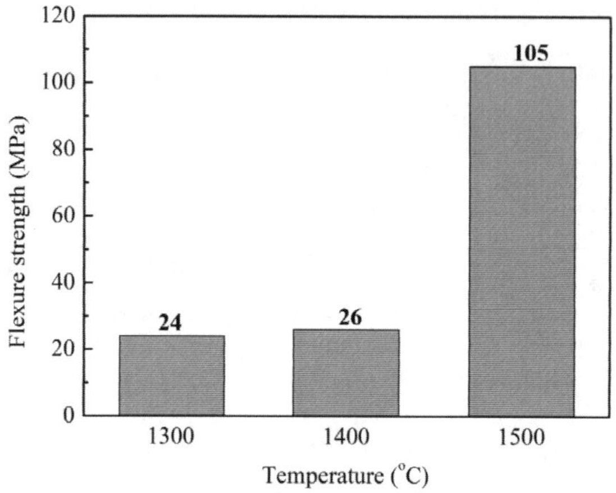


**Figure 6.** The flexure strength of alumina green body after sintering at different temperatures.

### *3.3 Printing accuracy analysis*

There are many thin-walled structures in the core model. Before printing the ceramic core of a complex structure, the limiting conditions of the photocuring process on thin-walled structures should be first explored. As shown in Figure 7 (a), three different thin-walled structures with wall thicknesses of 200 μm, 375 μm, and 500 μm are designed. When the wall thickness is 200 μm, the printed sample cannot be well-formed, as shown in the left diagram of Figure 7 (b), and the complete ring cannot be formed. It can achieve good forming when the wall thickness is 375 μm and 500 μm. The sample of model C was further observed by an optical microscope, as shown in Figure 7 (c). It was found that the thin-walled structure can be well formed and has high forming accuracy.

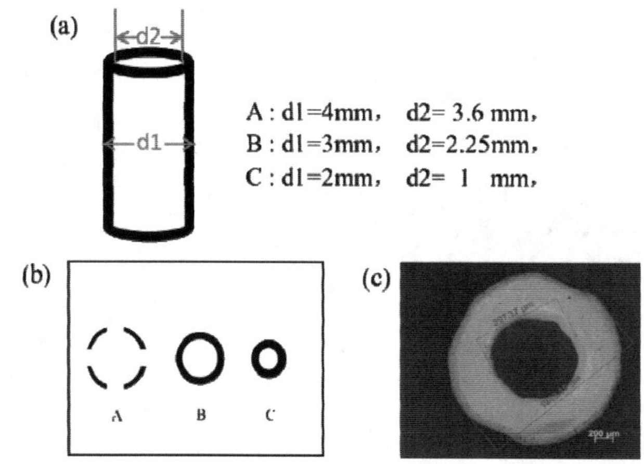


**Figure 7.** Three kinds of thin-walled models and printing effect diagram (a) Three thin-walled types of models, (b) Sample schematic diagram after printing, (c) Optical electron microscopy of Model C printed sample.

Before printing complex model cores, simple models should be printed to observe the effects of different structural designs, printing effects, and sintering results. As shown in Figure 8, there are seven components. The two cylinders of the alumina porcelain components are suspended. It can be seen from the physical map that although it can be printed, the interlayer bonding could be better, and there is deformation, indicating that the printing method has specific requirements for the shape and size of the suspended part. It can be seen from these components that the accuracy of 3D printing is very high, and the hollow circular hole has a high resolution.

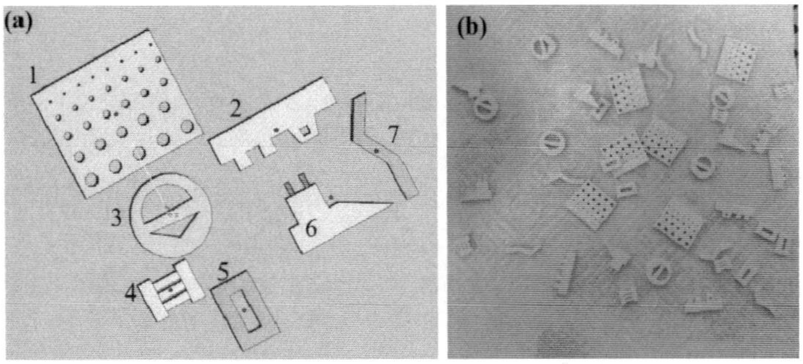


**Figure 8.** Model of typical core structure and printing effect diagram (a) model, (b) photo of the corresponding sample.

Figure 9 shows the morphology of the pore structure in detail. It can be seen that the difference between the printed apertures and the apertures designed for the model is very small. After printing, the pores of 300 μm shrink to 263 μm, and the pores of 600 μm shrink to 516 μm. By designing different pore size structures for printing, it is found that the pore size after printing is less different from the pore size designed by the model. Precise control of the delicate pores in the core can be achieved by the model compensation method.

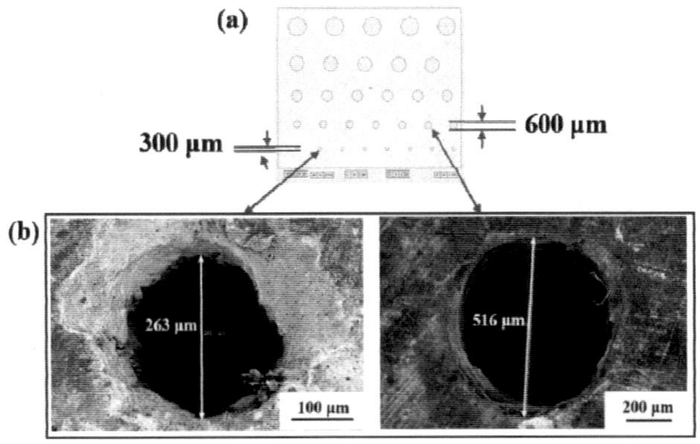


**Figure 9.** Models with different pore sizes and printing effect diagram (a) model, (b) scanning electron microscope diagram corresponding to different pore sizes.

## 4 Conclusions

In this paper, based on the DLP photocuring technology, alumina porcelain suspension of 50% by volume with low viscosity was successfully prepared. The printed body has a lower shrinkage rate after heat treatment at 1400°C. The porosity of the patulous orifice reached 48%. The flexural strength of the alumina porcelain component was 26 MPa, which met the requirements of the high-temperature compound metal blade for the porcelain core. At the same time, by designing and printing the typical and fine structure in the core model, the limiting conditions in the printing process are obtained, which lays a theoretical foundation for preparing high-precision complex structure ceramic cores.

## Acknowledgment

We acknowledge the technical support in material preparation and additive manufacturing from Jiangsu Quickdemos Intelligent Manufacturing Hightech Co., Ltd, China.

## References

[1] M. Gromada, A. Świeca, M. Kostecki, A. Olszyna, R. Cygan, Ceramic cores for turbine blades via injection moulding, Journal of Materials Processing Technology, 220 (2015) 107-112.

[2] E.-H. Kim, G.-H. Cho, Y. Yoo, S. Seo, Y.-G. Jung, Development of a new process in high functioning ceramic core without shape deformation, Ceramics International, 39 (2013) 9041-9045.

[3] Y.-M. Chen, R.J. Smales, K.H.-K. Yip, W.-J. Sung, Translucency and biaxial flexural strength of four ceramic core materials, Dental Materials, 24 (2008) 1506-1511.

[4] F. Bayindir, O. Ozbayram, Effect of number of firings on the color and translucency of ceramic core materials with veneer ceramic of different thicknesses, The Journal of Prosthetic Dentistry, 119 (2018) 152-158.

[5] S. Zakeri, M. Vippola, E. Levänen, A comprehensive review of the photopolymerization of ceramic resins used in stereolithography, Additive Manufacturing, 35 (2020) 101177.

[6] E. Zanchetta, M. Cattaldo, G. Franchin, M. Schwentenwein, J. Homa, G. Brusatin, P. Colombo, Stereolithography of SiOC ceramic microcomponents, Advanced Materials, 28 (2016) 370-376.

[7] S. Li, W. Duan, T. Zhao, W. Han, L. Wang, R. Dou, G. Wang, The fabrication of SiBCN ceramic components from preceramic polymers by digital light processing (DLP) 3D printing technology, Journal of the European Ceramic Society, 38 (2018) 4597-4603.

[8] B. Bae, Integrally cored ceramic investment casting mold fabricated by ceramic stereolithography, Materials Science and Engineering. University of Michigan, (2008).

[9] Bae C J, Kim D, Halloran J W. Mechanical and kinetic studies on the refractory fused silica of integrally cored ceramic mold fabricated by additive manufacturing[J]. Journal of the European Ceramic Society, 2019, 39 (2-3): 618-623.

[10] H. Li, Y. Liu, Y. Liu, K. Hu, Z. Lu, J. Liang, Influence of sintering temperature on microstructure and mechanical properties of $Al_2O_3$ ceramic via 3D stereolithography, Acta Metallurgica Sinica (English Letters), 33 (2020) 204-214.

[11] Gu Y, Duan W, Wang T, et al. Additive manufacturing of $Al_2O_3$ ceramic core with applicable microstructure and mechanical properties via digital light processing of high solid loading slurry[J]. Ceramics International, 2023.

# Precision cutting mechanics and mechanical performance testing on the spindle of machine tool

**Hongping Yang[1,2], Zhongwen Guo[1], and Yingjie Xia[1]**

[1]Tianshui Normal University, Tianshui, Gansu Province, 741000 China

[2]Corresponding author's e-mail: yanghp@tsnu.edu.cn

**Abstract**: The manufacturing industry is the foundation of national development. With science and technology developing, precision turning technology has gradually become one of the cutting-edge technologies of advanced manufacturing technology. In the paper, the turning mechanics model is established based on the principle of metal cutting, and the corresponding cutting force is calculated. Considering the cutting force, material weight, flange weight, and other loads, the analysis model of the CK61125 machine tool spindle is constructed based on the finite element method, and the corresponding maximum strain and modal solution are calculated. The results are that when the 30, 000 N load is applied, the largest strain of the spindle is 0.0013 mm/mm, and the largest stress is 265 MPa. When the 100, 000 N load is applied, the strain of the spindle largest is 0.0015 mm/mm, and the stress is 290 MPa. The four orders of vibration modes of the spindle are obtained, which are 215.3 Hz, 256.2 Hz, 436.33 Hz, and 974.5 Hz. By comparison with the experiment, the spindle vibration frequencies are 87.8 Hz and 625 Hz, while the spindle rotating testing speeds are 100 rpm, 500 rpm, and 800 rpm.

## 1. Introduction

The machinery manufacturing industry is an important symbol of the overall level of industrial development of a country, in which turning processing occupies a pivotal position and has been widely used in aviation, aerospace, shipbuilding, and medical industries [1]. Turning performance is a very important point that affects the surface workpiece quality and limits the productivity of the turning process. Cutting force is one of the important indicators of the workpiece elastic deformation, which is the workpiece distortion caused by the residual stress of the cutting surface. The processing vibration affects the machining accuracy and quality. It is of great significance to establish the cutting force model of the tool for studying the mechanism of precision turning [2]. Traditional turning force prediction models are mainly divided into the following three categories: The first is a complete prediction model based on a cutting experiment [3]; The second is a semi-empirical prediction model based on mechanical analysis [4]; The third is an analytical prediction model based on cutting theory and material properties. Ding et al. [3] established a new analytical prediction model of precision turning force based on a metal cutting mechanism and material constitutive relationship. Qin et al. [5] obtained the strain rate from the experiment of right-angle cutting based on the material constitutive relationship, and then extended it to the oblique cutting and established the milling force prediction model. Shamoto extended to bevel cutting and proposed a calculated bevel geometric angle method. Wang [6] proposed a parallel plane shear zone model for right-angle cutting. Hao [7] established the bevel-cutting model and obtained the relationship between shear flow angle and chip flow angle.

Based on the principle of cutting mechanics, cutting mechanics modeling is carried out. Considering the influence of cutting force and other factors on the machine tool spindle, the mechanical characteristics of the machine tool spindle are carried out. Combined with the field engineering test results, the working characteristics of the machine tool are analyzed.

## 2. Analytical modeling of turning force

In the process of metal cutting, the metal material undergoes a transition deformation under the action of the cutting tool. It is from elastic deformation to plastic deformation. Chips continuously flow out along the front tool face under the action of the main cutting force. The tool is mainly subjected to pressure and friction between the cut metal and the front tool face.

The force-cutting process analysis is shown in Figure 1. It is assumed that the machining process is a steady-state cutting process, and the force acting on the front tool face and the combined force acting on the shear surface is a pair of balanced forces. Generally, the cutting resultant force $F$ can be decomposed into the shear force $F_s$ along the cutting surface and the pressure of normal $F_n$ perpendicular to the cutting surface, as well as the friction force $F_u$ along the tool rake surface and the normal pressure $F_v$ perpendicular to the tool surface. The cutting force $F$ can be decomposed into the two forces. There are the tangential force $F_c$ along the cutting speed direction and the feed force $F_f$ along the cutting thickness direction.

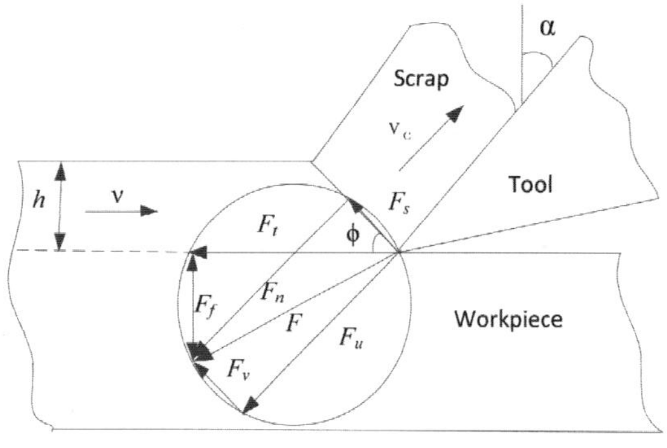


**Figure 1**. Schematic of cutting force.

It is assumed that the shear force is uniformly distributed. The force $F_s$ on the plane can be obtained by multiplying the shear flow stress.

$$A_s = \frac{A}{\sin\varphi} = \frac{bh}{\sin\varphi} \tag{1}$$

Where $A$ is the cutting layer area, $h$ is the cutting width, $b$ is the cutting depth, and $A_s$ is the shear plane area.

$$F_s = \tau_s A_s = \tau_s \frac{bh}{\sin\varphi} \tag{2}$$

Where $\tau_s$ is the stress of the shear flow on the plane for shear, and $\varphi$ is the angle for shear during the cutting process.

Based on the geometric relationship between the forces in Figure 1, we can derive the equation for $F_s$ and $F$.

$$F = \frac{F_s}{\cos(\varphi+\beta-\alpha)} = \frac{\tau_s bh}{\cos(\varphi+\beta-\alpha)\sin\varphi} \tag{3}$$

Where $\alpha$ is the nominal tool angle.

It is assumed that $\chi$ is the angle between the force $F$ and the plane for shear, that is, $\chi = \varphi + \beta - \alpha$, and $\chi$ angles vary freely at 45° with different values of the material.

$$F = \frac{\tau_s bh}{\cos(\chi)\sin\varphi} \tag{4}$$

According to the theory and experiment of material mechanics, the relationship between the true shear force $\tau$, strain $\varepsilon$, and yield strength $\tau_s$ during the metal cutting process is:

$$\tau = \tau_s \varepsilon^n \tag{5}$$

The practice has shown that in metal cutting, changes in the angle affect the deformation coefficient and relative slip. Figure 2 is a schematic diagram of shear deformation during metal cutting. During the cutting process, the cutting layer OHNM on the workpiece undergoes slip deformation to form OGPM. The concept of relative slip includes:

$$\varepsilon = \frac{\xi^2 - 2\xi\sin\alpha + 1}{\xi\cos\alpha} \tag{6}$$

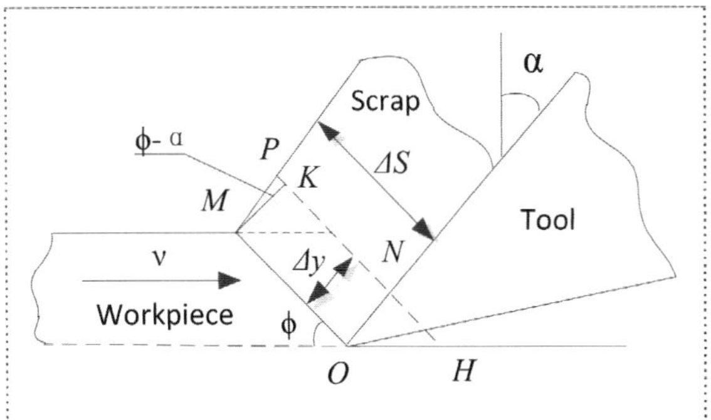


**Figure 2**. Schematic of cut deform.

From the relationship between the forces in Figure 2, it can be obtained that the force of Tangential $F_t$, the direction force $F_f$ , and the cutting force $F$ are expressed:

$$\begin{cases} F_t = F\cos(\beta - \alpha) = \dfrac{\tau_s bh\cos(\beta - \alpha)}{\sin\varphi\cos(\beta - \alpha + \varphi)} \\ F_f = F\sin(\beta - \alpha) = \dfrac{\tau_s bh\sin(\beta - \alpha)}{\sin\varphi\cos(\beta - \alpha + \varphi)} \end{cases} \tag{7}$$

The tool front angle α can be measured by three-dimensional confocal microscopy, the shear flow stress $\tau_s$ can be calculated by the Johnson-Cook shear flow stress model, and the friction angle $\beta$ can be calibrated by cutting experiments. The shear angle $\varphi$ in the cutting process is solved based on the minimum energy theory proposed by the merchant.

## 3. The dynamic characteristics and analysis principle of the spindle

The modal spindle system frequencies and vibration modes are obtained by using ANSYS. These parameters are very important for dynamic characteristics analysis.

The spindle structure is divided into finite units. The **K** is the spindle system stiffness, **M** is spindle system mass, $\mathbf{x}(t)$ is the spindle displacement, $\ddot{\mathbf{x}}(t)$ is the spindle total acceleration by each unit acceleration, and $\mathbf{F}(t)$ is the spindle concentrated force. Therefore, the spindle dynamic equilibrium equation is:

$$\mathbf{M}\ddot{\mathbf{x}}(t)+\mathbf{K}\mathbf{x}(t)=\mathbf{F}(t) \tag{8}$$

If $\mathbf{x}_0(t)$ is the amplitude and $\omega$ is the frequency, we can obtain:

$$\mathbf{x}(t)=\mathbf{x}_0(t)\cos(\omega t) \tag{9}$$

The homogeneous equation is:

$$\left(\mathbf{K}-\omega^2\mathbf{M}\right)\mathbf{x}_0=0 \tag{10}$$

When the spindle is vibration-free, each node amplitude is not zero, and the coefficient determinant is zero.

$$\left|\mathbf{K}-\omega^2\mathbf{M}\right|=0 \tag{11}$$

## 4. The characteristics analysis of spindle

### *4.1 The spindle system parameters*

We use the CK61125 horizontal CNC lathe spindle as an object. The spindle material is 45 steel that is quenched and tempered.

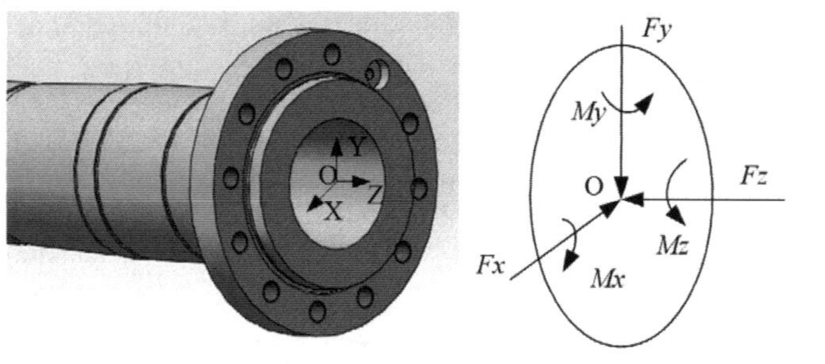


**Figure 3**. Schematic of simplified loading of spindle.

The spindle torque is transited by a motor through the synchronous belt wheel, and the torque value is $Mz=12$ N·m. The assembly speed range of the spindle is from 100 rpm to 800 rpm. While the spindle is working, the workpiece's mass and the cutting force are simplified to the end plane center O, as shown in Figure 3. The parameter values are $Fz=25$ kN, $Fx=15$ kN, and $My=10$ kN·m.

### *4.2. Stress characteristic modeling of spindle*

Taking the situation of processing machine tools, the spindle mechanical performances are analyzed. Its machining load is 3 tons and 10 tons, respectively.

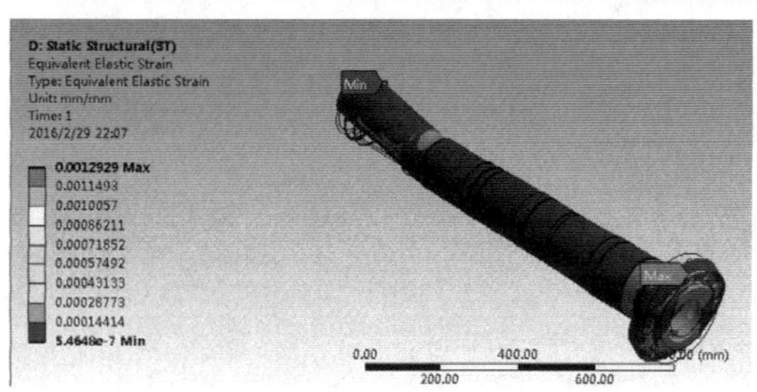


**Figure 4**. The stain for 30, 000 N.

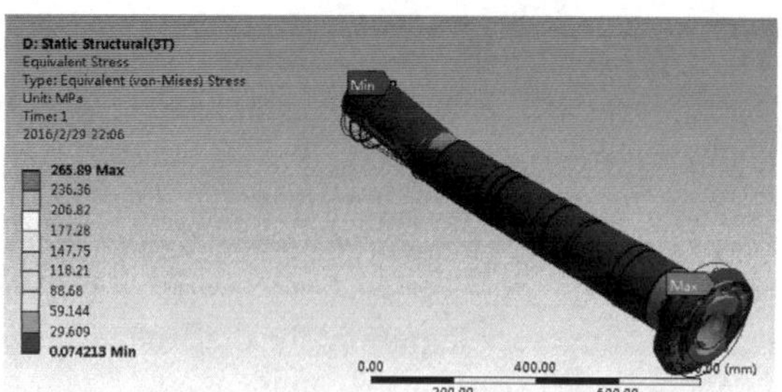


**Figure 5**. The stress for 30, 000 N.

From Figure 4, we can see that the spindle strain is the workpiece 30, 000 N of 0.0013 mm/mm. From Figure 5, the spindle maximum stress is 265 MPa. However, 45 steel allowable stress has a value of 355 MPa.

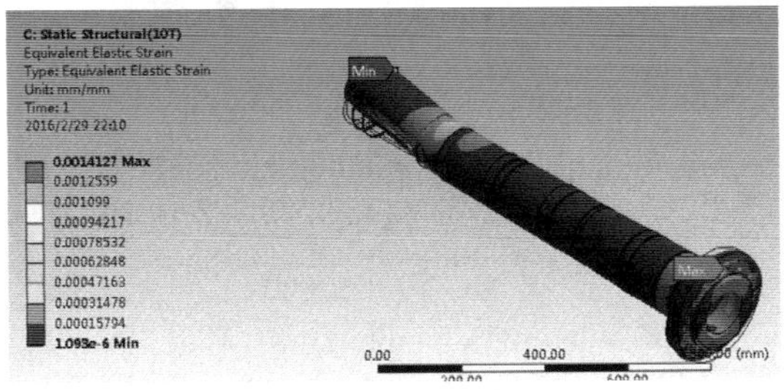


**Figure 6**. The stain for 100, 000 N.

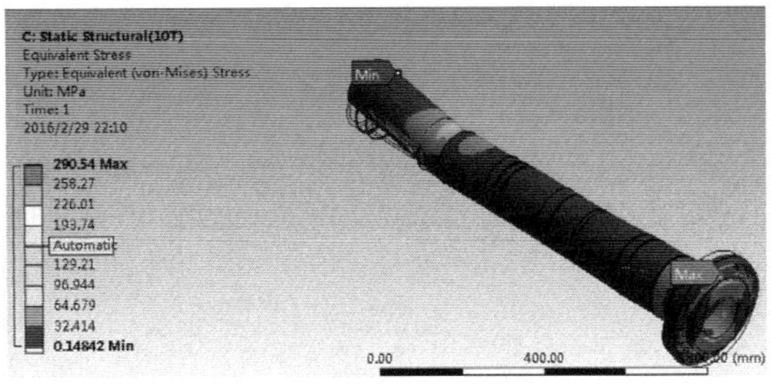


**Figure 7**. The stress for 100, 000 N.

From Figure 6, we can see that the spindle strain is the workpiece 100, 000 N of 0.0015 mm/mm. From Figure 7, the spindle maximum stress is 290 MPa. However, 45 steel allowable stress has a value of 355 MPa. The values are all less than the allowable stress of spindle materials.

### *4.3 Modal analysis of spindle*

The spindle model and calculation use ANSYS. The natural frequency and vibration mode for the spindle system are obtained, as shown in Table 1.

**Table 1**. The natural frequency (Hz).

| Modal order | 1 | 2 | 3 | 4 |
|---|---|---|---|---|
| Natural frequency | 215.3 | 256.2 | 436.33 | 974.5 |

## 5. The dynamic performance testing of the spindle system

To evaluate the spindle static and dynamic performance in the engineering application, we installed it on the CK61125 numerical control lathe. The spindle rotating speeds are 100 rpm, 500 rpm, and 800 rpm. The spindle vibration response signals are obtained through the sensor's axial and radial direction. The testing points are shown in Figure 8 and Figure 9.

**Figure 8**. Radial vibration testing of spindle.

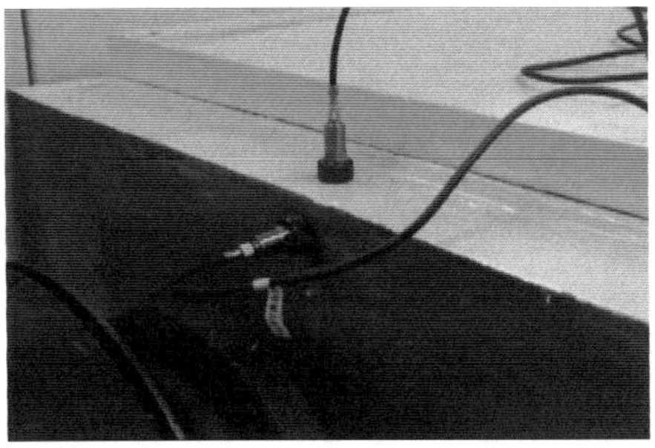

**Figure 9**. Axial and radial vibration testing of spindle.

**Table 2**. The testing vibration frequency parameters of the spindle.

| Spindle speed (rpm) | Radial frequency (Hz) | Axial frequency (Hz) |
|---|---|---|
| 100 | 87.8, 625 | 625 |
| 500 | 625 | 87.8, 625 |
| 800 | 625 | 87.8, 625 |

From Table 2, we can see that the frequency of the spindle box mainly is 87.8 Hz and 625 Hz.

## 6. Conclusion

For the CK61125 machine tool spindle parts, we analyze the spindle cutting force and strain under certain loads by using the finite element analysis method and analyze the spindle modal frequency, pointing out the weak spot of the spindle structure. Through the machine tool spindle under the condition of different speeds of analysis and testing of cutting work, the spindle vibration frequency is obtained. By comparison with the experiment and theoretical analysis, it validates the correctness of theoretical analysis.

## Acknowledgments

This research is supported by the Key Project of Science and Technology of Gansu Province (22YF7GE169), the University-Industry Support Plan Project of Gansu Province (2023CYZC-61), and the National Natural Science Foundation of China (51565053).

## References

[1] Ganesan G., et al. A Review of Cutting Tools for Ultra-precision Machining, Machining Science and Technology, 2022, 26 (6) 923-976.
[2] Mohsen S., Behrooz A., and Roza D. Machine Learning and Artificial Intelligence in CNC Machine Tools. Sustainable Manufacturing and Service Economics, January 2023, online.
[3] Ding Z. C., Yan S. Y., and Sun Y. W. Research on Cutting Force Modeling in Precision Turning Process. Mechanical Engineer, 2020, (11): 13-16+19.
[4] Wei X. L. and Hou Z. J. Experimental Study on Cutting Force in Turning Process. Journal of Hebei University of Technology (Natural Science Edition), 2017, 34 (3): 97-100.
[5] Qin L., Liu E., Xing H. W., et al. Research Progress on Stability Analysis of Turning Process. Journal of Harbin University of Science and Technology, 2019, 24 (01): 14-22.
[6] Wang A. Dynamic Modeling and Analysis of Turning Chatter. Lanzhou University of Technology, 2021.
[7] Hao W. X. Nonlinear Dynamics Analysis in Turning Process. Lanzhou Jiaotong University, 2023.

# Research on maintenance-free design and low friction characteristics of industrial robot sliding components

**Zhuomiao Zhu**[1], **Junde Guo**[2,3]

[1]Faculty of Engineering Sciences University College London, WC1E 7JE, UK
[2]School of Mechatronic Engineering Xi'an Technological University Xi'an, 710021 China

[3]Corresponding author's e-mail: gjd0119@163.com

**Abstract:** With the continuous progress of industrial automation, the application of industrial robots in various fields has become increasingly common, and ensuring their efficient and stable operation and reducing maintenance costs is crucial. This study used powder metallurgy technology, combined with multiple sintering-rolling processes and oil-immersed vacuum assistance, to successfully prepare Fe-Cu-Ni-Sn-graphite oil-immersed self-lubricating composite materials to meet this demand. In-depth microstructure and wear surface studies revealed that this porous oil-immersed self-lubricating composite material exhibits long-term low friction and high wear resistance and reduces the friction coefficient by 40% compared to untreated samples after a specific composite process. In addition, the material exhibits excellent friction performance in the high-temperature pin-on-disc friction and wear test machine. Even after continuous sliding for 24 hours, its friction coefficient remains low and stable. The study also found that its lubrication mechanism may be attributed to solid-liquid synergistic lubrication, thanks to the appearance of oil-graphite mixtures around the wear track. This design ensures high rigidity and reduced frictional loss, providing a strong reference for the design and optimization of sliding components of industrial robots, and is highly suitable for the widespread application of industrial robots.

## 1. Introduction

With the increasing depth of industrial automation, the application of industrial robots is becoming increasingly widespread [1–3]. The low friction and long lifespan characteristics of sliding components are crucial to ensure the stable operation of industrial robots and reduce maintenance costs. Traditional lubrication methods may be difficult to meet the long-term and efficient operation requirements, particularly in high-frequency and high-load working environments. The demand for low-friction materials to enhance the long-term serviceability of industrial robot joint components is becoming increasingly strong [4, 5]. Self-lubricating materials can maintain low friction characteristics without external lubrication or the need for lubricants, making them highly applicable in this field.

Iron-based self-lubricating materials prepared by powder metallurgy technology exhibit good self-lubricating properties, with the lubricating phase uniformly distributed in the material, demonstrating excellent self-lubricating performance [6–8]. In addition, these materials also possess excellent anti-friction and anti-wear properties while ensuring mechanical strength due to the uniformly distributed lubricating phase in their structure, such as graphite and molybdenum disulfide, providing continuous lubrication. The porous characteristics of powder metallurgy products provide conditions for storing lubricants, allowing the formation of microscopic lubricant reservoirs on the surface of sliding components, thus achieving a self-lubricating effect [9]. When lubricant consumption or wear occurs, the internal lubricant can replenish the surface, achieving lasting lubrication. Graphite, as a common solid lubricant, has a low friction coefficient, which enables porous materials combined with it to have outstanding anti-friction and anti-wear properties [10]. However, it remains a great challenge to maintain the unity of self-lubricating performance and mechanical strength. The rolling and secondary sintering methods are used to ensure both mechanical strength and long-term serviceability.

Traditional sliding components of industrial robots often require regular lubrication, leading to higher maintenance costs. In our early exploration, we sought to create these components using powder metallurgy technology. However, we encountered a challenge: the low density of porous P/M materials resulted in unsatisfactory strength and hardness under high load conditions [1–13]. To address this issue, inspired by applying porous materials in self-lubricating sliding components, we utilized the unique properties of oil-containing porous powder metallurgy and employed secondary sintering and rolling to enhance the frictional characteristics of the components. Furthermore, we added graphite during the material mixing process to optimize the lubrication effect to achieve synergistic lubrication between solid and liquid phases [14, 15]. This design not only ensures the strength of the components but also achieves lower frictional losses, providing a reference for the application of sliding components such as joints in industrial robots.

## 2. Experimental method

### *2.1 Raw materials*

**Table 1.** Specifications and participation ratio.

| Name | Particle size/μm | Purity/% | Appending proportion (wt%) |
|---|---|---|---|
| Copper powder | ≈80 μm | >99.5 | 33wt% |
| Iron powder | ≈50 μm | >99.5 | 42wt% |
| Tin powder | ≈85 μm | >99.5 | 6wt% |
| Nickel powder | ≈50 μm | >99.5 | 15wt% |
| Graphite powder | ≈25 μm | >99.5 | 4wt% |

The raw materials used in this experiment are high-purity copper powder, iron powder, tin powder, nickel powder, and graphite powder, with their main performance indicators and addition ratios shown in Table 1.

**Table 2.** Parameters of sintering and pressing processes.

| Name | Technological process | Sintering temperature | Rolling temperature | Secondary sintering temperature | Hardness (HRc) |
|---|---|---|---|---|---|
| 1 | Sintering | 800 °C | / | / | 7.2 |
| 2 | Sintering-Rolling | 800 °C | 650°C | / | 8.2 |
| 3 | Sintering-Rolling-Sintering | 800°C | 650°C | 650°C | 10.3 |

According to the proportions in Table 1, the five powders are mixed in a drum-type blender for 3 hours, adding 1% machine oil (N460) as an auxiliary material before mixing to aid in pore formation. The mixed powders are then pressed in a steel mold at a pressure of 350~400 MPa, held for 1 minute, demolded, and sintered in a hydrogen-protected tube resistance furnace at 800°C for 30~50 minutes. After sintering, the sample is cooled in the furnace to obtain Sample 1. Sample 2 is obtained by rolling the sintered sample at a temperature of 400°C, rolling speed of 0.08 m/s, and rolling precision of ±5 mm. Sample 3 is obtained by sintering-rolling-secondary sintering at a temperature of 650°C, with the same rolling process as Sample 2. After preparation, each sample undergoes vacuum oil impregnation at 120°C, using oil with a room temperature viscosity of 375 cps. Finally, the samples are precision machined to the target dimensions. Table 2 lists the parameters of the sintering and pressing processes.

### *2.2 Experimental method*

The surface morphology of the friction layer was observed using a scanning electron microscope (SEM, model JSM-6610), and XRD and EDS analyzed its composition. The surface of the samples was observed using an optical metallographic microscope, the porosity of the samples was estimated through binary processing, and the average pore size was calculated. The friction and wear performance of the material was studied on a high-temperature reciprocating friction and wear test machine, where the test piece was made of a developed material, the pin size was Ø5 mm, the test plate was made of 40Cr steel with a size of Ø40 mm, and a hardness greater than 25 HRc. The experiment was conducted at room temperature and in an atmospheric environment for 45 minutes, as shown in Figure 1, with a load of 3 MPa, a friction radius of 25 mm, a sliding time of 24 hours, and a speed of 0.8 m $s^{-1}$. Different loads and speeds were considered in the experiment, and the mass before and after wear was measured using an electronic balance, and the mass loss and wear rate were calculated. Hardness testing of sintered samples was performed on a Brinell hardness tester, using a 10 mm diameter hardened steel ball or tungsten carbide ball as the indenter, applying an experimental force of 62.5 kgf (approximately 613.3 Newtons) and pressing it into the surface of the tested material. An electronic balance was used to determine the density of the sample, and its mechanical properties at room temperature were evaluated using a universal materials testing machine.

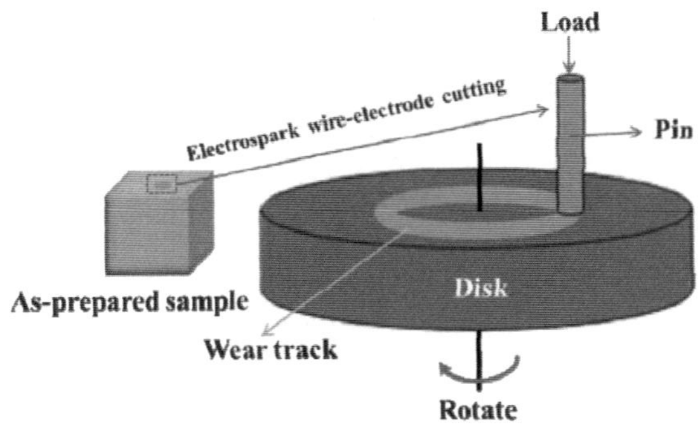


**Figure 1.** Schematic diagram of pin-on-disk tribometer and sample testing process.

## 3. Results and discussions

### *3.1 The microstructure and phase composition of composite materials*

As shown in Figure 2, the metallographic morphology of the Cu-Fe-Ni-Sn-Graphite sample varies with different preparation processes. For the sintered sample (a), the phase composition appears relatively dispersed, with various non-continuous phases. However, after rolling, the phase composition of sample b becomes more continuous. Most notably, sample c, after sintering-rolling-sintering, exhibits the most uniform and continuous phase distribution with significantly reduced porosity. With the increase in process steps, the microstructure of the sample gradually becomes more uniform and dense, and the distribution of each component becomes more continuous, which may be related to the compression during the rolling process and the solid-phase diffusion during the sintering process. Rolling is a process of deformation of materials by applying pressure. During this process, the material is compressed, leading to enhanced contact and bonding between powder particles, reducing porosity and voids. Additionally, sintering is a heat treatment process under high-temperature conditions that binds powder particles through solid-phase diffusion mechanisms, promoting atomic migration between particles and material redistribution, further enhancing the density of the sintered layer. Therefore, the combined effects of compression during rolling and solid-phase diffusion during sintering result in a more uniform and dense sample.

The XRD results shown in Figure 3 indicate that only the sintered Sample 1 and the sintered-rolled Sample 2 have a similar composition, existing mainly as a solid solution of Cu, Sn, Fe, and Ni, suggesting that rolling did not alter the composition of the samples. For Sample 3, after rolling and secondary sintering, there is almost no detection of the diffraction peak for Fe/(Fe, Ni, Cu) in the matrix. At the same time, the surface composition is mainly a Cu-Sn-Fe-Ni solid solution, indicating that the secondary sintering process increased the degree of alloying in the sample. These results show that the sintering-rolling-sintering composite process can effectively increase the alloying degree of the matrix and improve the material's hardness. This is mainly due to the significant grain refinement of the matrix after rolling deformation, which can effectively improve the strength of the composite material. The deeper reason is that rolling increases the dislocation density in the matrix, causing work hardening and increasing the strength and hardness of the matrix, which is consistent with the hardness trend in Table 2 [16–18].

**Figure 2.** Morphology of friction-sintered layer of Cu-Fe-Ni-Sn-Graphite samples under different preparation processes. (a) sintering; (b) sintering-rolling (c) sintering-rolling-sintering.

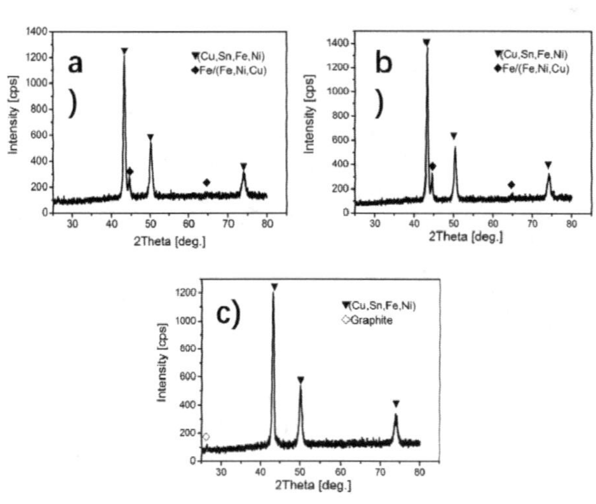


**Figure 3.** XRD patterns of Cu-Fe-Ni-Sn-Graphite samples prepared by different processes. (a) sintering; (b) sintering-rolling (c) sintering-rolling-sintering.

Figure 4 displays the EDS full spectrum results of three different sintering samples. As shown in Figure 4, the matrix compositions of the three samples all contain Fe, Cu, Ni, and Sn elements with similar proportions. The weight percentage and atomic percentage of Fe and Ni elements increase, while the weight percentage and atomic percentage of Cu decrease. This may be due to the precipitation of Cu during rolling and secondary sintering-rolling processes. After secondary sintering, the degree of alloying increases, and the graphite content subsequently increases, consistent with the XRD results.

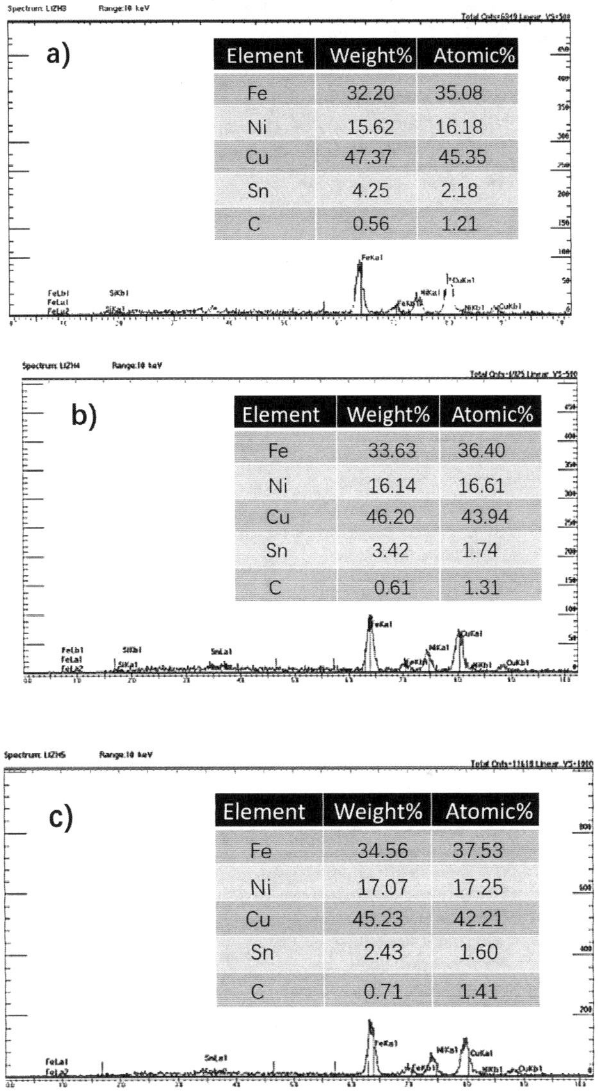


**Figure 4.** EDS results of a sintered layer of Cu-Fe-Ni-Sn-Graphite samples prepared by different processes. (a) Sintering; (b) Sintering-Rolling (c) Sintering-Rolling-Sintering.

Figure 5 shows the porosity of powder metallurgy samples prepared under three different process conditions: sintering, sintering-rolling, and sintering-rolling-sintering. The results indicate that the porosity decreases as the process steps increase. Compared to single sintering, the sintering-rolling and re-sintering process can effectively reduce the porosity of the material and increase its density, making it the preferred process for obtaining denser powder metallurgy components. This provides references for further optimization of powder metallurgy processes.

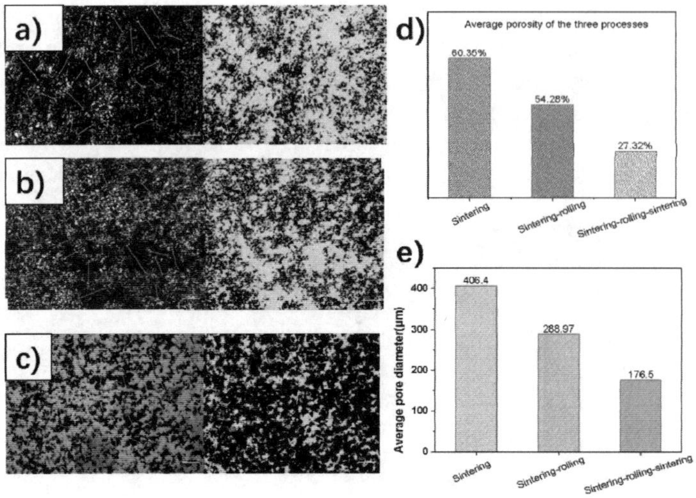


**Figure 5.** Mean porosity of Cu-Fe-Ni-Sn-Graphite under different preparation processes. (a) Sintering; (b) Sintering-Rolling; (c) Sintering-Rolling-Sintering; (d) Mean porosity values for the three processes (e) Mean pore size for the three processes.

### *3.2 Frictional wear performance*

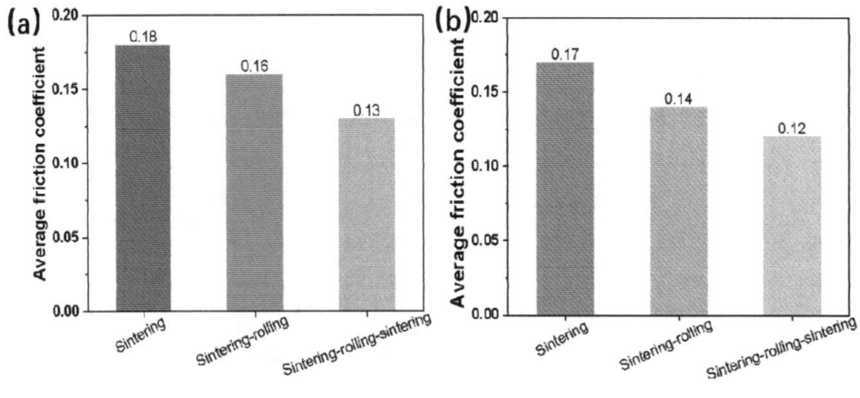


**Figure 6.** Friction performance of samples prepared by three different processes. (a) Average friction coefficient; (b) Average friction coefficient after running-in period.

According to Figure 4, friction coefficient testing of Cu-Fe-Ni-Sn-Graphite powder metallurgy samples prepared by three different processes revealed that the average friction coefficient of the sintering-rolling-secondary sintering process samples decreased from 0.18 to 0.13 compared to samples sintered only, indicating a 27% improvement in friction performance. After the running-in period, the friction coefficient of the sintering-rolling-secondary sintering process samples remained at 0.12, representing a 40% increase in friction performance compared to samples sintered only. This indicates that the sintering-rolling-secondary sintering process can significantly improve the anti-friction performance of the material. The reason is that the sintering-rolling-secondary sintering process improves the material's density and hardness, refines the grain size [19], promotes carbide generation, and enhances the wear resistance of the material [20]. This provides a reference for further optimizing the anti-friction performance of copper-iron-based self-lubricating composite materials. At the same time, rolling can promote more graphite particles to precipitate from the matrix surface, which is beneficial to enhance the lubrication performance of the samples, consistent with the EDS results in Figure 4 showing an increase in carbon content with secondary sintering and rolling [21].

*3.3 Abrasion mechanism*

To analyze the influence of different processes on the wear performance of materials, we observed the wear surface of the prepared materials through an optical metallographic microscope. As shown in Figure 7(a), the wear surface of the metal prepared by the sintering process only presents obvious furrows, with significant pores and a large amount of worn abrasive particles. As shown in Figure 7(b), the wear surface of the samples prepared by rolling and rolling-secondary sintering processes is smoother, with improved wear and less worn abrasive particles. Additionally, the samples of the rolling-secondary sintering process exhibit a more obvious metallic luster, which is associated with the increased hardness of the samples. The wear surface of the samples prepared by the sintering-rolling-sintering process is the smoothest, with no apparent worn abrasive particles, indicating a significant increase in material hardness and the best wear resistance. In general, with the improvement of the preparation processes, from sintering to sintering-rolling and then to sintering-rolling-sintering, the wear resistance of the metallic materials is significantly enhanced. This is mainly due to the denser material structure and finer grains achieved through rolling and secondary sintering, which substantially increases hardness and significantly improves wear resistance. This result is consistent with the friction performance results in Figure 6. After rolling, the porosity of the samples decreases, allowing better oil retention after immersion, preventing oil loss, and promoting solid-liquid synergistic lubrication. Additionally, after rolling, some fine graphite particles are extruded at the corresponding wear marks, forming a lubricating film and improving the lubrication performance of the samples. The metal carbides generated by secondary sintering also contribute to the improvement of wear resistance [20].

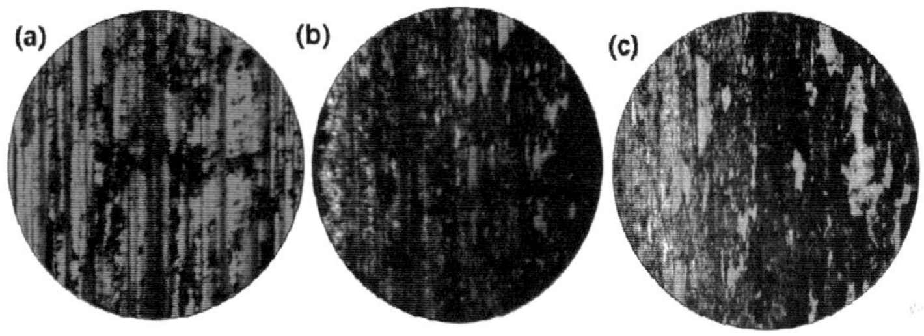


**Figure 7.** The wear morphology of three materials. (a) Sintering; (b) Sintering-rolling; (c) Sintering-rolling-sintering.

## 4 Conclusions

This study provides important references for the optimization design of sliding components in industrial robots. The developed materials, which are not only high in mechanical strength and low in friction loss but also highly applicable in the field of sliding components such as industrial robot joints, offer an effective approach to extending the service life of industrial robots and reducing maintenance costs. The main conclusions are as follows:

- This study utilized powder metallurgy technology to prepare Fe-Cu-Ni-Sn-Graphite porous self-lubricating composite materials with ultra-low friction properties capable of effectively storing lubricating oil.
- Through the sintering-rolling-secondary sintering process, the material's density and hardness were significantly improved, with an 18% increase in density and a 43% increase in hardness, providing a reference for optimizing the process of self-lubricating composite materials.
- Materials processed by the sintering-rolling-secondary sintering technique showed a 40% reduction in the average friction coefficient after the running-in period. They maintained a low and stable friction coefficient of 0.12 after continuous sliding for 24 hours, significantly better than samples processed only by sintering. This reveals the wear mechanism of the sliding interface of materials processed by sintering-rolling-secondary sintering, achieving solid-liquid synergistic lubrication with long-lasting lubricating properties.

**References**

[1] Analysis of the application of industrial robots in intelligent manufacturing. *Int. J. Front. Eng. Technol.* **4**, (2022).
[2] Evjemo, L. D., Gjerstad, T., Grøtli, E. I. & Sziebig, G. Trends in Smart Manufacturing: Role of Humans and Industrial Robots in Smart Factories. *Curr. Robot. Rep.* **1**, 35–41 (2020).
[3] Li, Z., Li, S. & Luo, X. An overview of calibration technology of industrial robots. *IEEECAA J. Autom. Sin.* **8**, 23–36 (2021).
[4] Adams, M. L. *Bearings: Basic Concepts and Design Applications*. (CRC Press, 2018). Doi: 10.1201/b22177.
[5] Li, M. *et al.* Super-wetting interfaces as a multiphase composite prototype for ultra-low friction. *Green Chem.* **24**, 7492–7499 (2022).
[6] Amigó-Mata, A., Haro-Rodriguez, M., Vicente-Escuder, Á. & Amigó-Borrás, V. Development of Ti–Zr alloys by powder metallurgy for biomedical applications. *Powder Metall.* **65**, 31–38 (2022).
[7] Ke, B. *et al.* Powder metallurgy of high-entropy alloys and related composites: A short review. *Int. J. Miner. Metall. Mater.* **28**, 931–943 (2021).
[8] Meignanamoorthy, M. *et al.* Microstructure, Mechanical Properties, and Corrosion Behavior of Boron Carbide Reinforced Aluminum Alloy (Al-Fe-Si-Zn-Cu) Matrix Composites Produced via Powder Metallurgy Route. *Materials* **14**, 4315 (2021).
[9] Guo, J. *et al.* Fabrication and tribological behavior of Fe-Cu-Ni-Sn-Graphite porous oil-bearing self-lubricating composite layer for maintenance-free sliding components. *Mater. Res. Express* **8**, 015801 (2021).
[10] Sun, J., Yan, K., Zhu, Y. & Hong, J. A High-Similarity Modeling Method for Low-Porosity Porous Material and Its Application in Bearing Cage Self-Lubrication Simulation. *Materials* **14**, 5449 (2021).
[11] Jia, H. *et al.* Hierarchical porous silicon structures with extraordinary mechanical strength as high-performance lithium-ion battery anodes. *Nat. Commun.* **11**, 1474 (2020).
[12] Ruan, H. *et al.* Efficacy of hierarchical pore structure in enhancing the tribological and recyclable smart lubrication performance of porous polyimide. *Friction* **11**, 1014–1026 (2023).
[13] Zhu, X. *et al.* Enhanced compressive strength of carbon aerogels with low density and high specific surface areas. *J. Porous Mater.* **29**, 1279–1285 (2022).
[14] Hejazi, V. & Nosonovsky, M. Wetting Transitions in Two-, Three-, and Four-Phase Systems. *Langmuir* **28**, 2173–2180 (2012).
[15] Wang, X. *et al.* Vegetable oil-based nanofluid minimum quantity lubrication turning: Academic review and perspectives. *J. Manuf. Process.* **59**, 76–97 (2020).
[16] Gao, C. *et al.* Effects of Strain Rate on Shear Bands and Texture Evolution in Grain-oriented Silicon Steel. *J. Phys. Conf. Ser.* **2268**, 012008 (2022).
[17] He, B. B. *et al.* High dislocation density–induced large ductility in deformed and partitioned steels. *Science* **357**, 1029–1032 (2017).
[18] Li, Z. *et al.* Enhancement of Strength by Multiple Rolling at Room Temperature in the 1045 Steel Prepared by Aluminothermic Reaction. *Steel Res. Int.* **90**, 1800556 (2019).
[19] Tarraste, M. *et al.* Ultrafine Cemented Carbides with Cobalt and Iron Binders Prepared via Reactive *In Situ* Sintering. *Solid State Phenom.* **320**, 176–180 (2021).
[20] Tymoshenko, Ya. G., Gadzyra, M. P., Davydchuk, N. K., Pinchuk, M. O. & Galyamin, V. B. Structurization of Steel Produced from Iron Doped with a Nanosized Silicon Carbide Composite Powder. *Powder Metall. Met. Ceram.* **60**, 323–330 (2021).
[21] Cofiño-Villar, A., Alvarez-Antolin, F., Asensio-Lozano, J. & Garcia-Garcia, M. Control over the Percentage, Shape and Size of the Graphite Particles in Martensitic White Castings Alloyed with Cr, Nb and Mg. *Materials* **12**, 185 (2019).

# Experimental and numerical study on flexural properties of glass fiber reinforced plastic sandwich plates

**Huijie Wang[1], Bei Zhang[1], Zehao Qi[1], Hong Wang[1] and Yongsheng Chai[1,2]**

[1]Defence Engineering Institute, AMS, PLA, Beijing 100850, China

[2]Corresponding author's e-mail: 2285657249@qq.com

**Abstract:** Due to their high specific strength and stiffness, glass fiber reinforced plastic (GFRP) sandwich plates are widely used in various fields such as construction, aerospace, shipbuilding, and military. This paper studies the bending performance of FRP sandwich plates with reinforcement through experimental and numerical simulation methods. The study shows that the bending performance of unidirectional and bidirectional reinforced plates under the same concentrated load is similar, and bidirectional reinforcement cannot significantly reduce the bending deformation of the sandwich plate. Under uniform load, the four-sided supported bidirectional reinforced sandwich plate can significantly reduce the deformation of the plate, and the deformation can be reduced by 40% compared to the unidirectional reinforced plate.

## 1. Introduction

The glass fiber reinforced plastic (GFRP) sandwich structure has high specific stiffness, high specific strength, and good heat and sound insulation capabilities, widely used in construction, aerospace, marine, military, and other fields. Typical sandwich structures mainly include honeycomb, foam, and other structures. The traditional honeycomb sandwich structure has high strength, but it is difficult to manufacture and maintain and high cost [1-2]; the production of foam sandwich plates is relatively simple, but the bending stiffness is relatively low. Research has shown that adding stiffeners to composite sandwich structures could effectively improve the structural performance of the sandwich plates [3-5]. Theoretical, experimental, and numerical methods are widely used in the research of the mechanical properties of composite sandwich plates [6-11]. With the development of numerical simulation technology, the numerical method could reduce the experimental work and improve the research efficiency of composite sandwich structure study [12-15]. To reveal the effects of reinforced forms and boundary conditions on the bending performance of GFRP sandwich plates, two kinds of reinforced forms of GFRP sandwich plates were designed, and through bending tests and numerical simulation research, the deformation laws of GFRP sandwich plates under different boundary conditions were studied.

## 2. Specimen preparation

The GFRP reinforced sandwich plate proposed in this paper uses GFRP as the skin, filled with phenolic foam as the interlayer, and a GFRP square tube as the stiffener. Phenolic foam and GFRP square tube are bonded by vinyl room temperature curing resin. Figure 1 is the structural diagram of the GFRP reinforced sandwich plate. The basic mechanical parameters of each part of the GFRP reinforced sandwich plate are: The elastic modulus of GFRP skin is 29.6 GPa, tensile strength is 483 MPa; the elastic modulus of the stiffener is 46 GPa, and the tensile strength is 670 MPa; the compressive strength of foam is 0.19 MPa, and the elastic modulus is about 16 MPa. The length, width, and height of the

GFRP reinforced sandwich plate specimens are 2000 mm, 1000 mm, and 50 mm, respectively. The skin thickness is 10 mm; the thickness of the phenolic foam filling layer is 30 mm. The side length of the stiffener cross section is 30 mm, and the wall thickness is 4 mm. Figure 2(a) shows the unidirectional stiffened model. Figure 2(b) shows the bidirectional stiffened model.

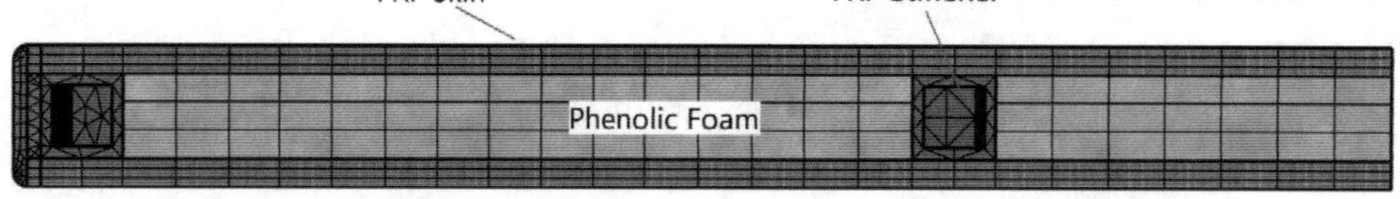


**Figure 1.** Structural diagram of GFRP reinforced sandwich plate.

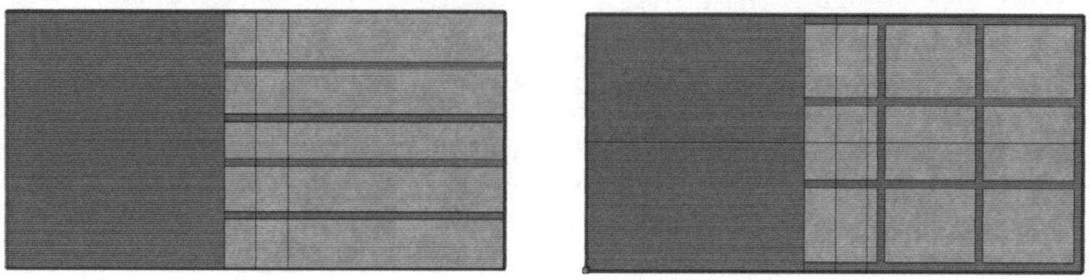

(a) unidirectional stiffened (b) bidirectional stiffened

**Figure 2.** GFRP reinforced sandwich plate model.

## 3. Flexural properties test

The three-point flexural test is carried out on the specimens of unidirectional stiffened sandwich plates. The loading device of the flexural test adopts the gradual loading mode of heavy objects, and the loading diagram of the three-point flexural test is shown in Figure 3. The concentrated load was gradually applied during the test, and the midspan displacement of the sandwich plate specimen was recorded.

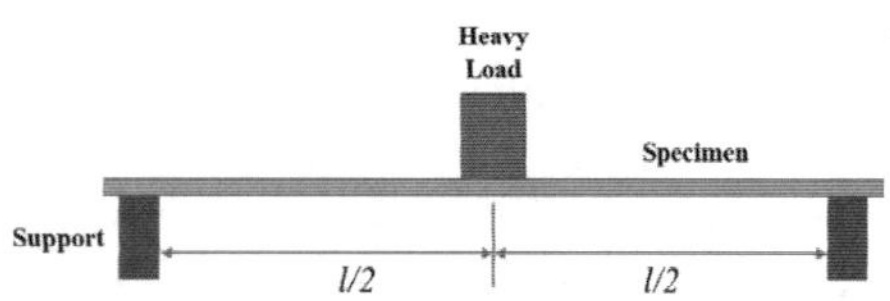


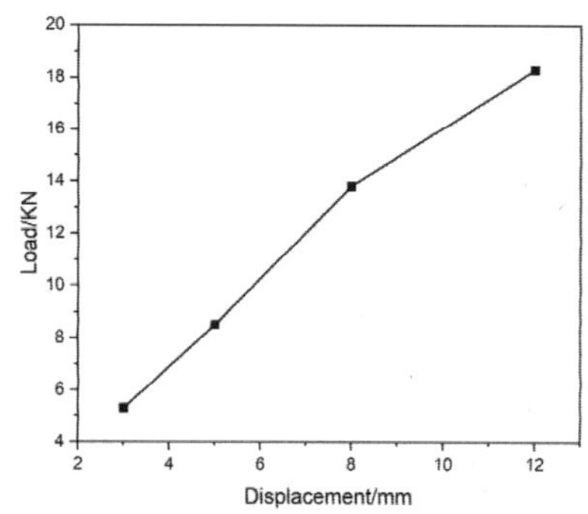


**Figure 3.** Sketch map of flexural test set-up. **Figure 4.** Deflection curve of a specimen.

Figure 4 shows the variation of mid-span displacement of unidirectional reinforced specimens with load. It shows that the displacement gradually increases with the increase of load. When the load is less than 13.8 kN, the mid-span displacement increases linearly with the load; when the load exceeds 13.8 kN, the slope of the load displacement curve reduces.

## 4. Numerical simulation of flexural test

This paper uses the finite element software COMSOL to simulate the bending of a reinforced sandwich plate. Firstly, the bending test and numerical simulation results of the unidirectional stiffened sandwich structure are compared. Based on verifying the reliability of the numerical method, the bending performance of a bidirectional stiffened sandwich structure is numerically simulated.

*4.1 Numerical model and parameters*
According to the symmetry principle, a quarter of the stiffened sandwich plate model is constructed to improve the calculation efficiency. A uniformly distributed surface load is applied within 125 mm from the symmetrical boundary on the upper surface of the right end of the model to replace the concentrated load in the test, as shown in Figure 5. The plate and support adopt the contact boundary.

Model parameters are as follows: The length is 1000 mm, the width is 500 mm, and the height is 50 mm. To improve computational efficiency, hexahedral structured grids are used for skin and foam, and triangular prism grids are used for stiffeners. Figure 5 shows the meshing results of the model.

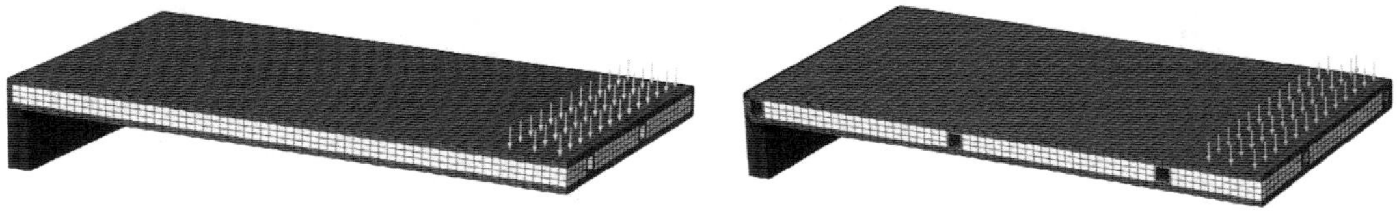

(a) unidirectional stiffened (b) bidirectional stiffened

**Figure 5.** Mash of GFRP reinforced sandwich plate model.

*4.2 Numerical simulation results of the unidirectional stiffened model*
In this paper, the flexural properties test of unidirectional stiffened sandwich plate under different loads were simulated respectively. The mid span displacement of the sandwich plate is listed in Table 1. The numerical simulation results are close to the experimental values, and the error is within 9%. Therefore, the numerical simulation method can effectively simulate the bending deformation behavior of the sandwich structure.

**Table 1.** The mid span displacement of the unidirectional stiffened sandwich plate.

| Load/kN | Mid span displacement /mm | | Error |
|---|---|---|---|
| | Test | Numerical | |
| 5.3 | 3 | 3.2 | 7% |
| 8.5 | 5 | 4.99 | 0% |
| 13.8 | 8 | 8.16 | 2% |
| 18.3 | 12 | 10.89 | 9% |

For example, the stress of different parts of the unidirectional reinforced sandwich structure under 18.3 kN is shown in Figure 6. It shows that the high stress is mainly distributed in the mid span of the skin and the upper and lower sides of the stiffener, and the maximum value is about 28.77 MPa, which is far less than the tensile strength. The stress of the foam filling layer is very small, and the maximum value is about 31.75 KPa.

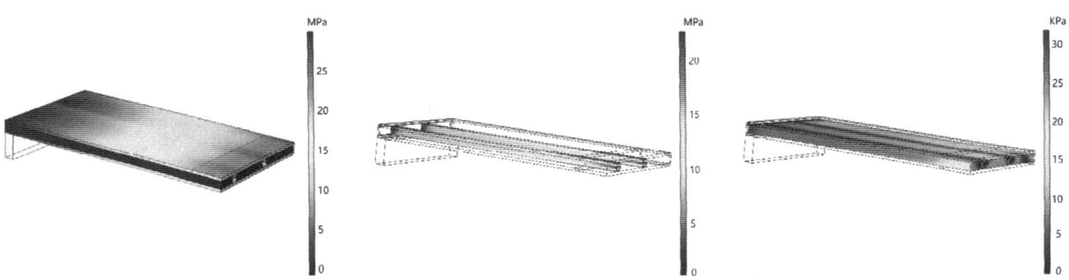


**Figure 6.** Von mises stress of unidirectional stiffened model under 18.30 kN.

*4.3 Numerical simulation results of the bidirectional stiffened model*
Aiming to reveal the influence of different reinforcement forms on the bending performance of sandwich plates, a numerical simulation method was used to simulate the bending deformation of bidirectional

reinforced sandwich plates under different loads. Table 2 lists the displacement at the middle of the sandwich plate under different loads. It shows that under the same load, the displacement at the middle of the plate span is close, and the difference is less than 4%. The bidirectional stiffened reinforcement mode in this paper cannot significantly improve the bending performance compared to the unidirectional stiffened mode.

**Table 2.** The mid span displacement of the stiffened sandwich plate.

| Load/kN | Mid span displacement /mm | | Error |
|---|---|---|---|
| | Unidirectional stiffened | Bidirectional stiffened | |
| 5.3 | 3.2 | 3.08 | 4% |
| 8.5 | 4.99 | 4.84 | 3% |
| 13.8 | 8.16 | 7.92 | 3% |
| 18.3 | 10.89 | 10.56 | 3% |

For example, the stresses of the bidirectional stiffened reinforced sandwich structure under 18.3 kN are shown in Figure 7. It shows that the stress of the middle part of the skin and the upper and lower sides of the stiffener of the sandwich plate is large, and the maximum value is about 29.86 MPa, which is far less than the tensile strength. The stress of the foam filling layer is very small, and the maximum value is about 34.47 KPa. The stress distribution is more uniform in the bidirectional stiffened model than in the unidirectional stiffened model.

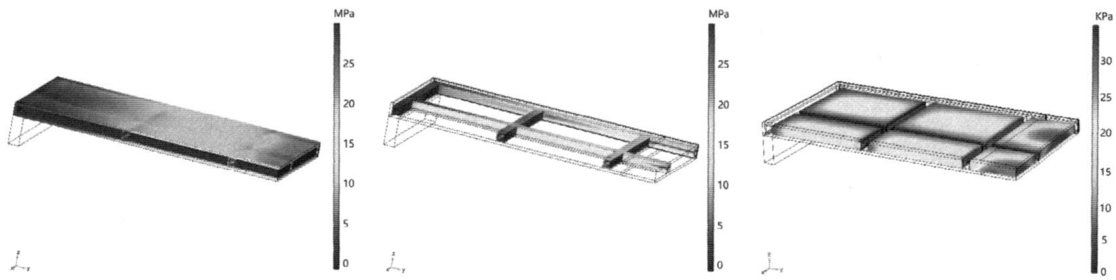


**Figure 7.** Von mises stress of bidirectional stiffened model under 18.30 kN.

## 5. Simulation of four sides supported stiffened sandwich plate

To study the influence of boundary conditions on plate deformation, two kinds of stiffened sandwich plates under the same uniform load in the state of four sides supported were studied by numerical simulation. Figure 8 shows the deformation of the stiffened model. The results show that for the four side supported model, the deformation of the center of the plate is the largest under the uniformly distributed load. The deformation of the bidirectional stiffened model is significantly less than that of the unidirectional stiffened model, and the deformation is reduced by about 40% under the same load conditions.

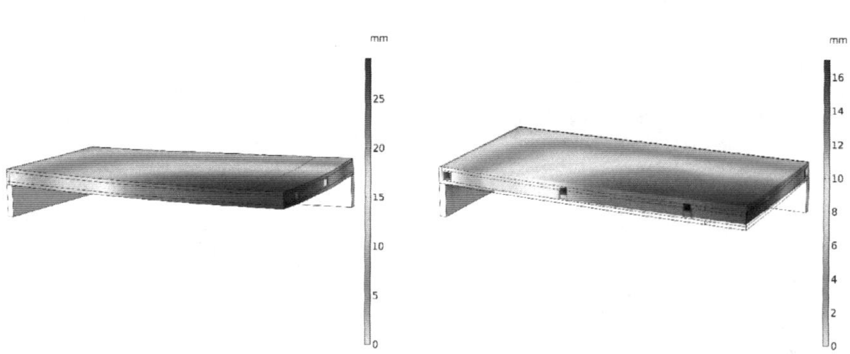


**Figure 8.** Deformation of the stiffened model with four sides supported.

Figure 9 shows that under the same load, the internal stress of the bidirectional stiffened reinforced model is much smaller than that of the unidirectional stiffened reinforced model, and the maximum stress is reduced by about 50%; the high stress area of the unidirectional reinforced model is mainly distributed around the stiffener, and the internal stress distribution of sandwich structure is quite nonuniform; the stress distribution of the bidirectional stiffened model is relatively uniform, and the high stress is mainly distributed on the skin surface and the side of the short stiffener. The above results show that under the condition of being supported on four sides, the mechanical properties of different parts of the sandwich plate can be effectively played by bidirectional reinforcement.

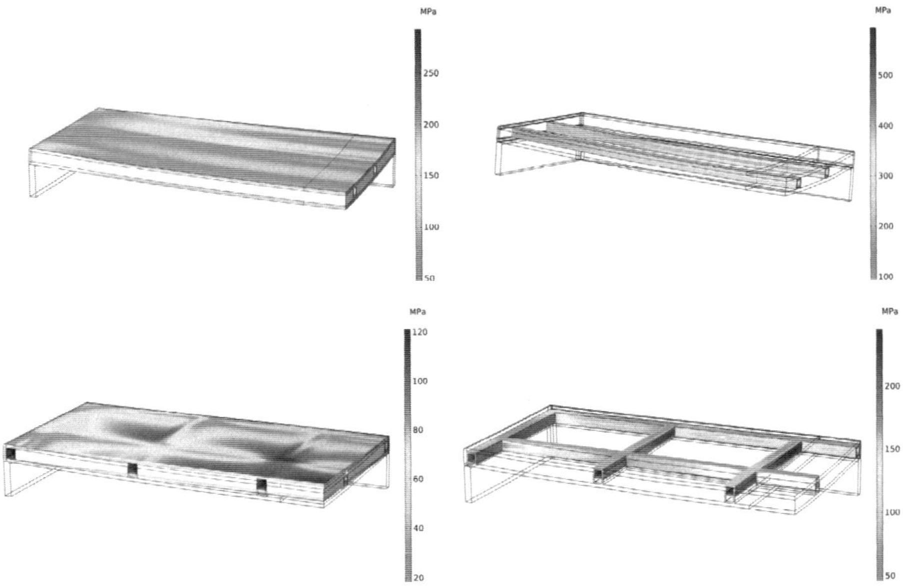


**Figure 9.** The stress of the stiffened model with four sides supported.

## 6. Conclusions

In this paper, two types of GFRP reinforced sandwich plates were constructed, and the flexural properties of the plate were studied through points test and numerical simulation. A numerical model was developed for a four side simple supported reinforced sandwich plate to investigate the effect of reinforcement type on deformation. The conclusions are as follows:

(1) With the three-point bending condition, the bidirectional stiffened reinforcement paper could not significantly improve the bending performance of the unidirectional stiffened.

(2) Under four sides simple supported boundary condition, the deformation of the bidirectional stiffened model is significantly less than that of the unidirectional stiffened model, and the deformation is reduced by about 40%.

(3) The stress distribution inside the two types of reinforced model is similar under a three-point bending condition; the internal stress of the bidirectional stiffened plate is about 50% of that of the unidirectional stiffened model under four side simple supported boundary condition.

## Reference

[1] Zhang, X., Huang, J., Wang, Y., et al. (2019) Numerical simulation and experimental research on resistance performance of honeycomb structure composite protective equipment. Protective Engineering 2019, 41(2): 24-30.

[2] Wu Q., Fang H., Gao F., et al. (2020) Experimental testing and numerical simulation on flexural behavior of composite grid panels [J]. Composites Science and Engineering, 1: 12-20.

[3] Kong L., Zhang H., Fang G., et al. (2017) Mechanical properties analysis of FRP sandwich plate frame structure with a large opening. Fiber Reinforced Plastics/Composites, 2017(6): 5-11.

[4] Wei, S., Zhai, Y., Mei, Z. (2014) Analysis of the dynamic natural characteristic's influencing factors to composite reinforced sandwich plate. Ship Engineering, 2014, 36(03): 99-102.
[5] Li. Q., Liu, W., Fang, H., et al. (2010) Study of trusscore polyurethane sandwich wall plate stiffened by pultruded I-beam, 2010, 37(7): 54-56.
[6] Chen H., Wang L., Liu Q., et al. (2017) Analysis of I-type fracture toughness of sandwich structure with GFRP skins and a foam core under coupling effects of temperature and loading. Journal of Nanjing Tech University, 39(01): 93-99.
[7] Zhang, B., He, H., Zhou, J., et al. (2019) Construction and failure analysis of ultra-light GFRP fluted-core sandwich protective structures. Composites Science and Technology, 173: 73-82.
[8] Triantafillou, T. C., Gibson, L. J. (1987) Minimum weight design of foam core sandwich panels for a given strength. Materials Science & Engineering, 95: 55-62.
[9] Hong J., Liu W., Fang H., et al. (2018) Stress analysis of composite sandwich panels under unidirectional bending [J]. Engineering Mechanics, 35(4): 41-51.
[10] Steeves, C.A., Fleck, N.A. (2004) Collapse mechanisms of sandwich beams with composite faces and a foam core, loaded in three-point bending. Part I: analytical models and minimum weight design. International Journal of Mechanical Sciences, 46, 561-583.
[11] Nazari, A., Hosseini-Toudeshky, H., et al. (2019) Experimental investigations on the sandwich composite beams and panels with elastomeric foam core. Journal of Sandwich Structures and Materials, 21, 865 - 894.
[12] Russo, A., Zuccarello, B. (2007) Experimental and numerical evaluation of the mechanical behaviour of GFRP sandwich panels. Composite Structures, 81, 575-586.
[13] Wang N., Xiao B., Xiao S. (2013) Analysis for bending strength of foam cored FRP sandwich plate. Ship & Ocean Engineering, 42(6): 46-48.
[14] Arruda, M. R. T., Garrido, M., et al. (2018) Numerical modelling of the creep behavior of GFRP sandwich panels using the Carrera Unified Formulation and Composite Creep Modelling. Composite Structures, 183: 103-113.
[15] Tang C., Chen X., Wang H. (2009) Analysis for mechanical properties of foam cored sandwich based on nonlinear FEM. Chinese Quarterly of Mechanics, 30(2): 273-280.

# Effect of ultrasonic surface rolling on the friction and wear properties of 8620H carburized steel at elevated temperatures

**Hao Zhong[1], Shengguan Qu[1,3], Wang Long[2], Rixin Deng[2], and Wenxue Zhou[2]**

[1]South China University of Technology, Guangzhou, Guangdong, 510640, China
[2]Liuzhou Arays Technology Co., LTD, Liuzhou, Guangxi, 545000, China

[3]Corresponding author's e-mail: qusg@scut.edu.cn

**Abstract:** The surface strengthening treatment of 8620H carburized steel was carried out by ultrasonic surface rolling (USR) process, and the samples with and without the treatment were subjected to friction and wear experiments at 25°C, 100°C, and 150°C. The experimental results showed that the samples with USR treatment had higher surface hardness, lower surface roughness, deeper hardened layer depth, and higher surface residual compressive stress than the samples without USR treatment. Meanwhile, the wear degree of the samples deteriorated as the temperature increased, but the anti-wear performance of the samples with USR treatment increased by 52.9%, 29.8%, and 14.9% compared with the samples without USR treatment at 25°C, 100°C, and 150°C, respectively, indicating that the USR treatment still had a better strengthening effect at high temperatures.

## 1. Introduction

AISI 8620H steel is an alloy structural steel renowned for its high hardness and strength, making it a widely used material in the manufacture of gears and bearings. Furthermore, it proves useful in manufacturing mechanical parts for challenging environments such as those with high temperatures and pressures. This steel exhibits good oxidation resistance at elevated temperatures and can endure a certain amount of thermal fatigue and creep [1]. However, improving the service life of parts under elevated temperatures is very important to reduce costs and improve efficiency, as the service life of parts under elevated temperatures is generally reduced. Ultrasonic surface rolling is a process that uses an ultrasonic tool to apply constant pressure to the surface of a part and vibrate at an ultra-high frequency. This process can increase the hardness of the parts' surface, reduce the surface roughness, and form a deeper hardened layer and higher residual stress on the parts' surface to achieve the effect of improving the wear resistance. Currently, numerous reports examine the friction and wear performance of steel with surface strengthening treatment. For instance, LI Y et al. [2] studied the effect of laser peening on the wear resistance of 20CrNiMo steel, and the results indicated that the wear resistance of the samples improves with an increasing laser impact overlap rate. Zhang Y L et al. [3] have investigated the influence of shot peening (SP) on the friction and wear properties of 17Cr2Ni2MoVNb steel, which showed that the SP-treated samples have better wear resistance. Hu X F et al [4] conducted a study on the influence of discrete surface laser quenching on the friction and wear performance of 25CrNi2MoV steel and found that the use of a discrete surface laser quenching process led to a substantial decrease in the wear volume of the samples by approximately 93.2%. As for the study of the wear resistance of steel at different temperatures, this also has a lot of research such as He Q [5] et al. and Emami M [6] et al. studied the wear resistance of 9Cr18Mo steel and 31CrMoV9 nitride steel at various high temperatures, respectively.

However, there is limited research on the investigation of friction and wear characteristics of steel at varying temperatures after USR treatment. Therefore, in this paper, we initially strengthened 8620H carburized steel samples utilizing ultrasonic surface rolling. Subsequently, we performed friction and wear experiments on the above samples with and without the USR treatment to investigate their friction and wear performance at 25°C, 100°C, and 150°C.

## 2. Preparation of the sample

The chemical composition of AISI 8620H steel is shown in Table 1 [7]. The material selected consisted of a 50 mm diameter bar. All samples were dis-shaped, measuring φ45 mm × 6 mm, cutting from this bar. The samples received a carburizing heat treatment, and the process is depicted in Figure 1. First, place the samples in a carburizing furnace set at 940°C for 3.5 hours, then cool to 850°C and keep for 0.5 hours. Thereafter, quenched in oil and maintained at 160°C for 2.5 hours before air-cooling and tempering. After the heat treatment process, the samples underwent uniform grinding on a surface grinder to eliminate the decarburization layer on the surface. Following that, an ultrasonic cleaner was employed to complete the pretreatment of the samples. Subsequently, the samples were cleaned and blown dry, resulting in a surface hardness of approximately 62 HRC after the completion of the pretreatment process.

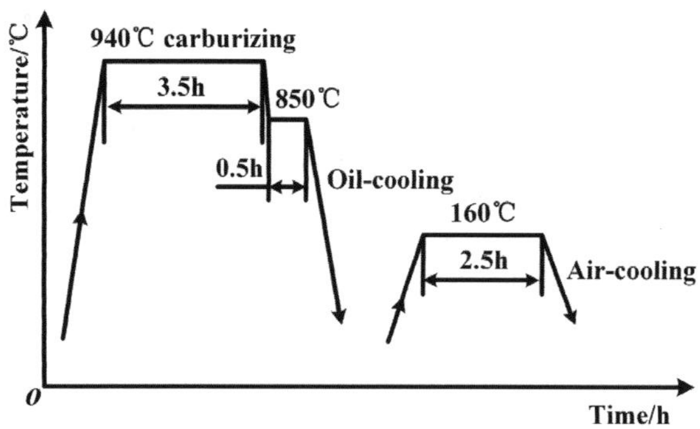


**Figure 1.** Carburizing heat treatment programmed for samples.

**Table 1.** Chemical composition of AISI 8620H steel (*ωt.*/%).

| C | Si | Mn | Cr | Ni | Mo | Al | P | S | Cu |
|---|---|---|---|---|---|---|---|---|---|
| 0.17 | 0.21 | 0.84 | 0.57 | 0.45 | 0.20 | 0.02 | 0.01 | 0.02 | 0.1 |

The ultrasonic surface rolling apparatus is utilized to conduct USR processing on the sample, and its processing principle is illustrated in Figure 2. The device's front end features a 10 mm diameter silicon nitride ceramic rolling ball with a hardness of approximately 95 HRC. An air compressor and an ultrasonic generator are employed to ensure a constant static pressure is exerted on the rolling ball, which carries out high-frequency vibration on the sample's surface whilst maintaining a constant pressure. The high-frequency vibrations cause grain refinement and plastic deformation of the sample's surface material. In USR processing, the sample is secured onto the machine tool and co-rotated, while the ultrasonic rolling device applies perpendicular pressure onto the sample's surface. The rolling device then moves reciprocally along the radial direction of the sample, permitting a specific range of multi-pass USR processing on its surface. During USR processing, heat can easily accumulate on the surface of contact. To prevent the ball and sample from reaching high temperatures, lubricating oil can continuously flow out from around the ball. This process helps avoid high temperatures and provides lubrication and cooling for the rolling ball and sample. The optimized processing parameters for the USR, derived from laboratory experimentation, are presented in Table 2 below.

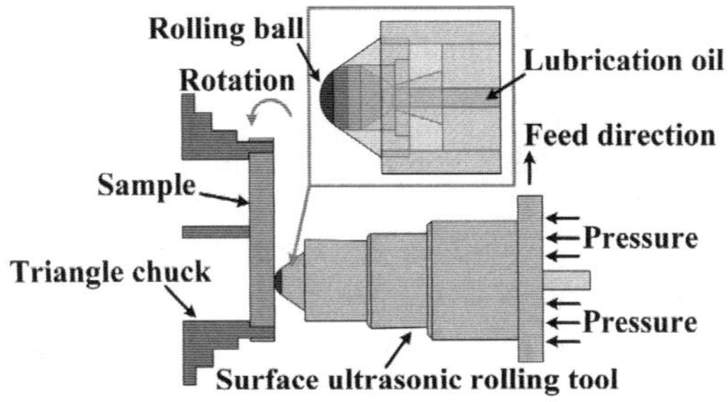


**Figure 2.** Schematic diagram of USR processing.

**Table 2.** Parameters of USR processing.

| Frequency/ (kHz) | Pressure/ (MPa) | Amplitude/ (μm) | Spindle speed/ (rpm) | Axial feed/ (mm/r) | Passes |
|---|---|---|---|---|---|
| 27.5 | 0.50 | 10 | 45 | 0.08 | 6 |

## 3. Friction and wear tests

The test at elevated temperatures of 8620H carburized steel was performed using the MMU-5G friction and wear tester. The counter-abrasion form was applied using the pin and disc contact mode, and the custom-made GCr15 ball-end pin was used as the counter-abrasion sample. The pin is 18 mm long, with a head diameter of 3 mm, and heat-treated to achieve a hardness of 60-62 HRC. The ball-end region was polished and ground to a surface roughness of Ra 0.8 μm. Figure 3 displays the testing machine and samples used in the experiment, the pin sample is mounted in the fixture holes located at the upper end of the machine, and the disc sample is fixedly mounted on the fixture at the lower end.

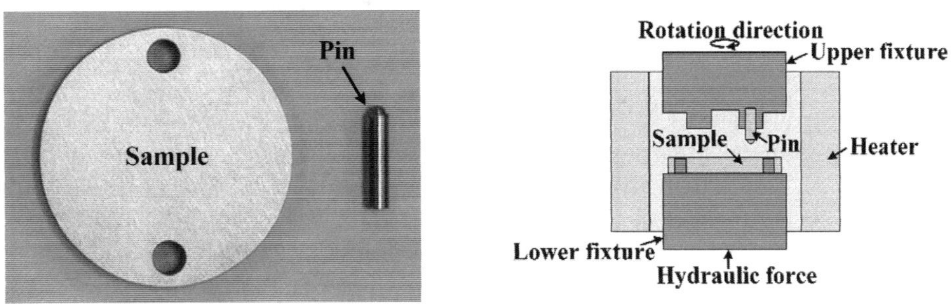


a) Samples b) Schematic diagram of MMU-5G tester

**Figure 3.** Experimental sample and schematic diagram of the tester.

At the start of the experiment, the testing machine's hydraulic spindle exerts a vertical upward load on the lower fixture. It then moves the disc sample upward, bringing the two samples in close contact. Subsequently, it applies a force to achieve and maintain the pre-determined load between the sample and the pin. Last, the heating furnace is activated to elevate the temperature. Once the heating furnace has reached its set temperature and remained stable for at least 5 minutes, the upper fixture's spindle will engage, causing the pin to rotate on the sample's surface. This enables the performance of friction and wear tests under high-temperature work conditions. The friction wear tester records real-time data on various test parameters including loading load, spindle speed, friction coefficient, test temperature, and test duration. The device automatically stops the test when the predetermined time limit is reached. Table 3 displays the friction wear test parameters applied in this study.

**Table 3.** Test parameters.

| Sample status | Hydraulic force/(N) | Spindle velocity /(r·min-1) | Wear time /(s) | Temperatures / (°C) |
|---|---|---|---|---|
| Untreated/USR | 300 | 60 | 3600 | 25/100/150 |

## 4 Results and discussion

*4.1 Ultrasonic surface rolling strengthening properties.*

Figure 4 displays the surface SEM morphology of the samples with and without USR treatment. It is clear that the surface of the sample without USR treatment has more noticeable and deep scratches, while after USR treatment, most of the scratches on the surface of the sample disappeared, and only a few deep scratches remain on the sample surface, which indicates that the sample surface can achieve a relatively flat effect under the USR parameters shown in Table 2, but compared with the effect of USR treatment in [8], there are still more unclosed scratches on the surface of this experiment, which does not achieve the effect of complete smoothing. This is due to the high hardness of the surface of the 8620H carburized sample, which reduces the ability of the USR processing to close the abrasive marks on the sample surface [9].

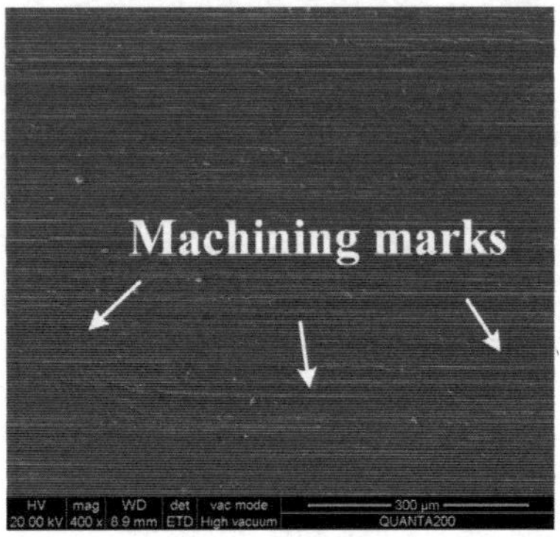


a) Untreated sample

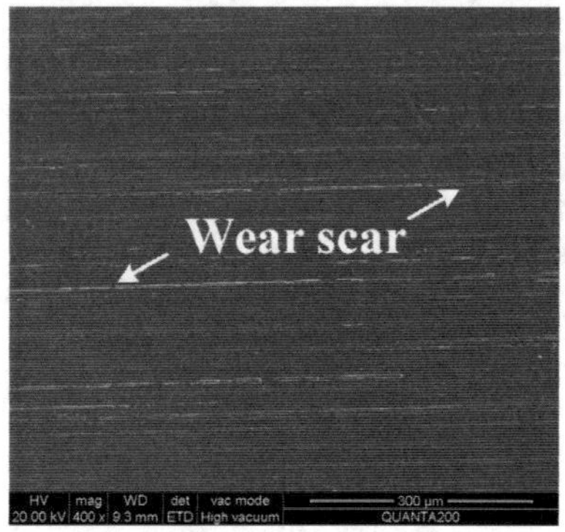


b) USR-treated sample

**Figure 4.** Surface morphology of samples.

Table 4 displays the surface hardness, surface roughness, and surface residual compressive stress of the samples with and without USR treatment. As depicted in the table, the sample's hardness augmented from 760 HV0.2 to 834 HV0.2, accompanied by a decline in its surface roughness from 0.798 to 0.406 and an elevation in its residual compressive stress from -207 MPa to -621 MPa. These results demonstrate that USR treatment can significantly enhance the surface properties of 8620H carburized steel, and thus increase the samples' resistance to wear and fatigue [10].

**Table 4.** Surface properties of different samples.

| | Untreated sample | USR-treated sample |
|---|---|---|
| Surface hardness /($HV_{0.2}$) | 760 | 834 |
| Surface roughness /(μm) | 0.798 | 0.406 |
| Surface residual compressive stress /(MPa) | -207 | -621 |

The cross-sectional organization of samples with and without USR treatment is shown in Figure 5, the organization of untreated sample mainly consists of slaty martensite and residual austenite, and the density distribution of the organization is relatively uniform in the cross-sectional, whereas the organization of the USR-treated sample produces a significant change in the range of 24 μm from the surface, in which the distance between martensite is significantly reduced. In this range, the martensite morphology becomes finer and shorter than that of the untreated sample, and the volume of residual austenite is significantly smaller or even disappears, indicating that USR treatment under this parameter can produce obvious organization refinement of the surface organization of the 8620H carburized steel, and enhance its surface hardness, thus improving its wear resistance.

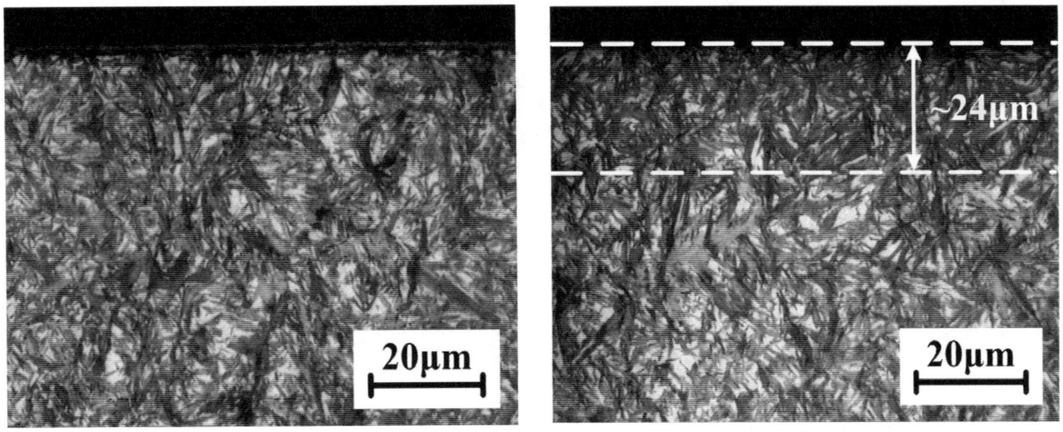


a) Untreated sample b) USR-treated sample

**Figure 5.** Cross-sectional organization of samples with and without USR treatment.

*4.2 Frictional performance*

Figure 6 shows the sliding friction coefficients of the samples with and without USR treatment at different temperatures. As shown in Figures 6(a), 6(b), 6(d), and 6(e), the friction coefficients of the samples at 25°C and 100°C increased slowly at the beginning of the experiment and then remained stable, indicating that the initial break-in phase lasted for a long time at these two temperatures. This is because the surface of the samples has a higher hardness at relatively low temperatures. At the beginning of contact between the two samples, the contact mode between the ball joint pin and the surface of the sample is the point contact mode. At this time, the friction force is small. As wear progresses, the contact form of the sample gradually changes to the surface contact mode, and the friction force also gradually increases, resulting in a longer time in the initial break-in phase. While at 150°C, as shown in Figure 6(c) and (f), the break-in phase is shorter and the range of fluctuations in the friction coefficient is significantly smaller compared to the samples at 25°C and 100°C, which demonstrates that the wear process of the samples is smoother under high-temperature conditions. This is because the sample reaches a low-temperature tempering temperature at 150°C, causing a gradual decrease in its hardness. At the same axial load, the surface material is also more susceptible to abrasion, resulting in a reduced friction force and less sudden changes. In addition, comparing Figures 6(a), 6(b), 6(c) with Figures 6(d), 6(e), 6(f), it can be seen that the fluctuation range of the coefficient of friction of the samples with USR treatment is smaller than that of the samples without USR treatment at these three temperatures, which suggests that the USR treatment can make the wear process of the samples smoother because the USR processing has greatly reduced the roughness of the sample surface, which in turn suppressed the possibility of the friction to be subjected to sudden changes.

Figure 7 shows the average friction coefficients in the steady-state of the samples with and without USR treatment at various temperatures, and the results demonstrate that the average friction coefficients in the steady-state of the samples with and without USR treatment at 25°C, 100°C, and 150°C, are 0.521, 0.498, 0.450 and 0.496, 0.454, 0.438 respectively. This shows that the friction coefficients of the samples at various temperatures were reduced after USR treatment, indicating that the USR treatment can significantly enhance the friction performance of the samples at various temperatures. Moreover, as the temperature increases, the average coefficient of friction in a steady state decreases gradually. This implies that under high-temperature conditions, the friction force produced by the samples is lesser than that under low temperatures. Such a decrease may be attributed to the reduction in surface hardness of the samples when exposed to high temperatures or the development of an adhesive layer on the sample surface.

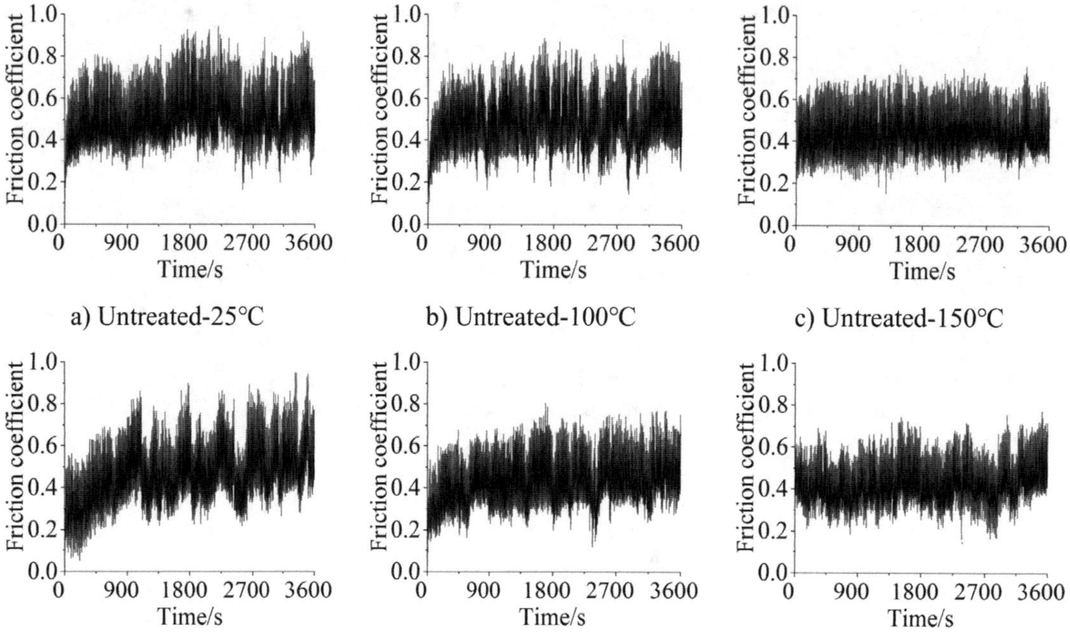


a) Untreated-25°C b) Untreated-100°C c) Untreated-150°C

d) USR-25°C e) USR-100°C f) USR-150°C

**Figure 6.** Sliding friction coefficients of samples at various temperatures.

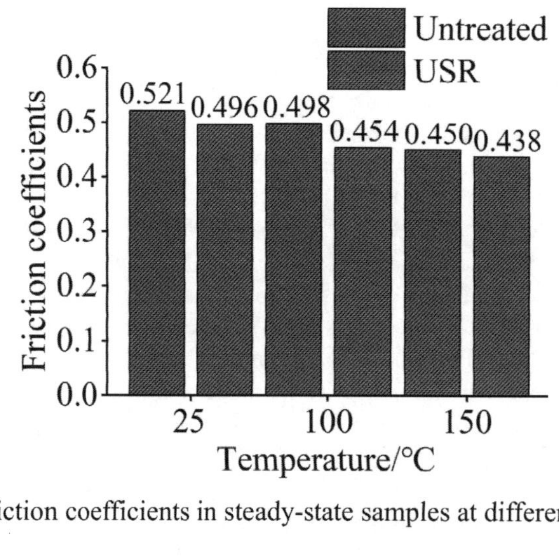


**Figure 7.** Average friction coefficients in steady-state samples at different temperatures.

*4.3 Wear performance*

The 3D morphology of sliding wear of samples with and without USR treatment at various temperatures is shown in Figure 8, from which the wear area and depth of wear on the surfaces of the samples increase gradually with the increase in temperature, which indicates that the samples experience a higher degree of wear as the temperature increases. This is because as the temperature increases, the surface hardness of the sample diminishes, which reduces the specimen's ability to resist wear. In addition, the ability of the sample to resist crack propagation diminishes as the temperature increases, which may also contribute to the greater degree of wear on the sample. Comparing Figures 8(a), 8(b), and 8(c) with Figures 8(d), 8(e), and 8(f), the wear degree of samples with USR treatment at various temperatures is reduced compared to untreated samples, which is most obvious at ambient conditions.

Figures 8 (a1), 8(b1), 8(c1) and 8 (d1), 8(e1), 8(f1) show the corresponding wear contours of the samples with and without USR treatment at three temperatures, respectively, and it can be seen from the contour plots that the average depth of the wear pits of the samples with USR treatment is smaller than that of the samples without USR treatment at all three temperatures. Calculate the area enclosed by the contour curve and the straight line with the depth of the abrasion mark of 0 and integrate in the circumferential direction to find out the wear volume of the sample under various working conditions. As shown in Figure 9, the wear volumes of samples with and without USR treatment are 0.129 $mm^3$, 0.219 $mm^3$, 0.355 $mm^3$, and 0.274 $mm^3$, 0.312 $mm^3$, 0.417 $mm^3$ at 25°C, 100°C, and 150°C, respectively. The wear volumes of the samples at 25°C, 100°C, and 150°C after USR treatment are reduced by 52.9%, 29.8%, and 14.9%, indicating that the USR treatment can significantly enhance the wear performance of the samples at various temperatures, but its enhancement effect decreases with the rise of temperature.

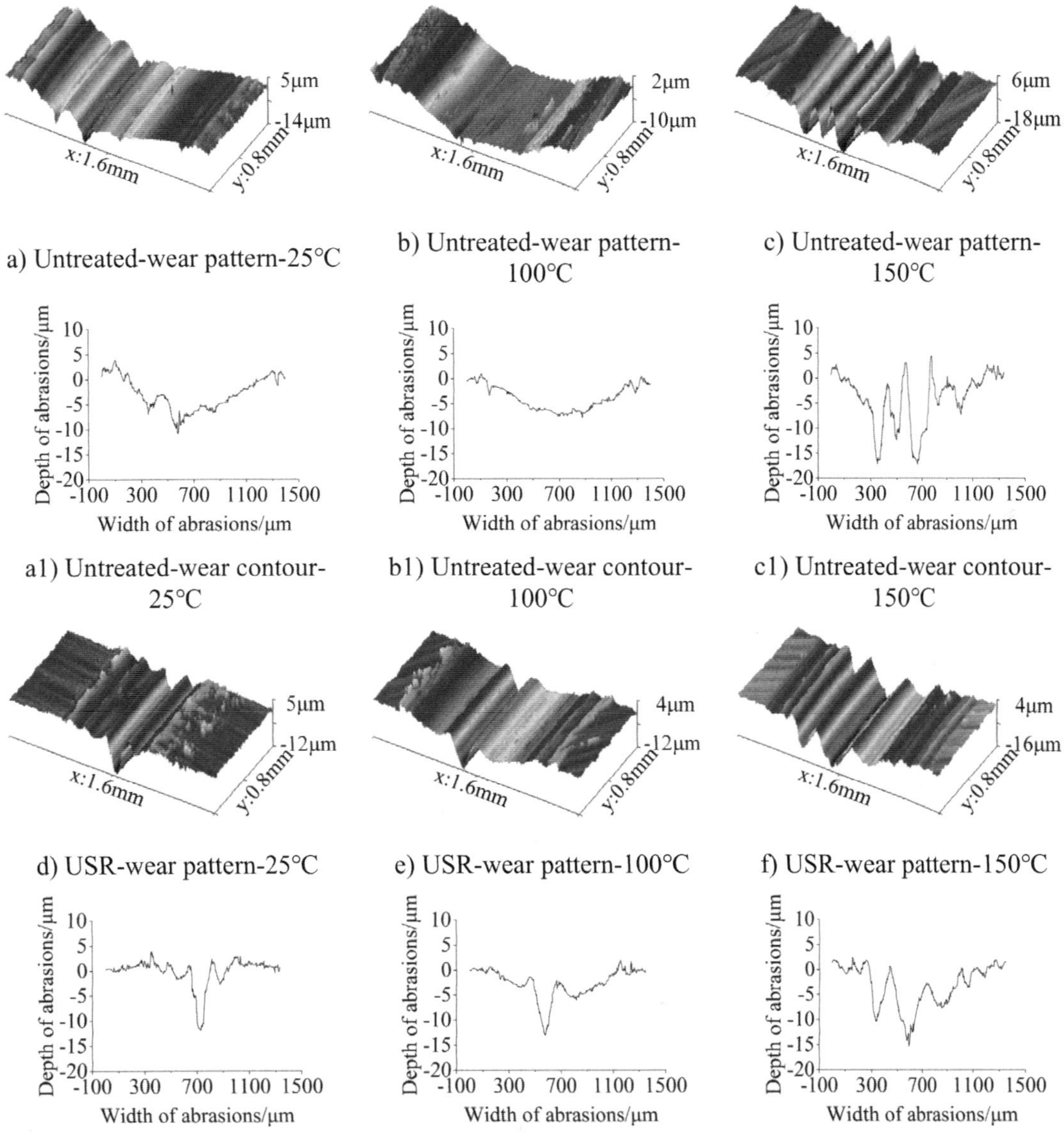


a) Untreated-wear pattern-25°C
b) Untreated-wear pattern-100°C
c) Untreated-wear pattern-150°C
a1) Untreated-wear contour-25°C
b1) Untreated-wear contour-100°C
c1) Untreated-wear contour-150°C
d) USR-wear pattern-25°C
e) USR-wear pattern-100°C
f) USR-wear pattern-150°C
d1) USR-wear contour-25°C
e1) USR-wear contour-100°C
f1) USR-wear contour-150°C

**Figure 8.** 3D morphology of sliding wear of samples at different temperatures.

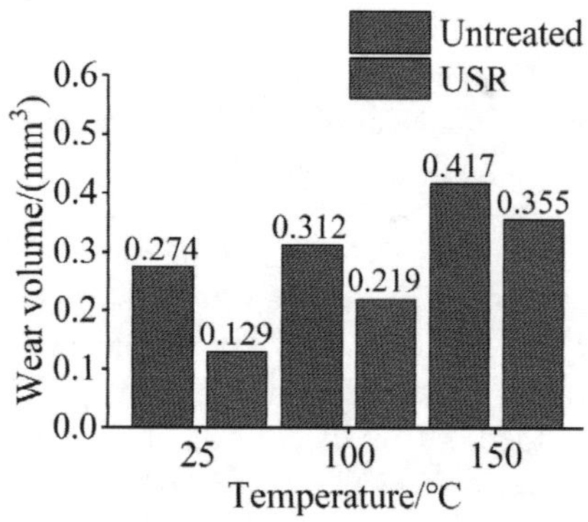


**Figure 9.** Wear the volume of samples at different temperatures.

*4.4 Analysis of wear mechanisms*

The wear morphology of the samples without USR treatment at various temperatures is shown in Figures 10(a), 10(b), and 10(c), from which the wear surfaces of the samples at the three temperatures have a lot of adhesive layers and ploughings, which indicates that the samples have obvious adhesive wear and abrasive wear mechanisms at the three temperatures. The EDS elemental analyses of the adhesive layer of the samples at these three temperatures are shown in Figure 10(a1), 10(b1), and 10(c1), respectively, which show that the adhesive layer contains a high concentration of oxygen, indicating that there is a significant oxidative wear phenomenon for the untreated samples at all three temperatures. This is because the wear surface of the sample is in contact with the air and thus oxidation occurs. However, comparing Figures 10(a), 10(b), and 10(c), the area of spalling pits on the sample surface expands with the increase of temperature, and at 150°C, the surface of the sample has a long strip of spalling cracks and the phenomenon of delamination occurs, indicating that the phenomenon of fatigue wear occurring in the untreated sample is more pronounced with the increase of temperature. This is because as the temperature increases, the surface hardness of the sample decreases and the residual compressive stress on the surface is slowly released, so the ability of the sample to resist crack extension also decreases and the spalling pit gradually becomes larger.

Figures 10(d), 10(e), and 10(f) show the wear morphology of the samples with USR treatment at various temperatures. As shown in the figure, the wear surfaces of the samples have obvious adhesive layers and ploughings at all three temperatures, which indicates that the samples with USR treatment still have the mechanisms of adhesive wear and abrasive wear at all three temperatures, but compared with the samples without USR treatment, the area of adhesive layers on the surfaces and the amount of the plowings are smaller at all three temperatures. This is because the USR treatment improves the hardness of the sample surface, which produces less hard abrasive debris during the experiment. Therefore, less hard abrasive debris adheres to the sample surface, and it is also more difficult for the hard abrasive debris to produce significant plowing of the sample surface. Figures 10(d1), 10(e1), and 10(f1) show the EDS elemental analysis of the adhesive layer in Figures 10(d), 10(e), and 10(f), respectively, from which all the adhesive layers have a certain content of oxygen elements, but compared to the samples without treatment, the oxygen content overall decreased, which indicates that the USR treatment has improved the samples' resistance to oxidative wear.

In addition, the figures show the presence of small spalling pits on the surface of samples with USR treatment at 25°C and 100°C. At 25°C, powdering was observed around the spalling pit on the sample surface, whereas at 100°C, the entire spalling pit was heavily powdered and a large number of powdered particles were observed around the spalling pit. At 150°C, the number and area of the spalling pits on the surface increase, and cracks appear around them. This highlights the presence of fatigue wear phenomenon on the USR samples at all three temperatures, with the degree of fatigue wear increasing as the temperature rises. This is because even though the samples with USR treatment exhibit an increase in surface hardness and residual compressive stress, the efficacy of this strengthening effect declines as the temperature rises. Consequently, under high-temperature working conditions, the surface of the samples with USR treatment still exhibited large spalling pits. However, comparing Figures 10(a), 10(b), and 10(c), the fatigue spalling pit area of the USR samples at all three temperatures is smaller than that of the samples without USR treatment, which indicates that the fatigue wear resistance of the samples at various temperatures can be enhanced through USR treatment.

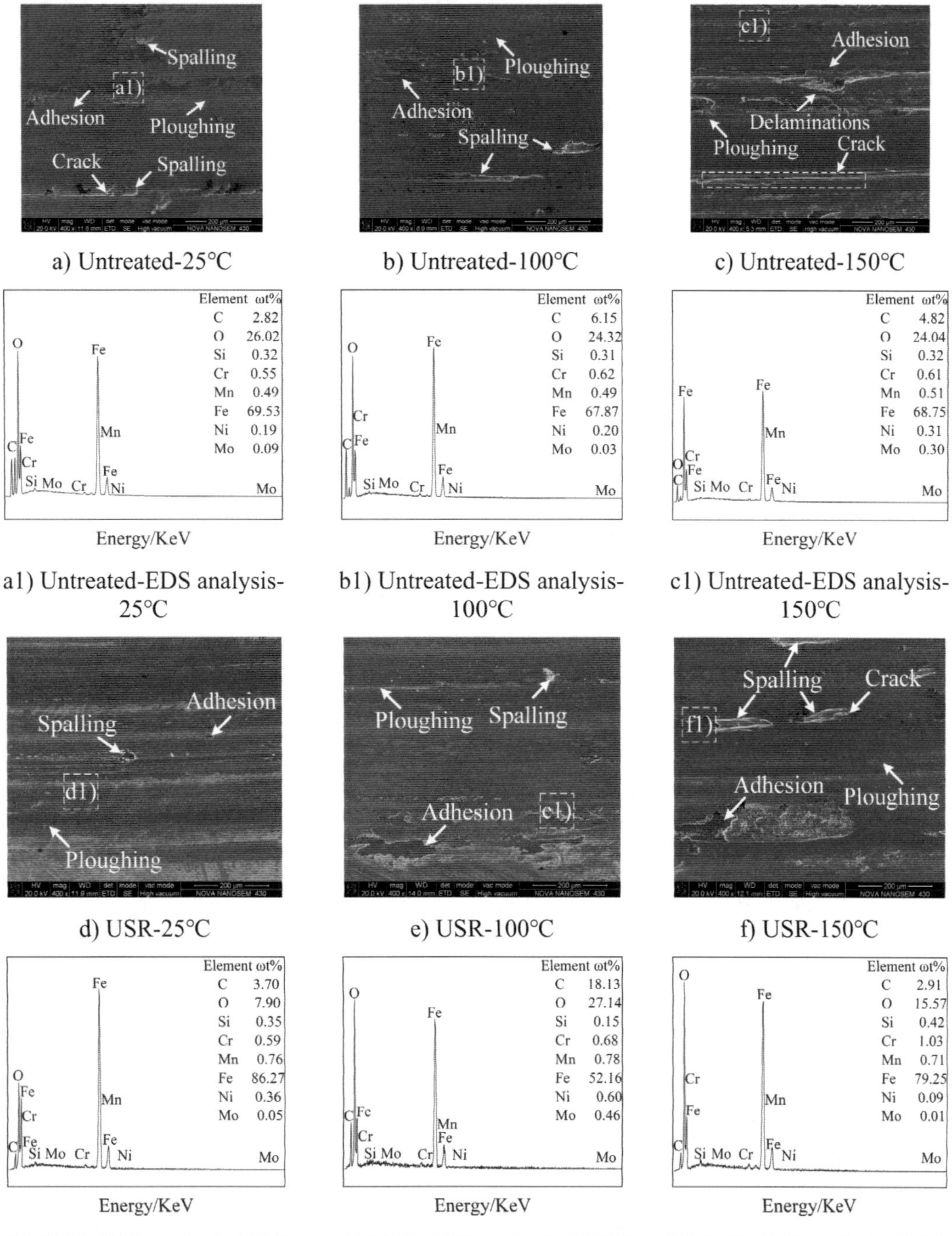


a) Untreated-25°C b) Untreated-100°C c) Untreated-150°C

a1) Untreated-EDS analysis-25°C b1) Untreated-EDS analysis-100°C c1) Untreated-EDS analysis-150°C

d) USR-25°C e) USR-100°C f) USR-150°C

d1) USR-EDS analysis-25°C e1) USR-EDS analysis-100°C f1) USR-EDS analysis-150°C

**Figure 10.** Wear patterns of samples with and without USR treatment at various temperatures.

## 5 Conclusions

The USR treatment was employed to strengthen the surface of 8620H carburized steel. To evaluate the effectiveness of this treatment, high-temperature abrasive wear experiments were conducted on samples with and without USR treatment at temperatures of 25°C, 100°C, and 150°C. The findings are presented below.

(1) The USR treatment can enhance the hardness and residual compressive stress of the sample, decrease roughness, and refine grain organization, ultimately bolstering sample wear resistance.

(2) The wear volume of the samples rose as the temperature increased. However, the wear volume of the samples with USR treatment decreased by 52.9%, 29.8%, and 14.9% at temperatures of 25°C, 100°C, and 150°C, respectively compared to the samples without USR treatment.

(3) The degree of adhesive wear, abrasive wear, oxidative wear, and fatigue wear of the samples with and without USR treatment increased as temperature rose, but the degree of occurrence of the above wear mechanisms in the samples with USR treatment was reduced as compared to that in the samples without USR treatment.

## Acknowledgments

This work was supported by the Liuzhou Science and Technology Plan Project [No. 2021AAD0103]; and the Zhuhai industrial core research project [No. 2220004002348].

## References

[1] Sankaran R, Rajamani D, Natarajan S and Thirugnanasambantham KG (2017). Sliding wear behavior and its mechanisms of carbonitrided AISI 8620 steel at 100°C under unlubricated conditions. Surface Engineering, Communication, 33:42-48.

[2] Yang L I and Xu P (2012). Investigation on tribological behaviors of surface micro-forming 20CrNiMo steel under laser shock processing. Laser Technology, Communication, 36(6): 814-817.

[3] Zhang Y L, Lai F Q, Qu S G, Ji V, Liu HP, and Li X Q (2020). Effect of shot peening on residual stress distribution and tribological behaviors of 17Cr2Ni2MoVNb steel. Surface & Coatings Technology, Communication, 386: 125497.

[4] Hu X F, Jia S Y, Lai F Q, Jiang L Z, Li X Q, and Qu S G (2021). Investigation on the parameters optimization and sliding wear behaviors under starved lubrication of discrete laser surface hardened 25CrNi2MoV steel. Tribology International, Communication, 163: 107176.

[5] He Q, Li A, Qu W H, Zhang Y, Tao W, and Kong L H (2018). Investigation on Friction and Wear Properties of High-Temperature Bearing Steel 9Cr18Mo. Materials Research-ibero-american Journal of Materials, Communication, 21: 3.

[6] Emami M, Ghasemi H M and Rassizadehghani J (2010). High-temperature tribological behavior of 31CrMoV9 gas nitride steel. Surface Engineering, Communication, 26: 168-172.

[7] Shao W, Yi M, Tang J Y and Sun S Y (2020). Prediction and Minimization of the Heat Treatment Induced Distortion in 8620H Steel Gear: Simulation and Experimental Verification. Chinese Journal of Mechanical Engineering, Communication, 35: 1-12.

[8] Zhang Y L, Huang L M, Lu F., Qu S G, Ji V, Hu X F and Liu HP (2021). Effects of ultrasonic surface rolling on fretting wear behaviors of a novel 25CrNi2MoV steel. Materials Letters, Communication, 284: 128955.

[9] Wang J T, Qu S G, Lai F Q, Hu X F, Deng Y Q and Li X Q (2021). Synergistic effects of a combined surface modification technology on rolling contact fatigue behaviors of 20CrMoH steel under different contact stresses. International Journal of Fatigue, Communication, 153: 106487.

[10] Duan Y, Qu S G, Jia S Y and Li X Q (2021). Effects of ultrasonic surface rolling processing on microstructure and wear properties of high-carbon high-chromium steel. Surface & Coatings Technology, Communication, 422: 127531.

# Effect of cooling temperature on microstructure and mechanical properties of Ti microalloyed steel

**Xin Zheng[1,2], Yujian Zuo[1], Hongxin Tian[1], and Yingshuo Wang[1]**

[1]Angang Steel Company Limited, Liaoning, 114000, China

[2]Corresponding author's e-mail: 261158216@qq.com

**Abstract.** Ti is a relatively inexpensive alloying element. Ti micro-alloying enables steel to have higher strength, which has already been applied to steel products in many domains, including engineering machinery. In this work, the effects of cooling temperature upon the mechanical properties and microstructure of Ti microalloyed steel were researched by OM, TEM, tensile and impact tests. The consequences show that as the final cooling temperature (FCT) decreased, the steel plate microstructure changed from PF + P + B (less) to GB + PF (less)+ P (less), and the grain size decreased obviously. As the temperature further decreased to 607°C, the intragranular dislocation tangles and subgrains increased significantly. The size and number of large TiC precipitates in the steel plate decreased with the decrease of the final cooling temperature. Additionally, the mean size of precipitates decreases gradually. As the FCT varied between 672°C and 645°C, the steel plate strength and low temperature toughness were significantly improved. As the temperature is further reduced to 607°C, the strength increase was reduced. Due to the improvement of grain uniformity, the low temperature toughness was further enhanced to 156 J.

## 1. Introduction

In mining and mineral processing, chemical metallurgy, construction and transportation industries, high-strength steel is extensively applied to the manufacture of engineering machinery and vehicle structural parts. At present, in the domain of engineering machinery, high-strength steel mainly uses Nb microalloying as the main technical means to ensure strength and toughness. However, with the intensification of market competition, the cost of steel needs to be further reduced, alloy reduction has become the focus of research, and Ti has been widely considered as a cheaper microalloying element. In recent years, the influence of Ti on phase transition, precipitates and properties have been widely studied [1-3].

The cooling process exerts a prominent effect on precipitates and properties of titanium-containing steels. The distribution and morphology of precipitated phases in as-cast steel (0.08C-0.2Mo-0.37Ti) were investigated at disparate cooling rates [4]. These results showed that as the subcooling degree increases, the size of TiC particles decreases, and the TiC nucleation location grain boundaries shift to the grain interior. García-Sesma et al. [5] investigated the influence of final cooling temperature (FCT)/coiling temperature on the organization and properties of Nb microalloyed steels by adding different contents of Ti. These results illustrated that the steel hardness increased significantly with increasing Ti content, which was mainly due to vast quantities of small-sized (< 3 nm) precipitated phases in the steel. Meanwhile, the coiling temperature affects the steel strength by influencing the level of precipitation strength.

To further reduce the production cost of steel, the micro-alloying element Ti was added to study the influence of FCT on phase transformation, precipitation and mechanical capacities of steel plates.

## 2. Test materials and methods

Steel ingots are melted in one lab vacuum induction melting stove with ingot dimensions of 120 mm × 120 mm × 250 mm. Table 1 shows the chemical component. The ingot was heated in a muffle stove at 1200°C for 1 h. The plate was rolled on a φ550 two-roll test mill, using a two-stage rolling method. As for the intermediate billet, its thickness reached 48 mm, while the thickness of the rolled steel plate reached 12 mm. After rolling, the steel plate entered laminar cooling. After cooling, the steel plate was immediately put into the preheating muffle furnace and slowly cooled to room temperature. Table 2 is the test process. The tensile test was carried out on a Zwick / Roell Z600 tensile tester, and the sample was a φ10 ratio sample. The -20°C KV2 impact test was carried out on the SANS ZBC2602 impact tester, and the sample size was 10 mm × 10 mm × 55 mm. The metallographic structure was observed under Zeiss optical microscope. The fine structure and carbon extraction replica precipitates were observed under the Tecnai $G^2$20 transmission electron microscope.

**Table 1.** Chemical composition (mass fraction/%).

| C | Si | Mn | S | P | Als | Ti |
|---|---|---|---|---|---|---|
| 0.08 | 0.15 | 1.76 | 0.002 | 0.008 | 0.03 | 0.06~0.16 |

**Table 2.** Test process.

| Number | Yield Strength /MPa | Tensile Strength /MPa | Elongation /% | -20°C Impact energy /J |
|---|---|---|---|---|
| 1# | 572 | 653 | 19.5 | 53 |
| 2# | 668 | 772 | 21 | 118 |
| 3# | 692 | 807 | 19 | 156 |

## 3. Experimental results and analysis

### *3.1 Microstructure*

Figure 1 displays the test steel microstructure. It illustrates that the grain size declines significantly as the FCT decreases. By comparing Specimens 1 #, 2 # and 3 # of the steel plates, the microstructures vary from PF + P + B (less) to GB + PF (less) + P (less) when the FCT is lowered from 672°C to 645°C. As temperature further decreases, the microstructure does not change significantly and remains GB + PF (less) + P (less), indicating that 607 ~ 645°C is in granular bainite transformation. According to the fine structure of the test steel shown in Figure 2, when the FCT is 672°C, the grain boundaries of Sample 1 # are relatively flat, with lamellar organization at the boundaries of the grain and low dislocation density in grains. While the FCT was lowered to 645°C, Sample 2# showed a large quantity of irregularly shaped grain boundaries with vast quantities of dislocations distributed within the grains. While the FCT was lowered to 607°C, a large number of dislocation entanglements and high-density dislocation subgrains appeared in the microstructure of Sample 3#. Also, subgrain boundaries of slaty bainite ferrite were observed. The dislocation density of Specimen 3# significantly exceeded that of Specimens 1 # and 2 #.

As the FCT decreases, the dislocation density increases and the dislocation content increases significantly. This is because temperature affects the thermal motion of atoms, in turn, the driving force of dislocation motion is affected. If the temperature is higher, the driving force of dislocation motion will be greater, and the speed of dislocation motion will be faster. Under the same rolling conditions, as the FCT of the steel plate decreases, the lower the slow cooling temperature is, the lower the driving force of dislocation motion within the grain is, because more dislocations are pinned and entangled in the precipitated phase to be retained [6]. At the same time, the rate of dislocation movement from the grain to the grain boundaries slowly decreases due to the decrease in driving force, and more dislocations are retained inside the grain.

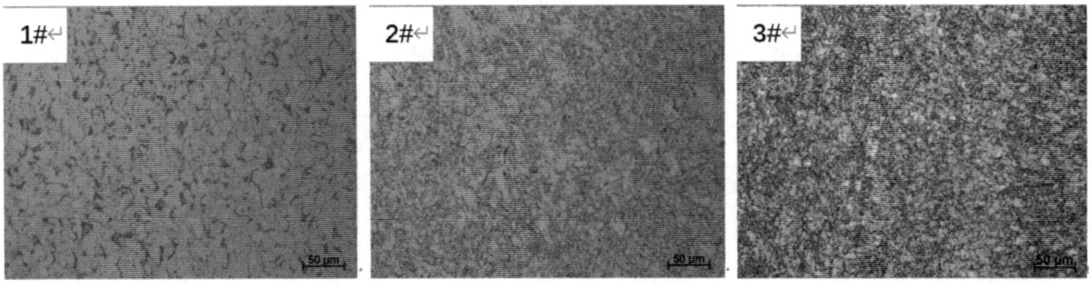


**Figure 1.** Microstructure of test steels.

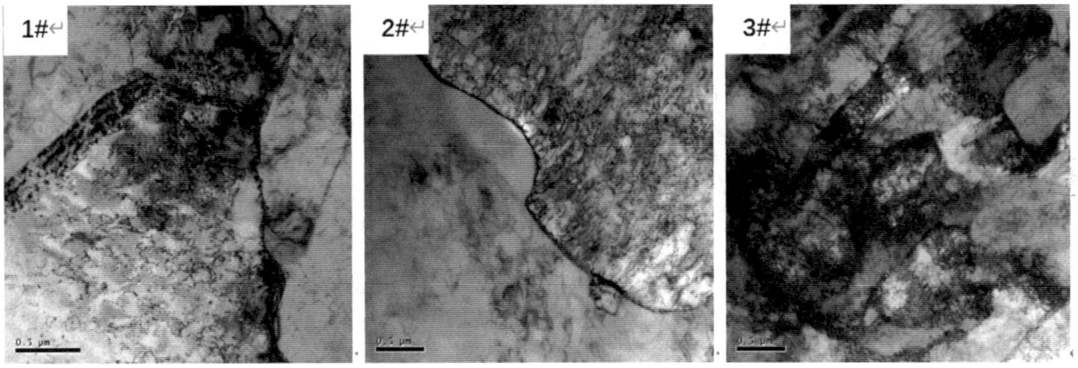


**Figure 2.** High magnification microstructure of test steels.

*3.2 Precipitates*

The morphology of the precipitated phases of the test steel is shown in Figure 3. Besides, the TiC precipitated phases of those test steels are spherical at different FCTs. When the FCT is 672°C, the mean size of the TiC precipitated phase within Sample 1 # is about 18 nm, containing a large-size precipitated phase with a diameter of 30 nm. When the FCT was lowered to 645°C, the average size of TiC precipitated phases in Samples 2 # and 3 # was about 11 nm, with large-size precipitated phases with a diameter of about 15 nm. When the FCT was lowered to 607°C, the diameters of the large-size precipitates were still about 15 nm. Although the larger-sized precipitated phase still exists in the steel plate, the quantity is greatly reduced, and the mean size reaches about 9 nm. As the FCT decreases, the size and the quantity of the larger TiC precipitated phase in the steel plate decrease, and the mean size of the precipitated phase decreases step by step. This is because when the FCT is lower, and the degree of austenite subcooling is greater, the driving force of precipitated phase nucleation will be greater, the rate of precipitated phase nucleation will be higher, and the size of the precipitated phase will be smaller. Synchronously, if the slow cooling temperature of the cooled steel plate is lower, the amount of dislocations will retain greater. These large amounts of dislocations provide a lower energy nucleation location for the precipitated phase, which further increases the nucleation rate of the precipitated phase [7]. In addition, the lower the slow cooling temperature of the steel plate is, the smaller the range of atomic diffusion is, and the smaller the number of atoms available for the formation and growth of the precipitated phase is. The increase in the nucleation rate exacerbates the competition for the atoms needed for the precipitation process, bringing about a decline in the size of the grain and the average grain size of the large-size precipitated phase.

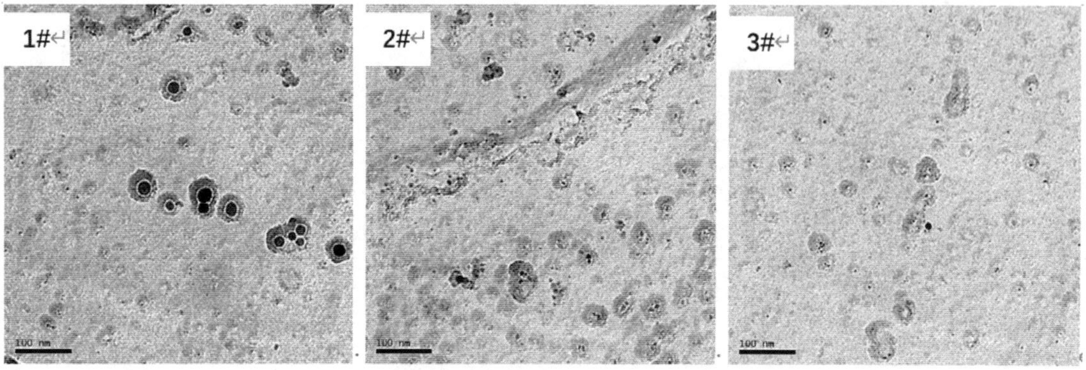


**Figure 3.** Precipitated phases of test steels.

*3.3 Mechanical properties*

In terms of the tested steels, their mechanical capacities are shown in Table 3. When the FCT is lowered from 672°C to 645°C, the yield strength varies between 572 MPa and 668 MPa, and the tensile strength varies between 653 MPa and 772 MPa. Besides, the impact energy at 20°C is increased from 53 J to 118 J, which is a significant improvement in the mechanical properties. This is on account of the FCT reduction. In terms of the steel plate, the grain size is greatly refined, the size of the precipitation phase is reduced, and the dislocation content increases, which enhances the fine grain reinforcement, precipitation reinforcement and dislocation reinforcement, so that the strength of the steel plate is greatly improved. At the same time, grain refinement is capable of effectively improving the crack extension work, and the reduction of the size of the precipitated phase can reduce its harm to the toughness of the steel plate, thus improving the low-temperature toughness. When the FCT continues to reduce to 607°C, the yield strength continues to increase to 692 MPa, tensile strength increases to 807 MPa, and the strength increase is reduced. This is due to the reduction in the FCT from 645°C to 607°C. The reduction in grain refinement and precipitation phase refinement of the steel plate provides a limited strength increment. The grain refinement was more uniform in Specimen 3 # as compared to Specimen 2 # and hence the impact energy was further increased to 156 J at -20°C.

**Table 3.** Mechanical properties of test steels.

| Number | Yield Strength /MPa | Tensile Strength /MPa | Elongation /% | -20°C Impact energy /J |
|---|---|---|---|---|
| 1# | 572 | 653 | 19.5 | 53 |
| 2# | 668 | 772 | 21 | 118 |
| 3# | 692 | 807 | 19 | 156 |

## 4. Conclusion

In this paper, the impact of cooling conditions upon the properties of Ti microalloyed steel was detected by analyzing microstructure and precipitation phases. The conclusions of this research are listed.

1) With the decrease of FCT, the grain size decreases significantly. When the FCT was lowered from 672°C to 645°C, the steel plate organization changed from PF + P + B (less) to GB + PF (less) + P (less). As the temperature deeply decreases to 607°C, the phase composition does not change significantly, but the intracrystalline dislocation entanglement and subgranular grains increase significantly.

2) As the FCT decreases, the size and number of larger TiC precipitated phases in the steel plate decrease, and the mean size of the precipitated phases is reduced step by step.

3) When the FCT was lowered from 672°C to 645°C, the strength and toughness improved significantly. As the temperature is deeply lessened to 607°C, the increase in strength decreases. The low temperature toughness was further increased to 156 J due to the improvement of grain uniformity.

Journal of Physics: Conference Series **2720** (2024) 012030 doi:10.1088/1742-6596/2720/1/012030

**Reference**

[1] Xu, G., Gan, X., Ma, G., Luo, F., Zou, H. (2010) The development of Ti-alloyed high strength microalloy steel. Materials & Design, 31: 2891-2896.

[2] Gaier, M., Lin, H. T., Farhat, Z. N., Plucknett, K. P. (2021). Precipitation hardenable TiC-steel cermets. Wear, 477: 203804.

[3] Kostryzhev, A. G., Killmore, C. R., Yu, D., Pereloma, E. V. (2020). Martensitic wear resistant steels alloyed with titanium. Wear, 446: 203203.

[4] Chen, T., Ji, C., Zhu, M. (2020) Effect of cooling rate on the nucleation and growth of large TiC particles in Ti-Mo steel. Journal of Alloys and Compounds, 823: 153650.

[5] García-Sesma, L., López, B., & Pereda, B. (2019) Effect of coiling conditions on the strengthening mechanisms of Nb microalloyed steels with high Ti addition levels. Materials Science and Engineering: A, 748: 386-395.

[6] Xu, Y., Zhang, W., Sun, M., Yi, H., Liu, Z. (2015) The blocking effects of interphase precipitation on dislocations' movement in Ti-bearing micro-alloyed steels. Materials Letters, 139: 177-181.

[7] Li, C. N., Li, X. L., Yuan, G., Misra, R. D. K., Kang, J., Wang, G. D. (2016) Precipitation behavior and mechanical properties of a hot rolled Ti-bearing dual phase steel. Materials Science and Engineering: A, 673: 213-221.

# The microstructure and properties of high-strength structural steel for fruit auxiliary picking equipment

**Lulu Feng[1], Yanjun Yang[2], Wenfeng Zhang[3], Weiwen Qiao[4], and Xuefeng Tian[1,5]**

[1]School of Advanced Manufacturing, Guangdong Songshan Polytechnic, Shaoguan 512126, China
[2]Training Center, Guangdong Songshan Polytechnic, Shaoguan 512126, China
[3]School of Intelligent Manufacturing, Guangdong Polytechnic of Water Resources and Electric Engineering, Guangzhou 510000, China
[4]Jiangsu Huaneng Cable Co., Ltd., Gaoyou, Jiangsu, 225613 China

[5] Corresponding author's e-mail: 2361682703@qq.com

**Abstract:** The microstructure and mechanical properties of high-strength structural steel used in the fruit-picking equipment were studied by using a metallographic microscope, a tensile tester, and an impact toughness tester. The results revealed that the structure of the experimental steels was a mixed microstructure of ferrite and pearlite, with an average grain size of ferrite of 12 μm. The yield strength was between 515-540 MPa and the tensile strength was 635-645 MPa. The impact toughness at room temperature reached over 200 J. The physical properties of the studied steels completely fulfilled the requirements of the high-strength structural steel for fruit auxiliary picking equipment. The new method cut down the tempering heat treatment process and decreased production costs.

## 1. Introduction

The application of the fruit auxiliary picking equipment in the agricultural and rural fruit planting industry has become the focus of research both at home and abroad. The fruit auxiliary picking equipment is agricultural machinery with a rotating mechanical arm device. The mechanical arm of this equipment has to bear a huge load, therefore, it requires the use of high-strength structural steel with a yield strength of 460 MPa and a tensile strength of over 630 MPa [1-3]. This kind of steel for fruit auxiliary picking equipment is usually prepared by incorporating ultra-low carbon content (0.04%-0.07% wt.) and is produced by using the thermo-mechanical rolling process (TMCP) combined with the tempering heat treatment process to improve the quality of the produced steel [4]. However, this production process has certain drawbacks, such as a long process duration and a high cost. In this context, a novel type of high-strength structural steel for use in the fruit auxiliary picking equipment was designed in the present study with a medium carbon content of 0.16%-0.18% wt [5-6]. The chemical composition, production manufacturing technique, structure, and physical properties of the steel of fruit auxiliary picking equipment with different thickness specifications (8, 10, 12, and 14 mm) were investigated. Certain micro-alloy elements, such as Nb, V, and Ti, were added to refine the grain during the smelting process, which also improved the quality of the steel. In the production process, the advanced rolling process (CR) and cooling process(CC) were adopted. The produced steel plate could be used directly after rolling without the tempering treatment, which cuts down the tempering heat treatment process and reduces the production cost. The present study would serve as a reference basis for the production of such kinds of steel.

## 2. Experimental materials and methods

The molten iron pretreatment, converter steelmaking, LF furnace refining, RH furnace vacuum degassing, dynamic soft reduction, and electromagnetic stirring technology were employed to control the production of the cc slab during the test steel smelting. The slab is 200 mm in thickness, and its chemical composition is shown in Table 1. The steel samples with different thicknesses of 8, 10, 12, and 14 mm were rolled by using different rolling procedures, then subjected to accelerated cooling, cooling bed cooling, trimming process, and quality inspection, and finally were stored until use. The flowchart for the production process is presented in Figure 1. The rolling manufacturing process was roughing and finishing stages: rough rolling and finishing rolling. The heating temperature of rough rolling was controlled at 1, 240°C. The heat preservation was performed for 180 min. After the descaling process, rough rolling was conducted at 980-1, 040°C. After the completion of rough rolling, the temperature control stage commenced. After the completion of temperature control, the finishing rolling was conducted at 960-830°C. After the completion of finishing rolling, the accelerated cooling process was performed by using the relevant device and water for cooling. The cooling was performed at a rate of 20-25°C until the final cooling temperature of 600°C was reached. Subsequently, the natural air cooling was performed on a cooling bed to reach the temperature of 300°C, then cooled to room temperature in the air. The entire rolling process is illustrated in Figure 2.

**Table 1.** Chemical composition analysis of experimental steels (%).

| Thickness | C | Si | Mn | P | S | Alt | Nb | V | Ti | Fe. |
|---|---|---|---|---|---|---|---|---|---|---|
| 8 | 0.18 | 0.38 | 1.41 | 0.007 | 0.002 | 0.046 | 0.032 | 0.061 | 0.017 | Bal. |
| 10 | 0.16 | 0.34 | 1.38 | 0.009 | 0.005 | 0.043 | 0.032 | 0.058 | 0.013 | Bal. |
| 12 | 0.17 | 0.38 | 1.37 | 0.008 | 0.004 | 0.041 | 0.034 | 0.061 | 0.014 | Bal. |
| 14 | 0.16 | 0.34 | 1.39 | 0.008 | 0.003 | 0.043 | 0.031 | 0.056 | 0.015 | Bal. |

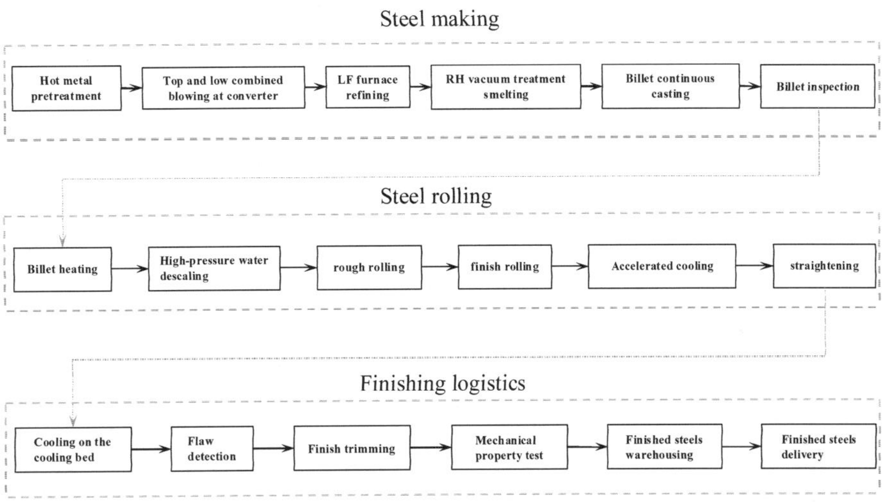


**Figure 1.** Production flowchart of steel plate.

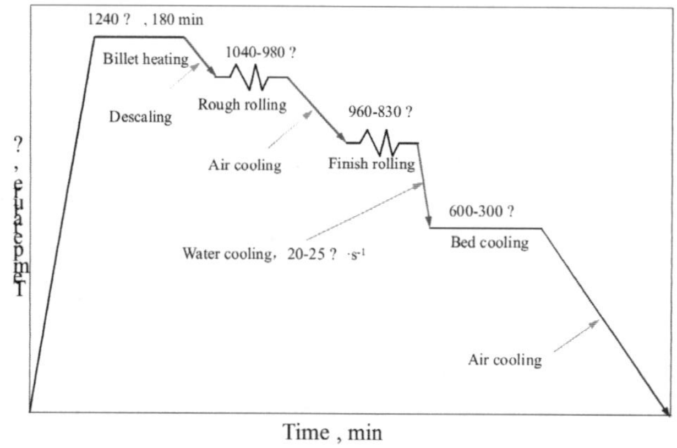


**Figure 2.** The rolling process of steel plate.

## 3. Microstructure and morphology of test steels

The structure of the test steels is depicted in Figure 3. As visible in the figure, the metallographic microstructure of the test steel comprised black pearlite and white ferrite [7-8]. The black pearlite was distributed between the white ferrite in strips. The ferrite grain size of 8 mm in thick test steels was relatively small, and the average ferrite grain size was 12 μm. With the increase in the thickness of the experimental steel, the ferrite grain size also increased, while the ferrite grain level decreased. The statistical analysis was performed next. As depicted in Figure 3(a), the ferrite grain size reached Grade 14 for a thickness of 8 mm, Grade 13.5 for a thickness of 10 mm, and Grade 13 for thicknesses of 12 mm and 14 mm.

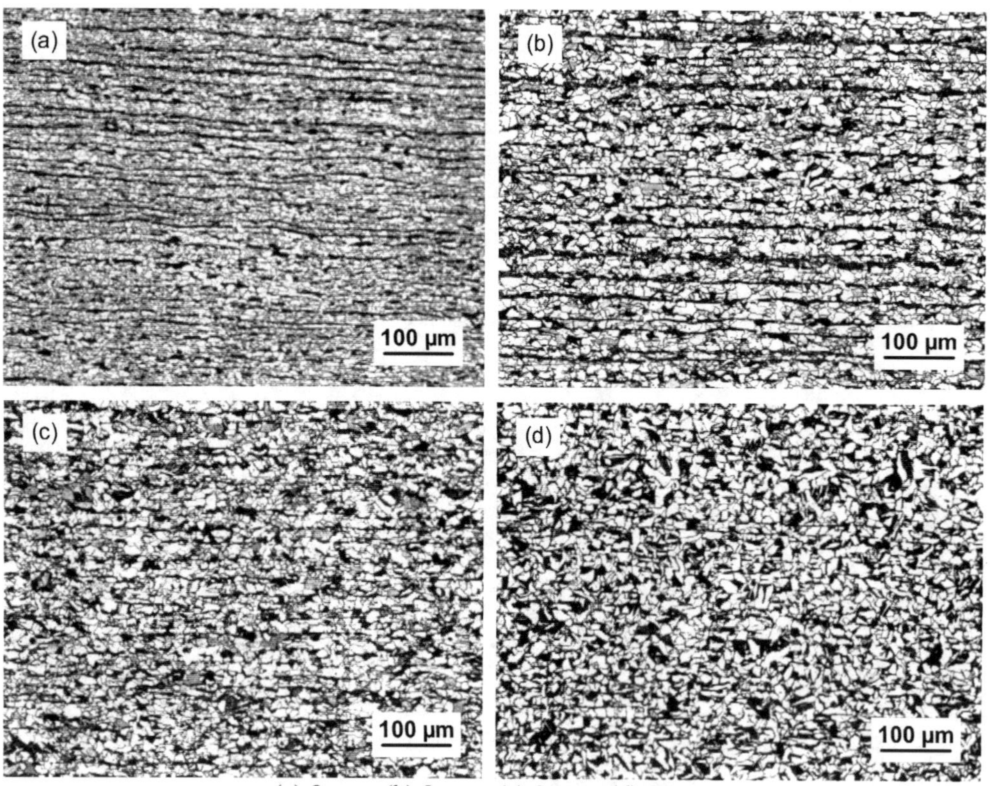


(a) 8 mm; (b) 8 mm; (c) 8 mm; (d) 8 mm

**Figure 3.** Microstructure and morphology of test steel.

## 4. Mechanical properties of test steel

The yield strength of the steel was predicted and estimated by Formula (1) [9-10] provided below.

$$\sigma_s = \sigma_o + \sigma_A + \sigma_D + \sigma_\rho + \sigma_{IN} + kd^{-\frac{1}{2}} \tag{1}$$

Where $\sigma_s$ denotes the yield strength; $\sigma_0$ denotes the lattice friction resistance; $\sigma_A$ denotes the solid solution strengthening; $\sigma_D$ denotes the location strengthening; $\sigma_\rho$ denotes the precipitation strengthening; $\sigma_{IN}$ denotes the strengthening caused by the interstitial atoms; $k$ denotes the strengthening coefficient; $d$ is the average grain diameter; $\sigma_0$ was approximately 50 MPa for the general low-alloy steel; $\sigma_A$ was solution-enhanced, and the calculation formula is provided below:

$$\sigma_A = \sum_i k_i(x_i) \tag{2}$$

The calculated $\sigma_A$ was approximately 43.1 MPa.

In the formula, $\sigma_D$ denotes the dislocation strengthening, and the yield stress was then estimated through the general formula for calculating dislocation density, which is provided below:

$$\sigma_D = \alpha\mu b\rho^{\frac{1}{2}} \tag{3}$$

In the case of a cubic crystal, $\alpha$ is 0.5, while for a closely arranged hexagonal crystal, $\alpha$ is 1.1. The calculated $\sigma_D$ was approximately 123.2 MPa.

In the formula, $\sigma_\rho$ denotes the precipitation strengthening and was estimated by using Formula (4) provided below:

$$\sigma_\rho = 0.298(\frac{\mu b}{l})\ln(\sqrt{2/3}\cdot\frac{d}{b}) \tag{4}$$

We assumed that the dislocation density $\rho = 1.5 \times 10^8$ mm$^2$, the shear modulus $\mu = 8 \times 10^4$ MPa, Burgess vector $b = 2.5\times10^{-7}$ mm, the particle diameter $d = 20$ nm, and the precipitation spacing $l = 250$ nm. The calculated value $\sigma_\rho$ was approximately 99.6 MPa.

At $\sigma_{IN} = 9.4 \times 10^4 \times$ f, the value of f was $10^{-3}$, and the calculated value $\sigma_{IN}$ was approximately 94.3 MPa.

When the contribution of grain refinement was considered, $k = 20$ N/mm$^{-3/2}$ was used, and the average grain size was $d \approx 12$ μm. The calculated value $kd^{-\frac{1}{2}}$ was then approximately 115.2 MPa.

After the above estimation, the yield strength was determined to be approximately 525 MPa, which was equivalent to the measured value presented in Table 2.

The determined mechanical properties are shown in Table 2. The yield strength was between 515 and 540 MPa, and the tensile strength was between 635 and 645 MPa. The steel fulfilled all requirements for the steel used in the fruit auxiliary picking equipment, exhibiting excellent impact toughness, with the impact energy value reaching over 200 J.

**Table 2.** Mechanical properties of experimental steel.

| Thickness | σs (MPa) | σb (MPa) | δ (%) | Akv (J) (at 20°C) |
|---|---|---|---|---|
| 8 | 535 | 640 | 28 | 235, 234, 260 |
| 10 | 530 | 640 | 29 | 220, 252, 245 |
| 12 | 540 | 645 | 26 | 260, 235, 245 |
| 14 | 515 | 635 | 27 | 255, 255, 245 |

## 5. Conclusions

(1) The structure of the experimental steel comprises a mixture of ferrite and pearlite. The grain size of ferrite is between Grades 13 and 14, and the average grain size of ferrite is 12 μm.

(2) The yield strength of the experimental steel is between 515 and 540 MPa, and the tensile strength is between 635 and 645 MPa. The steel exhibits excellent impact toughness at room temperature, with the impact energy value reaching over 200 J.

(3) The tested steel is designed with medium carbon content and produced by rough rolling and finishing the rolling process. The physics properties of the studied steel completely fulfill the requirements of the high-strength structural steel for fruit auxiliary picking equipment. The new method cuts down the tempering heat treatment process and decreases production costs.

**Acknowledgments**

This work was financially supported by Guangdong Higher Education Science and Technology Service Rural Revitalization Research Project (2022ZDZX4082, 2023ZDZX3068), Shaoguan Municipal Science and Technology Funding Plan Project (210728104530543), Guangdong Provincial Education Science Planning Project (2022GXJK493).

## References

[1] Han Y. Q., Sun B. G., Zhang P., Qi Y., and Zheng X.B. (2023). Research progress of agricultural picking robot, *Forestry Machinery & Woodworking Equipment*, 51 (04) 4-8. DOI: 10.13279/j.cnki.fmwe.2023.0045.

[2] Gou Y. M., Yan J. W., Zhang F. G., Sun C. Y., and Xu Y. (2023). Strawberry image processing based on machine vision, *J. Computer Engineering and Applications*, 59 (09) 13-26. DOI: 10.3778/j.issn.1002-8331.2209-0183.

[3] Zhang Y. Y., Zhou J. L., Hu H. R., and Wang J. P. (2022). Research, analysis and development trend of fruit and vegetable picking end-effector, *Forestry Machinery & Woodworking Equipment*, 50 (11) 4-10+17. DOI: 10.13279/j.cnki.fmwe.2022.0162.

[4] Jansto S. G. (2016), High Quality TMCP Production and Metallurgy of Niobium Bearing Steels. *Mater Sci Forum*, 879:820-825. DOI:10.4028/WWW.SCIENTIFIC.NET/MSF.879.820.

[5] Maisuradze M. V.and Björk T. (2022) Microstructure and mechanical properties of high-strength steel with improved machinability. *Metallurgist*, 66(3-4):391-402. DOI:10.1007/S11015-022-01341-Y.

[6] Sisodia R. and Gáspár M. (2021). Experimental assessment of microstructure and mechanical properties of electron beam welded S960M high strength structural steel, *Manufact Letters*, prepublish. DOI:10.1016/J.MFGLET.2021.05.004.

[7] Zhou H., Tang J.C., Zhou X.J., Wang H.J., Chen Z.X., Long Y.l. and Ai P.H. (2023) Clamping experiment and simulation for the picking of heavy string-typed fruit. *Journal of Agricultural Mechanization Research*, 45(06):68-75. DOI: 10.13427/j.cnki.njyi.2023.06.013.

[8] Chen X.Y., Feng L.L., Song Z.Q., QIAO W.W., Li P.P. and Wu K.M. (2021) Effect of critical quenching on impact toughness of ultra-thick Hydropower Steel. Journal of Materials and Metallurgy, 20(03):211-216. DOI: 10.14186/j.cnki.1671-6620.2021.03.009.

[9] Saastamoinen A., Kaijalainen A., Nyo T. T., Suikkanen P., Porter D. and Kömi J.. (2019) Direct-quenched and tempered low-C high-strength structural steel: The role of chemical composition on microstructure and mechanical properties. *Mater Sci and Eng: A*, 760: 346-358. DOI:10.1016/j.msea.2019.06.018.

[10] Chatterjee D. (2021). Effect of repeated warm rolling cold rolling and annealing on the microstructure and mechanical properties of AISI 301LN grade austenitic stainless steel, *Mater Today: Proceedings*, 46 (20) 10604-10611. DOI:10.1016/J.MATPR.2021.01.341.

# Vector ducted propeller aerodynamic modeling and verification testing

**Peng Tang[1,2], Chuangxin Zhao[1], Sen Mao[1], Jiajie Tian[1] and Shizhe Liang[1]**

[1]AVIC Chengdu Aircraft Industrial (Group) Co. Ltd., Chengdu 610091, China

[2]Corresponding author's e-mail: tp_study@hotmail.com

**Abstract** The tail-sitter aircraft has the advantages of both high-speed cruise and capability of the take-off and land vertically, high-performance control laws usually rely on the accurate flight mechanics model, while its flight aerodynamic modeling is a complex problem. This paper analytically develops an aerodynamic model of its propelling system based on first principles, including three components of a propeller, a vector duct, and eight slipstream vanes. This model can give the propeller's induced velocity, power, thrust, and moment, also considers the augmented effect of ducted, and the lift and drag forces of vanes. Moreover, a propelling system test bed is built to measure the forces and moments along the body frame, which is composed of the tail-sitter prototype, power, communication link, upper system, and servo drive device. In addition, experimental results show the effectiveness of the model in estimating the thrust and torque with different angles of control vanes, the measured thrust is consistent with the calculated values, and the steering performance of each vane is further examined. The moments exerted on the two axes are linear with the vanes angles, and the symmetry is also well, which is convenient for flight controller design for this type of tail-sitter aircraft.

## 1. Introduction

In recent years, unmanned aircraft vehicles (UAVs) have been developing rapidly due to the capabilities of executing tasks in remote and complex environments. The vertical take-off and landing (VTOL) aircraft has comprehensive performance in cruise and hover among various UAVs, especially in scouting, mapping, and emergency rescue applications. As a type of VTOL, tail-sitter UAV has the benefits of high-speed cruise, which satisfies higher flight envelope requirements. Various tail-sitter prototypes thus are developed to explore the integrated design and control technology. For instance, R. Hugh developed a "T-wing" UAV and conducted tests in different flight modes [1]. A "V-bat" prototype UAV with a ducted propeller was designed, and flight verification was also tested [2]. The advanced flight control technique is the key point for this type of UAV to achieve high performance, especially in the transition stage. In this stage, system dynamics shows strong nonlinear characteristics, and the aerodynamics analysis is a challenging problem. Several controllers have been proposed to cope with this situation. An anti-disturbance control method for tail-sitter UAVs was provided in [3-5]. Lyu et al. [6] proposed an adaptive controller for a tail-sitter with two rotors. In general, independent control laws are designed to hover and level flight by switching the two modes in a simple logic. In [7], an integrated controller consisting of an observer and an adaptive law was presented which can adapt the full envelope. High-performance control laws usually rely on the accurate flight mechanics model, therefore, researchers always dedicated to creating a high-fidelity dynamic model of VTOL aircraft. The way of identification is often used on this type of aircraft, for instance, Sun et al. [8] built a digital flight model for the

"Caipirinha" tail-sitter based on wind tunnel data, and model accuracy had also been verified. This paper aims to explore the aerodynamic characteristics of the tail-sitter's propelling system, the rest is arranged as follows. In the next section, the aerodynamic model is given based on the first principle including the propeller, duct, and control vanes. The third section introduces the prototype test bed. The truest test is also conducted considering different angles of control vanes. Finally, a brief conclusion is given.

## 2. Aerodynamic modeling

### *2.1. Propeller force*

There are many ways to calculate the propeller thrust, the following thrust model is introduced and the duct is taken into account independently [2, 9]. The velocity at the disk is given as

$$v_b = (u - u_w) + \frac{2}{3}\omega_p r_p \left(\frac{3}{4}K_t\right) \quad (1)$$

Where $\omega_p, r_p$ are propeller's angular velocity and radius respectively, $K_t$ is the twist at the three quarters of blade, $\mathbf{V_b} = [u_b, v_b, w_b]^T$ is the body speed measured along three body axis, and $\mathbf{V_w} = [u_w, v_w, w_w]^T$ is corresponding wind speed. Therefore, the thrust generated by propeller can be calculated by

$$T = \frac{1+K_{\text{aug}}}{4}(v_b - v_i)\omega_p r_p^2 \rho a_0 b_p c_p \quad (2)$$

Where $\rho$ is air density, $a_0$ denotes the lift curve slope, $b_p, c_p$ is the number and chord of the blade. $K_{\text{aug}}$ denotes the thrust augmentation factor result from the duct, and $K_{\text{aug}} \in [0\ 1]$. The $v_i$ represents the induced velocity, and is given by

$$v_i = \frac{T}{2\rho\pi r_p^2 v_f} \quad (3)$$

$v_f = \sqrt{(v - v_w)^2 + (w - w_w)^2 + (u - u_w + v_i)^2}$ represents the far-field velocity. It is worth noting that there is a lack of explicit expression of thrust and induced velocity, which are usually found by numerical solution iteratively. In general, the convergence of solving process is fast and can enter a lowest error bound within about 10 steps.

After obtaining the thrust, induced velocity and angular velocity, the induced power is given by

$$P_{\text{i}} = \frac{v_i T}{1+K_{\text{aug}}} \quad (4)$$

Therefore, the reaction moment generated by the propeller can be represented as $M_{\text{i}} = P_{\text{i}}/\omega_p$.

### *2.2. Ducted fan*

The ducted fan can enlarge the thrust of propelling system, and also generate significant drag. The airspeed magnitude at the top of duct is given by

$$\text{V}_0 = \sqrt{u_{\text{duct}}^2 + v_{\text{duct}}^2 + w_{\text{duct}}^2} \quad (5)$$

Where $\boldsymbol{V}_{\text{duct}}$ is the relative airspeed at the mouth of duct, which is represented as

$$u_{\text{duct}} = u - u_w$$
$$v_{\text{duct}} = v - v_w + r l_{\text{duct}} \quad (6)$$

$w_{\text{duct}} = w - w_w + q l_{\text{duct}}$ $q, r$ are the body angular rates along the Y axis and Z axis, and $l_{\text{duct}}$ is the length between the mouth of the duct and the gravity center of the whole body. Let us define the ducted fan's attack angle $\alpha_{\text{duct}}$ and sideslip $\beta_{\text{duct}}$ as follows:

$$\alpha_{\text{duct}} = \tan^{-1}\left(\frac{\sqrt{v_{\text{duct}}^2 + w_{\text{duct}}^2}}{u_{\text{duct}}}\right),\ \beta_{\text{duct}} = \tan^{-1}\left(\frac{v_{\text{duct}}}{w_{\text{duct}}}\right) \tag{7}$$

The momentum drag generated by duct is given by [10]

$$f_d = K_{\text{duct}}\rho\pi A r_{\text{duct}}^2 (V_0 \cos(\alpha_{\text{duct}}) + v_i) V_0 \sin(\alpha_{\text{duct}}) \tag{8}$$

The $r_{\text{duct}}$ is the inner radius of the duct. Naturally, the moment along the body axis is calculated as

$$M_{\text{duct}} = \begin{bmatrix} -f_d \sin(\beta_{\text{duct}}) l_{\text{duct}} \\ -f_d \cos(\beta_{\text{duct}}) l_{\text{duct}} \\ 0 \end{bmatrix} \tag{9}$$

### 2.3. *Slipstream vanes*

The body frame is described as a right-handed coordinate system. Each slipstream vane deflection will generate the lift and drag forces as

$$\begin{aligned} f_{L,i} &= \frac{1}{2}\rho v_v^2 S_v \delta_i C_L \\ f_{D,i} &= \frac{1}{2}\rho v_v^2 S_v \delta_i C_D \end{aligned} \tag{10}$$

Where $\delta_i$ is the actual deflection angle of the *i*-th slipstream vane, which is deemed positive in a clockwise rotation. $v_v = u_{\text{duct}} + v_i$ is the airflow velocity over the corresponding slipstream vane, $S_v$ is the area of vane, $C_D$ and $C_L$ is the coefficient of vane's drag and lift respectively. Then, the total drag force aligned with the Z-axis is given as

$$f_{Z,\text{vane}} = \sum_{i=1}^{8} |f_{D,i}| \tag{11}$$

The forces aligned with X and Y axis are

$$\begin{aligned} f_{X,vane} &= -\sum_{i=1}^{8} cos\left(\frac{\pi(i-1)}{4}\right) f_{L,i} \\ f_{Y,vane} &= \sum_{i=1}^{8} sin\left(\frac{\pi(i-1)}{4}\right) f_{L,i} \end{aligned} \tag{12}$$

The moment exerted on the propelling system body is expressed as

$$M_{\text{vane}} = \begin{bmatrix} -d_{vane} f_{Y,vane} & d_{vane} f_{X,vane} & \frac{L_{vane}}{2}\sum_{i=1}^{8} f_{D,i} \end{bmatrix}^T \tag{13}$$

$L_{\text{vane}}$ and $d_{\text{vane}}$ are the lengths from the center of duct, and center of gravity of the body to the vane's pressure center respectively.

## 3. Thrust test

In this section, the propelling system test bed is built to measure the forces and moments along the body frame, as presented in Figure 1, it is composed of the tail-sitter prototype, power, communication link, upper system, and servo drive device.

Firstly, we consider that there is no slipstream vane deflection, as shown in Figure 2, and it is observed that the measured thrust values are consistent with that calculated by the model, when the propeller's angular velocity reaches about 6000 rpm, the thrust measured is a little over 800 N. Steering performance of each vane is further examined, Figure 3 shows the measured moments generated by the first and second vanes at 50% and 55% throttle rate of engine.

The measured moment values along the X and Y axis are proportionate to the vane's deflection angle roughly, while is not perfectly symmetrical in opposite rotational direction for the Vane 2, it results from the nonlinear and coupling aerodynamic characteristics of vanes. Similarly, force test experiments further validate the handling performance of the 7th and 8th control vanes in Figure 4, it can be seen that the moments exerted on the two axes are linear with the vane's angles, and the symmetry is also well, which is convenient to flight controller design for this type of tail-sitter aircraft.

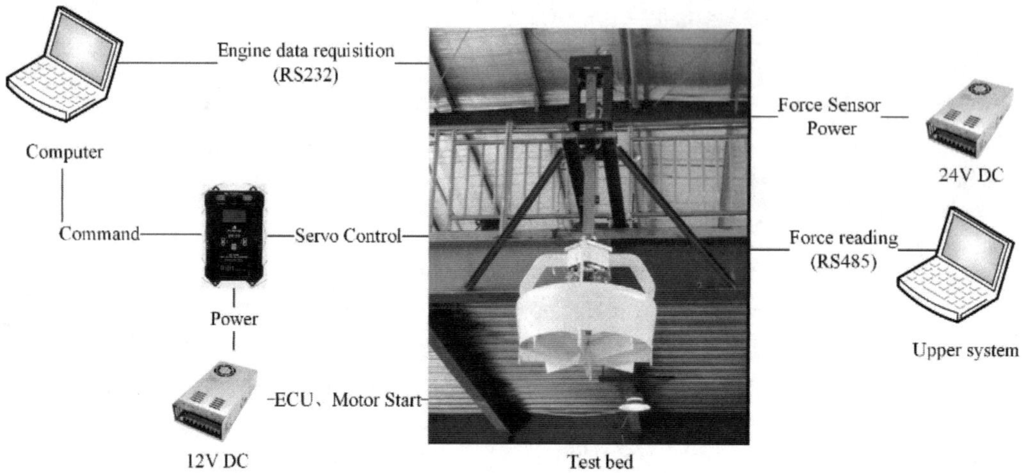


**Figure 1.** Propelling system test bed.

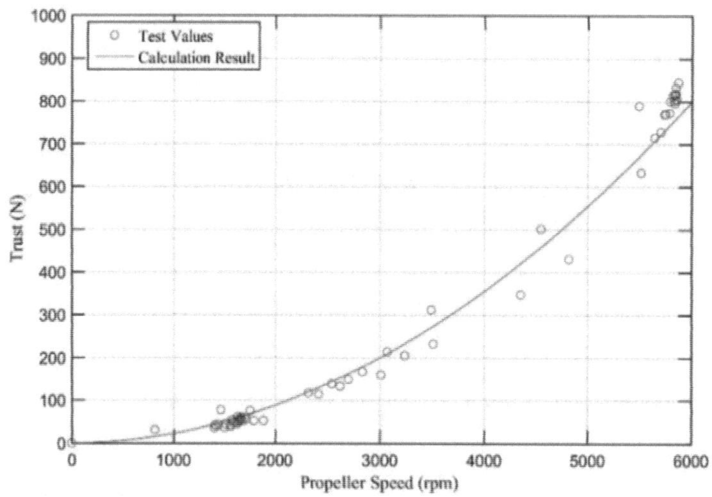


**Figure 2.** Trust test comparison.

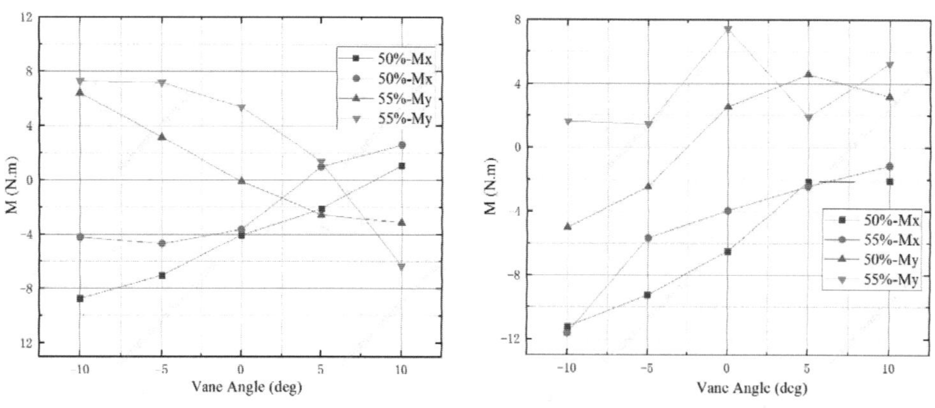


(a) Vane 1 (b) Vane 2

**Figure 3.** Control vanes performance.

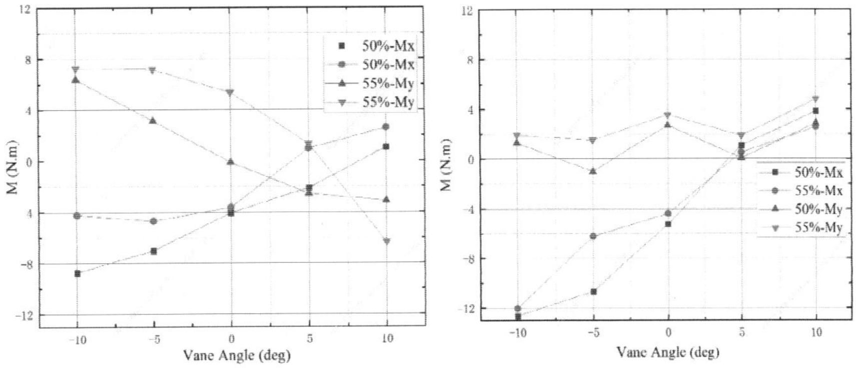


(a) Vane 7 (b) Vane 8

**Figure 4.** Control vanes performance.

## 4. Conclusion

This paper analytically develops an aerodynamic model for a type of vector ducted propeller, the components of the propeller, ducted fan, and slipstream vanes are considered. Then, a prototype with a test system is constructed to evaluate the force and moment performance, the measured results verify the effectiveness of the model in estimating the thrust, and the handling capability test of the propelling system also shows a satisfactory result.

## Reference

[1] Stone R H. Anderson P, Hutchison C, et al. Flight testing of the tail-sitter unmanned air vehicle [J]. *Journal of Aircraft*, 2008, 45(2): 673-685.

[2] M. E. Argyle, R. W. Beard and S. Morris, The Vertical Bat tail-sitter: Dynamic model and control architecture, *2013 American Control Conference*, Washington, DC, USA, 2013, 806-811.

[3] Lustosa L R, Defay F, Moschetta J M. Global Singularity-Free Aerodynamic Model for Algorithmic Flight Control of Tail Sitters [J]. *Journal of Guidance, Control, and Dynamics*, 2019, 42(2): 303-316.

[4] Dalwadi, N., Deb, D., Kothari, M. and Ozana, S.. Disturbance observer-based backstepping control of tail-sitter UAVs [J]. Actuators, 2021, 10(6): 119.

[5] Meng, Fanliang, Congwei Zhao, Qianzhang Ren, and Xinhua Wang. Research of Tail-Sitter VTOL UAV in Transition Process Based on an Improved L1 Adaptive Control Method. *6th International Symposium on Autonomous Systems (ISAS)*, pp. 1-6. IEEE, 2023.

[6] Lyu XM, Zhou JN, Gu hw, et al. Disturbance observer based hovering control of quad-rotor tail-sitter VTOL UAVs using $H_\infty$ synthesis [J]. *IEEE Robotics and Automation Letters*, 2018, 2(4): 870 – 882.

[7] CHENG Zihuan, PEI Hailong. A full envelope flight controller for ducted fan tail sitter vertical take-off and landing. *Control Theory & Applications*, 2021, 38(11): 1863 – 1873.

[8] Sun J, Li B, Shen, L, et al. Dynamic Modeling and Hardware-in-loop simulation for a Tail-sitter Unmanned Aerial Vehicle in Hovering Flight [J]. *AIAA modeling and simulation technologies conference*, 2017: 0811.

[9] Argyle ME. Modeling and Control of a Tail-sitter with a Ducted Fan. Brigham Young University, 2016.

[10] E. L. Tobias, Development of Modeling and Simulation Tools for Analysis of Ducted Fan Aircraft, Master of Science, Pennsylvania State University, 2008.

# Study on the effect of annealing rate on the impact resistance of the H-ZLaF50 vehicle plate glass

**Wentao Jia[1,2], Jiaxin Liu[1], Yaohong Wang[1], Guanpeng Yao[1], Xiaofang Gao[1], Xuezhou Fang[1], Xiaoyong Kou[1] and Guangyi Xu[1]**

[1]Hubei New Huaguang Information Materials Co., Ltd., Xiangyang 441000, China

[2]Corresponding author's e-mail: jiawt@hbnhg.com

**Abstract**. The H-ZLaF50 glass has attracted much attention in the field of automotive glass applications due to its excellent mechanical properties. The effect of the annealing rate on the impact resistance of the H-ZLaF50 glass was studied. High-precision infrared stress meter was used to test the stress distribution of glass under different annealing rates and the influence of annealing rate on the stress distribution was found. Secondly, the impact resistance of the H-ZLaF50 glass samples was evaluated by the test method of electric falling ball impact resistance. The maximum impact energy of the H-ZLaF50 glass samples was 0.120 J. The results show that, compared with other annealing rates, the annealing rate of 1°C/min effectively improves the impact resistance of the H-ZLaF50 glass, which makes it more widely used in the automotive flat glass market.

## 1. Introduction

As the material of the interior display protective cover plate, after solving the problem of the safety of the vehicle gauge, the vehicle cover glass has been applied in batches in many models for its high hardness (not easy to be scratched), excellent optical performance (clearer image) and good tactile feeling, and has become one of the mainstream choices of the vehicle display cover plate material, with broad market prospects [1-3]. Compared with the mobile phone cover glass, the size of the vehicle cover glass is larger, the equipment investment is large, the process maturity is insufficient, and the research space is larger. On the premise that the optical properties of vehicle glass are satisfied, the mechanical properties of vehicle glass are also the focus of many researchers during the process of vehicle running [4, 5].

H-ZLaF50 vehicle cover glass produced by NHG company belongs to heavy lanthanum firestone glass with high refractive index ($n_d$=1.80400) and low dispersion ($v_d$=0.017262). The glass has high transmissibility, relatively low specific gravity, and relatively low cost. It is widely used in single-lens, monitoring and projectors, and other optical systems of general materials. It can effectively meet the design needs of customers for cost-effective optical systems. At the same time, the glass has high hardness, good impact resistance, and excellent chemical stability, which is more suitable for use in harsh environmental conditions, especially in the current automotive glass industry, where it is widely used [6]. Due to the extremely high requirements of safety performance in the vehicle industry, the glass itself needs to have high toughness and will not cause secondary damage to the human body due to fragmentation. In the process of preparing the onboard cover glass, in order to eliminate the internal stress in the glass, the onboard cover glass needs to be annealed, and the speed of the annealing rate has a certain influence on the mechanical properties of the glass [7-9]. At present, the most common testing methods for the mechanical properties of vehicle glass include the three-point bending test, Knoo hardness test, and simply supported beam impact test. At present, there are few pieces of research on the impact resistance performance of vehicle glass falling balls [10]. Therefore, this paper will focus on the effect of the annealing rate on the internal stress distribution and impact strength of the H-ZLaF50 onboard glass.

## 2. Materials and methods

In this experiment, the coarse annealed H-ZLaF50 vehicle-mounted glass produced by NHG Company was used as the research object. Glass samples after cutting and polishing, were processed into sizes of 40×40×5 mm cube sample. According to the transition temperature of the H-ZLaF50 onboard glass ($T_g$=688°C), the annealing temperature of the onboard glass was determined to be 660°C. The annealing furnace temperature was raised from room temperature to 600°C. After the sample was placed in the annealing furnace, the annealing rate was set at 16°C/min, 12°C/min, 8°C/min, and 4°C/min. To study the properties of glass samples at lower annealing rates, the annealing rates were set as 3°C/min, 2°C/min, and 1°C/min, respectively. The samples obtained at different annealing rates were labeled as CG-16, CG-12, CG-8, CG-4, CG-3, CG-2, and CG-1. When the annealing furnace temperature dropped to room temperature, the mechanical properties of the vehicle glass under different annealing rates were tested.

The energy-saving Muffle furnace (GM. M8/12A) manufactured by Sigma was used for annealing, with a maximum power of 4 kW. A high-precision infrared stress meter (Exicor 150ATH-1550 nm) was used to measure the stress distribution of different samples. The laser wavelength was 1550 nm. The electric falling ball impact test machine (ELB-DDLQ-15) produced by Alliplatinum Company was used to test the impact strength of the vehicle glass samples. Each group of samples was tested 5 times, and then the average value was obtained. A magnetic carbon steel ball was selected, with a weight of 8.36 g and a diameter of 11 mm.

## 3. Test results and discussions

In the process of application of vehicle cover glass, the internal stress of the glass has a great influence on the mechanical properties of the glass. In the process of preparing glass samples, the internal stress in glass can be effectively eliminated by annealing treatment of different degrees. Figure 1 shows the real photos of the CG-4 sample after annealing at 4°C/min.

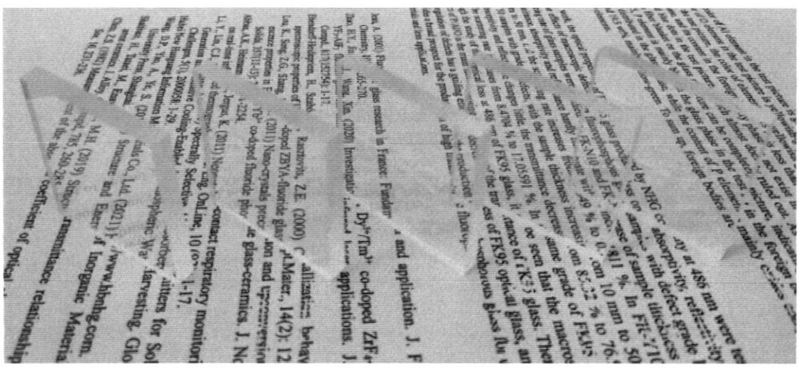

**Figure 1.** Photograph of CG-4 samples after annealing at 4°C/min.

An ideal optical glass is isotropic, but internal stresses can occur when the temperature inside and outside the glass is not consistent during annealing, or when the temperature is not the same throughout the annealing furnace. To further study the stress distribution of each group of glass samples, a high-precision infrared stress meter was used to test the samples at different annealing rates. The stress distribution in the middle of CG-16 and CG-12 samples is small, while the stress at the edge is large, as shown in Figure 2. With the decrease in the annealing rate, the stress distribution in the middle of the CG-8 sample is relatively uniform. When the annealing rate is 4°C/min, the stress distribution of the whole sample is more uniform, and there is no phenomenon of high side stress. The comparison shows that with the decrease in annealing rate, the average value of stress birefringence inside the plate glass is 17.72 nm/cm, 13.32 nm/cm, 5.51 nm/cm, and 1.84 nm/cm. The structure shows that the glass sample CG-4 has the least stress birefringence, and the lower annealing rate can effectively eliminate the internal stress inside the glass so that the stress distribution inside the vehicle glass is more uniform.

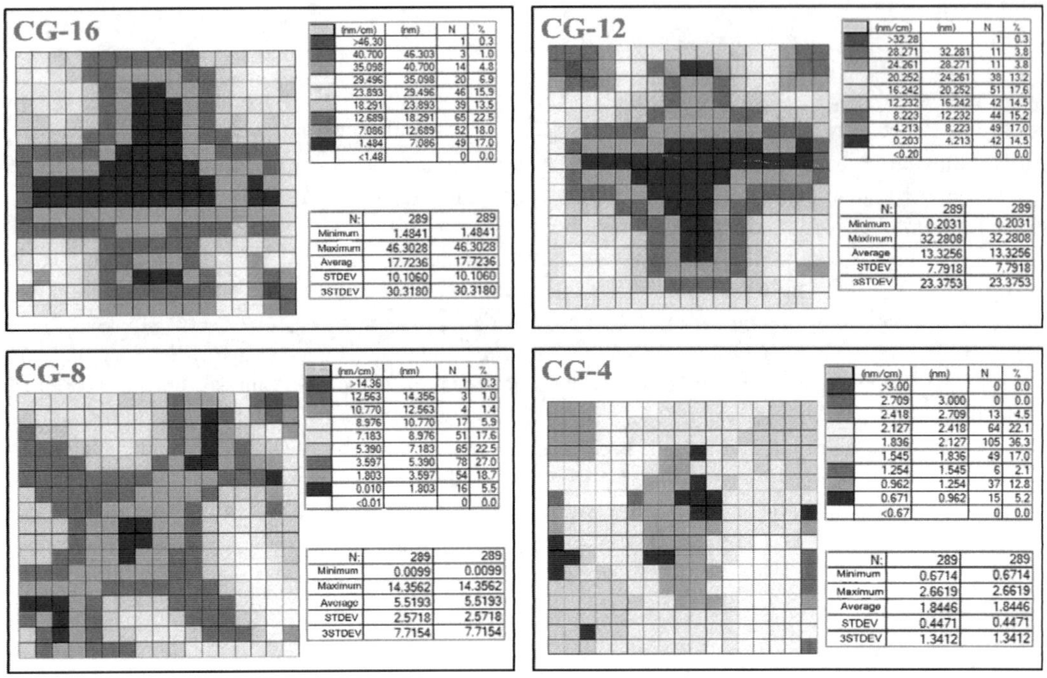


**Figure 2.** Stress distribution of CG-16~ CG-4 plate glass samples at different annealing rates.

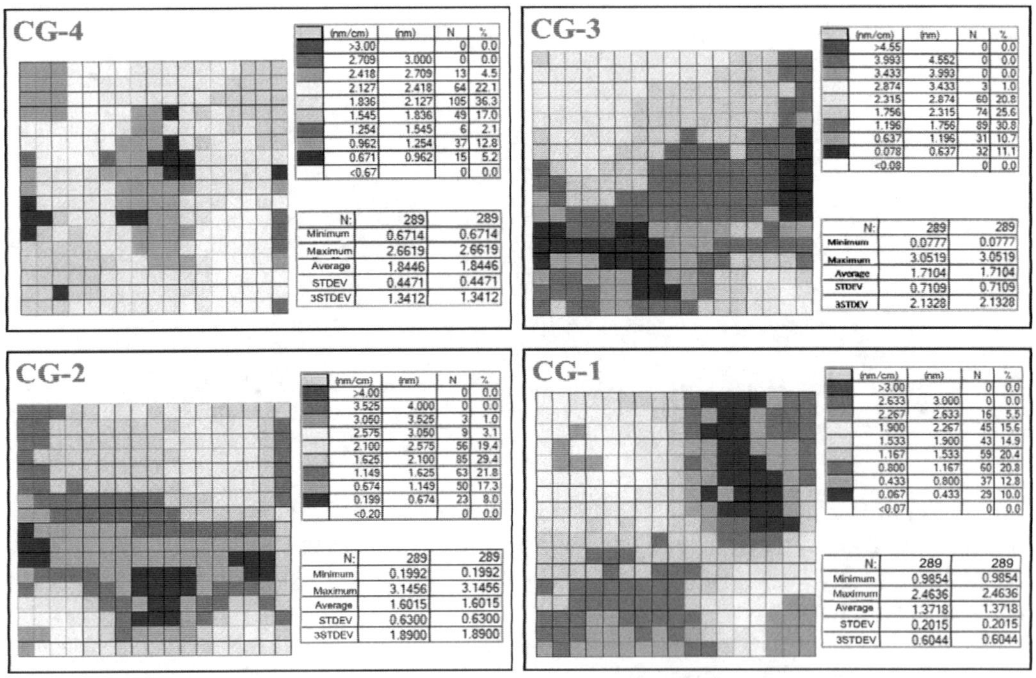


**Figure 3.** Stress distribution of CG-4 ~ CG-1 plate glass samples at different annealing rates.

To further verify the effect of a lower annealing rate on the stress birefringence of the plate glass, the samples were annealed at 3°C/min, 2°C/min, and 1°C/min respectively. According to the test results in Figure 3, the average stress birefringence of CG-4, CG-3, and CG-2 samples is 1.84 nm/cm, 1.71 nm/cm, and 1.60 nm/cm. The results show that the change trend of stress birefringence is not great and almost tends to be stable with the annealing rate decreasing successively. When the annealing rate of the sample is 1°C/min, the stress birefringence of the CG-1 sample reaches 1.37 nm/cm, indicating that the lower annealing rate can eliminate the internal stress in the sample to the greatest extent.

To study the falling ball impact strength of vehicle cover glass, an automatic falling ball impact test machine was used to test the samples obtained at different annealing rates. Before testing, a black marker was used to draw 2 diagonal lines along the test surface to find the middle of the sample. The red laser projection point of the device coincided with the impact-resistant point when the sample was placed. In the process of testing, the sample was dropped from a height of 1000 mm, marked "○" if the surface was not damaged, then the test height was further increased by 50 mm, and the tests were carried out successively. If the sample rose to a certain height, the impact of glass fracture after the falling ball fell or cracked was marked as "×"; The height above the falling ball failure sample was not tested and was marked as "-". Table 1 shows the experimental results of the falling ball impact resistance of the glass sample of the vehicle cover plate.

**Table 1.** Test results of falling ball impact resistance of glass samples of vehicle cover plate.

| Height (mm) | CG-16 | CG-12 | CG-8 | CG-4 | CG-3 | CG-2 | CG-1 |
|---|---|---|---|---|---|---|---|
| 1200 | ○ | ○ | ○ | ○ | ○ | ○ | ○ |
| 1250 | ○ | ○ | ○ | ○ | ○ | ○ | ○ |
| 1300 | ○ | ○ | ○ | ○ | ○ | ○ | ○ |
| 1350 | × | ○ | ○ | ○ | ○ | ○ | ○ |
| 1400 | — | × | ○ | ○ | ○ | ○ | ○ |
| 1450 | — | — | × | × | × | × | ○ |
| 1500 | — | — | — | — | — | — | × |
| 1550 | — | — | — | — | — | — | — |
| 1600 | — | — | — | — | — | — | — |

According to the test results in Table 1, the CG-16 ~ CG-1samples of the vehicle cover plate were damaged when falling at the height of 1350 mm, 1400 mm, 1450 mm, 1450 mm, 1450 mm, 1450 mm, and 1500 mm, respectively. In order to accurately determine the energy, when the sample was damaged, the height value of the sample was subtracted 50 mm, and then the gradient of 10 mm was used to carry out experiments successively to obtain the most accurate impact height of the test sample. The test results are shown in Table 2.

Table 2 shows the experimental results of the impact resistance of the second falling ball of the vehicle cover glass. It can be seen from the test that the onboard cover plate samples CG-16 ~ CG-1 were damaged when falling at the height of 1330 mm, 1370 mm, 1410 mm, 1430 mm, 1430 mm, 1440 mm, and 1460 mm respectively. Although the test results in Table 2 are slightly lower than the height values in Table 1 when the glass is damaged, the test values in Table 2 are more accurate. The impact resistance energy of the vehicle-mounted cover glass sample can be obtained by putting the height value and the mass of the steel ball into Equation (1):

$$E = mgh \qquad (1)$$

In Equation (1), m is the mass of the steel ball, and the unit is Kg; h is the distance m between the bottom of the steel ball and the sample; g is the unit of gravitational acceleration 9.8 m/s2; E is the impact energy and the unit is J [11].

**Table 2.** Experimental results of second falling ball impact resistance of vehicle cover glass sample.

| Height (mm) | CG-16 | CG-12 | CG-8 | CG-4 | CG-3 | CG-2 | CG-1 |
|---|---|---|---|---|---|---|---|
| 1300 | ○ | – | – | – | – | – | – |
| 1310 | ○ | – | – | – | – | – | – |
| 1320 | ○ | – | – | – | – | – | – |
| 1330 | × | – | – | – | – | – | – |
| 1340 | – | – | – | – | – | – | – |
| 1350 | – | ○ | – | – | – | – | – |
| 1360 | – | ○ | – | – | – | – | – |
| 1370 | – | × | – | – | – | – | – |
| 1380 | – | – | – | – | – | – | – |
| 1390 | – | – | – | – | – | – | – |
| 1400 | – | – | ○ | ○ | ○ | ○ | – |
| 1410 | – | – | × | ○ | ○ | ○ | – |
| 1420 | – | – | – | ○ | ○ | ○ | – |
| 1430 | – | – | – | × | × | ○ | – |
| 1440 | – | – | – | – | – | × | – |
| 1450 | – | – | – | – | – | – | ○ |
| 1460 | – | – | – | – | – | – | × |

Through calculation, the impact strength of glass samples CG-16, CG-12, CG-8, and CG-4 is 0.109 J, 0.112 J, 0.116 J, and 0.117 J, respectively. The impact strength of glass samples CG-3, CG-2, and CG-1 is 0.117 J, 0.118 J, and 0.120 J, respectively. By comparison, the impact resistance values of glass samples are the same at the annealing rate of 4°C/min~3°C/min. With the annealing rate decreasing from 3°C/min to 1°C/min, the impact resistance of glass samples increases from 0.117 J to 0.120 J. The glass sample CG-1 has the highest impact strength. The test results show that the internal stress of the onboard glass is effectively eliminated and the impact strength is further improved at the annealing rate of 1°C/min.

## 4. Conclusion

In this paper, the H-ZLaF50 vehicle plate glass was annealed with different annealing rates. A high-precision infrared stress meter was used to measure the stress distribution in the glass sample. With the annealing rate decreasing from 16°C/min to 1°C/min, the stress birefringence decreases from 17.72 nm/cm to 1.37 nm/cm. The H-ZLaF50 vehicle plate glass annealing rate is 1°C/min, the internal stress inside the glass is the lowest, and the maximum value of the falling ball impact strength is 0.120 J. The results show that the H-ZLaF50 plate glass produced by NHG has a wide range of application prospects in the field of plate material.

## References

[1] Dawid B, Stanisław B, Wojciech W. (2023) Analysis of passenger car crash with a cable barrier installed with anti-glare screens on a horizontal convex road curve with 400 m radius, Int J Impact Eng, 173, 104486: 1-5.

[2] Jing C, Nick B, Yang S. (2021) The influence of mobile phone location and screen orientation on driving safety and the usability of car-sharing software in-car use, Int J Ind Ergonom,84, 103168: 1-4.

[3] Hilkka G, Tuomo K. (2020) Impacts of Touch Screen Size, User Interface Design, and Subtask Boundaries on In-Car Task's Visual Demand and Driver Distraction, Int J of Hum-Comput St,142, 102467: 1-3.

[4] Basudev S, Jae R, Chan G. (2022) Industrial recycling of end-of-life vehicle windshield glass by mechanical beneficiation and complete recovery of polyvinyl butyral, J Clean Prod, 334, 130192: 1-3.

[5] Karrouchi0 M, Messaoudi A, Kassmi K. (2023) Intelligent driver monitoring system: An Internet of Things-based system for tracking and identifying the driving behavior, Comput Stand Inter, 84, 103704: 1-3.
[6] http://www.hbnhg.com.
[7] Song M, Li Y, Wu G, He H. (2011) The effect of annealing on the mechanical properties of a ZrAlNiCu metallic glass, J Non-Cryst Solids. 357(3): 1239-1241.
[8] Li J, Wang G, Yi J. (2022) Annealing effect on mechanical properties and strain localization of a Zr-based metallic glass under shear, Intermetallics,151, 107714: 1-2.
[9] Huang F, Xie W, Li S. (2020) Effect of annealing environment on the microstructure and mechanical property of CrWN glass molding coating, Surf Coat Tech, 383, 125281: 1-2.
[10] Önder A, Robinson M. (2018) Harmonised method for impact resistance requirements of E-glass fiber/unsaturated polyester resin composite railway car bodies, Thin-Wall Struct, 2018, 131: 151-164.
[11] Klaus W, Bhola N. (2015) On the potential energy in a gravitationally bound two-body system, New Astron, 34: 250-252.

# Effect of intercritical quenching and tempering on microstructure and properties of DP1180

**Zhiwei Zou [1], Bowen Jiang [1], Liejun Li [1,4], Jianping Ren [2], and Jietao Dai [3]**

[1]Guangdong Provincial Key Laboratory of New Metal Materials Preparation and Forming, South China University of Technology, Guangzhou 510640, China
[2] Guangzhou JFE Steel Sheet Co., Ltd., Guangzhou 510000, China
[3]School of Mechanical and Electrical Engineering, Guangzhou University, Guangzhou 510006, China

[4] Corresponding author's e-mail: liliejun@scut.edu.cn

**Abstract**: Advanced high-strength steels for automobiles are attracting more and more attention due to the need to conserve energy, reduce emissions, and develop lightweight designs. DP1180 is set to become the mainstay of automotive steel, offering a good combination of strength and plasticity. In this paper, the effect of the quenching temperature on the microstructure and mechanical properties of the DP1180 steel has been studied by means of scanning electron microscopy (SEM) and mechanical tests. The results show that the DP1180 steel had the highest product of strength and plasticity equal to 15.7 GPa·%, quenched at 825°C and tempered at 430°C for 5 minutes, the ultimate tensile strength (UTS) is about 1, 342 MPa, and the total elongation is 11.7%. This study offers a dependable theoretical guideline for the optimization of process parameters for the production of DP1180 steel.

## 1. Introduction

With the development of the automotive industry and various energy shortages, higher requirements have been put forward for automotive steel [1]. Not only do they need to meet normal usage requirements, but they also need to achieve energy conservation and emission reduction to a certain extent. Lightweight is one of the effective means to reduce emissions, and achieving lightweight requires the development and use of high-strength automotive steel. After three generations of evolution, the third generation of advanced automotive steel DP1180 exhibits ferrite and martensite as its main microstructure with high tensile strength, comprehensive mechanical properties, and crash reliability, which plays an important role in promoting automotive lightweight [2-4]. Through the study of the phase transformation law of DP1180 steel in the continuous cooling process, scholars have concluded the product differences under different cooling rates [5]. It has shown that the quenching and tempering process in the intercritical zone can improve the mechanical properties of DP980 steel [6]. Although DP1180 steel has a high strength, its poor plastic toughness makes it difficult to be widely used in actual production. This paper aims to explore the impact of critical zone quenching and tempering on the mechanical properties of DP1180 steel and attempts to improve its plasticity problem and enhance its practicality. The main work content is as follows:

(1) We perform thermal simulation experiments on DP1180 steel samples by using the Gleeble525 thermal simulator to determine the phase transformation temperature.

(2) We investigate the microstructural and mechanical properties of the samples by subjecting them to intercritical different temperature zones during quenching and tempering processes.

(3) We assess and compare the performance of components subjected to heat treatment processes with those that have not undergone any heat treatment process (cold-rolled DP1180 steel), specifically during the quenching and tempering in the critical zone.

## 2. Experimental materials and methods

The experimental steel is a cold-rolled DP1180 steel plate provided by a steel company, and the specific chemical composition of the cold-rolled DP1180 steel is shown in Table 1. As can be seen from the table, the basic composition of DP1180 steel is Carbon and Manganum, and a certain amount of Chromium and Molybdenum is added to improve the hardenability [7].

**Table 1.** The main chemical composition of the DP1180 steel (wt. %).

| C | Si | Mn | P | S | Al | N | Cr | Ni | Mo | V |
|---|---|---|---|---|---|---|---|---|---|---|
| 0.17 | 0.2 | 1.4 | 0.01 | 0.01 | 0.08 | 0.01 | 0.22 | 0.06 | 0.02 | 0.08 |

To determine the approximate range of phase transformation temperature of DP1180 steel, three sets of experiments were conducted by using a Gleeble525 thermal simulation testing machine with a thermal expansion device, as shown in Figure 1. The $Ac_1$, $Ac_3$, $M_s$, and $M_f$ temperatures were determined respectively, as shown in Table 2.

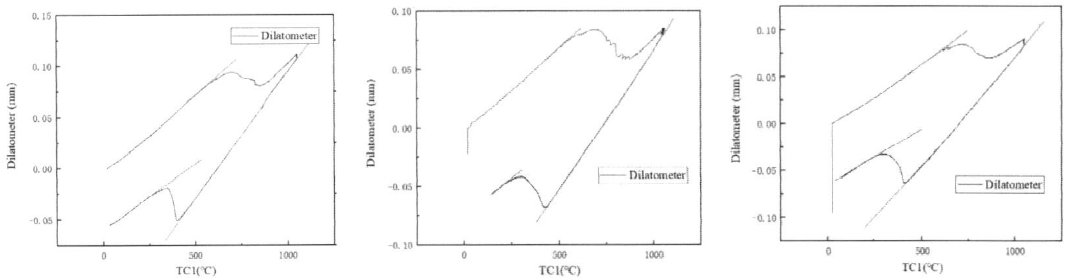


(a) Sample 1; (b) Sample 2; (c) Sample 3

**Figure 1.** Thermal expansion curve of three sets of DP1180 steel.

Based on the analysis of the above experimental data combined with the literature, the temperatures of the $Ac_1$, $Ac_3$, $M_s$, and $M_f$ temperatures were finally determined to be 610°C, 836°C, 456°C, and 250°C, respectively. According to this data, the temperatures required for future experiments were determined.

**Table 2.** The phase transition temperatures of DP1180 steel at different quenching processes.

| Group | $Ac_1$ (°C) | $Ac_3$ (°C) | $M_s$ (°C) | $M_f$(°C) |
|---|---|---|---|---|
| 1 | 600 | 845 | 463 | 271 |
| 2 | 601 | 825 | 450 | 242 |
| 3 | 630 | 838 | 457 | 266 |

The DP1180 steel was heated by an OTF-1500X open tubular furnace. The samples were cut from the cold rolled steel plate and were carried out on quenching and tempering treatment in the critical zone. The specific heat treatment process is shown in Figure 2 below. Thus, the quenching temperatures were selected at 825°C and 850°C respectively, followed by water quenching. The tempering temperature was set at 430°C, lasting for 4 minutes. Our study aims to explore the effects of different quenching temperatures on the microstructural and mechanical properties of DP1180 steel. In this paper, the original cold-rolled DP1180 steels were marked as Sample 1, the steels quenching at 825°C were marked as Sample 2, and the steels quenching at 850°C were marked as Sample 3.

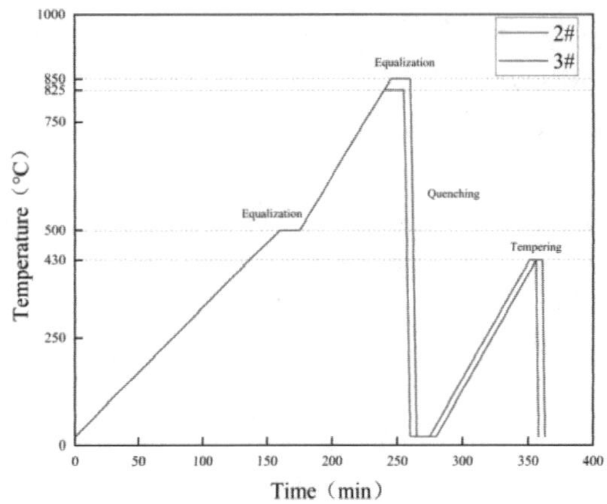


**Figure 2.** Heat treatment process routes of DP1180 steel.

In this study, optical microscopy (OM) and scanning electron microscopy (SEM) were used to observe the microstructure of DP1180 steel under different heat treatment conditions. The test samples were taken from the middle of the original cold-rolled steel. Then we ground with 200, 600, 1, 200, 1, 500, and 2, 000 mesh sandpaper, and then polished mechanically by a polishing agent with particle sizes of 1.5 μm and 1.0 μm, respectively. Finally, samples were corroded with nitric acid alcohol solution (volume fraction of 4%), lasting for 3-10 s.

Tensile samples were cut from the middle of the original cold-rolled steel in the rolling direction (RD) with a total length of 100 mm, a parallel segment length of 29 mm, and a gauge width of 6.12 mm. The specific dimensions of the tensile samples are shown in Figure 3. The tensile samples were polished to smooth surfaces before tensile tests. At least three groups of tensile tests were conducted for every sample to obtain an average value and ensure reliability.

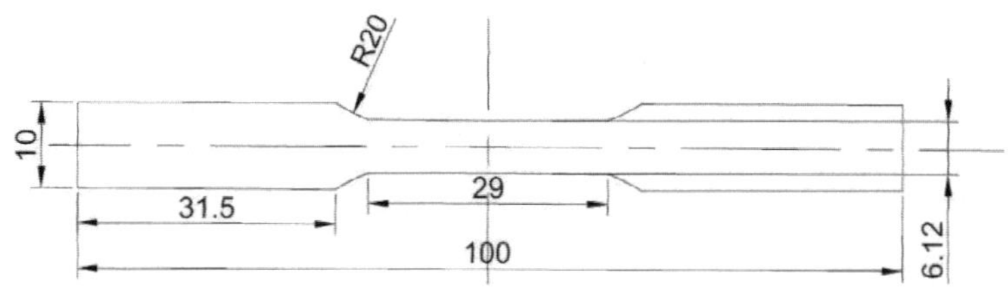


**Figure 3.** Dimension diagram of DP1180 tensile specimen.

## 3. Results and discussion

### *3.1 Effect of quenching temperatures on microstructure*

Figure 4 shows the microstructure of all samples. Figures 4 (a) and (d) show that the microstructure of cold-rolled steel consists of ferrite and island pearlite. However, owing to the cold-rolling process, the island pearlite was compressed and elongated with some fine broken grains. Figures 4 (b) and (e) show the microstructure of steel quenching at 825°C. According to the existing literature, we can find that the microstructure exhibits acicular bainite and lumpy lath martensite with a small amount of fine carbide precipitation [9]. These two phases are refined in comparison with the former. As shown in Figures 4 (c) and (f), when quenching at 850°C, the original cold-rolled steel shows an obvious refinement. When quenching at 825°C, some of the acicular bainite and lath martensite have grown, exhibiting a significant non-uniform grain size distribution.

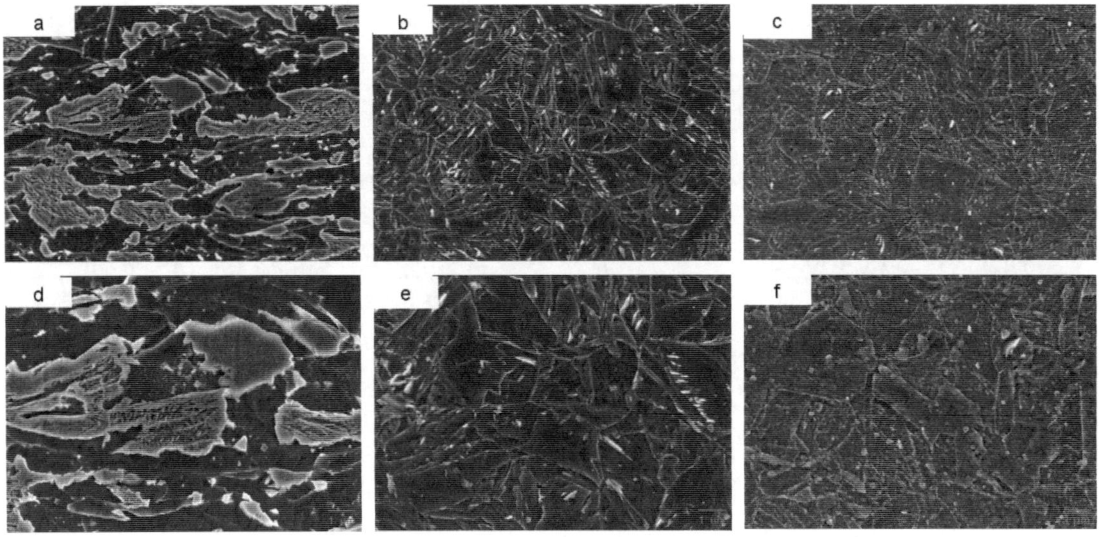


(a) Cold-rolled DP1180, 5000X; (b) 825°C-430°C, 5000X; (c) 850°C-430°C, 5000X;
(d) Cold-rolled DP1180, 10000X; (e) 825°C-430°C,10000X; (f) 850°C-430°C, 10000X

**Figure 4.** SEM micrographs of DP1180 steel at different quenching temperatures.

*3.2 Effect of quenching temperatures on mechanical properties*
Table 3 shows the total elongation (TE), ultimate tensile strength (UTS), and product of strength and elongation of the original cold-rolled steel and steels at different quenching temperatures of 825°C and 850°C. We can learn from Table 3 that as the quenching temperature increases, the ultimate tensile strength (UTS) decreases from 1, 425 MPa at the original cold-rolled state to 1, 342 MPa at 825°C and 1, 248 MPa at 850°C. However, we can also see that the total elongation (TE) increases from 9.9% of the original cold-rolled state to 11.9% at 825°C and 11.7% at 850°C as the quenching temperature increases. The product of strength and elongation is determined by ultimate tensile strength (UTS) and total elongation (TE), as shown in Table 3. When the quenching temperature reaches 825°C, the value of Sample 2 (15.97%) is significantly higher than that of Sample 1(14.11%) and Sample 3 (14.60%), which means that the comprehensive properties and the applicability of the material are greatly improved.

Figure 5 shows the engineering stress-strain curves of the three samples. Based on the above data, we consider that because the cold-rolling deformation causes the dislocation entanglement and jam, the microstructure produces a work hardening effect, which is also the reason why the original cold-rolled samples have higher ultimate tensile strength (UTS) and lower total elongation (TE). However, as the temperature increases, the microstructure undergoes a recovery phenomenon, and the diffusion ability of atoms inside the material is enhanced. When the atoms are rearranged into an equilibrium state, most of the residual stress in the cold-rolled steel is eliminated. At the same time, the entanglement of dislocations in the matrix will loosen and open, and at this time, the displacement can not only slip but also climb. Finally, by controlling the critical zone holding temperature and time, as well as the cooling rate, new fine and uniform phase structures of martensite and bainite can be generated through phase transformation, resulting in steel with high strength and good plasticity. However, as for the different quenching temperatures, the two groups of quenching samples also show different microstructural and mechanical properties, mainly due to the temperature excess of $Ac_3$ during the quenching process. The phase transformation of some austenite has finished ahead of time and continues to grow, and the effect of fine grain strengthening is reduced, which makes the microstructural and mechanical properties of Sample 3 superior to Sample 1 and inferior to Sample 2.

**Table 3.** Mechanical property indexes of different quenching temperatures under intercritical quenching and tempering process.

| Sample | Ultimate tensile strength (MPa) | Total Elongation (%) | Product of strength and elongation (GPa·%) |
|---|---|---|---|
| Sample 1 (Cold-rolled DP1180) | 1, 425 | 9.9 | 14.11 |
| Sample 2 (825°C-430°C) | 1, 342 | 11.9 | 15.97 |
| Sample 3 (850°C-430°C) | 1, 248 | 11.7 | 14.60 |

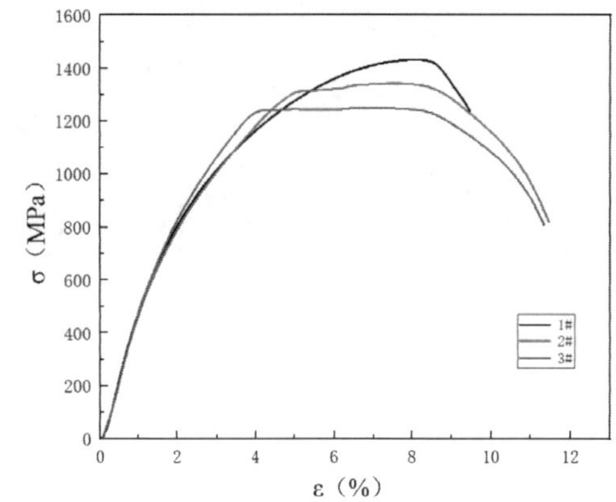


**Figure 5.** Engineering stress-strain curves of industrial DP1180 steel and experimental DP1180 steel at different quenching processes.

## 4. Conclusions

The impact of varying quenching temperatures on the microstructural and mechanical properties of DP1180 steel was investigated. The primary findings can be summarized as follows:

(1) The DP1180 steel exhibits martensite-bainite dual-phase microstructure under different quenching and tempering processes. It is worth mentioning that the two-phase microstructure in steel is finer and uniformly quenched at 825°C and tempered at 430°C.

(2) The quenching temperature has the potential to influence both the ultimate tensile strength (UTS) and total elongation (TE) of the DP1180 steel. The ultimate tensile strength (UTS) at 825°C is higher than that at 850°C, and the total elongation (TE) is slightly better than the latter. The comprehensive mechanical properties (expressed in the product of strength and elongation) are both superior to those of cold-rolled steel.

(3) The DP1180 steel has the highest product of strength and plasticity, i.e., 15.97 GPa·%, quenched at 825°C and tempered at 430°C for 5 minutes, which offers a dependable theoretical guideline for the optimization of the process parameters for the production of DP1180 steel.

## References

[1] Tasan C. C., Diehl M., Yan D., et al. (2015). An overview of dual-phase steels: advances in microstructure-oriented processing and micromechanically guided design [J]. Annual Review of Materials Research, 45 (1): 391-431.

[2] Shang Q., Guo X., Wang G., et al. (2019). Study on the structure and properties of the laser-tailored welded joint of ultra-high-strength duplex steel DP1180 [J]. Thermal Processing Technology, 48 (07): 66-68.

[3] Li Y., Huang Z., Li D., et al. (2019). Advanced high-phase steel forming limit model [J]. Journal of Plastic Engineering, 26 (05): 256-262.

[4] Wang X., Li J., et al. (2017). Research and development and technical innovation of automotive board product of Hegang Group [J]. Steel, 52 (08): 70-74.

[5] Feng X., Yan S., Xing C., Sun X., Zhao N., and Zhao X. (2021). DP1180 phase transition law in the continuous cooling transition process of dual-phase steel for automobiles [J]. Physical and Chemical Inspection (physical volume), 57 (05): 35-38.
[6] Fan Y., Li L., Peng Z., et al. (2022). Effect of intercritical quenching on microstructure and properties of DP980 steel [C]. Journal of Physics: Conference Series. IOP Publishing, 2, 390 (1): 012009.
[7] Wang T. and Huang N. (2014). Application analysis of high-strength steel in the white body material of a small car [J]. Shanghai Automobile, (04): 56-58.
[8] Zhang W., Guo H., Zhou H., et al. (2005). Superplastic stretching behavior of coarse crystalline LC4 alloy [J]. Thermal Processing Process, (02): 18-20.
[9] Xu N., Wang L., Hu J., et al. (2023). Enhancement of the yield strength of intercritically annealed Q&P steel via bainite-based quenching and partitioning treatment [J]. Journal of Materials Research and Technology, 27: 3, 996-4, 004.

# Analysis and prevention of copper pipe corrosion

**Jinpeng Xie [1, 2], Yulin Teng [1], Li Zhang [1] and Hui Li [1]**

[1] Gansu Electric Power Research Institute Technology Center Co., Ltd., Gansu, Lanzhou 730070, China

[2] Corresponding author's e-mail: 1101178341@qq.com

**Abstract:** The failure of the locomotive heat exchanger due to pitting and corrosion of the T2 heat exchange copper tube is used as the engineering application background. The pitting corrosion and influencing factors of T2 copper tu were studied by weight loss method, SEM, and EDS. The optimal ratio of compound corrosion inhibitors (borax, sodium nitrite, sodium silicate, and benzotriazol) of T2 copper tube was obtained. The results show that the main reason for the pitting corrosion of T2 copper pipe is the local rupture of cuprous oxide protective film on the copper surface, resulting in the copper at the bottom of the pit as the anode and the $Cu_2O$ oxide film on the surface as the cathode to form a corrosion cell so that copper dissolves and pitting in the pit. The rupture of the $Cu_2O$ oxide film is related to excessive carbon-containing residue and the deposition of dirt (foreign matter) on the surface, mechanical attack, and defects of the material itself. The optimal ratio of compound corrosion inhibitor in 2 L pure aqueous solution is as follows: Borax content is 4.4 g, sodium silicate content is 4.23 g, sodium nitrite content is 4 g, and benzotriazole content is 0.2 g.

## 1. Introduction

The corrosion and fouling deposition of the heat exchanger will seriously reduce the heat exchange efficiency, causing the failure or even shutdown of the pipes in the heat exchanger increasing operating costs. The materials that make up the heat exchanger are mainly copper and its alloys, and the corrosion of copper and its alloys can be attributed to uniform corrosion, pitting corrosion, crevice corrosion, and selective corrosion [1-2]. It has been found that the corrosion of copper and its alloys is mainly determined by the solubility of $O_2$ and $CO_2$ in the solution, pH, temperature, microorganisms, corrosive environment (in the presence of $SO_2$ and $H_2S$), composition of the electrolyte solution, flow rate, and microstructure potential difference (different metallurgy), as well as the content of other elements in copper and its alloys [3-16].

## 2. Materials and testing

### *2.1 Test material*

The following is the selection and processing of specimens. T2 copper tube was selected, the chemical element was mainly Cu (copper content>99.9%), and the T2 copper tube was processed into $\Phi(10\pm0.03)$ mm$\times$ $(0.6\pm0.03)$ mm, and the length was $(50\pm0.5)$ mm.

Borax, sodium silicate, sodium nitrite, benzotriazole (BTA), and sodium chloride were selected, as shown in Table 1.

**Table 1.** Chemical reagents required for the experiment.

| Chemical agent | Molecular formula | Purity | Source |
|---|---|---|---|
| Borax | $Na_2B_4O_7$ | AR | Tianjin Kaitong Chemical Reagent Co., Ltd |
| Sodium silicate | $Na_2SiO_7 \cdot 9H_2O$ | AR | Yantai Shuangshuang Chemical Co., Ltd |
| Sodium nitrite | $NaNO_3$ | AR | Yantai Shuangshuang Chemical Co., Ltd |
| Benzotriazole | $C_6H_5N_3$ | AR | Shanghai Zhongqin Chemical Reagent Co., Ltd |
| Sodium chloride | NaCl | AR | Sinopharm Chemical Reagent Co., Ltd |

*2.2 Test setup*
The test device is a hanging piece dynamic immersion corrosion instrument, and the schematic diagram is shown in Figure 1. The test device comprises a constant-temperature water bath, a corrosion specimen, and an electric stirrer. The device is an RCC series rotary corrosion pendant tester, dimensions: 1200×650×1100 mm, the number of test hanging pieces: 10 sets in 2 groups, temperature control range: (20-180)±1°C, and speed: (40~160)±2 r/min.

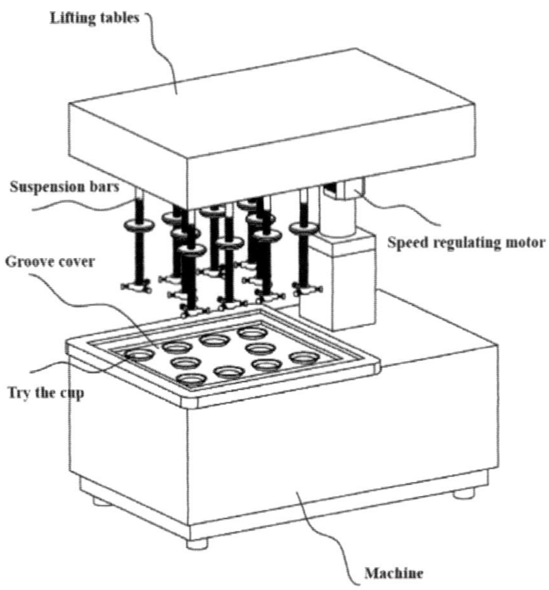


**Figure 1.** Corrosion tester.

The test beaker was calibrated with a 2 L volumetric flask, and the graduation mark was placed on the mural on the outside of the beaker to facilitate the addition of distilled water during the test.

Preparation of simulated fluid: According to Table 1, we weigh the drug required for the test as for the volumetric flask and place it in a 2 L beaker to let it completely dissolve after it is completely dissolved.

After the temperature of the test solution reaches the test temperature, we start the rotation system, adjust the speed to 75 r/min, and start the timing.

When the test running time reached 72 h, the rotation system was stopped, the specimen was taken out, and the appearance of the specimen was observed. Then, it was placed on a clean filter paper, blotted dry with the filter paper, placed in a desiccator for more than 4 hours, and finally weighed (accurate to 0.1 mg) and recorded the data.

*2.3 Test the pilot project*
(1) Corrosion weight loss: The degree of metal corrosion is measured by measuring the change in the quality of T2 copper before and after corrosion, that is, the weight loss method. The calculation is as follows:

$$v = \Delta G_1 /(t \cdot A) \tag{1}$$

Where v is the corrosion rate, and the unit is mg/(h·m$^2$); ΔG1 is the change in the mass of the specimen before and after corrosion, and the unit is mg; A is the surface area of the specimen, and the unit is m$^2$.

Scanning electron microscope morphology analysis and elemental analysis of corrosion products: Specimens with large weight loss and serious corrosion were selected to analyze the microscopic morphology by scanning electron microscope, and the microscopic morphological characteristics of different corrosion areas were observed and studied. The composition of corrosion products was further analyzed by electron microscope with EDS to study the pitting phenomenon and corrosion mechanism of specimens.

Compound corrosion inhibitor orthogonal test: According to TB/T 1750-2006 "In Coolant for Internal Combustion" No. 3 formula materials, borax, sodium silicate, sodium nitrite, benzotriazole (BTA), borax, sodium silicate, sodium nitrite, benzotriazole (BTA) and temperature were selected as test factors. According to the coolant temperature range (40-80°C) provided by the operation and maintenance party and the chloride ion content (10 mg/L) during water quality testing, the orthogonal test table $L_{18}$ ($3^7$) of 7 factor 3 levels was established. The test was carried out according to the orthogonal test table in Table 2.

We establish non-quantitative indexes, as shown in Table 3, and conduct an orthogonal analysis of the test results based on weight loss and morphological condition indexes.

**Table 2.** Orthogonal test table for test.

| Serial number | Factor | | | | |
|---|---|---|---|---|---|
| | Temperature (°C) | Borax (g) | Sodium silicate (g) | Sodium nitrite (g) | Benzotriazole (g) |
| 1 | 40 | 3.60 | 4.23 | 3.60 | 0.18 |
| 2 | 40 | 4.00 | 4.70 | 4.00 | 0.20 |
| 3 | 40 | 4.40 | 5.17 | 4.40 | 0.22 |
| 4 | 40 | 3.60 | 5.17 | 4.40 | 0.20 |
| 5 | 40 | 4.00 | 4.23 | 3.60 | 0.22 |
| 6 | 40 | 4.40 | 4.70 | 4.00 | 0.18 |
| 7 | 60 | 3.60 | 4.23 | 4.00 | 0.20 |
| 8 | 60 | 4.00 | 4.70 | 4.40 | 0.22 |
| 9 | 60 | 4.40 | 5.17 | 3.60 | 0.18 |
| 10 | 60 | 3.60 | 4.70 | 4.40 | 0.18 |
| 11 | 60 | 4.00 | 5.17 | 3.60 | 0.20 |
| 12 | 60 | 4.40 | 4.23 | 4.00 | 0.22 |
| 13 | 80 | 3.60 | 4.70 | 3.60 | 0.22 |
| 14 | 80 | 4.00 | 5.17 | 4.00 | 0.18 |
| 15 | 80 | 4.40 | 4.23 | 4.40 | 0.20 |
| 16 | 80 | 3.60 | 5.17 | 4.00 | 0.22 |
| 17 | 80 | 4.00 | 4.23 | 4.40 | 0.18 |
| 18 | 80 | 4.40 | 4.70 | 3.60 | 0.20 |

**Table 3.** Scoring standard.

| Appearance after corrosion | Bright | Generally bright | Slight discoloration | Slightly discolored | Slight discoloration | Discolor ation | Corrode | Severe corrosion |
|---|---|---|---|---|---|---|---|---|
| Score | 6 | 8 | 10 | 12 | 14 | 16 | 18 | 20 |

## 3. Trial analysis

### *3.1 Weight loss test analysis*

A multi-index analysis was carried out by establishing the weightless index and morphological change as the indicators to measure corrosion, and Table 4 and Figure 2 were obtained. As you can see from Table 4 and Figure 2, whether it is the index of weightlessness or the index of morphological change, the influence of borax on these two indicators is great, considering that for the index of weightlessness, borax is larger than the temperature k value. The morphology change index is not as accurate and stable as weightlessness, so it can be determined that borax is the first influencing factor, followed by temperature. It can be seen that sodium silicate is the third factor, sodium nitrite is the fourth factor, and benzotriazole is the fifth factor. The order of the corrosion degree of T2 copper pipe specimens from large to small is borax > temperature> sodium silicate> sodium nitrite >benzotriazole.

**Table 4.** Multi-index analysis of different factors on the T2 copper sheet.

| Index | | Factor | | | | |
|---|---|---|---|---|---|---|
| | | Temperature (°C) | Borax (g) | Sodium silicate (g) | Sodium nitrite (g) | Benzotriazole (g) |
| Quality Status | K1 | 2.07 | 3.97 | 4.07 | 4.63 | 4.33 |
| | K2 | 7.26 | 7.81 | 7.26 | 4.10 | 3.97 |
| | K3 | 6.50 | 4.04 | 4.50 | 7.09 | 6.63 |
| | k1 | 0.69 | 1.32 | 1.36 | 1.54 | 1.443 |
| | k1 | 2.42 | 2.60 | 2.42 | 1.37 | 1.32 |
| | k1 | 2.17 | 0.44 | 1.50 | 2.36 | 2.21 |
| Range R1 | | 1.73 | 2.16 | 1.06 | 1.00 | 0.89 |
| Topography | K1 | 94.00 | 104.00 | 100.00 | 100.00 | 98.00 |
| | K2 | 100.00 | 94.00 | 96.00 | 96.00 | 100.00 |
| | K3 | 106.00 | 102.00 | 104.00 | 104.00 | 104.00 |
| | k1 | 31.33 | 34.67 | 33.33 | 33.33 | 32.67 |
| | K2 | 33.33 | 31.33 | 32.00 | 32 | 33.33 |
| | K3 | 35.33 | 34 | 34.67 | 34.67 | 34.67 |
| Range R2 | | 2.00 | 3.34 | 2.67 | 2.67 | 2.00 |

It can be seen that the change in the concentration of borax in the solution has the greatest impact on the weight loss of the T2 copper tube, which is related to the hydrolysis of borax as a strong alkali and weak acid in the water, the aqueous solution of borax is alkaline and the borax solution is a pH buffer solution so that the metal ions (such as $Cu^{2+}$, $Ca^{2+}$, $Mg^{2+}$, $Fe^{2+}$, etc.) in the solution react with acid ions (such as $CO_3^{2-}$, $SiO_3^{2-}$, $PO_4^{3-}$) to form dirt deposited on the metal surface. The pH of the solution and the $Cu^{2+}$ in the solution are almost unchanged. The decrease of concentration accelerates the corrosion of T2 copper tubes, the effect of temperature on T2 copper tube specimens accelerates the above-mentioned scaling reaction, and although sodium silicate is a strong alkali and weak salt, a large number of polysilica micelles are obtained by hydrolysis in water. The micelles are easy to combine with Ca2+, Mg2+ and Fe2+. The result of the interaction between sodium nitrite and dissolved oxygen in the solution is related to the density of the oxide film on the surface of the T2 copper tube. Benzotriazole can chelate metal ions (such as Cu2+, Ca2+, Mg2+, Fe2+, etc.). Hence, the concentration of metal ions (such as $Cu^{2+}$, $Ca^{2+}$, $Mg^{2+}$, $Fe^{2+}$, etc.) in the solution is constant, which inhibits the corrosion of T2 copper pipe to a large extent.

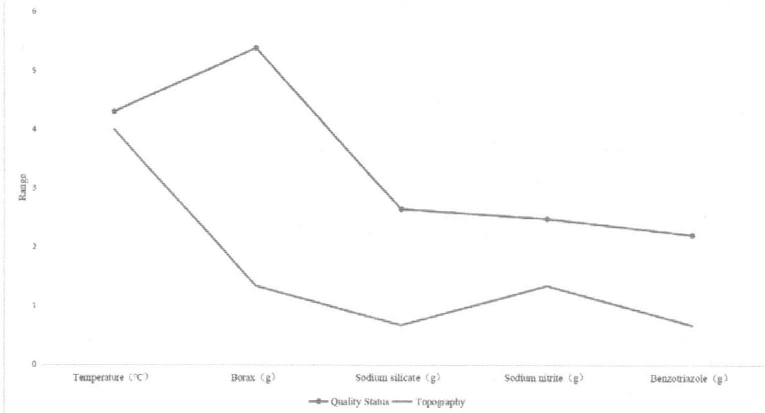


**Figure 2.** Range comparison of weight loss index and quality condition index.

Figure 3 plots the distribution of parameter k in the orthogonal table, representing the distribution of different factor levels. It shows the weight loss index and morphology change index of the T2 copper tube with factors. As can be seen from Figure 3, considering that weight loss is more important and reliable for measuring the severity of corrosion of T2 copper tube specimens than measuring the corrosion status of T2 copper specimens through the morphological changes observed by the naked eye, the K value is not much different for the morphology change index, so the optimal ratio of compound corrosion inhibitor in 2 L pure aqueous solution is obtained. The optimal temperature is 40°C. The worst temperature is 60°C. The best borax content is 4.4 g. The worst is 4 g. The best sodium silicate content in the solution is 4.23 g, with a worst of 4.7 g. Sodium nitrite in solution was best at 4 g and 4.4 g at worst. Benzotriazole at best 0.2 g and worst at 0.22 g. This is shown in Figure 4.

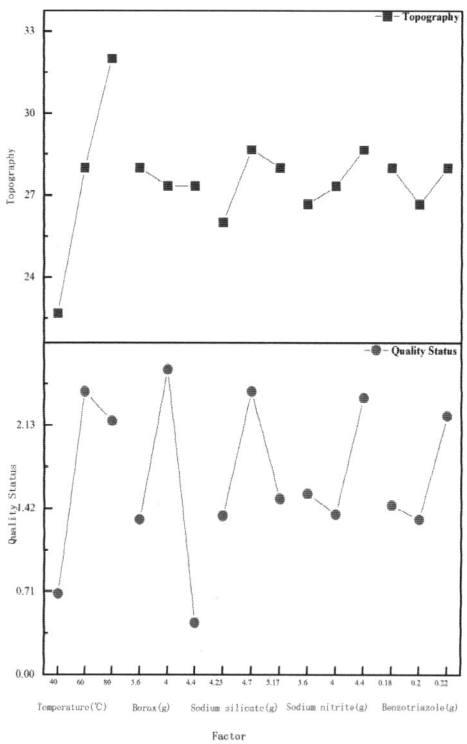


**Figure 3.** Change of T2 copper tube loss weight index and quality condition index with factors.

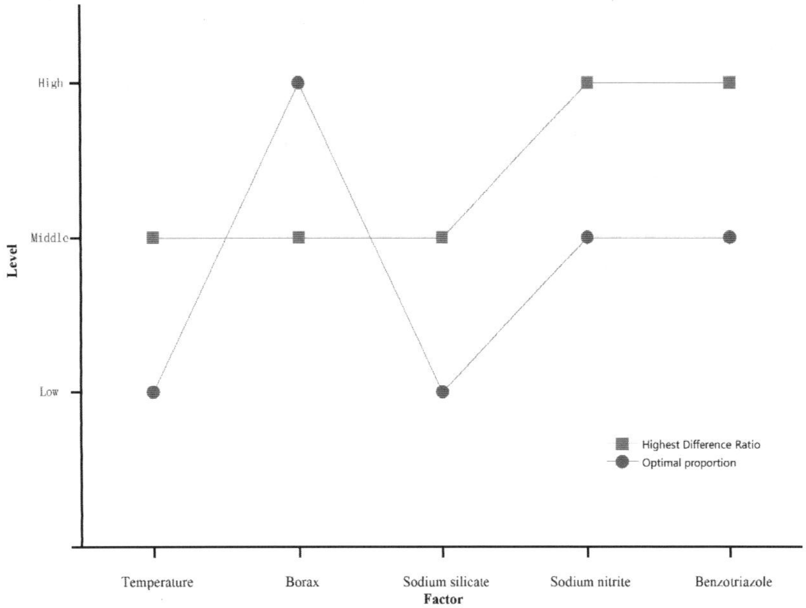


**Figure 4.** T2 copper pipe proportion.

*3.2 Microscopic analysis*

As shown in Figure 5, the left picture is the topography of the No. 9 T2 copper tube specimen at 200 times magnification, from which it can be seen that there is a disordered circular small white spot on the surface of the specimen and one of the round small white dots is magnified and observed. The morphology of the small white spot at 2000 times is shown on the right. From the right figure, it can be seen that there are circular and irregular pits on the surface of the specimen and lumpy materials are observed in the circular pits. An EDS analysis is carried out on the corroded part of the right figure, and the analysis results are shown in Figure 6 and Table 3.

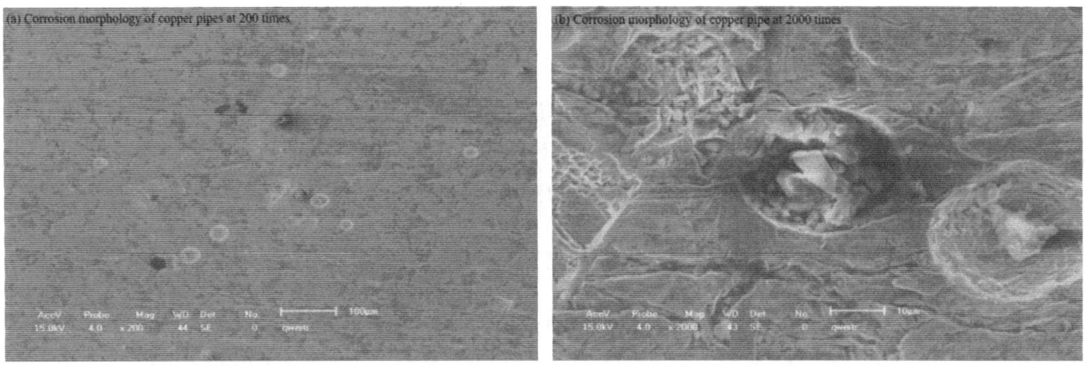


**Figure 5.** SEM appearance of T2 copper tube after corrosion.

Table 4 shows that the elemental composition of T2 copper pipe specimens before and after corrosion is very different. Mg, Al, K, Ca, and Fe elements not present in the copper pipe substrate appear in 34 places, and the content of O is high, indicating that oxidation has occurred in this area, resulting in corrosion. The inner side of the corrosion pit is lighter than the other parts, and the element composition at the bottom (37 places) is almost the same as that of the copper pipe substrate. In addition, Si appears in each area of the corrosion specimen, and the content of Si on the inner surface of the corrosion pit is significantly higher than in other regions.

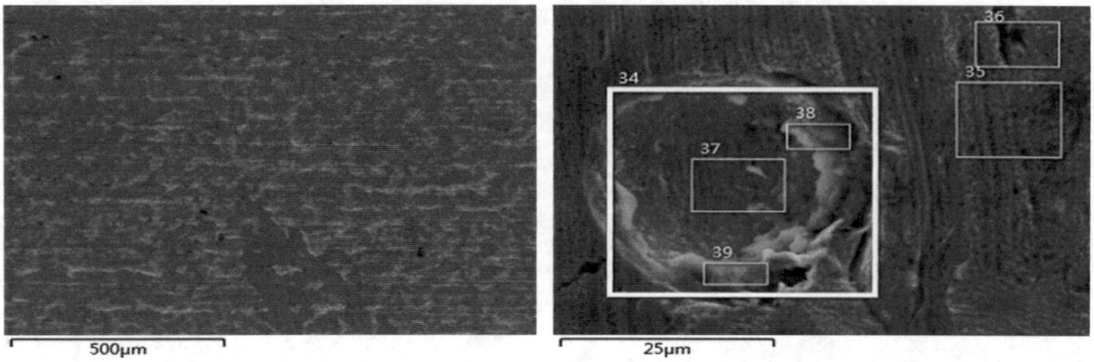


**Figure 6.** Morphology of T2 copper tube blank (left) and corroded (right).

**Table 5.** Element and element content of the test piece in the designated area after corrosion.

| Region | Atomic percentage | | | | | | | | | |
|---|---|---|---|---|---|---|---|---|---|---|
| | C | O | Cu | Si | Mg | Al | K | Ca | Fe | P |
| Blank | 47.30 | 5.95 | 46.75 | / | / | / | / | / | / | / |
| 34 | 37.54 | 20.79 | 34.63 | 3.25 | 1.05 | 1.15 | 0.10 | 1.01 | 0.48 | / |
| 35 | 34.04 | 11.82 | 51.49 | 1.65 | / | / | / | / | / | / |
| 36 | 37.78 | 12.55 | 47.51 | 1.96 | / | / | / | / | / | 0.20 |
| 37 | 23.28 | 6.56 | 68.83 | 1.33 | / | / | | / | / | / |
| 39 | 45.01 | 38.80 | 6.39 | 5.26 | 0.97 | 1.04 | 0.30 | 0.88 | 1.36 | / |

As can be seen from Figures 5, 6, and 4, the cause of pitting on the copper surface can be attributed to a local crack in the cuprous oxide protective film, which can be caused by mechanical damage or by defects in the material itself (including the deposition of impurities on the material). Table 5: The content of the Cu element is high in the 37 area at the bottom of the pit, while the content of the O element is very low. Around the pit (area 39), the Cu element is low, and the O is high. It can be seen that the copper at the bottom of the pit is used as the anode. The cuprous oxide film on the surface is used as the cathode, forming an electric couple that causes the copper to dissolve in the pit, and the corrosion mechanism is as follows:

$$Cu \rightarrow Cu^{2+} + 2e^{-} \tag{2}$$

$$H_2O + 0.5O_2 + 2e^{-} \rightarrow 2OH^{-} \tag{3}$$

$$2H^{+} + 2e^{-} \rightarrow H_2 \uparrow \tag{4}$$

$$2Cu^{2+} + 2(OH)^{-} + CO_3^{2-} \rightarrow CuCO_3 \cdot Cu(OH)_2 \tag{5}$$

The water ionization in the solution yields $H^+$ and $OH^-$, while the $OH^-$ produced in the cathode region is consumed by $Cu^{2+}$ precipitation. The formation of $CaCO_3$, $SiO_2$, $Al_2O_3$, and $Fe_2O_3$ deposition is promoted due to the increase of the pH value of the solution around the corrosion pit, which explains why the Cu element around the corrosion pit decreases. In contrast, the content of the O element and these "unpresent" elements, such as Mg, Al, K, Ca, and Fe, increases, and the salt scale deposited around the pit further thickens.

## 4. Conclusion

(1) The factors affecting the corrosion degree of T2 copper pipe specimens from large to small are as follows: borax > temperature> sodium silicate> sodium nitrite > benzotriazole.

(2) The optimal ratio of corrosion inhibitors in 2 L pure aqueous solution was 4.4 g of borax, 4.23 g of sodium silicate, 4 g of sodium nitrite, and 0.2 g of benzotriazole.

(3) The main reason for the T2 copper pipe is the local rupture of the cuprous oxide protective film on the copper surface, which causes the copper at the bottom of the pit to the anode and the cuprous oxide film on the surface as the cathode to form electrochemical corrosion. It makes the copper dissolve in the pit and cause pitting. The rupture of the oxide film is related to the deposition of excessive carbon-containing residue and dirt (foreign matter) on the surface, mechanical attack, and defects of the material itself.

(4) It is recommended to closely monitor the water quality and temperature during the heat exchange operation to avoid impurities mixed in the aqueous solution, which will impact the copper tube of the heat exchanger and replace the unqualified hot water transfer solution in time.

## References

[1] Hana F, Amel S, Amel D, et al (2024) A comparative study of novel synthesized sulfamide compounds: Electrochemical, morphological, XPS, and theoretical investigations on copper corrosion inhibition in 1.0 M HCl [J]. Journal of Molecular Liquids, 394 123781-.

[2] Ahmed F, Arafat T, O.K. A, et al (2024) Electrochemical, spectroscopic, kinetic and surface analysis of the inhibitory performance of Alcian blue dye for copper corrosion in sulfuric acid solution[J]. International Journal of Electrochemical Science, 19(1):100429-.

[3] Renhao L, Xinyu H, Fangyuan W, et al (2024) Enhancing performance in copper corrosion inhibitors through molecular structural modifications: Mechanisms, design, and future pathways [J]. Journal of Molecular Liquids, 394 123750-.

[4] Yeongil S, Jongmin J, Youngwook P, et al (2023). Modeling of step height reduction with temperature function in copper CMP [J]. Journal of Mechanical Science and Technology, 37 (12): 6213-6220.

[5] Gong Bo, Yu Haijing, Liu Rui (2022) Equipment Management and Maintenance, (10): 81-82.

[6] Chen Xudong, Cui Kangping, Hai Zibin (2022) Research progress on Cu_2O photocorrosion mechanism and inhibition method [J]. Applied Chemical Industry, 51(05): 1461-1465+1471.

[7] Wang Chaoping (2022) Corrosion fracture analysis and improvement measures of tube bundle of hydrogenated heat exchanger [J]. Petrochemical Corrosion and Protection, 39(02): 54-58.

[8] Li Siyuan (2021) Study on corrosion behavior of Cu-Ti alloy in S~(2-) polluted seawater under different heat treatment conditions [D]. Taiyuan University of Technology.

[9] Guo Jian (2021) Study on mechanical behavior and corrosion characteristics of Ni/Cu layered composites [D]. Dalian University of Technology.

[10] Li Hao, Hui Weijun, Wang Jiaojiao (2020) Effect of Cu element on the corrosion behavior of 42CrMoV high-strength bolt steel in simulated atmospheric corrosive environment [J]. Materials Protection, 53(11): 13-20.

[11] Huang Songpeng (2020) Study on corrosion behavior of copper alloys protected by BTA in atmospheric environment [D]. University of Science and Technology of China.

[12] Lv Wenting (2020) Synergistic effect of alloying elements Cu and Cr in the process of local corrosion of low-alloy steel [D]. University of Science and Technology Beijing.

[13] Liu Jiaxin, Zhou Zili, Cao Zhongqiu (2019) Study on corrosion electrochemical properties of Cu-50Co alloy in HCl solution [J]. Liaoning Chemical Industry, 48(10): 960-962.

[14] Li Peng, Li Chengtao, Dong Shuai (2018) Effect of Cu~(2+) on the corrosion behavior of materials in seawater [J]. Corrosion and Protection, 39(11): 843-848+853.

[15] Wang Xiaoyan, Cao Guoliang (2014) Comparison of pitting resistance of several typical seawater steels [J]. Total Corrosion Control, 28(02): 63-67.

[16] Liu Hong, Shen Yuemin (2009) Research on the application of phytic acid in metal corrosion [J]. Applied Chemical Industry, 38(11): 1701-1702.

# Design and implementation of facility intelligent carbon dioxide incubator control system based on POA optimized fuzzy PID

**Qiangke Luo[1], Zeneng Zhao[1], Xingjiang Chen[2], Huanbo Zhang[1] and Yong Liu[1,3]**

[1]School of Biological and Chemical Engineering, Zhejiang University of Science and Technology, Zhejiang Province, Hangzhou; 310023, China
[2]Guizhou Academy of Tobacco Science, Guizhou Province, Guiyang; 550001, China

[3]Corresponding author's e-mail: 116047@zust.edu.cn

**Abstract:** To solve the problem that traditional carbon dioxide incubators are difficult to accurately control the required carbon dioxide environment to the detriment of experimental cultivation, and given the difficulties of time lag, nonlinearity, time-varying, and poor adaptive ability of the incubator control system, this study designs a set of intelligent carbon dioxide incubator control system suitable for laboratory cultivation. The system is based on the POA (Pelican optimization algorithm) to optimize the fuzzy PID (Proportional Integral Derivative) control to regulate the opening degree of the solenoid valve to accurately control the carbon dioxide concentration in the incubator to the set range. In this study, the simulation and comparison tests of POA-based optimized fuzzy PID control, PID control, and fuzzy PID control are carried out by MATLAB/Simulink, and it is found that the overshooting amount of the system is reduced by 11.6%, the regulating time is shortened by 69.3 s, and the steady state error is reduced by 0.1% compared with that of the traditional fuzzy PID control. It can be obtained through practical tests that the accuracy, response, and stability of the optimized fuzzy PID control of the pelican algorithm are better than those of the fuzzy PID control method, and it has better robustness. Therefore, the carbon dioxide incubator control system has high practical application value.

## 1. Introduction

A carbon dioxide incubator is a device that can be used to cultivate plant and animal cells, tissues, organs, bacteria, fungi, and other types of biological samples, and it is characterized by the ability to control the carbon dioxide concentration, humidity, temperature and other parameters in the culture environment [1] to provide the most suitable culture environment. Carbon dioxide incubators can also provide a high concentration of carbon dioxide to meet the needs of cell division, differentiation, and reproduction of various biological samples [2]. Precise control of carbon dioxide concentration is critical to the accuracy and reliability of many biological experiments. Unstable or inappropriate carbon dioxide concentration may affect the growth, metabolism, signaling, and other important processes of organisms, which ultimately affects the interpretation of experimental results and the reliability of data [4]. Hence, ensuring accurate control of carbon dioxide levels within the incubator is a critical concern that requires attention during experimental cultivation.

The time lag, nonlinearity, time-varying, and poor adaptive ability of the $CO_2$ incubator control system adversely affect the precise regulation of $CO_2$ in the incubator, so it is crucial to develop an efficient $CO_2$ control strategy for the incubator [3]. Currently, carbon dioxide control is mainly controlled by PID and fuzzy PID control strategies. Wang et al. [5] explored an effective control logic system for automotive air conditioning utilizing PID control. They created a dynamic simulation model for transcritical $CO_2$ air conditioning to enhance the control system's stability. However, it's worth noting that their approach is currently limited to the simulation phase. Qi et al. [6] developed a cost-effective and integrated greenhouse carbon dioxide measurement and control system using fuzzy PID control, which offers practicality and efficiency. Shuai and Lin [7] devised a carbon dioxide control system that combines BP neural networks with fuzzy control. This innovative approach addresses challenges seen in existing control methods, eliminating the necessity for extensive data inputs. However, it's worth noting that their approach remains in the simulation phase.

In summary, PID control and fuzzy PID control can solve the system time lag, nonlinearity, time-varying, and other issues. Carbon dioxide has a certain degree of regulation to a certain extent, but some of the research is still in the simulation stage, and the use of group intelligence algorithm optimization fuzzy PID control research is relatively small.

To enhance control precision and system adaptability, this study presents a novel intelligent carbon dioxide incubator control system. This system leverages a fuzzy PID controller optimized by the POA. The POA emulates pelican behaviors and hunting strategies, enabling continuous iterations to determine the optimal fuzzy PID parameters. This iterative process seeks the global optimal solution, ultimately facilitating precise control of carbon dioxide concentrations within the incubator, ensuring they remain within the specified range.

## 2. Incubator carbon dioxide control system

### *2.1. System architecture and design*

The carbon dioxide control system for the incubator is depicted in Figure 1, comprising four distinct modules: the detection module, execution module, environment module, and control module.

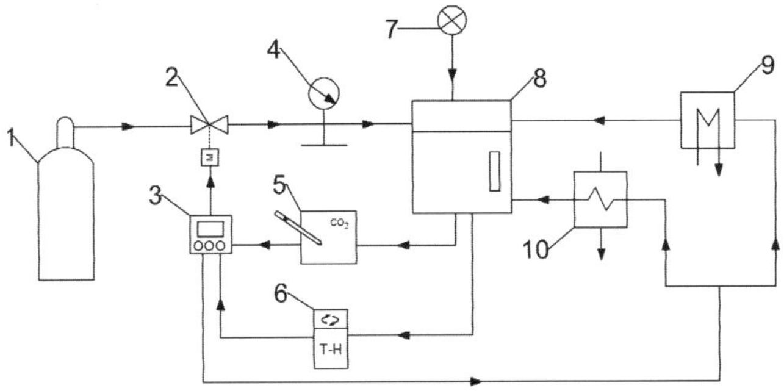


**Figure 1.** Overall structure design of the $CO_2$ control system.

1. gas injection cylinder 2. solenoid valve 3. STM32 controller 4. flow meter 5. carbon dioxide sensor 6. temperature and humidity sensor 7. ultraviolet lamp 8. incubator 9. cooler 10. Heater.

### *2.2. Carbon dioxide concentration C modeling and analysis*

The carbon dioxide concentration model adopted in this study is the classical first-order hysteresis model. This model accounts for the relationship between the inflow of carbon dioxide into the tank and the flow rate, which is determined by the opening of the solenoid valve, denoted as F(t). This opening is established according to the principles of mass conservation once the system attains equilibrium.

$$\frac{d(C(t)V_T)}{dt} = q_0 C_0 F(t) + q_1 C_1 + q_2 C(t) \tag{1}$$

$V_T$—Volume of the incubator, L; $C_0$—Carbon dioxide concentration into the incubator, %; $C_1$—Carbon dioxide concentration in the air entering the incubator, %;C(t)—Carbon dioxide concentration in and out of the incubator, %; $q_0$—Flow rate of carbon dioxide into the incubator, L/min; $q_1$—Flow rate of air into the incubator, L/min; $q_2$—Flow rate of carbon dioxide out of the incubator, L/min; $t_0$—Lag time, s; t—System working time, s

Figure 2 illustrates the dynamic model of the carbon dioxide control system. The control process adheres to the principle of mass conservation, ensuring that the total mass of carbon dioxide entering the incubator and the mass of carbon dioxide exiting the incubator are combined. Also, it equals the total amount of carbon dioxide within the incubator. This equilibrium is achieved through the adjustment of the solenoid valve's opening degree.

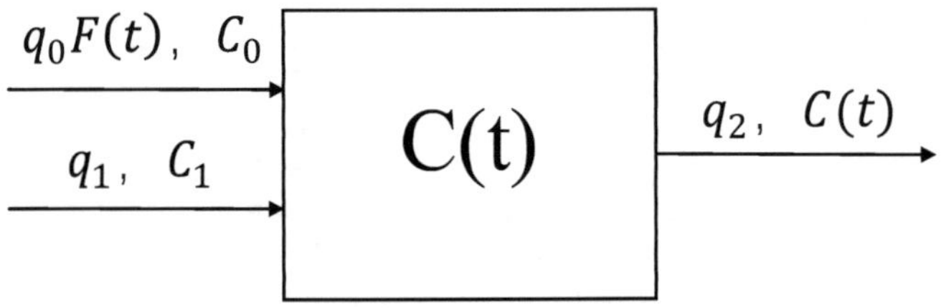


**Figure 2.** Dynamic model of the $CO_2$ control system.

Because of the minimal concentration of carbon dioxide in the air, its influence on this control system is negligible, resulting in:

$$\frac{d(C(t)V_T)}{dt} = q_0C_0F(t) + q_2C(t) \tag{2}$$

Equation (2) is obtained by taking the Laplace transform:

$$C(s) = \frac{C_0q_0F(s)}{V_Ts+q_2} \tag{3}$$

From Equation (3), the control response is characterized as a first-order linear system. Through the actual test, $C_0$=5%, $q_0$=3.74 L/min, $V_T$=150L, $q_2$=2.90 L/min, and $t_0$=0.57 s. Bringing the measured data into Equation (3), the transfer function of the system can be derived as:

$$\frac{C(s)}{F(s)} = \frac{C_0q_0}{V_Ts+q_2} = \frac{1.465}{3103s+1}e^{-0.57s} \tag{4}$$

## 3. POA-optimized fuzzy PID control system design

### *3.1. Pelican optimization algorithm*

The Pelican Optimization Algorithm is a population-based optimization technique that draws inspiration from the hunting behaviors of pelicans. It mimics the two main phases of pelican hunting: surveying prey and exploiting it. Through iterative updates, the algorithm converges towards the best solution [7].

*3.1.1. Initialization.* The initialization expression for the pelican population is as follows:

$$x_{i,j} = l_j + rand \cdot (u_j - l_j), i = 1,2,\dots,N, j = 1,2,\dots,m \tag{5}$$

$x_{i,j}$ — Location of the $j^{th}$ dimension of the $i^{th}$ pelican; N — Pelican populations; m — Dimension of the solution problem; rand — Random numbers in the range of [0,1]; $u_j$、$l_j$ — The solution problem's j dimension has upper and lower bounds.

In the Pelican Optimization Algorithm, the pelican population can be represented using the following matrix:

$$X = \begin{bmatrix} X_1 \\ \vdots \\ X_i \\ \vdots \\ X_N \end{bmatrix}_{N\times m} = \begin{bmatrix} x_{1,1} & \cdots & x_{1,j} & \cdots & x_{1,m} \\ \vdots & \ddots & \vdots & & \vdots \\ x_{i,1} & \cdots & x_{i,j} & \cdots & x_{i,m} \\ \vdots & & \vdots & \ddots & \vdots \\ x_{N,1} & \cdots & x_{N,j} & \cdots & x_{N,m} \end{bmatrix}_{N\times m} \tag{6}$$

*$X$ —Population matrix of pelicans; $X_i$ —Location of the $i^{th}$ pelican*

Within the Pelican Optimization Algorithm, the objective function of a specific problem is employed to compute the objective function value for each pelican. These individual objective function values are subsequently organized into a vector, collectively forming the objective function values for the entire population of pelicans.

$$F = \begin{bmatrix} F_1 \\ \vdots \\ F_i \\ \vdots \\ F_N \end{bmatrix}_{N\times 1} = \begin{bmatrix} F(X_1) \\ \vdots \\ F(X_i) \\ \vdots \\ F(X_N) \end{bmatrix}_{N\times 1} \tag{7}$$

F —Vector of objective functions for pelican populations; $F_i$ —The objective function's value for the $i^{th}$ pelican.

*3.1.2. Approaching prey (survey phase).* In the initial stage, the Pelican Optimization Algorithm focuses on locating its prey and subsequently navigating towards a predefined area. By emulating the pelican's approach to its target, this algorithm effectively scans the search space. This approach leverages the algorithm's ability to explore diverse regions within the search space. Within the POA algorithm, the prey's location is generated randomly within the search space. This randomization enhances the algorithm's exploration capacity, making it adept at solving diverse search problems. Its mathematical expression is shown below:

$$x_{i,j}^{P_1} = \begin{cases} x_{i,j} + \text{rand} \cdot (p_j - I \cdot x_{i,j}), F_p < F_i \\ x_{i,j} + \text{rand} \cdot (x_{i,j} - p_j), \quad \text{else} \end{cases} \tag{8}$$

$x_{i,j}^{P_1}$ —Considering the $j^{th}$ dimension of the $i^{th}$ pelican's position after the first-stage update.; I —Random integers of 1 or 2; $p_j$ —The position of the $j^{th}$ dimension of the prey's location.; $F_p$ —The objective function value of the prey.

In the POA algorithm, a new position for a pelican is chosen only when it leads to an improvement in the objective function value at that particular position. This updating mechanism, commonly known as an "effective update," guarantees that the algorithm stays within or moves towards optimal or higher-performing regions, avoiding transitions to suboptimal areas. Its mathematical expression is shown below:

$$X_i = \begin{cases} X_i^{P_1}, & F_i^{P_1} < F_i \\ X_i, & else \end{cases} \tag{9}$$

$X_i^{P_1}$ ——The updated position of the $i^{th}$ pelican.; $F_i^{P_1}$ ——The objective function value calculated using the new position of the $i^{th}$ pelican after the first-stage update

*3.1.3. Horizontal flight (development phase).* During the second stage of the POA algorithm, as the pelicans reach the water's surface, they employ a strategy involving the spreading of their wings, lifting the fish, and subsequently storing the captured prey in a throat pocket. This surface-based flying tactic employed by pelicans enables them to capture more fish within a designated attack area. By modeling this specific behavioral process, the POA algorithm enhances its capacity to converge toward more favorable positions within the prey area. Consequently, this improvement bolsters the algorithm's ability to perform local search and exploit nearby regions. From a mathematical perspective, the algorithm examines positions in proximity to the pelican to facilitate convergence toward better positions. The mathematical expression is shown below:

$$x_{i,j}^{P_2} = x_{i,j} + R \cdot \left(1 - \frac{t}{T}\right) \cdot (2 \cdot rand - 1) \cdot x_{i,j} \tag{10}$$

$x_{i,j}^{P_2}$ —Considering the $j^{th}$ dimension of the $i^{th}$ pelican's position after the second-stage update; R — Random integers of 0 or 2; t — Current number of iterations; T — The highest or maximum number of iterations allowed or specified.

### *3.2. Fuzzy PID controller design*

The fuzzy PID controller comprises two main components: the fuzzy controller[9] and the PID controller [10]. The block diagram representing this configuration is depicted in Figure 3. This controller takes the system error $e$ and the rate of change of the error $e_c$ as the inputs of the controller, which are fuzzified and defuzzified to obtain the correction quantities $\Delta K_p$, $\Delta K_i$, and $\Delta K_d$, which are used as the three parameters of the PID controller.

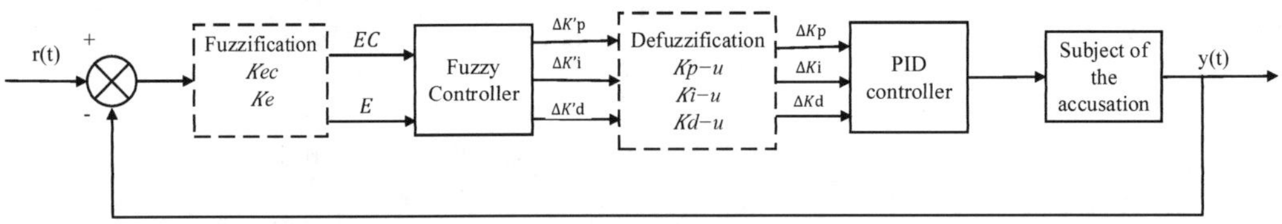


**Figure 3.** Block diagram of Fuzzy PID control principle.

The equation for PID is

$$u(t) = K_p(e(t) + \frac{1}{T_i}\int_0^t e(t)dt + T_d\frac{de(t)}{dt}) \tag{11}$$

u(t) — Controller output; e(t) — The deviation quantity is the difference between the given quantity and the output quantity; $K_p$ — Constant of proportionality; $T_i$ — Integration time constant, i.e., parameter $K_i = \frac{K_p}{T_i}$; $T_d$ — Differential time constant, i.e. parameter $K_d = \frac{K_p}{T_d}$

The PID three-parameter correction equation is

$$\begin{cases} K_p = K_{p0} + \Delta K_p \\ K_i = K_{i0} + \Delta K_i \\ K_d = K_{d0} + \Delta K_d \end{cases} \tag{12}$$

Where $K_p$, $K_i$, and $K_d$ are the corrected proportional, integral, and differential coefficients, and $K_{p0}$, $K_{i0}$, and $K_{d0}$ are the proportional, integral, and differential parameters before correction, respectively.

The fuzzy PID controller is given by

$$u(t) = (K_{p0} + \Delta K_p)(e(t) + \frac{1}{K_{i0}+\Delta K_i}\int_0^t e(t)dt + (K_{d0} + \Delta K_d)\frac{de(t)}{dt}) \tag{13}$$

The error $e_c$, the rate of change of error $e_c$ and the corrections $\Delta K_p$, $\Delta K_i$, $\Delta K_d$ of the carbon dioxide concentration collected by the carbon dioxide sensor from the set value are transformed into the fuzzy linguistic variables E, $\Delta E/\Delta t$, and $K_p$, $K_i$ and $K_d$ through the blurring process, respectively. The quantization domains of the corresponding fuzzy linguistic variables are all defined as [-3,3], and the fuzzy subsets used in this system are defined as follows: {NB (negative big), NM (negative medium), NS (negative small), Z (zero), PS (positive small), PM (positive medium), and PB (positive big)}. Fuzzification of these subsets is accomplished through triangular membership functions, and the system utilizes the Mamdani inference mechanism with the center of gravity defuzzification. Detailed fuzzy rules can be found in Table 1.

**Table 1.** Table of fuzzy rules.

| $K_p/K_i/K_d$ | | $\Delta E/\Delta t$ | | | | | | |
|---|---|---|---|---|---|---|---|---|
| | | NB | NM | NS | Z | PS | PM | PB |
| | NB | PB/NB/PB | PB/NB/PM | PM/NM/PS | PM/NM/PM | Z/Z/Z | Z/Z/Z | Z/Z/Z |
| | NM | PB/NB/PS | PM/NB/PB | PM/NM/PS | PS/PS/NS | PS/NS/PS | Z/Z/NS | NS/Z/Z |
| | NS | PM/NM/PS | PM/NS/PS | PS/NS/PS | PS/NS/Z | Z/Z/Z | NS/PS/NS | NS/PS/Z |
| *E* | Z | PM/NS/Z | PM/NB/PS | PS/NS/NS | Z/Z/NS | NS/PS/NS | NS/PM/NS | NM/PM/PS |
| | PS | NS/NM/Z | PS/NS/Z | PS/Z/NS | PS/PS/PS | PM/PS/PM | PM/PM/PS | PB/PM/PB |
| | PM | PS/Z/PM | Z/Z/Z | NS/PS/NS | NM/PS/NM | NM/PM/NM | NM/PS/NM | NB/PB/NB |
| | PB | Z/Z/PB | Z/Z/Z | NM/PS/NM | NM/PM/NM | NM/PM/NM | NB/PS/NB | NB/PB/NB |

*3.3. POA-optimized fuzzy PID control*

Through the POA optimization algorithm to optimize the quantization factors $K_{ec}$, $K_e$ and the proportional factors$K_{p-u}$, $K_{i-u}$, $K_{d-u}$ of the fuzzy PID, the optimization is searched within the range of meeting the set requirements to find the optimal proportional factor and quantization factor, so that the system regulation does not only rely on the experience of the experts, but also has the ability of self-adjustment [8]. The system's control characteristics and self-adaptive capabilities are enhanced through the schematic block diagram illustrated in Figure 4.

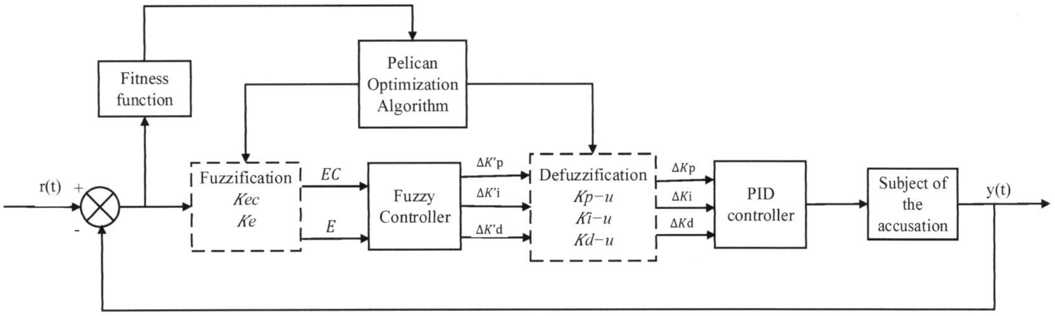


**Figure 4.** Schematic structure of optimized fuzzy PID controller with pelican algorithm.

To ensure response time, overshoot, and stability in the control system, the fitness function is established using the Absolute Error Integral performance index (ITAE):

$$J_{ITAE} = \int_0^{+\infty} t|e(t)|dt \tag{14}$$

## 4. Analysis of test results

A Simulink simulation diagram has been developed within MATLAB/Simulink [11] to evaluate the control performance of the carbon dioxide incubator control system using the fuzzy PID controller optimized by the Pelican Optimization Algorithm, as illustrated in Figure 5. Furthermore, a dedicated incubator device has been designed for experimental purposes, and its physical configuration is presented in Figure 6.

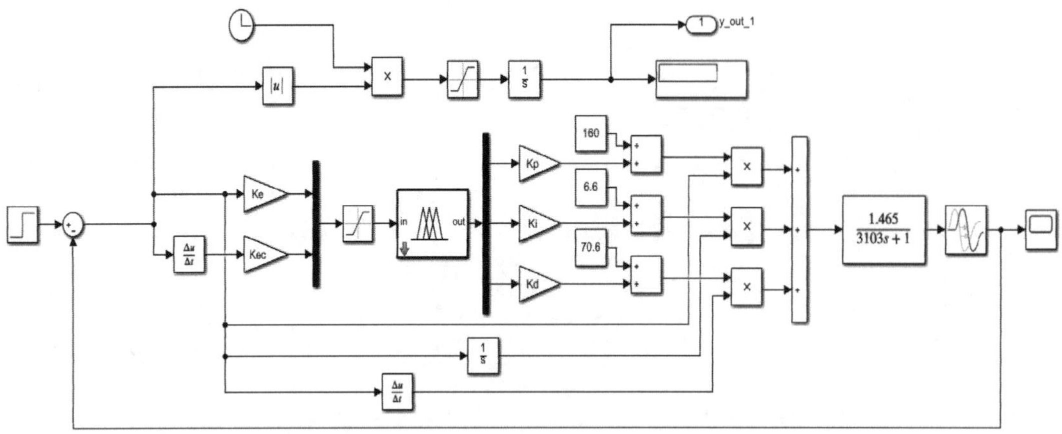


**Figure 5.** Control model diagram of carbon dioxide incubator based on POA Fuzzy PID.

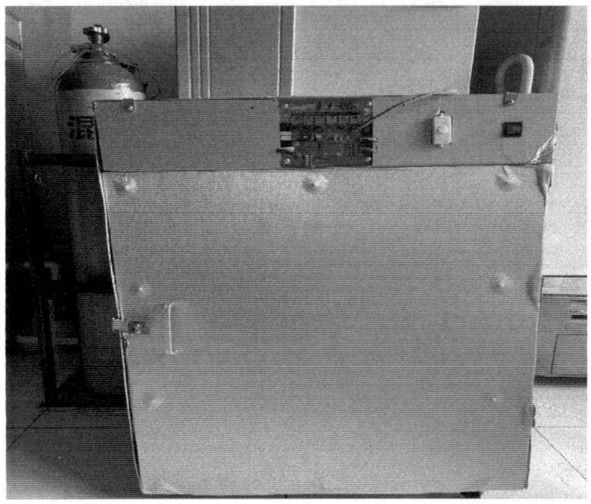

**Figure 6.** Carbon dioxide incubator control system test set.

A comparative analysis has been conducted to assess the suitability of the Pelican algorithm for optimizing fuzzy PID control. This analysis includes a comparison with both traditional PID control and conventional fuzzy PID control through simulations. The results of these simulations are illustrated in Figure 7, and a detailed comparison is provided in Table 2.

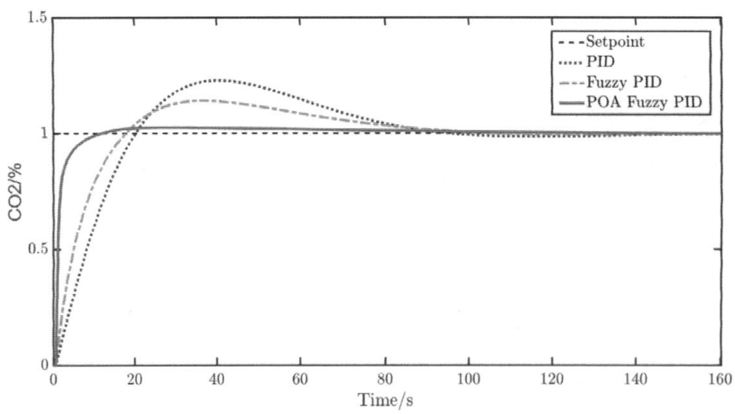


**Figure 7.** Comparison of control model simulation curve results.

**Table 2.** Comparison of simulation results of control model for $CO_2$ concentration.

| Control method | Rising time /s | Overshoot /% | Adjustment time /s | Steady state error /% |
|---|---|---|---|---|
| PID | 20.3 | 23.0 | 86.4 | 0.160 |
| Fuzzy PID | 17.9 | 14.1 | 87.6 | 0.100 |
| POA Fuzzy PID | 11.7 | 2.5 | 18.3 | 0.002 |

The comparison presented in Figure 7 and Table 2 demonstrates the significant precision achieved by traditional PID control, fuzzy PID control, and Pelican algorithm-optimized fuzzy PID control. Among these control methods, it is evident that the Pelican algorithm-optimized fuzzy PID control stands out with the highest level of accuracy, boasting an impressively low steady-state error of just 0.002%. The data analysis underscores the benefits of the Pelican algorithm optimized fuzzy PID control: it reduces rise time by 8.6 s and 6.2 s, curbs overshooting by 20.5% and 11.6%, and trims regulation time by 68.1 s and 69.3 s. The optimized fuzzy PID control using the Pelican algorithm has demonstrated remarkable improvements in system performance. It substantially reduces overshooting, shortens both rise and regulation times, and enhances control accuracy, achieving an impressive steady-state error of just 0.002%. This performance surpasses that of both traditional PID control and fuzzy PID control. In contrast to the other methods, the Pelican algorithm optimized fuzzy PID control considerably diminishes overshooting, trims response times, and bolsters stability and interference resistance. This marks a substantial enhancement in system performance. In conclusion, the Pelican algorithm's optimized fuzzy PID control emerges as highly suitable for the carbon dioxide concentration control system. It effectively meets performance and precision demands, ensuring accurate carbon dioxide concentration management.

## 5. Conclusion

(1) To address the challenges posed by nonlinearity, time-varying behavior, and delays in controlling carbon dioxide concentration within a carbon dioxide incubator used for experimental cultivation, we developed a mathematical model for the carbon dioxide concentration control system. Subsequently, we optimized this system using the Pelican algorithm to implement precise regulation of carbon dioxide concentration.

(2) We conducted simulations and comparisons involving the Pelican algorithm-optimized fuzzy PID control, fuzzy PID control, and PID control. The analysis outcomes demonstrate significant performance improvements achieved by the fuzzy PID control when optimized with the Pelican algorithm. When contrasted with traditional PID control, it demonstrates a 20.5% reduction in overshooting, a 68.1 s decrease in regulation time, a 0.16% improvement in steady-state error, and precise carbon dioxide concentration regulation within the target range. This not only enhances the steady-state accuracy of the system but also renders it highly suitable for incubator carbon dioxide control systems.

## Acknowledgments

Upon the conclusion of this paper, I wish to extend my heartfelt gratitude to my advisor, Liu Yong, for his invaluable guidance and unwavering support. Additionally, I deeply appreciate the financial assistance provided by the Science and Technology project of Guizhou Province Company, China Tobacco Corporation (No. 2021XM11). Lastly, I want to express my thanks to everyone who has contributed to this endeavor.

## Reference

[1] Gu Z and Li J. (2007) Principle and Selection of Carbon Dioxide Incubator. *Life Science Instruments*, (6): 49-51.
[2] YORK A. (2021) Adapting to high levels of $CO_2$. *Nature Reviews Microbiology*, 19(7): 405-405.
[3] Song S, Jiao Y. (2003) Overview of Carbon Dioxide Usage and Production Status. *Chemical Technology Market*, (12): 12-15.

[4] Ou Z, Peng C. (2003) Research progress on the effects of high carbon dioxide concentration on plants. *Journal of Tropical Subtropical Botany*, (2): 190-196.
[5] Wang J, Sun X, Fang J, et al. (2020) Dynamic Performance simulation of multi-PID Control for Trans-critical $CO_2$ Vehicle Air Conditioning. *Journal of Xi'an Jiao Tong University*, 54(8): 168-176.
[6] Qi H, Niu X, and Zheng C. (2019) Greenhouse carbon dioxide concentration measurement and control system. *Journal of Jilin University (Information Science Edition)*, 37(2): 141-147.
[7] Shuai Z, and Lin W. (2006) Greenhouse carbon dioxide control based on fuzzy neural network. *Modern Computer*, (9): 7-10.
[8] TROJOVSKÝ P, and DEHGHANI M. (2022) Pelican Optimization Algorithm: A Novel Nature-Inspired Algorithm for Engineering Applications. *Sensors*, 22(3): 855.
[9] Su M, Chen L, and Lin Hao. (2004) Fuzzy PID Control and its MATLAB Simulation. *Modern Machinery*, (4): 51-55.
[10] Wang L, Song W. (2004) PID control. *Automation Instrument*, (4): 3-8.
[11] Kudinov Y , Kolesnikov V , Pashchenko F , et al. (2017) Optimization of Fuzzy PID Controller's Parameters. Procedia Computer Science , 103618-622.

# Research on hydrogen induced cracking (HIC) performance of vessel steels

**Yanqiu Fan[1, 2,4], Changwen Ma[1, 2, 3], Shaopo Li[1, 2], Wenhua Ding[1, 2], Yang Liu[1, 3] Zhanjun Li[1, 3] and Rensheng Chu[1, 3]**

[1] Research Institute of Technology, Shougang Group Co., Ltd., Beijing, Beijing, 100043, China
[2]Beijing Engineering Research Center of Energy, Beijing, Beijing, 100043, China
[3]Beijing Key Laboratory of Green Recyclable Process for Iron and Steel Production Technology, Beijing, Beijing, 100043, China

[4]Correspondence: fanyq@shougang.com.cn

**Abstract:** Hydrogen-induced cracking performance is quite important for vessel steels. But for this type of medium carbon steel, studies on the effect of the dislocation density and distribution condition of carbides on hydrogen-induced cracking performance were insufficient. In this study, the tested steel was treated with three different heat-treatment processes: scanning electron microscopy (SEM), hydrogen-induced cracking tests (HIC), and transmission electron microscope (TEM), which were used to reveal the mechanism of affecting hydrogen-induced cracking performance. The results showed that microstructure, dislocation density, and distribution condition of carbides were the significant key to the hydrogen-induced cracking performance of tested steels. As the tempering temperature grew from 450°C to 650°C, the density of dislocation gradually went down from $9.71/10^{10}$ $cm^{-1}$ to $2.97/10^{10}$ $cm^{-1}$. And the shape of carbides transformed from bar into fine particles, and the area fraction increased from 37% to 44%. More uniformly distributed carbides can enhance hydrogen-induced cracking resistance because it can promote trapped hydrogen atoms uniformly distribute, reduce the local hydrogen pressure and prohibit crack initiation and propagation. In other words, a tempering temperature above 550°C would be beneficial for the hydrogen-induced cracking performance of vessel steels.

## 1. Introduction

Anti-hydrogen-induced cracking vessel steels were usually used in the petrochemical industry and coal chemical industry, especially in oil refining, desulphurization, and sewage treatment equipment [1]. Under a wet $H_2S$ environment, pressure vessels were more likely to emerge hydrogen-induced cracks, and it had become one of the most prominent technical problems. Hydrogen-induced cracks is the phenomenon that hydrogen atoms produced from corrosion in a wet $H_2S$ environment entered into steel, and accumulated near inclusions and segregation zone, and when the pressure surpassed the crack initiation limit, cracks would appear in the steel [2].

Anti-hydrogen-induced cracking performance was a significant indicator for evaluating vessel steels. Some researchers found that many aspects, such as chemical composition, microstructure, non-metal inclusions, and other defects can affect anti-hydrogen-induced cracking performance [3-6]. Ai et al. proposed a new model to discover that segregation is the key to determining the slab quality and shrinkage cavity and inclusions resulting from segregation are easy to become crack initiation sites [7]. Besides, hydrogen trapping sites including reversible trapping sites and irreversible trapping sites were the most important factors in HIC susceptibility. And as the most common reversible hydrogen trapping sites, dislocation density would be quite different under various heat-treatment processes [8-11]. So in this paper, the hydrogen-induced cracking performance of vessel steels under different heat treatment processes was studied to discover the mechanism of affecting its hydrogen-induced cracking performance.

## 2. Experimental material and methods

The tested slabs were melted in an induction furnace and forged into a small ingot with the size of 125 (thickness)*125*250 mm. The main elements that were added to steels were carbon (C), silicon (Si), manganese (Mn), and aluminum (Al). And impurity elements such as phosphorus (P) and sulfur (S) should be strictly controlled according to the content range in Table 1.

**Table 1.** Chemical compositions of the experimental steel(mass fraction/%).

| C | Si | Mn | P | S | Alt |
|---|---|---|---|---|---|
| 0.15~0.25 | 0.3 | 0.5~1.5 | 0.008 | 0.006 | 0.03 |

The tested plates were cooled from 800°C to room temperature at the rate of 20°C/s and then were cut into three smaller plates. Then they were quenched at 900°C and tempered at 450°C, 550°C and 650°C for 120 min, individually. And they were marked as specimens NO.1, NO.2, and NO.3 in turn.

Metallographic specimens were made from the three plates, and the cross-section surfaces were polished and corroded by a 4% nitric alcohol solution. Microstructures were watched by using scanning electron microscopy (SEM), which model is FEI QUANTA 200E with the scanning parameters: 200~250 kV, 20-150× magnification. Also, thin slices were cut from the tested plates with the size of 10*10*0.3 mm, and then ground to 40 μm and observed by transmission electron microscope (TEM). In the TEM experiment, a JEM 2100 model equipment was used to obtain substructure, dislocation distribution condition, and so on. Besides, the hydrogen-induced cracking (HIC) test was conducted following the standard NACE TM 0284-2003, to evaluate the exact level of the hydrogen corrosion resistance of specimens.

## 3. Results and discussions

### *3.1 Mechanical property*

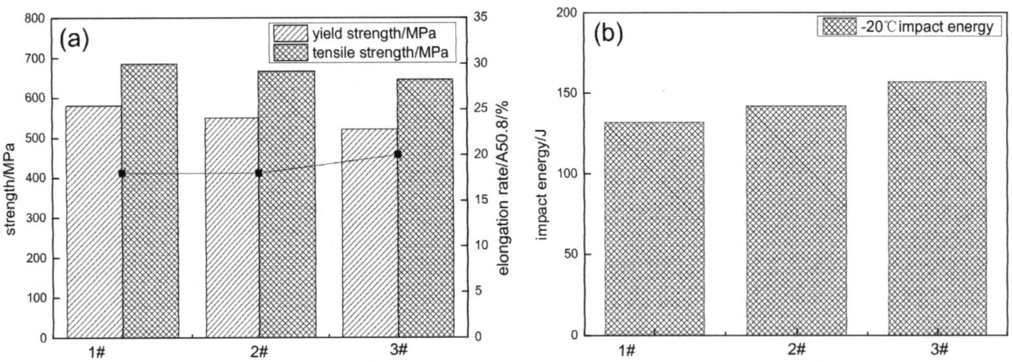


**Figure 1.** Mechanical property.
(a) tensile test (including strength and elongation rate); (b) impact test under -20°C.

The mechanical property of specimens were shown in Figure 1. As shown in Figure 1(a), the yield strength and tensile strength both gradually went down. For specimen NO.1, when the tempering temperature was 450°C, the yield strength and tensile strength were 580 MPa and 685 MPa. When the tempering temperature grew to 550°C, the yield strength and tensile strength decreased to 550 MPa and 667 MPa (specimen NO.2), individually. And for specimen NO.3 tempered at 650°C, the yield strength and tensile strength declined to 521 MPa and 645 MPa, individually. There was a slight change in elongation rate, with 18%, 18%, and 20%, individually. Besides, for the three specimens, impact energy under -20°C gradually grew from 132 J to 142 J, and then to 157 J (Figure 1(b)).

### *3.2 Microstructure*

The microstructure morphology of the three tempered steels were displayed in Figure 2. On the whole, the tempering temperature had a great influence on microstructure morphology, including the size, shape, and distribution condition of carbides. From Figure 2(a), it can be seen that when the tempering temperature was 450°C, the main microstructure type was tempered lath bainite, and the carbides were unevenly distributed on the matrix in the shape of a band. While, when the tempering temperature reached 550°C (specimen NO.2), the microstructure transformed into granular bainite, and the carbides uniformly distributed on the matrix with a small ball shape, as shown in Figure 2(b). Meanwhile, at a tempered temperature of 650°C (specimen NO.3), the microstructure was still mainly granular bainite, the number of precipitated carbides was greater and the size was larger, and the shape was still mainly sphere (Figure 2(c)).

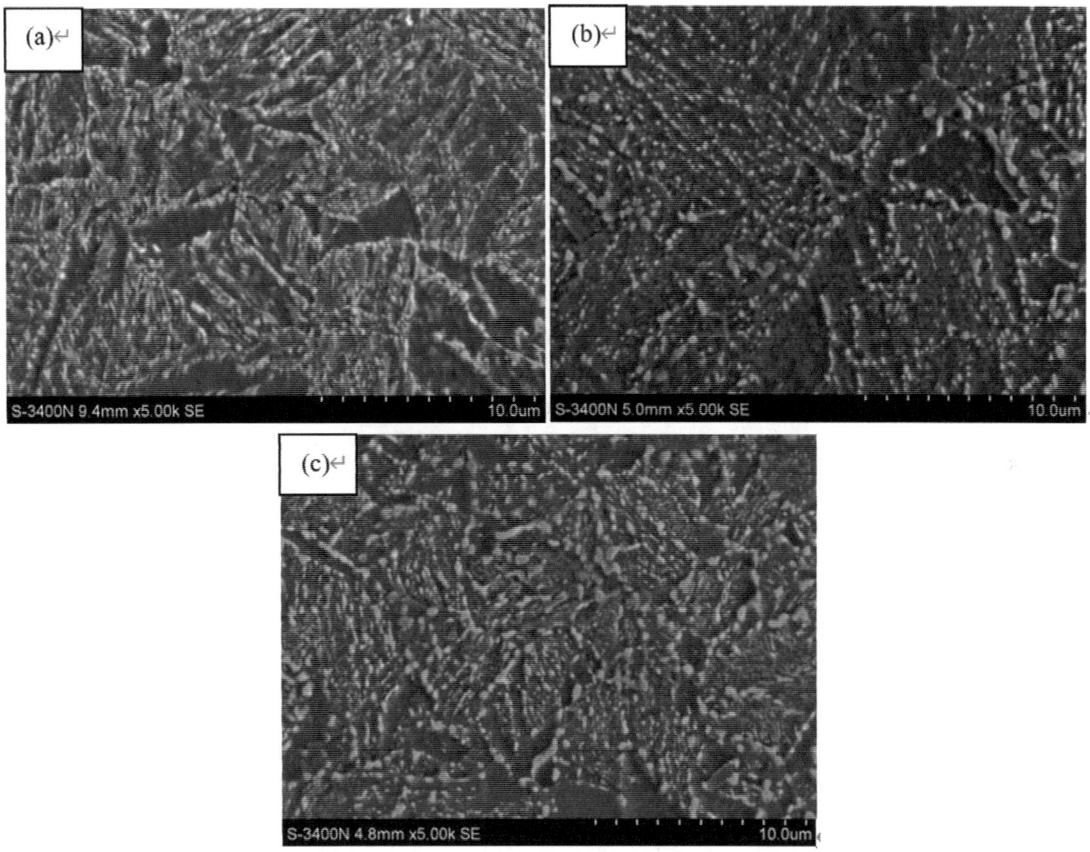


**Figure 2.** Microstructure morphology of three tempered sheets of steel.
(a) specimen NO.1; (b) specimen NO.2; (c) specimen NO.3.

Statistical results of carbides distribution were listed in Table 2. Tempered at 450°C (specimen NO.1), the average size of precipitated carbides was 0.1789 μm, and the area fraction was 37%. For specimen NO.2, the average size of precipitated carbides was 0.2387 μm, and the area fraction increased to 44%. When the tempering temperature grew to 650°C (specimen NO.3), the average size of precipitated carbides and area fraction increased markedly, with 0.3083 μm and 72%, individually.

Overall, as the tempering temperature increased, the average size of precipitated carbides and area fraction gradually dropped.

**Table 2.** Statistical results of carbides distribution.

| Specimens | NO.1 | NO.2 | NO.3 |
|---|---|---|---|
| Average size/μm | 0.1789 | 0.2387 | 0.3083 |
| Area fraction/ % | 37 | 44 | 72 |

*3.3 HIC test results*

The hydrogen-induced cracking (HIC) test results were exhibited in Table 3. The calculated value of CLR, CTR, and CSR of specimen NO.1 were 22.54%, 8.71%, and 5.25%. The figure for CLR, CTR, and CSR of specimens NO.2 and NO.3 were all 0%. In other words, no cracks were found in specimens NO.2 and NO.3. It was discovered that when steel was tempered at 450°C, the anti-hydrogen cracking performance can not meet standard requirements. And when tempering temperatures were 550°C (specimen NO.2) and 650°C (specimen NO.3), the steel can have excellent hydrogen-induced cracking resistance.

**Table 3.** Hydrogen-induced cracking (HIC) test results.

| Specimens | Sections | CLR/% | CTR/% | CSR/% |
|---|---|---|---|---|
| NO.1 | 1 | 16.38 | 7.29 | 5.29 |
| | 2 | 27.78 | 10.53 | 6.17 |
| | 3 | 23.46 | 8.31 | 4.29 |
| | Average | 22.54 | 8.71 | 5.25 |
| NO.2 | 1 | 0 | 0 | 0 |
| | 2 | 0 | 0 | 0 |
| | 3 | 0 | 0 | 0 |
| | Average | 0 | 0 | 0 |
| NO.3 | 1 | 0 | 0 | 0 |
| | 2 | 0 | 0 | 0 |
| | 3 | 0 | 0 | 0 |
| | Average | 0 | 0 | 0 |

*3.4 Dislocation*

TEM results of three tested steels were shown in Figure 3. When steel was tempered at 450°C (specimen NO.1), it can be seen that a large number of dislocations were distributed on the matrix, and most of the carbides were in the shape rod or band, and they mostly precipitated on grain boundary among laths. For specimen NO.2 (tempered at 550°C), the amount of dislocation that can be watched greatly decreased, the shape of carbides transformed to granular, and the size was smaller. As the tempering temperature was 650°C, it can be observed from Figure 3(c) that the number of dislocations further fell, and the size of precipitated carbides turned larger while the number was smaller.

For the three specimens, the dislocation density was $9.71/10^{10}$ $cm^{-1}$, $5.62/10^{10}$ $cm^{-1}$, and $2.97/10^{10}$ $cm^{-1}$, individually. As the tempering temperature grew from 450°C to 550°C to 650°C, the density of dislocation gradually descended. Combined with the HEDE theory and HELP theory, hydrogen easily concentrated at local stress concentration locations and reduced energy that dislocation mobility needed. And dislocation with hydrogen moved to the crack tip and resulted in hydrogen gathering at crack tips, thus increasing the hydrogen-induced cracking sensitivity [12]. And see in figure 4.

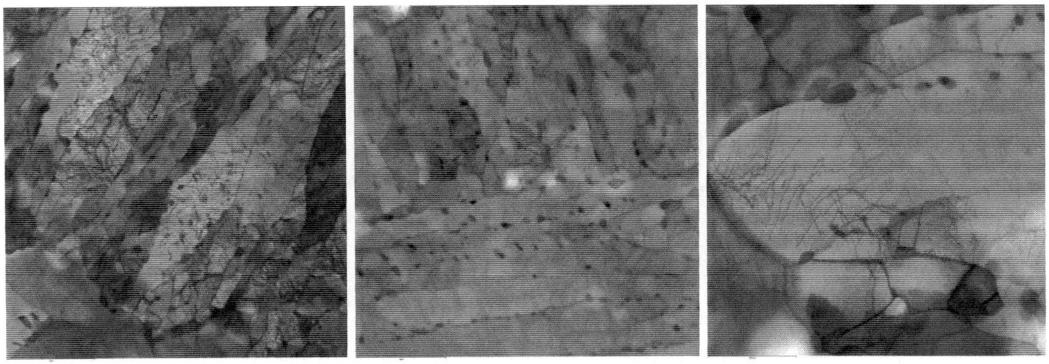

**Figure 3.** TEM results of three tested sheets of steel.
(a) specimen NO.1; (b) specimen NO.2; (c) specimen NO.3.

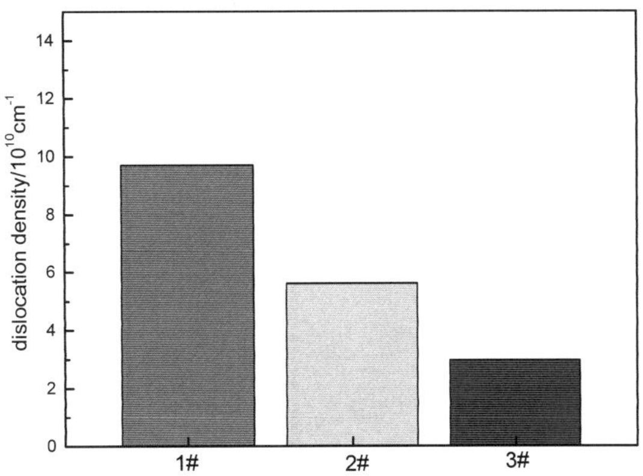


**Figure 4.** Dislocation density.
(a) specimen NO.1; (b) specimen NO.2; (c) specimen NO.3.

*3.5 Analysis*

Besides dislocation, the distribution condition of carbides was also an important factor in enhancing anti-hydrogen cracking resistance. Generally, non-uniformly distributed carbides in the shape of a bar is harmful to the toughness. When carbides are transformed into the shape of a particle, it would be beneficial for toughness. From the above results, when the tempering temperature was 550°C and 650°C, carbides gradually distributed in the shape of particles, and it can be beneficial for improving the hydrogen-induced cracking resistance.

Also, as irreversible hydrogen traps, uniformly distributed carbides can enhance hydrogen-induced cracking resistance [13]. As shown in Figure 5, after proper tempering techniques, the greater number of carbides that are uniformly distributed on the matrix can trap more hydrogen atoms, significantly reduce the content of diffusible hydrogen atoms at crack tips, and lower hydrogen partial pressure near cracks. Besides, when the main crack nearby carbides are generated, hydrogen is easily diffused with dislocation mobility to the elastic stress zone under stress effects. Under the dual effects, it was likely to induce hydrogen cracks at these sites, and main cracks propagate ahead and merge with micro-crack, and in this way, main cracks can extend ahead through continuous nucleation and merging mechanism. While, in specimen NO.2 and specimen NO.3, lower dislocation density can prevent hydrogen mobility and micro-cracks were not easily emerge at crack tips, thus inhibiting main cracks propagating ahead. Meanwhile, uniformly distributed hydrogen atoms can promote trapped hydrogen atoms evenly distribution, and it can effectively lower local hydrogen pressure resulting from hydrogen atoms concentration.

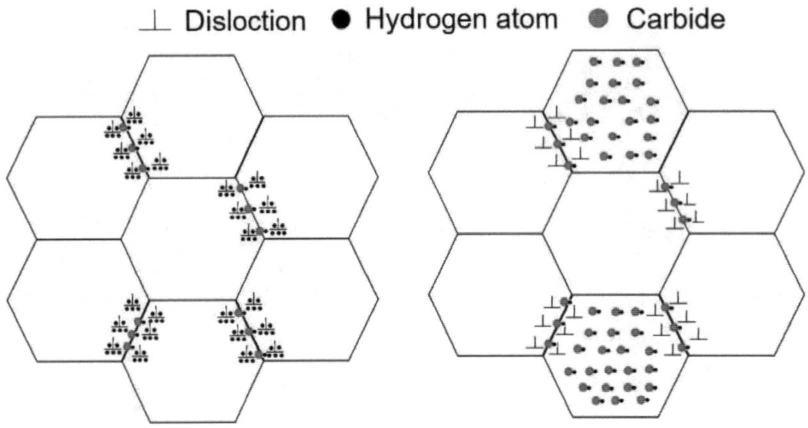


**Figure 5.** Schematic diagram of the effect of uniformly distributed carbides on the distribution of hydrogen atoms after tempering at the proper temperature.

Undoubtedly, when the tempering temperature were 550°C and 650°C, excellent hydrogen-induced cracking resistance performance was mainly because of lower dislocation density and a larger number of uniformly distributed carbides, and these factors relieved hydrogen atom concentrate at local sites, then reduced local hydrogen pressure and were hard to emerge cracks. As a result, it can promote excellent hydrogen-induced cracking resistance.

## 4. Conclusions

In this study, different heat treatment process was tested on the vessel steel, which was quenched at 900°C and tempered at 450°C, 550°C and 650°C for 120 min, individually. Microstructure, dislocation, and carbides were all studied, and the underlying kinetics affecting hydrogen-induced cracking performance were discussed. Some findings were listed as follows:

(1) As the tempering temperature increased from 450°C to 650°C, the yield strength and tensile strength gradually decreased, and there were minor changes in elongation rate and impact energy.

(2) The distribution condition of carbides exhibited great differences after various tempering temperature processes. When the tempering temperature went up from 450°C to 550°C, the shape of carbides transformed from bar into fine particles, and the area fraction climbed from 37% to 44%. And tempered at 650°C, the amount of precipitated particle carbides was greater and the size was larger, and the area fraction soared to 72%.

(3) When tempering temperatures were 550°C (specimen NO.2) and 650°C (specimen NO.3), no cracks were found and the steel had excellent hydrogen-induced cracking resistance. When steel was tempered at 450°C, the anti-hydrogen cracking performance can not meet standard requirements.

(4) As tempering temperature grew from 450°C to 650°C, the density of dislocation gradually declined from $9.71/10^{10}$ $cm^{-1}$ to $2.97/10^{10}$ $cm^{-1}$. And it can reduce the possibility of hydrogen atoms moving to crack tips, and then relieve the hydrogen embrittlement susceptibility.

(5) As an important factor in influencing hydrogen-induced cracking resistance, uniformly distributed carbides can enhance hydrogen-induced cracking resistance because of trapping hydrogen atoms. And it can promote hydrogen atoms distribution more uniformly, reduce the local hydrogen pressure, and prohibit crack initiation and propagation.

## References

[1] Long Y., Zhang J.M. (2020) Development of Anti-sour Super-thick Plate Q345R(R-HIC) for Pressure Vessel. Metal Materials and Metallurgy Engineering, 48:14–19.

[2] Xiao C, Zhang J.M., Luo Y.Z. (2013) Control of macrosegregation behavior by applying final electromagnetic stirring for continuously cast high carbon steel billet [J]. Journal of Iron and steel research, 20(11):13-20.

[3] Ghosh G., Rostron P., Panday A. (2018) Hydrogen induced cracking of pipeline and pressure vessel steels: a review. Engineering Fracture Mechanics, 199: 609-618.
[4] Haq A.J., Muzaka K., Dunne D.P., Calka A., Pereloma E.V. (2013) Effect of microstructure and composition on hydrogen permeation in X70 pipeline steels. Int J Hydrogen Energy 38(5):2544–2556.
[5] Xue H.B., Cheng Y.F. (2011) Characterization of inclusions of X80 pipeline steel and its correlation with hydrogen-induced cracking. Corros Sci., 53(4):1201-1208.
[6] Lukito H., Szklarska-Smialowska Z. (1997) Susceptibility of medium-strength steels to hydrogen-induced cracking. Corros Sci., 39(12):2151-2169.
[7] Chong T., Kumar S.B., Lai M.O., Loh W.L. (2014) Effects of wet $H_2S$ containing environment on mechanical properties of NACE grade C-Mn steel pipeline girth welds. Engineering Fracture Mechanics, 131: 485-503.
[8] Bai P., Zhao H., Zheng S., Chen C. (2015) Initiation and developmental stages of steel corrosion in wet $H_2S$ environments. Corrosion Science, 93: 109-119.
[9] Zhang X., Li Q., Simon M., Zheng G., Tan Y. (2022) Effect of $H_2S$ content on relative permeability and capillary pressure characteristics of acid gas/brine/rock systems: a review. Journal of Rock Mechanics and Geotechnical Engineering, 14:2003-2033.
[10] Jiwon J., Ramadan A., Rida E. (2022) Hydrogen embrittlement of high-strength API carbon steels in $H_2S$ and $CO_2$ containing environments. Journal of Natural Gas Science and Engineering, 104: 1-12.
[11] Fragiel, A.Serna, S.Malo-Tamayo, J.Silva, P.Campillo, B.Martinez-Martinez, E.Cota, L.Staia, M. H.Puchi-Cabrera, E. S.Perez, R. (2019) Effect of microstructure and temperature on the stress corrosion cracking of two micro alloyed pipeline steels in $H_2S$ environment for gas transport. Engineering Failure Analysis, 105: 1055-1068.
[12] Pessu F., Barker R., Chang F., Chen T., Neville A. (2021) Iron sulphide formation and interaction with a corrosion inhibitor in H2S-containing environments. Journal of Petroleum Science and Engineering, 207:109152-109165.
[13] Wang Y.F., Hu S.Y., Tian Z.Q., Cheng G.X., Zhang J.X. (2021) Structures of hydrogen-induced cracks under different hydrogen concentrations for a high strength steel. Engineering Failure Analysis, 128:105587-105595.

# Research on the high-speed impact of projectiles on reinforced concrete equivalent targets

**Hongli Qi[1,2], Xinying Wang[1], Yi Qin[1] and Qiye Wen[1]**

[1]Shenyang Ligong University, Shenyang, Liaoning Province, 1100000, China

[2]Corresponding author's e-mail: 857851505@qq.com

**Abstract.** Due to the prolonged production cycle of reinforced concrete target plates and the need for a series of preparations and tests before use, it is essential to study the equivalent relationship between reinforced concrete target plates and steel target plates. Numerical simulations of the process of projectiles vertically penetrating reinforced concrete target plates were conducted using LS-DYNA software. Models for both reinforced concrete and steel targets were established, and the equivalent thickness conversion relationship between the two was obtained. Simulation results indicated an inverse proportional relationship between projectile residual kinetic energy and target thickness. Curve fitting using Origin software was performed to obtain the conversion relationship between the thicknesses of reinforced concrete target plates and steel targets. The simulation results were validated, showing an error in concrete target plate thickness conversion of less than 10%. The research results provide theoretical support and data references for the equivalent study of reinforced concrete target plates and steel target plates.

## 1. Introduction

Concrete material reinforced with steel bars is widely used in both military and civilian applications, including command centers, bunkers, bridges, and buildings. Concrete, as a composite material, is typically composed of a mixture of cement, mortar, aggregate (crushed stone), and water. It exhibits high compressive strength but lower tensile strength. However, with the addition of an appropriate amount of steel reinforcement, it can enhance the overall structural load-bearing capacity and ductility.

The concrete target plate often requires a lengthy curing period during the manufacturing phase, during which the concrete strength gradually increases [1]. The strength of concrete is influenced by various factors such as materials, mix proportions, temperature, humidity, and manual compaction [2]. For reinforced concrete target plates, the reinforcement ratio and the selection of steel bars also significantly affect the concrete strength. Testing to meet experimental requirements typically takes a considerable amount of time. Therefore, establishing an equivalent target for reinforced concrete is necessary to replace actual reinforced concrete targets in ballistic resistance studies. Considering factors such as usage versatility, cost, and material strength, 45# steel is chosen as the equivalent target material for relevant experimental research, replacing reinforced concrete. This approach aims to achieve savings in labor, time, and economic costs while ensuring research progress.

Numerical simulation studies on high-speed penetration of medium-caliber projectiles into reinforced concrete target plates of varying thickness were conducted using LS-DYNA dynamics software. A finite element model for the penetration of projectiles into reinforced concrete target plates was established, and the reliability of the model was verified. The study involved analyzing the damage characteristics of reinforced concrete targets during the penetration process and determining the equivalent relationship between target plates of different thicknesses and steel targets.

## 2. Target equivalent principle

The target plate equivalence principle is mainly divided into the following four types, namely, the material strength equivalence principle, the ultimate penetration speed equivalence principle, the remaining penetration depth equivalence principle, and the target body absorption kinetic energy equivalence principle [3]. The principle of equivalent kinetic energy absorbed by the target, that is, the kinetic energy loss of the projectile during the process of penetrating the target plate is an equivalent index and measurement standard [4]. This principle can reflect the value of the remaining kinetic energy after the projectile penetrates the target plate. The remaining kinetic energy can be considered a criterion to measure whether the target body is equivalent, the equivalence principle of target absorption kinetic energy is selected as the equivalence principle in this article. Expressed as:

$$\mathrm{E_P} = \frac{1}{2} m_p v_0^2 - \frac{1}{2} m_p v_r^2 \tag{1}$$

In the equation: $m_p$ is the mass of the projectile; $v_0$ is the initial velocity of the projectile; $v_r$ is the remaining velocity of the projectile.

This paper uses LS-DYNA to carry out numerical simulation research on the vertical penetration of projectiles into reinforced concrete target plates and equivalent steel targets. The principle of equivalent target body absorption kinetic energy is selected to conduct equivalent target body research, that is, projectiles with the same material properties and the same speed penetrate vertically into steel targets with different thicknesses $h_s$ and reinforced concrete target plates with different thicknesses $h_c$. During the numerical simulation calculation process, assuming that the impact of projectile abrasion and deformation is not considered, the kinetic energy lost by the projectile during the penetration process can be calculated by the above formula when the projectile mass, projectile initial velocity, and projectile residual velocity are known. get. When the residual velocity of the projectile is the same, the steel target with thickness $h_s$ and the reinforced concrete target plate with thickness $h_c$ are considered equivalent targets.

## 3. Numerical simulation of projectile penetration into reinforced concrete target plates

Based on the target energy absorption equivalent principle, dynamic simulations of high-speed projectile penetration into reinforced concrete target plates were conducted using LS-DYNA software. The Lagrange dynamic algorithm was employed, taking into account material failure and erosion contact. The projectile, steel reinforcement, and steel target utilized the JOHNSON-COOK constitutive model, while concrete was modeled using the KCC constitutive model[5]. Material failure was defined using the *MAT_ADD_EROSION keyword, and erosion contact was defined using the *CONTACT_ERODING _SURFACE_TO_SURFACE keyword.

### *3.1 Three-dimensional model and material parameter setting*

The projectile and target plate have symmetry, in order to save calculation time, 1/4 model is used for simulation calculation. The oval solid projectile is selected to simulate the projectile, with a diameter of 90 mm and a length of 250 mm. Modeling and meshing are shown in Figure 1. The cross section of the concrete target 1/4 model and steel target 1/4 model is 400 mm × 400 mm, the steel bars are cylinders with a diameter of 6 mm, and the interval between steel bars is 80 mm. The modeling is carried out in a common node way, and the interval between each layer of steel mesh is 50 mm, so as to ensure that the reinforcement ratio of reinforced concrete target is constant at 1.5%. The modeling and mesh division are shown in Figure 2, global symmetry constraints are carried out, and boundary constraints are set outside the target. The simulation time step is set to 0.69 s.

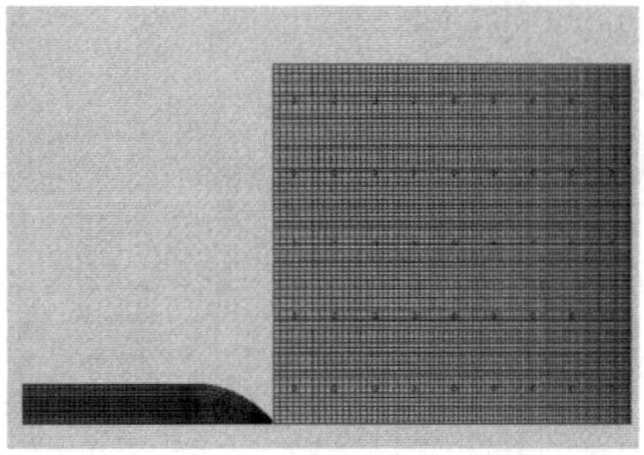

**Figure 1.** 1/4 projectile model and mesh generation.

The projectile material is 30CrMnSiA steel, the concrete material is C40 concrete, the reinforcement material is Q235 steel, and the steel target material is 45 # steel. The material properties are shown in Table 1 and Table 2. [6-7]

**Table 1.** Material parameter.

| Material | density (kg/m$^3$) | Elastic modulus/GPa | Poisson ratio | A/MPa | B/MPa |
|---|---|---|---|---|---|
| 30CrMnSiA | 7830 | 207 | 0.3 | 1587 | 382.5 |
| Q235 | 7800 | 200 | 0.33 | 293.8 | 230.2 |
| 45 # steel | 7800 | 200 | 0.3 | 507 | 320 |

**Table 2.** Concrete material parameters.

| density (kg/m$^3$) | PR | Ft | A0 | A1 | A2 |
|---|---|---|---|---|---|
| 2314 | 0.19 | 0.2011 | 0.01342 | 0.44636 | 1.78 |
| Lambda | B1 | OMEGA | A1F | Lambda1 | Lambda2 |
| 100 | 1.6 | 0.5 | 0.4417 | 0 | 8×10-6 |
| Lambda3 | Lambda4 | Lambda5 | Lambda6 | Lambda7 | Lambda8 |
| 2.4×10-5 | 4×10-5 | 5.6×10-5 | 7.2×10-5 | 8.8×10-5 | 3.2×10-4 |

Under the numerical simulation condition, the projectile penetrates the reinforced concrete target with different thicknesses and a reinforcement ratio of 1.5% at the velocity of 1000 m/s, and the residual velocity after the projectile penetrates the target is obtained. In the process of numerical simulation, it is found that there is no failure grid, that is, the mass of the projectile does not change during penetration. In this paper, the thickness of a reinforced concrete target is equivalent to that of a steel target when the residual velocity of the projectile is basically the same and the mass of the projectile is constant. Then the projectile penetrates steel targets with different thicknesses at the velocity of 1000 m/s, and the residual velocity of the projectile is obtained. The specific working conditions of projectile penetrating reinforced concrete targets are shown in Table 3.

**Table 3.** Working condition setting.

| Working condition | 1 | 2 | 3 | 4 | 5 |
|---|---|---|---|---|---|
| Thickness of reinforced concrete target plate/cm | 20 | 30 | 40 | 45 | 50 |

*3.2 Analysis of penetration results*

In order to verify the rationality of the numerical simulation method adopted in this paper, the experimental condition of the projectile penetrating concrete target at 130 m/s in Xu's experiment is numerically simulated [8], and the results are compared with the experimental data in the literature. From the numerical simulation results, the peak acceleration of the projectile is about $1.79 \times 10^4$ g, and the penetration depth is 16.9 cm. In Xu's experiment, the peak acceleration of the projectile is about $1.6 \times 10^4$ g, and the penetration depth is 17 cm.

Using LS-DYNA software for dynamic simulation, the acceleration curves of projectiles with different thicknesses are obtained, as shown in Figure 2, which is basically consistent with the trend of projectile acceleration time history curve in Huang's simulation experiment [9]. And the same laminar cracking phenomenon occurs in the penetration process as Wang's experiment[10].

When the thickness of the target is less than 45 cm, the whole mass of the target is smaller. The kinetic energy distribution in the part of the target increases with the increase of the thickness, and the kinetic energy lost by the projectile in the penetration process also increases with the increase of the thickness of the target. When the thickness of the target plate is greater than 45 cm, the compression wave generated by the contact between the projectile and the target will be transmitted to a longer distance, and the transmission time of the compression wave will also take longer.

When the warhead completely invades the target plate, the energy transmitted by the projectile to the target plate will be more widely dispersed in all positions of the target plate, resulting in the acceleration of the projectile unable to reach the maximum when the cross-sectional area of the projectile invasion is maximum. The time that the projectile is affected by resistance increases with the increase of target thickness, which is reflected in the velocity of the projectile, that is, when the projectile finally penetrates the target, the velocity is lower. In terms of energy loss, the larger the target thickness, the more the projectile energy loss. The residual velocity of the projectile is shown in Table 4.

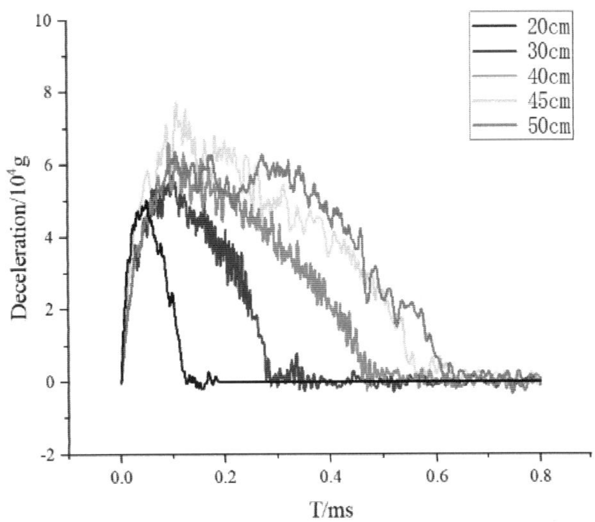


**Figure 2.** Acceleration time history curve of projectile reinforced concrete target plate.

## 4. Equivalent relationship between reinforced concrete target and steel target

### *4.1 Projectile penetration into an equivalent steel target*

Compared with projectile penetrating reinforced concrete targets, reinforced concrete appears to surface craters, cracks, shedding, and spallation. These phenomena do not occur when the projectile penetrates the steel target, and a smooth round hole will be left in the center of the steel target when the projectile penetrates the target completely. Based on the residual velocity of projectile penetrating reinforced concrete target, the projectile penetrates steel targets with different thicknesses at a speed of 1000 m/s, and the residual velocities of projectile penetrating steel targets with different thicknesses are obtained. Table 4 lists the residual velocities of projectile penetrating steel targets which are roughly consistent with Table 4. The time history curve of projectile acceleration is shown in Figure 3.

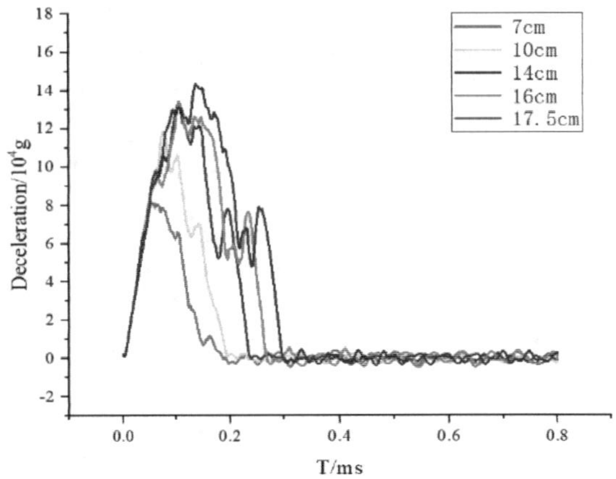


**Figure 3.** Acceleration time history curve of projectile penetrating steel target.

**Table 4.** Residual velocity of projectile.

| Thickness of reinforced concrete target /cm | Residual velocity /(m/s) | Thickness of steel target/cm | Residual velocity /(m/s) |
|---|---|---|---|
| 20 | 922 | 7 | 922 |
| 30 | 890 | 10 | 888 |
| 40 | 814 | 14 | 818 |
| 45 | 780 | 16 | 787 |
| 50 | 743 | 17.5 | 742 |

*4.2 Equivalent relationship between reinforced concrete target plate and steel plate*

According to the equivalent thickness of the reinforced concrete target plate and steel plate at the velocity of 1000 m/s obtained by the above numerical simulation, it is considered that the target plate corresponding to the same residual velocity of the projectile in Table 4 is equivalent.

According to the discrete data of the target thickness of the two different materials, the equivalent thickness conversion curves of the two materials are established by fitting with Origin software. Taking the thickness of the reinforced concrete target plate as an independent variable and the equivalent target thickness as a dependent variable, the equivalent equation between them at the velocity of 1000 m/s is obtained after fitting. This equation is approximate to the cubic equation, and the equivalent equation is:

$$y = -0.0003x^3 + 0.0302x^2 - 0.6884x + 10.888 \quad (2)$$

*4.3 Equivalent relation verification*

Based on the equivalent equation obtained by fitting the equivalent target thickness data, the conversion between 25 cm and 35 cm reinforced concrete target plate and equivalent steel target thickness is verified, and the equivalent steel target thickness of 25 cm and 35 cm reinforced concrete target plate is obtained by numerical simulation. Compared with the steel target thickness calculated by the equivalent equation, the error of the conversion result is less than 10%, which is shown in Table 5.

**Table 5.** Verification of equivalent target thickness transformation relationship.

| Reinforced concrete target thickness/cm | Steel target simulated thickness/cm | Steel target calculated thickness/cm | Error rate |
|---|---|---|---|
| 25 | 8 | 7.8 | 1% |
| 35 | 12.5 | 10.9 | 8% |

## 5. Conclusion

According to the principle of equivalent kinetic energy absorbed by the target, the equivalent relationship between reinforced concrete target and steel target is established. The equivalent thickness between the reinforced concrete target and steel target is obtained from the residual velocity of the projectile obtained by numerical simulation. And the thickness conversion relationship between the reinforced concrete target and steel target is obtained by fitting, which is verified again by numerical simulation. The error between calculated results and simulation results is less than 10%, which provides a reference for the study of reinforced concrete targets and equivalent targets.

## References

[1] Gao Xiangling, Yan Yingying, Li Jie. Variation law of compressive strength of concrete under general atmospheric environment [J]. Journal of Civil Engineering, 2015, 48 (01): 19-26. DOI: 10.15951/ J.tmgcxb.2015. 01.003.

[2] Wu Yongchong, Zhang Yudai. Causes and prevention of cracks in cast-in-place concrete [J]. Guangdong Building Materials, 2023, 39 (09): 135-136 +123.

[3] Hou Junchao, Wang Chunguang, Deng Dezhi, et al. Study on equivalent thickness of high-speed projectile penetrating concrete target [J]. Journal of Ordnance Engineering, 2021, 42 (02): 9-14.

[4] Cao Bing.Experimental study on 603 target against EFP penetration into equivalent target [J]. Explosive Materials, 2007 (01): 36-39.

[5] Murray Y D.Users manual for LS-DYNA concrete material model 159:FHWA-HRT-05-062[R].Colorado:Springs,2007.

[6] CHRISTIAN H,JURGEN S.Comparison of the RHT Concrete Material Model in LS-DYNA and ANSYS AUTODYN [C]//11th European LS-DYNA Conference.Salzburg,Austria,2017.

[7] THOMAS B,WERNER R.The RHT concrete model in LS DYNA[C]//8th European LS-DYNA Users Conference.Strasbourg,French2011.

[8] Xu Wenzheng, Wang Jingyu, Lu Zhen, et al. Study on overload characteristics of elastic projectile penetrating concrete target [J]. Vibration and Shock, 2010, 29 (05): 91-95 +156 +242. DOI: 10.13465/J.cnki.jvs.2010.05. 027.

[9] Huang Chenglong, Chen Yeqing, Li Shutao, et al. Influence of Impact Point on Penetration Depth of Reinforced Concrete [J/OL]. Applied Mathematics and Mechanics: 1-20 [2023-09-11]. http://kns.cnki.net/kcms/detail/50.1060.O3.20230709.1213.002.html.

[10] Wang Meng, Ding Yubo. Numerical simulation of projectile penetrating thick reinforced concrete target at high speed [J]. Science, Technology and Engineering, 2022, 22 (16): 6506-6514.

# Study on the microstructure and mechanical properties of VW73B alloy at high extrusion ratio

**Xiaowei Wei [1, 2, 3], Yongjun Li [1, 2, 4], Minglong Ma [1, 2], Xinggang Li [1, 2], Guoliang Shi [1, 2], Jiawei Yuan [1, 2], Zhaoqian Sun[1, 2], and Kui Zhang [1, 2]**

[1]State Key Laboratory of Nonferrous Metals and Processes, GRINM Group Corporation Limited (GRINM), Beijing 100088, China
[2]GRIMAT Engineering Institute Co., Ltd., Beijing 101407, China
[3]General Research Institute for Nonferrous Metals, Beijing 100088, China

[4]Corresponding author's e-mail: lyj@grinm.com

**Abstract:** The microstructure and mechanical characteristics of VW73B prepared by a high extrusion ratio (70:1) were studied. High-strength plastic VW73B alloy containing long-period structural phase (LPSO) was obtained by extrusion process. OM, SEM, TEM, and EBSD were used to investigate the relationship between microstructure and mechanical characteristics of VW73B. The findings indicate that VW73B presents complete dynamic recrystallization after deformation, and there are 14H-LPSO phases, SFs, and a large amount of β′ phase, which are precipitated after aging treatment. The UTS, YS, and EL of VW73B in peak aging state are 406 MPa, 339 MPa, and 9%, respectively. The β′ phase, LPSO phase, poor texture, and fine grain strengthening of VW73B contribute to its exceptional characteristics.

## 1. Introduction

Owing to its low density, high specific strength, and green environmental protection, magnesium alloy, the lightest metal structural-functional material, is utilized extensively in the aerospace, transportation, and 3C product industries, and other fields [1]. More exacting specifications have been made for the mechanics, particularly the mechanical characteristics of magnesium alloys due to the rapid advancement of the military manufacturing field. Strength and plasticity can be increased by adjusting the alloy composition and optimizing the deformation process [2].

A beneficial method for enhancing the magnesium alloy's qualities is alloying. The most efficient Rare Earth (RE) elements to enhance the alloy's overall mechanical qualities are Gd and Y. Zn elements added to Mg-RE, which can produce an LPSO structural phase, which is a strengthening phase. The LPSO structural phase's bending and deformation mechanism can help to coordinate the magnesium alloy's plasticity and enhance its toughness [3]. Mg-RE-Zn alloy has great potential in achieving reasonable matching of alloy strength and plasticity.

The microstructure and characteristics of magnesium alloy are significantly impacted by the extrusion ratio. Guan et al. [4] investigated the characteristics of Mg-Sm-Zn-Zr at three different extrusion ratios of 6.9, 10.4, and 17.6. The results demonstrate that as the extrusion ratio increases, the alloy's DRX fraction rises along with the size of the grains, and basal texture decreases. Wang et al. [5] explored the impact of extrusion ratios of 7.6, 12.5, and 26.1 on Mg-2.5Nd-0.5Zn-0.5Zr. The DRX grain size initially reduces and subsequently increases with a rise in extrusion ratio, while the texture strength rises from 6.7 to 19.4. A higher extrusion ratio can provide a larger strain, which is conducive to obtaining a higher percentage of dynamically recrystallized grains and enhancing the alloy's general qualities.

At present, rare studies have been done on the ultra-high extrusion ratio method of creating Mg-RE-Zn-Zr alloys. We prepared Mg-7.0Gd-3.0Y-1.0Zn-0.5Zr (VW73B) by an extrusion ratio of 70:1, and the alloy was aged. The connection among the structure and characteristics was established by analyzing the alloy's microstructure, texture, degree of recrystallization, and precipitation equality. Additionally, the mechanical properties of VW73B were examined.

## 2. Experiment

VW73B alloy ingot with a large size of Φ400×1, 000 mm was prepared by die casting. The true composition of VW73B was determined by ICP as Mg-6.79Gd-3.36Y-1.12Zn-0.57Zr (wt.%). The homogenization of the as-cast alloy was carried out at 480°C×12 h+520°C×40 h. The small ingot with Φ120×150 mm was taken along the edge of the large ingot for hot extrusion deformation. Before extrusion, the ingot is preheated at 460°C for 4 h, the extrusion temperature is selected at 460°C, the rate of extrusion is 0.4 mm/s, and the extrusion cylinder's diameter is 130 mm. Finally, the extrusion bar with Φ15.5 mm was obtained. The extruded alloy was aged at 200°C. The microstructure, precipitated phase, dynamic recrystallization degree, and texture of the alloy were examined using OM, SEM, EBSD, and TEM, and the mechanical characteristics of the VW73B were tested.

## 3. Results and discussion

The OM of extruded VW73B alloy is shown in Figure 1. It is discovered that the extrusion process breaks both the lamellar and block-shaped LPSO structural phases. The larger extrusion ratio provides higher strain energy for the occurrence of recrystallization, and the microstructure distribution is relatively uniform, which can be found from the longitudinal cross-section microstructure in Figure 1 (b). The LPSO phase is distributed along the extrusion direction, and the alloy is almost completely dynamic recrystallization.

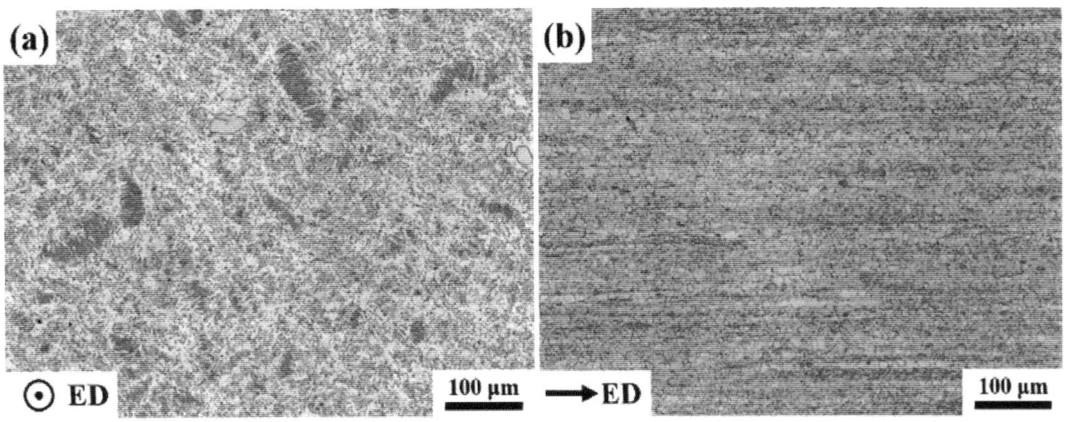


(a) Cross section; (b) Longitudinal section.

**Figure 1.** OM structure of extruded VW73B alloy.

Figure 2 shows the SEM of the extruded VW73B alloy. Following extrusion, the LPSO is twisted, fractured, and dispersed along the grain boundary, which encourages the occurrence of dynamic recrystallization. Lamellar LPSO is a dynamic precipitate in the extrusion process, distributed in the grain. The block-shaped and lamellar LPSO is fragmented and scattered throughout the matrix, as seen in the cross-section of Figure 2 (a). In Figure 2 (b), the block-shaped LPSO is elongated in the direction of extrusion, and the fractured LPSO is often twisted to coordinate deformation under the action of extrusion stress.

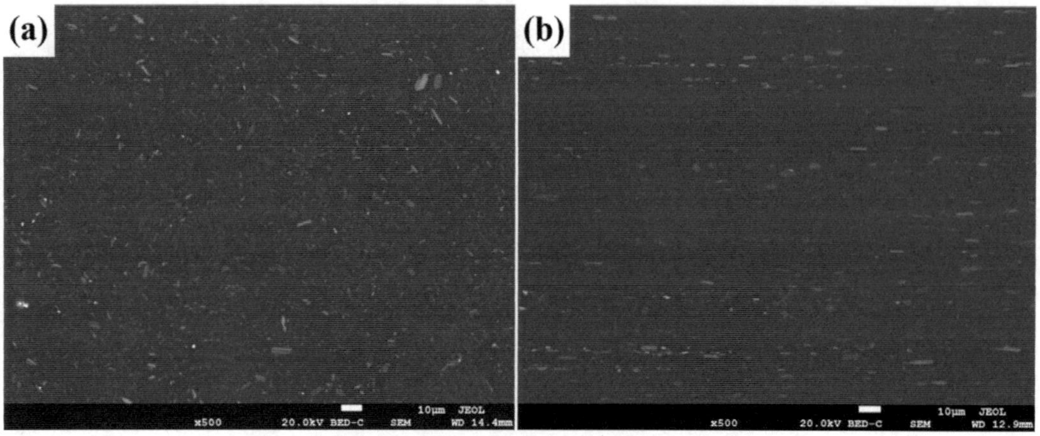


(a) Cross section; (b) Longitudinal section.

**Figure 2.** SEM photos of extruded VW73B alloy.

The precipitated phase in extruded VW73B alloy was analyzed by TEM. Figure 3 (a) shows the TEM open-field phase. Lamellated LPSO is twisted, and plastic deformation can be coordinated during the deformation process, as shown by the red arrow. The short rod-like phase is shown by the yellow arrow, with a small size and a length of about 1 μm. The SAED result is shown in Figure 3 (c), which can be determined to be 14H-LPSO. The lamellar structure is shown in Figure 3 (a) in the red line area, the SAED map in the direction of $[11\bar{2}0]$ is shown in Figure 3 (d), and the results show that it is a stratified fault (SFs). From Figure 3 (b), the topography of the SF structure can be more clearly seen with varying lengths.

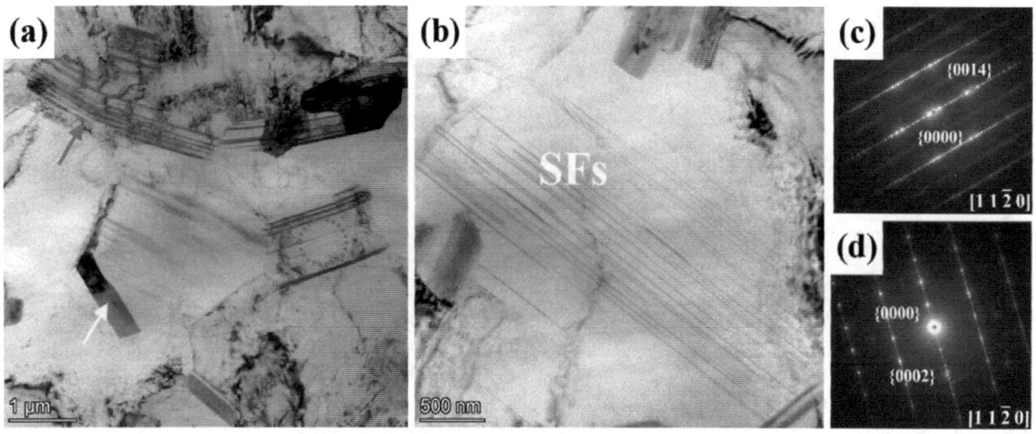


(a-b) TEM bright field image; (c-d) SAED map.

**Figure 3.** TEM results of extruded VW73B alloy.

The EBSD analysis of VW73B magnesium alloy was carried out. The IPF diagram can be seen in Figure 4 (a). Figure 4 (b) shows the grain size distribution diagram of VW73B. The orientation graphic presented it obviously under the extrusion ratio of 70:1, the alloy underwent complete dynamic recrystallization, the orientation distribution was relatively random, and the average grain size was 7.0 μm. The distribution of grain boundary orientation difference is shown in Figure 4 (c). The proportion of large angle grain boundaries LAGBs (>15°) is 93.8%, which is relatively high, so it can be concluded that a large extrusion ratio promotes the occurrence of dynamic recrystallization. Figure 4 (d) is the KAM diagram of the superposition size angle grain boundary. The LAGBs (15°~100°) are represented by the black line, while the small angle grain boundaries (2°~15°) are represented by a white line. The overall dislocation density in the DRX region is small.

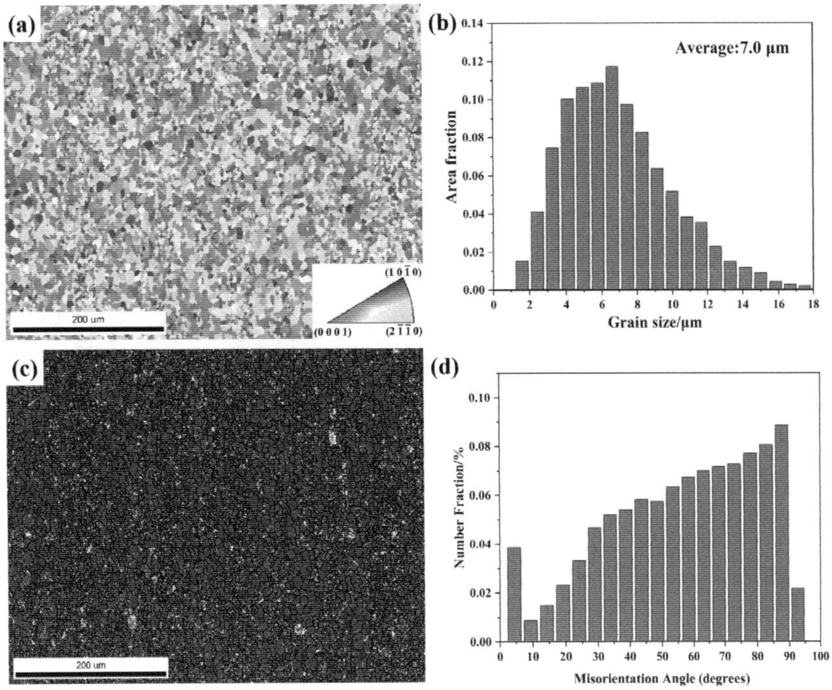


(a) IPF diagram; (b) Grain size distribution map; (c) Distribution map of grain boundary orientation difference; (d) KAM diagram.

**Figure 4.** EBSD analysis of extruded VW73B alloy.

The texture of VW73B alloy has been analyzed. Figures 5 (a) and (b) show the polar and inverse polar diagrams of VW73B magnesium alloy, and the texture strength is 3.603. Combined with the OM results in Figure 1 and EBSD analysis in Figure 4, the alloy is in a complete dynamic recrystallization state, consisting of fine equilateral crystals, whose grain orientation is more random, so the texture strength is weak.

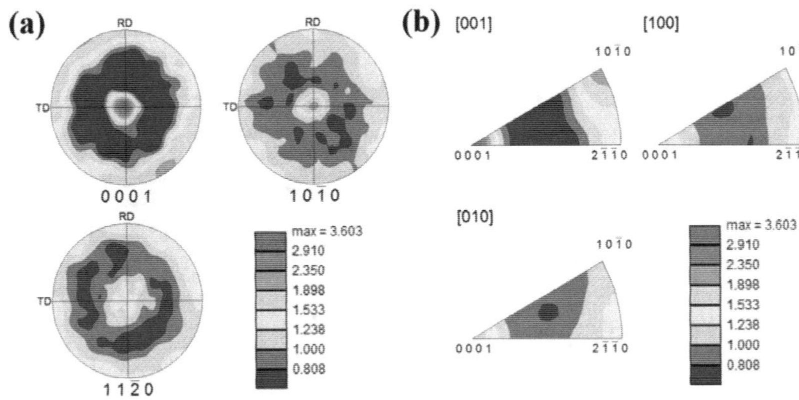


(a) Polar diagram; (b) Inverse polar diagram

**Figure 5.** Texture analysis of extruded VW73B alloy.

The aging hardness curves of the VW73B alloy at 180°C, 200°C, and 220°C were studied, and the results are shown in Figure 6. With the aging time, the hardness value showed a rising trend, and after reaching the peak value, the hardness value was maintained. Within 0-16 h, the changes in hardness values at the three temperatures were similar, mainly because the precipitated phase was uniformly precipitated at the early stage of aging. After 16 h, the precipitation at a low temperature of 180°C is less, while the aging precipitation at 220°C coarsens with the aging time. At these two temperatures, the hardness value is lower. Therefore, the aging temperature of VW73B alloy at 200°C was selected, and the hardness reached a peak of 114.4 HV after aging at 200°C for 80 h. Therefore, the extruded VW73B alloy was aged at 200°C for 80 h.

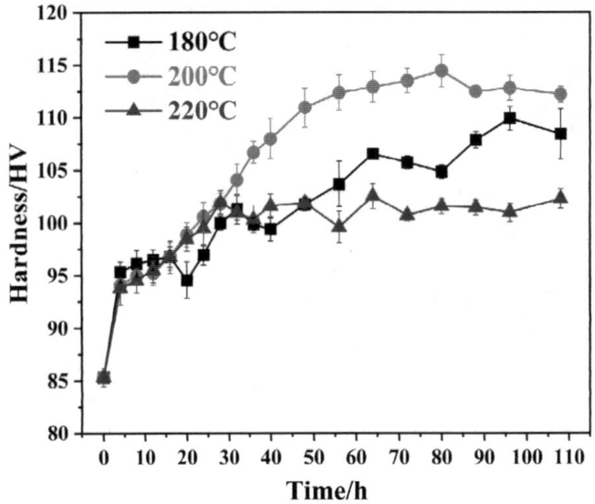


**Figure 6.** Hardness curves of VW73B alloy at different aging temperatures.

Figure 7 shows the TEM results of peak-aging VW73B alloy. TEM bright field images and HAADF-STEM images are displayed in Figures 7 (a) and (b), respectively. After aging treatment, a large number of uniformly distributed precipitated phases are deposited inside the matrix. The corresponding electron diffraction patterns in the [0001] direction and [11$\bar{2}$0] direction are displayed in Figures 7 (c) and (d). After calibration, the phase is the β' phase precipitated by aging. β' phase presents a parallel fringe in the matrix, precipitates equivalently along the <11$\bar{2}$0>$_\alpha$ direction, and shows an orientation relationship of 120° angle between each other [6].

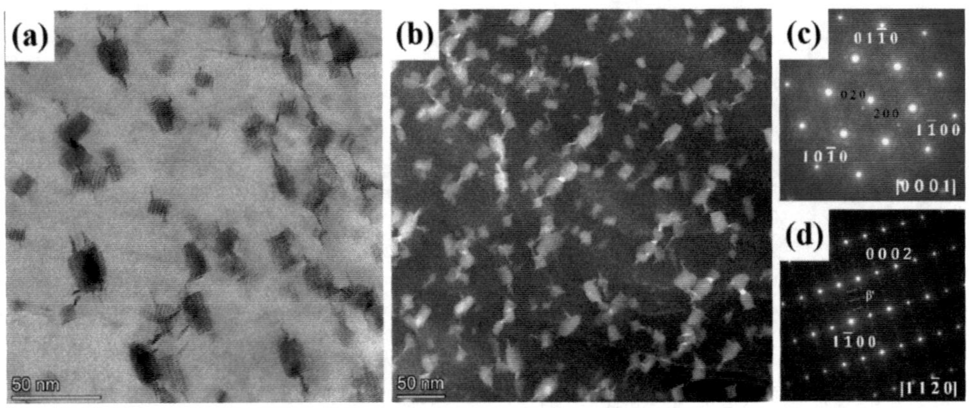


(a) TEM bright field image; (b) HAADF-STEM images; (c) SAED map in [0001] direction; (d) SAED map in direction [11$\bar{2}$0].

**Figure 7.** TEM analysis of peak-aging VW73B alloy.

The mechanical characteristics of aged and extruded VW73B alloys were examined. VW73B alloy shows good precipitation strengthening response, with extrusion UTS of 316 MPa, YS of 243 MPa, and EL of 16%. After aging, the UTS and YS of VW73B alloy reached 406 MPa and 339 MPa, and the EL was 9.0%. After aging treatment, a good deal of uniformly dispersed β' phases were formed within the alloy matrix, which played a good strengthening role in the matrix. As a precipitation strengthening phase, β' phase can significantly strengthen VW73B alloy's toughness and can nail the grain boundary to inhibit the slip of the grain boundary, so that compared with the extruded alloy, the plasticity is reduced.

## 4. Conclusions

In this paper, large-scale VW73B is prepared, and the microstructure and mechanical characteristics of VW73B were studied after the deformation of the ultra-large extrusion ratio (70:1).

(1). After extrusion deformation, LPSO in VW73B alloy is dispersed alongside the extrusion direction and violently shattered. Lamellar LPSO is dispersed in the matrix.

(2). Complete dynamic recrystallization occurred approximately after the extrusion of VW73B alloy. Its texture strength was weak, orientation distribution was relatively random, and the grain size was 7.0 μm.

(3). VW73B alloy has a good aging strengthening effect, and the dispersing nanoscale β' phases can be precipitated in the peak aging stage. The mechanical characteristics of VW73B in an aging state are best, with UTS of 406 MPa, YS of 339 MPa, and EL of 9.0%.

**Acknowledgments**

This work has been achieved by funding from China's National Key Research and Development Program (No. 2021YFB3701000).

**References**

[1] Jeong H. T. and Kim W. J. (2022). The hot compressive deformation behavior of cast Mg-Gd-Y-Zn-Zr alloys with and without LPSO phase in their initial microstructures [J]. Journal of Magnesium and Alloys, 10 (10): 2, 901-2, 917.

[2] Aalipour Z., Zarei-Hanzaki A., Moshiri A., Abedi H. R., Waryoba D., Kisko A., and Karjalainen L. P. (2022). Strain dependency of dynamic recrystallization during thermomechanical processing of Mg-Gd-Y-Zn-Zr alloy [J]. Journal of Materials Research and Technology, 18: 591-598.

[3] Ramezani S. M., Zarei-Hanzaki A., Salandari-Rabori A., Abedi H. R., Minarik P., Máthis K., and Horváth Fekete K. (2021). Unraveling the effect of deformation-induced phase transformation on microstructure and micro-texture evolution of a multi-axially forged Mg-Gd-Y-Zn-Zr alloy containing the LPSO phase [J]. Journal of Materials Research and Technology, 15: 2, 088-2, 101.

[4] Guan K., Ma R., Zhang J. H., Wu R. Z., Yang Q., and Meng J. (2021). Modification of microstructures and tensile properties of Mg-Sm based alloy via extrusion ratio [J]. Journal of Magnesium and Alloys, 9 (3): 1, 098-1, 109.

[5] Wang S. H., Pan J. Q., Xie W. Z., Yang J. L., Zhang W. C., and Chen W. Z. (2023). Effects of Extrusion Ratio on the Microstructure, Texture, and Mechanical Properties of Mg-2.5Nd-0.5Zn-0.5Zr Alloy Sheets [J]. Journal of Materials Engineering and Performance, 32 (11): 4, 834-4, 845.

[6] Du Z. W., Peng Y. G., Teng H., Cao Z., Zhang K., Ma M. L., Han X. L., Li T., Che C., and Li Y. J. (2023). Formation and growth of precipitates in a Mg-7Gd-5Y-1Nd-2Zn-0.5Zr alloy aged at 200°C [J]. Journal of Magnesium and Alloys, 11(7): 2, 326-2, 339.

# Laser cladding of Ni60/WC composite coatings on Ti6Al4V alloy and its effect on microstructure and properties

**Mingjia Feng**[1,2]

[1]School of Mechanical and Civil Engineering, Jilin Agricultural Science and Technology University, Jilin, 132101, China

[2]Corresponding author's e-mail: 491327847@qq.com

**Abstract.** Among its strong properties, titanium alloy is corrosion resistant, low density, and performs well at low temperatures. Aerospace, marine, automobile, armored, and biomedical equipment utilize it extensively. Titanium alloy also has some problems when used in some special environments, such as poor wear resistance and inadequate resistance to oxidation. By laser coating titanium alloy with Ni60/WC composite coating, the performance of titanium was better. To investigate impact of processing on cladding layer microstructure and properties, three different energy densities of 33.3 J/mm$^2$, 46.7 J/mm$^2$, and 53.3 J/mm$^2$ were used for experiments. Between substrate coatings, there is strong metallurgical bonding at three laser energy densities. The coating's microhardness first rises and then falls with increasing energy density. The performance of friction and wear coatings is superior to that of the substrate. Laser energy density at 33.3 J/mm$^2$ has the best wear resistance coating.

## 1. Introduction

Titanium alloy is a kind of alloy metal made of titanium and other metals. Among its strong properties, titanium alloy is corrosion resistant, low density, and performs well at low temperatures. At low temperatures and ultra-low temperatures, it can still maintain its mechanical properties, good biocompatibility, and other special properties. It has become an important structural material since its inception. It has been more and more used in aerospace, marine, automotive, armored, and biomedical equipment [1-3]. Although titanium alloy has many special properties, it also has some disadvantages. For example, it has poor tribological properties and cannot meet the wear resistance under strong load, so its application is limited [4]. At present, the common method to solve the limited application of titanium and titanium alloy is surface modification technology, which includes shot peening, electrochemical treatment, and laser cladding [5].

Shot peening treatment is to impact numerous 0.25-1 mm hard steel balls on the surface of the material at a certain speed, plastic deformation occurs on the surface, and a grain-refined surface modification layer can be obtained [6,7]. The hardness of titanium and titanium alloys and the wear resistance strengthened by shot peening have also been improved.

Electrochemical treatment is developed based on chemical reactions occurring on the surface characteristics of the alloy of titanium. Anodic oxidation is the electrochemical oxidation of metals and alloys, and it is a common method for preparing different oxide films on metal materials [8,9].

Laser cladding has high metallurgical bonding strength, reduces material cost and energy consumption, does not need to contact the workpiece during processing, and is green and pollution-free [10,11]. It is the most widely used surface coating technology in recent years and numerous academics researched this subject. Weng et al. [12] showed that TC4 was improved by NiCoCrAlY. Laser cladding produced a coating twice as hard as Ti-6Al-4V. Mahamood et al. [13] studied how laser velocity affected Ti-6Al-4V/TiC composites. Results demonstrate that the wearability of Ti-6Al-4V/TiC composites rises as scanning speed rises, however, scanning speed has a specific upper limit value. Only under the appropriate process parameters, titanium alloy can obtain excellent coating. Liu et al. [14] prepared Ni60 + 50 % WC and deloro22 powder backing + (Ni60 + 50%WC) two kinds of wear-resistant composite coatings on TC4 alloy. The outcomes show that resistance to wear, hardness and bonding strength of Ni60 + 50 % WC coating are better than those of deloro22 powder backing + (Ni60+50%WC) coating, which may be due to the thick backing of deloro22 powder.

The primary process parameters that impact the performance of surface coating are the energy of the laser, scanning speed, point diameter, etc. Because the impact of these parameters on coating quality is related, multiple factors on the cladding layer are analyzed the influence, and the laser energy density is introduced as a comprehensive examination index. The expression is :

$$E = \frac{P}{Dv} \tag{1}$$

Where $E$ is the laser energy density (J/mm$^2$); $P$ stands for power (W); $D$ stands for spot diameter (mm); $v$ stands for scanning speed (mm/s).

In paper, pre-coated powder Ni60+WC laser cladding Ti-6Al-4V was used, with three various laser energy densities of 33.3 J/mm$^2$, 46.7 J/mm$^2$, and 53.3 J/mm$^2$ were introduced to investigate performance of cladding layer.

## 2. Experimental materials and procedures

### *2.1. Experimental material*

The Ti6Al4V alloy with size of 50 × 50 × 5 mm was accepted as an experimental base. To begin the test, the oxide layer was removed by grinding at different mesh emery paper first, then cleaned and dried with anhydrous ethanol. The chemical composition of Ti6Al4V alloy is 5.8-6.3 wt% Al, 3.5-4.2 wt% V, 0.05-0.2 wt% C, 0.15 wt% Fe and Ti balance. The cladding materials used in the experiment are Ni60 (composition: 4.31 wt% Fe, 3.2 wt% B, 4 wt% Si, 15.9 wt % Cr, 0.78 wt% C, and Ni balance) and WC mixed powder, in which the mass fraction of WC is 35 %.

### *2.2. Preparation of experimental samples*

The laser equipment used in the laser experiment is DL-HL-T200 cross-flow $CO_2$ laser with 3 mm beam focus size. The whole experiment is acted upon under argon protection gas. In the experiment, the pre-placed cladding powder was 1 mm, and the cladding coating was prepared by using three different laser energy densities of 33.3 J/mm$^2$, 46.7 J/mm$^2$, and 53.3 J/mm$^2$.

### *2.3. Microstructure and performance testing*

Metallographic structure of cross-sectional microstructure of the sample after corrosion in the corrosive solution was observed by a metallographic microscope. Microstructure elements of cladding coatings were analysed by JSM-6490LV SEM. With an overload of 500 g and a residence duration of 10 s, the MH-500 microhardness instrument was utilized to evaluate the degree of hardness of coating layer. The coating friction and wear weight loss were trialed MGW-02 high-speed reciprocating friction tester.

## 3. Results and discussion

### *3.1. Microstructure*

Figure 1 depicts metallographic structure of specimens in different laser energy. The diagram demonstrates the strong metallurgy bond, as well as presence of a clear metallurgy bond line. The breadth of the heat-impacted zone narrows as the laser energy density increases. There are a large number of hard phase WC unmelted particles at low energy density, while the hard phase particles at high energy density are small and less, indicating that WC is decomposed at high laser energy density.

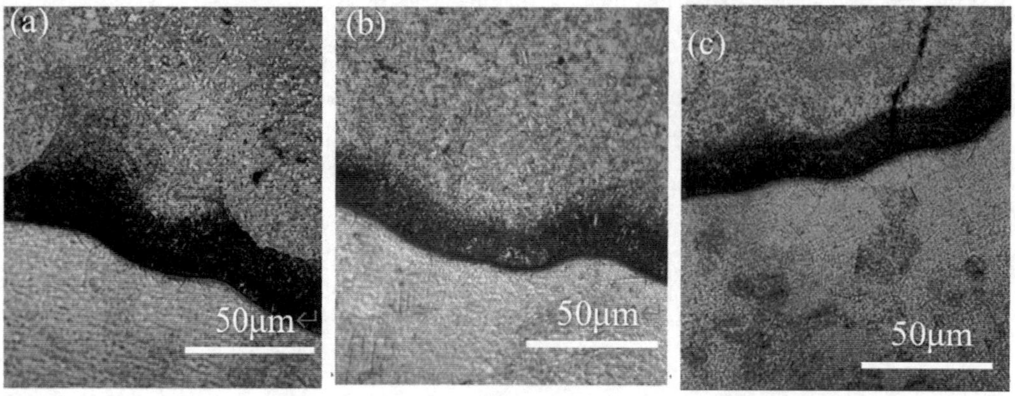


**Figure 1.** Metallographic structure (a) E = 33.3 J/ mm$^2$; (b) E = 46.7 J/ mm$^2$; (c) E = 53.3 J/ mm$^2$.

Because the powder of the three cladding samples is the same, but the energy density is different, the scanning electron microscope microstructure of the three samples is similar. The middle area of the coating has a certain amount of hard phase and the matrix diffuses into the cladding layer in Figure 2. Table 1 shows the EDS elemental composition analysis of Figure 2. It can be seen that some of the matrix Ti diffuses into the cladding layer, and there are Ni, W, and C elements.

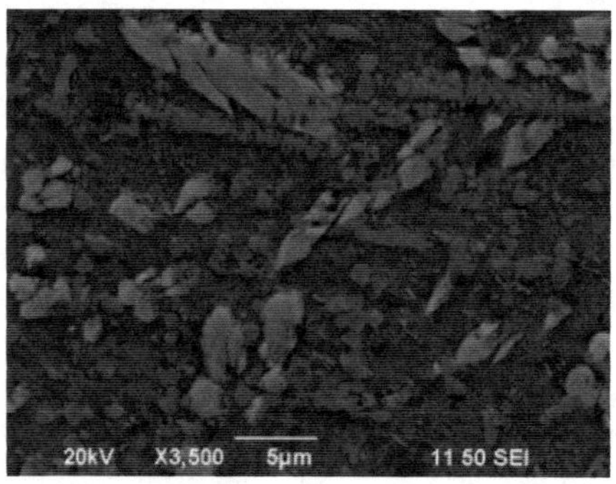


**Figure 2.** The middle area of the cladding layer.

**Table 1.** EDS results of middle region of cladding coating.

| Element | Atomic % |
|---|---|
| Carbon | 28.135 |
| Aluminum | 3.738 |
| Silicon | 3.506 |
| Titanium | 32.865 |
| Chromium | 5.496 |
| Iron | 5.035 |
| Nickel | 18.561 |
| Tungsten | 2.663 |

### *3.2. Microhardness*

Figure 3 illustrates the microhardness of specimens at various energy densities. As energy density increases, microhardness increases then decreases. If energy density is too large, energy absorbed by workpiece will increase, which will increase the decomposition of the WC hard phase. Meanwhile, the temperature gradient of cladding layer will increase, resulting slow cooling rate and reduced surface hardness. Average microhardness of cladding layer of three cladding samples is 1141.63$HV_{0.5}$,

1201.34HV$_{0.5}$, and 1017.79HV$_{0.5}$, respectively. The average hardness of the substrate is 364.78HV$_{0.5}$. The cladding coating's hardness is 2.79 to 3.29 times greater than that of the Ti6Al4V alloy used as the substrate.

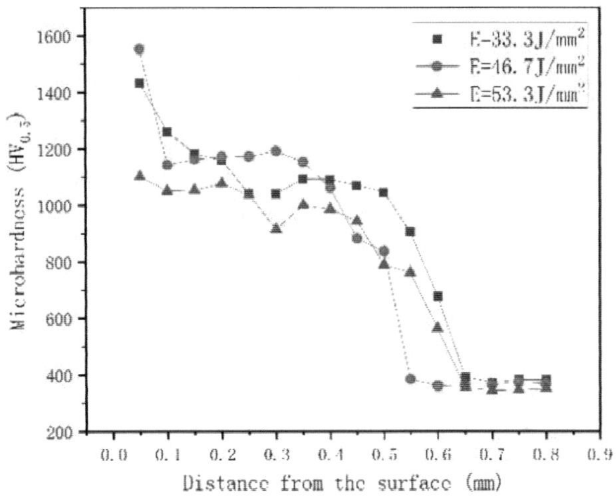


**Figure 3.** Microhardness curves of the coatings.

### *3.3. Tribological properties*

Figure 4 shows the laser energy density of 33.3 J/mm$^2$, 46.7 J/mm$^2$, 46.7 J/mm$^2$, and 53.3 J/mm$^2$ coating coefficient and loss weight. From Figure 1(a), coating coefficient of energy density of 53.3 J/mm$^2$ is the smallest. The intensity of the laser is 33.3 J/mm$^2$, which may be due to the dense and uniform coating structure under this energy density. From Figure 1(b), it is evident that the weight loss of coating under three energy densities is 4.8 mg, 6 mg, and 5.1 mg, respectively, which is much smaller than the weight loss of the substrate 27.6 mg. Weight loss increases initially, then declines as laser energy density increases.

Figure 5 shows wear morphology of substrate and the coating different energy densities. The figure shows that surface of substrate has the largest worn bandwidth, while the coating has a smaller worn bandwidth than the substrate. The coating's wear width with energy density 33.3 J/mm$^2$ has least wear width while the coating's wear width of 46.7 J/mm$^2$ has biggest wear width, which corresponds to the result of wear loss weight, showing that wear of the coating with energy 33.3 J/mm$^2$ is better than that of others. The wear morphology of the substrate has deep furrows and plastic deformation, and there is certain adhesive wear.

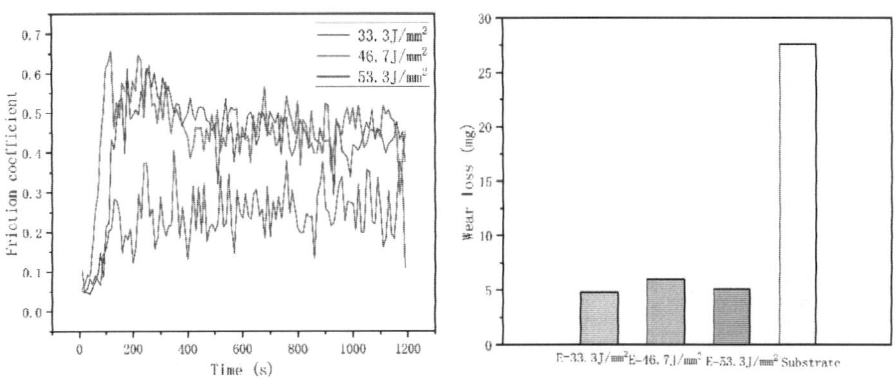


(a) (b)

**Figure 4.** Friction coefficient curves (a) and weight loss (b) of coating.

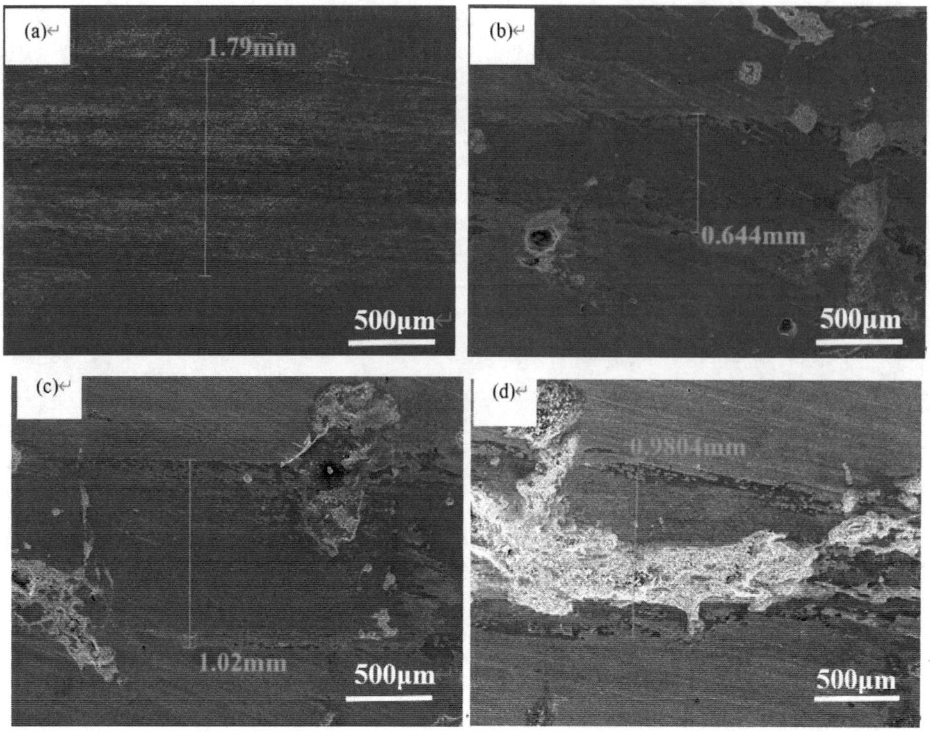


**Figure 5.** Worn morphology of Ti6Al4V (a) 33.3 J/mm$^2$ (b) 46.7 J/mm$^2$ (c) 53.3 J/mm$^2$ (d).

## 4. Conclusions

For this paper, the laser-clad Ni60/WC coated composites were created on Ti6Al4V alloy pre-laid powder method with different laser energy densities of 33.3 ,46.7 and 53.3 J/mm$^2$. The following is results.

(1) By applying three different energy densities to the surface of the Ti6Al4V alloy, Ni60/WC composite layers were created. The outcomes display that the samples have fine metallurgical bonding and good metallurgical bonding at three different speeds, and the heat breadth of the impacted zone reduces as the laser increases.

(2) The microhardness with varying energy levels, the outcomes display that microhardness increases then decreases with energy density increasing. The maximum hardness is 46.7 J/mm$^2$. The mean microhardness is 1201.34$HV_{0.5}$, which is 3.29 times that of the matrix.

(3) Under three energy densities, the cladding coating's wear qualities are superior to those of the base, and the loss of wear is minimal. The coating with laser energy of 33.3 J/mm$^2$ has smallest wear loss and the smallest wear width, showing better wear resistance.

(4) Comparing the three cladding samples from the hardness and tribological properties of coating, comprehensive performance with energy density 33.3 J/mm$^2$ is better.

## References

[1] Zhu YY, Liu D, Tian XJ, Tang HB, Wang HM. (2014) Characterization of microstructure and mechanical properties of laser melting deposited Ti–6.5Al–3.5Mo–1.5Zr–0.3Si titanium alloy. Mater Des, 56:445–53.

[2] An Q, Chen J, Tao Z, et al. (2020) Experimental investigation on tool wear characteristics of PVD and CVD coatings during face milling of Ti6242S and Ti-555 titanium alloys - ScienceDirect. International Journal of Refractory Metals and Hard Materials, 86:1–38.

[3] Sharma D, Mohanty S, Das AK. (2020) Surface modification of titanium alloy using hBN powder mixed dielectric through micro-electric discharge machining. Surface and Coatings Technology, 381, 1–47.
[4] Budinski KG. (1991) Tribological properties of titanium alloys. Wear, 151(2): 203–217.
[5] Ma Q, Li Y, Wang J, Liu K. (2015) Investigation on cored-eutectic structure in Ni60/WC composite coatings fabricated by wide-band laser cladding, J. Alloy. Compd. 645: 151–157.
[6] Xu SY, Cai Q, Li G, et al. (2022) Effect of scanning speed on microstructure and properties of TiC/Ni60 composite coatings on Ti6Al4V alloy by laser cladding. Optics & Laser Technology, 154:1–12.
[7] Thomas M, Lindley TC, Rugg D, Jackson M. (2012) The effect of shot peening on the microstructure and properties of a near-alpha titanium alloy following high temperature exposure, Acta Materialia, 60 (13-14), 5040–5048.
[8] Ortiz AL, Tian JW, Shaw LL, & Liaw PK. (2010) Experimental study of the microstructure and stress state of shot peened and surface mechanical attrition treated nickel alloys. Scripta Materialia, 62(3), 129–132.
[9] Zaraska L, Gawlak K, Gurgul M, Gilek D, Kozieł M, Socha RP, & Sulka GD. (2019) Morphology of nanoporous anodic films formed on tin during anodic oxidation in less commonly used acidic and alkaline electrolytes. Surface and Coatings Technology. Surface and Coatings Technology, 362, 191–199.
[10] Zhu LD, Xue PS, Lan Q, et al. (2021) Recent research and development status of laser cladding: A review.Optics & Laser Technology, 138, 1–26.
[11] Liu Y, Ding Y, Yang L, et al. (2021) Research and progress of laser cladding on engineering alloys: A review.Journal of Manufacturing Processes, 66(1): 341–363.
[12] Weng F, Chen CZ, Yu HJ. (2014) Research status of laser cladding on titanium and its alloys: A review. Materials and Design, 58: 412–425.
[13] Mahamood RM, Akinlabi ET, Shukla M, Pityana S. (2013) Scanning velocity influence on microstructure, microhardness and wear resistance performance of laser deposited Ti6Al4V/TiC composite. Mater Des, 50:656–666.
[14] Liu JG, Yang JH, Wang GS, et al. (2022) Microstructure and Wear Resistance of Laser Cladding WC Reinforced Ni Based Composite Coating on TC4 Titanium Alloy, Rare Metal Materials and Engineering. 51(8):2907–2914.

# Study on the effect of microstructure change of 20CrNiV5 high strength and toughness steel on hardenability

**Dongyuan Li [1, 2], Shupeng Song [1, 2,4], Wenbin Xia [3], Yong Zhou [3], Ze Lu [1], and Run Wu [1]**

[1]State Key Laboratory of Refractories and Metallurgy, Wuhan University of Science and Technology, Wuhan 430081, China
[2]Faculty of Materials, Wuhan University of Science and Technology, Wuhan 430081, China
[3]Hengyang Valin Steel Tube (Group) Co., Ltd., Hengyang 421001, China
[4]Corresponding author's e-mail: spsong@wust.edu.cn

**Abstract:** High-pressure and high-load drilling environment in the deep soil requires high strength and toughness of the oil well pipe, so the oil well pipe steel must be heat-treated, and the hardenability of the steel is particularly important in manufacturing. In this paper, Gleeble-3500 thermal simulator is used to measure the CCT curves of 20CrNiV5 steel, and the influence of microstructure on hardenability is studied. Deformation inhibits the martensitic transformation of the steel and leads to the decrease of its hardenability; the inhomogeneity of the steel pipe after hot forming also has an important impact on the hardenability, and the multiphase microstructure of bainite and ferrite after hot rolling leads to austenitic heating. Austenite grains are small and uneven, which reduces the hardenability of the steel. The microstructure of the inner and outer layers of the hot-formed steel pipe is smaller than that of the middle layer. After heat treatment, its smaller austenite results in weak hardenability.

## 1. Introduction

In recent years, oil and gas development has shifted to deep oil and gas fields with harsh environments [1]. The environment under high pressure and high load conditions requires oil well pipes to have high strength and toughness. An appropriate heat treatment process is a necessary condition for obtaining high strength and toughness, and hardenability is an important basis for formulating the heat treatment process. Good hardenability can ensure uniform surface and inner properties [2-4], so the effect of microstructure hardenability on steel properties is particularly significant.

The hardenability of steel mainly depends on its critical cooling rate, which is affected by many factors such as composition, heat treatment process, and austenite grain size [5-7]. The main research methods mainly include the end quenching and CCT curve methods. Alloying elements can effectively improve the hardenability of steel [8-12], high heating temperature and long holding time in austenitizing can shift the C curve to the right and improve the hardenability [13,14], larger austenite grains also improve hardenability [15]. In recent years, controlled rolling and cooling technology has been used in steel production to diversify the microstructure of steel. It is necessary to study the influence of various microstructures on the hardenability of steel. Therefore, based on measuring the CCT curves of 20CrNiV5 steel, the influence of hot-forming microstructure on its hardenability has been studied.

## 2. Experimental process and method

Experimental steel is 20CrNiV5, a kind of low alloy high strength steel, and its chemical composition is shown in Table 1. The steel cut from the hot-rolled slab is homogenized heat treated, and then the dynamic and static CCT curves are measured by Gleeble-3500 thermal simulator. When measuring static CCT curves, the steel was heated to 1150°C at 5°C/s and held for 5 min for austenitization, then cooled to 890°C at 10°C/s and held for 50 seconds, and then cooled to room temperature at 0.1, 0.3, 1, 5, 10, 20, 25, 30, 40, 50, 80°C/s. Dynamic CCT measurement parameters are the same, except that it is cooled from 1150 °C to 890°C, and then 20% compression was performed at a rate of 10°C/s after holding for 15 s. After holding for 35 s, it is also cooled to room temperature at the same cooling rate as the static CCT test.

**Table 1.** Composition of test steel (mass fraction, %).

| C | Si | Mn | Mo | P | S | Cu | Ni | Cr | Al |
|---|---|---|---|---|---|---|---|---|---|
| 0.2 | 0.21 | 0.41-0.5 | ≤0.8 | ≤0.02 | ≤0.01 | 0.15-0.2 | 0.05-0.08 | ≤0.5 | ≤0.03 |

After thermal simulation, the samples were cut along the thermocouple welding point to make a metallographic sample and then eroded with a 4% nitric acid alcohol solution. The microstructure was analyzed by metallographic microscope (OM) and FEI Nova 400 field emission electron microscope (SEM), and the hardness was measured by an HV-1000 A microhardness tester. The phase transformation point was measured from the thermal simulation expansion curve of each sample by tangent method, and the dynamic and static CCT curves were drawn by combining the microstructure and microhardness, and the critical cooling rate of quenching was obtained. To compare the structural characteristics of different layers, the samples were cut from the outer, middle, and inner layer of the hot-formed steel pipe, and the thermal simulation test of the critical cooling rate was carried out. According to GB/T 6394-2017 (the Metal average grain size determination method) [16], the grain size of the original austenite, the grain size, and the volume fraction of ferrite and bainite were determined.

## 3. Results and discussions

### *3.1. Critical cooling rate of 20CrNiV5 steel*

Figure 1 shows the OM ($a_1$-$d_1$) and SEM ($a_2$-$d_2$) images of static CCT after thermal simulation with different cooling rates. The microstructure of steel with a cooling rate of 0.1°C/s is ferrite and pearlite. At 1°C/s, the microstructure is mainly bainite, and there is trace ferrite. At 5°C/s, the microstructure of the steel is bainite. When the cooling rate ≥ 20°C/s, the microstructure is martensite. Figure 2 shows the OM and SEM images of dynamic CCT at 25°C/s cooling rate, which is full martensite.

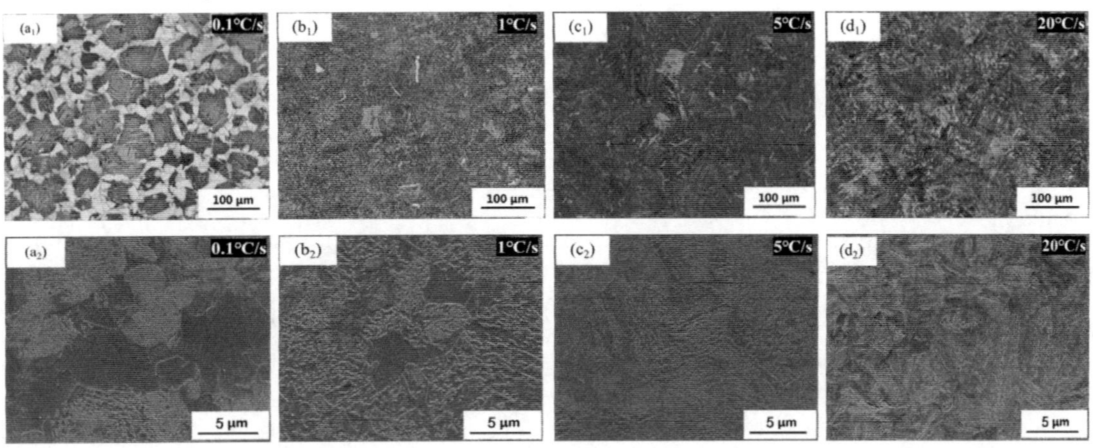


**Figure 1.** Metallographic microstructure ($a_1$-$d_1$) and SEM morphology ($a_2$-$d_2$) of static CCT samples under different cooling rates.

**Figure 2.** Metallographic microstructure (a) and SEM morphology (b) of dynamic CCT sample under critical cooling rates (25°C/s).

Expansion curves, microstructure, and hardness results of the dynamic and static CCT experiments were obtained, and the summary results are listed in Table 2. And the dynamic and static CCT results are drawn in Figure 3 using different curves.

**Table 2.** Static and dynamic (left/right) CCT phase transition start and end temperature.

| Cooling rate(°C/s) | $(F+P)_s$ | $(F+P)_f$ | $B_s$ | $B_f$ | $M_s$ | $M_f$ | Microstructure type and microhardness (HV0.05) |
|---|---|---|---|---|---|---|---|
| 0.1 | 763/778 | 639/651 | | | | | F(146)+P(284)/F(132)+P |
| 0.3 | 745/765 | 670/674 | 553/559 | 390/394 | | | F(185)+P+B(265)/F(204)+P+B |
| 1 | | | 562/574 | 385/391 | | | B(287)+paucityF/B(283)+F(183) |
| 5 | | | 565/576 | 381/390 | | | B(334)/B(341)+paucity F |
| 10 | | | 558/566 | | | 315/300 | B(370)+M(429)/B(338)+M |
| 15 | | | 542/552 | | | 264/252 | M(435)+paucityB/M(459) +paucity B |
| 20 | | | 545/553 | | | 269/273 | M(458)/M(487)+paucity B |
| 25 | | | | | 447/398 | 258/251 | M(472)/M(465) |
| 30 | | | | | 372/360 | 259/255 | M(436)/M(477) |
| 40 | | | | | 375/367 | 263/259 | M(459)/M(458) |
| 50 | | | | | 366/363 | 267/267 | M(488)/M(499) |
| 80 | | | | | 370/365 | 269/245 | M(475)/M(482) |

It can be seen from Table 2 that the start temperature of ferrite, pearlite, and bainite phase transition measured by dynamic CCT experiment is higher than that measured by static CCT, while the start temperature of martensite phase transition measured by dynamic CCT experiment is lower than that measured by static CCT. The microstructure at a low cooling rate of 0.1°C/s ~ 0.3°C/s is mainly ferrite and pearlite, and the microhardness value is about 132 ~ 284 HV0.05. When the cooling rate increases to 1 °C/s ~ 15 °C/s, the microstructure of the steel is mainly composed of bainite and martensite, and the microhardness value is about 334 ~ 459 HV0.05. In addition, when the cooling rate is further increased to 20 °C/s ~ 80 °C/s, the microstructure is mainly martensite, and the microhardness value is the largest, about 458 ~ 499 HV0.05.

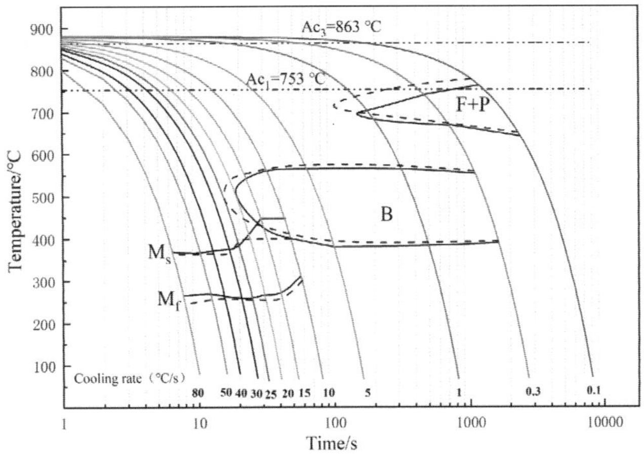


**Figure 3** Static (solid line) and dynamic (dotted line) CCT curves of test steel.

By comparing the dynamic and static CCT curves, it can be concluded that 0.1°C/s to 1°C/s is the static CCT ferrite and pearlite phase zone, and the dynamic CCT transition temperature range shift to the upper left, the static CCT bainite phase zone extends from 0.3°C/s to 20°C/s, while the dynamic CCT phase zone extends to 25°C/s. The critical quenching cooling rate of static CCT is 20°C/s, while that of dynamic CCT is 25°C/s.

The deformation makes the ferrite and pearlite start to phase transition temperature increase, the bainite phase transition zone expands, and the $B_s$ temperature increase. At the same time, the deformation increases the critical quenching cooling rate and reduces the $M_s$ point temperature. This is because the thermal deformation of austenite will increase the crystal defects and strengthen the austenite microstructure, so that the martensite non-diffusion shear mechanism is blocked, resulting in a decrease in hardenability [17,18].

### *3.2. Microstructure of hot-formed steel pipe and its hardenability at the critical cooling rate*

Fig.4 shows the original microstructure and quenched microstructure of the hot-formed steel pipe. The original microstructure is composed of ferrite and bainite. The grain sizes of bainite lath and ferrite are slightly larger in the middle layer, and slightly smaller in the inner and outer edge layers. For the quarter layer, the bainite lath average size is 1.29 μm, and the ferrite grain average size is 10.54 μm, which is larger than the edge layer and smaller than the middle layer. Corresponding quantitative statistics are shown in Table 3. This is because the inner and outer edges are cooled by water, and the cooling rate is faster than that of the middle layer so that the grains are refined. In the inner and outer edge layers, the volume fraction of bainite is more than that in the middle layer, but for the ferrite, this situation is the opposite. This is also related to the cooling rate. When the cooling rate is lower, it is beneficial to the ferrite transformation. The increase in the cooling rate will inhibit the diffusion of iron and carbon atoms, thus reducing the phase transformation of ferrite and promoting the bainite transformation [19].

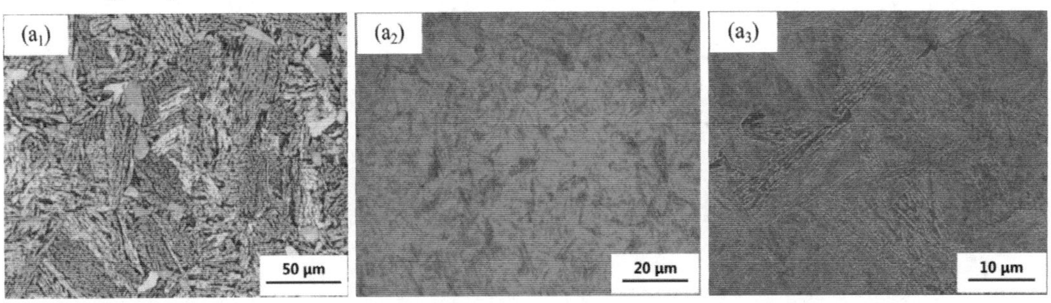


Microstructural changes of the outer edge layer
($a_1$) Original; ($a_2$) After quenching; ($a_3$) SEM observation after quenching

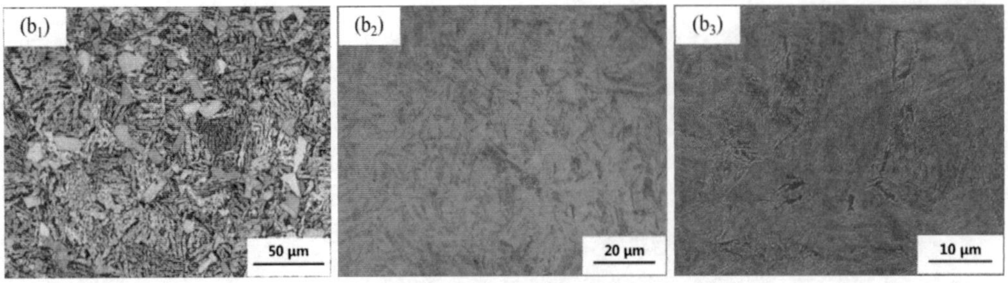


Microstructural changes of the Quarter layer
(b1) Original; (b2) After quenching; (b3) SEM observation after quenching

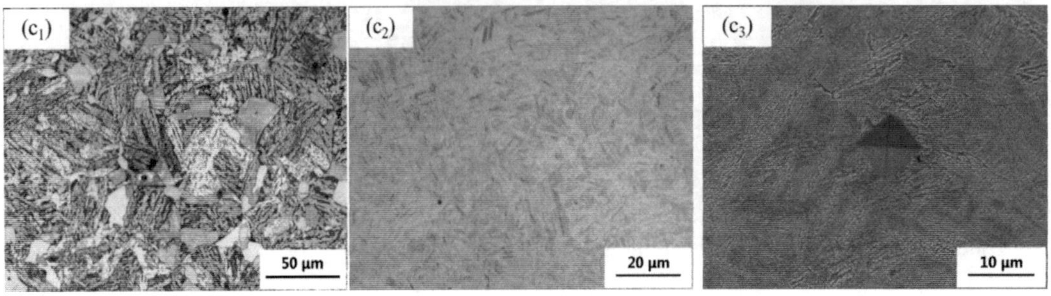


Microstructural changes of the middle layer
(c1) Original; (c2) After quenching; (c3) SEM observation after quenching

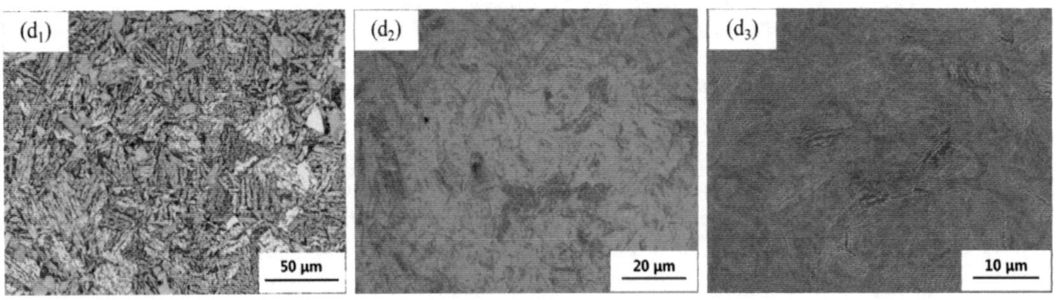


Microstructural changes of the inner edge layer
(d1) Original; (d2) After quenching; (d3) SEM observation after quenching

**Figure 4.** Original microstructure of hot forming steel and microstructure and SEM morphology after thermal simulation.

The microstructure of hot-forming steel after thermal simulation is mainly martensite, but based on the metallographic and SEM images, it can be seen that there are trace amounts of bainite and ferrite, and the bainite and ferrite in the inner and outer edge layers are more than those in the middle layer. After that, the inner edge layer sample has the weakest hardenability, and the corresponding cooling rate was increased to 26°C/s, full martensite was obtained under this thermal simulation condition. This microstructure is not completely hardenable, and its hardenability is weaker than that of steel after homogenization heat treatment.

**Table 3.** Quantitative statistics of the initial microstructure of hot-forming steel.

| Site | Ferrite | Bainite | Ferrite | Bainite lath | Austenite |
|---|---|---|---|---|---|
| | Volume fraction/vt.% | | Average grain size/μm | | |
| Outer edge layer | 9.11 | 90.89 | 9.07 | 1.12 | 5.66 |
| Outer quarter layer | 11.35 | 88.93 | 10.54 | 1.29 | 6.04 |
| Middle layer | 12.14 | 87.86 | 11.25 | 1.45 | 6.36 |
| Inner edge layer | 9.48 | 90.52 | 8.84 | 1.05 | 5.51 |

Figure 5 shows the OM images of the original austenite grain boundary which is corroded by picric acid. The austenite grain size of the steel after homogenization heat treatment is larger than that of the hot-forming steel, and the uniformity is better. The average grain size of the quarter layer is 6.04 μm, which is larger than 5.66 μm of the edge layer and smaller than 6.36 μm of the intermediate layer. The uniformity of each layer of the hot-forming steel is the same.

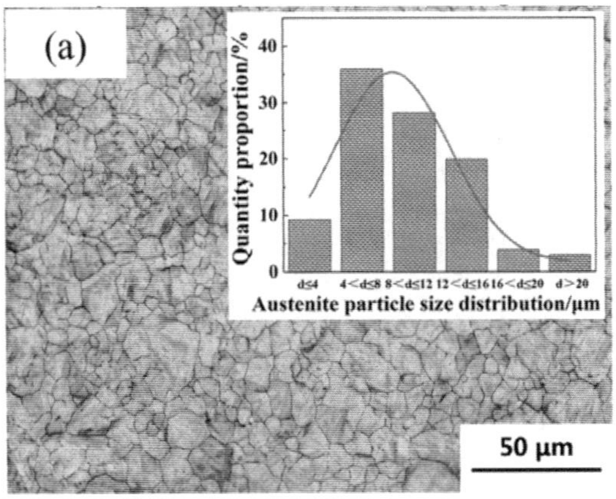


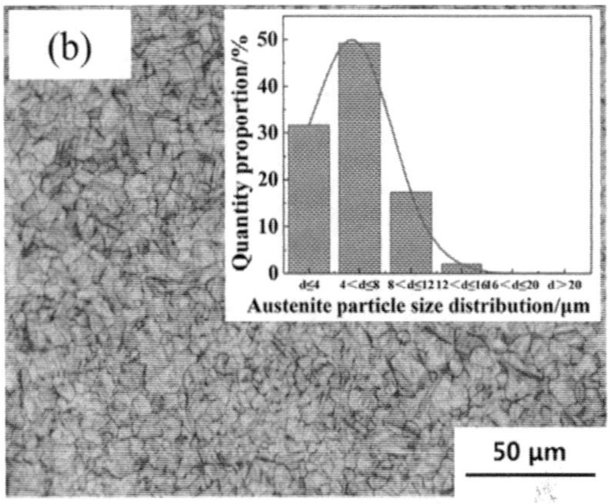


(a) Steel after homogenization heat treatment (b) Outer edge layer

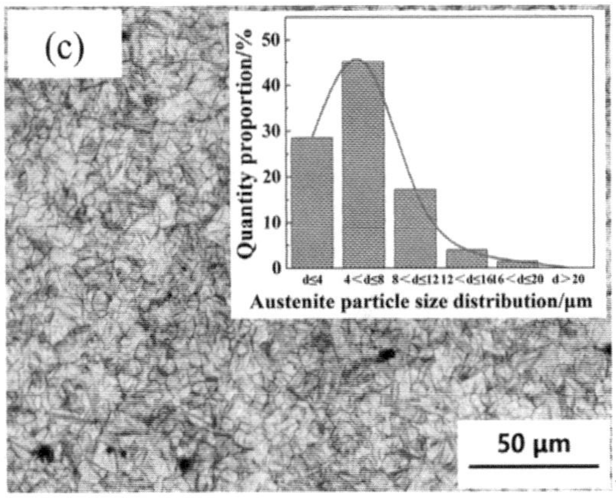


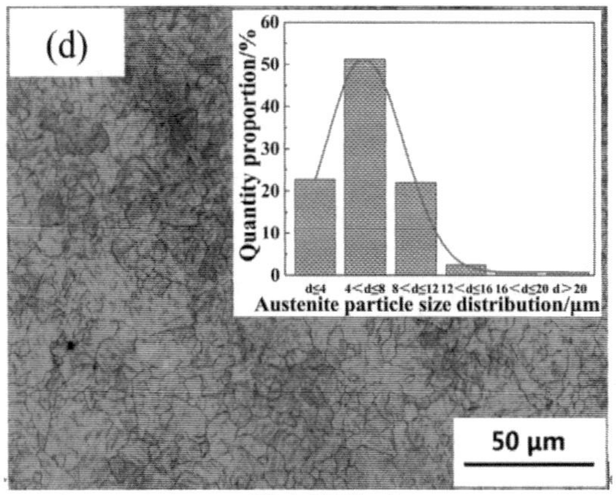


(c) Outer quarter layer (d) Middle layer

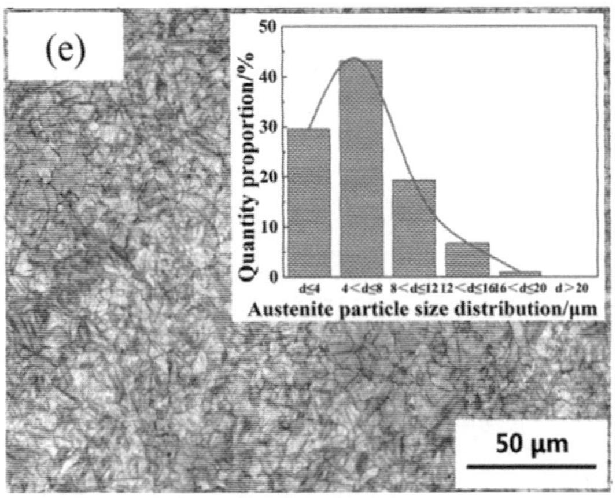


(e) Inner edge layer

**Figure 5.** The original austenite grain boundaries of steel and hot-forming steel after homogenization heat treatment.

The hardenability of hot-forming steel is weaker than that of steel after homogenization heat treatment. This is because the microstructure of hot-forming steel is ferrite and bainite, and austenite preferentially nucleates at the phase boundary. The non-uniformity of the multi-phase microstructure leads to more nucleation cores of austenite, and ferrite will promote the austenitization of bainite, and the austenite grains generated by final heating are relatively fine and uneven. Due to the low nucleation rate, the austenite grain size formed by thermal simulation heating is large and uniform. The larger the grain size is, the smaller the grain boundary area per unit volume is, and the fewer nucleation sites for diffusion-type transformation, thereby delaying the supercooled austenite transformation; the more uniform the microstructure is, the lower the nucleation rate of austenite transformation, and thereby increasing Supercooled austenite stability. Both factors shift the C curves to the right, reducing the critical cooling rate of steel and improving hardenability [20,21].

The hardenability of the inner and outer edge layers of hot-forming steel is weaker than that of the middle layer, and the hardenability of the quarter layer is weaker than that of the middle layer but stronger than that of the edge layer. This is due to the thinner bainite lath size of the inner and outer edge layers, resulting in more nucleation sites per unit area, thereby increasing the austenite nucleation rate. And the ferrite in the inner and outer edge layers is also small, and the heating transformation will produce structural heredity, these all lead to the formation of relatively small austenite grains, reducing the steel hardenability [22]. And the energy dispersive spectroscopy results also show no significant difference in layered elements, so the grain size is the core reason for the different hardenability of hot-formed steel after layered quenching.

## 4. Conclusion

Based on the results and discussions presented above, the conclusions are obtained as below:

(1) Dynamic CCT critical quenching cooling rate (25°C/s) is greater than the static CCT critical quenching cooling rate (20°C/s). Deformation inhibits martensitic transformation and increases the critical quenching cooling rate, resulting in a decrease in hardenability.

(2) Microstructure of hot stamping steel is bainite and ferrite, and it is not uniform. After heating, the austenite grains are relatively fine and uneven. After homogenization heat treatment, the microstructure of the steel is uniform tempered sorbite, and the austenite grains generated after heating are large and relatively uniform. Therefore, the hardenability of hot-forming steel is weaker than homogenization heat treatment ones.

(3) Grain size of bainite lath and ferrite in hot forming steel is slightly larger in the middle layer, slightly smaller in the inner and outer edge layers, and the bainite lath size and ferrite grain size of the quarter layer are larger than those of the edge layer but smaller than those of the middle layer. These lead to the smaller austenite grain size formed by recrystallization in the inner and outer edge layers after heating and holding, and it weaker hardenability of the edge layer.

(4) The original microstructure of steel has a decisive influence on its hardenability. The performance and toughness of unhardened steel components are insufficient and even fail in advance in use. Therefore, systematic research in this area has an important guiding role in the production and processing of steel. The follow-up research should focus on the effect of the original microstructure on the impact behavior of steel after heat treatment.

## References

[1] Chang, Y.W., Wang, Z.Q., Liu, B.L., et al. (2018) Characteristics and enlightenment of global oil and gas development. J. International Petroleum Economics, 26(12):58-66.
[2] Long, J.L. (2018) Research on the hardenability of CM690 anchor steels. D. Shandong University.
[3] Luan, Y., Tang, Y.F. (1996) Review of standards for steels for hardenability (H steel). J. Metallurgical Standardization and Quality, (10):21-23.
[4] Liu, X.G., Wu, Y., Sun, W., Wang, K.Z. (2016) Experimental study on CCT curve and hardenability of 200~250 km/h EMU axle steel. J. Journal of Railway Science, 38(05): 42-45.

[5] Wang, X.K., Guo, Z.Q. (2016) Control of narrow hardenability of 20CrMnTiH gear steel for cars. J. Hebei Metallurgy, (11): 10-13.
[6] Bian, S. (2012) Research on hardenability of 25CrMo axle steel used in high-speed rail. D. Shenyang University of Technology.
[7] Jia, X.S., Zhang Z.W., Huang, Y., et al. (2017) Simulation and analysis on chemical composition optimization of large semi-autogenous mill liner. J. Metal Heat Treatment, 42(05) :1-8.
[8] Kobayashi J, Ina D, Yoshikawa N, et al. (2012) Effects of the addition of Cr, Mo and Ni on the microstructure and retained austenite characteristics of 0.2% C–Si–Mn–Nb ultrahigh-strength TRIP-aided bainitic ferrite steels. J. ISIJ international, 52(10): 1894-1901.
[9] Wang Z, Qi J, Liu Y. (2019) Effect of silicon content on the hardenability and mechanical properties of link-chain steel. J. Journal of Materials Engineering and Performance, 28: 1678-1684.
[10] Huang S, Yu Y S, Wang Z Q, et al. (2022) Crystallographic insights into the role of nickel on hardenability of wear-resistant steels. J. Materials Letters, 306: 130961.
[11] Huang, S., Wu, B.B., Wang, Z.Q., et al. (2019) EBSD study on the significance of carbon content on hardenability. J. Materials Letters, 254(1):412-414.
[12] Ling, X., Zhu, P., Li, B.P., et al. (2023) Effect of Ti on the hardenability of carbon structural steel. J. Physical Testing, 41(01):12-15.
[13] Kokosza A, Pacyna J. (2015) Influence of austenitising temperature on kinetics of phase transformations in medium carbon TRIP steel. J. Materials Science and Technology, 31(7): 803-807.
[14] Potapov A I, Malikov I T, Urazov V I, et al. (2010) Effect of the chemical composition and austenitizing conditions on the hardenability of 35GR steel. J. Russian Metallurgy (Metally), 2010(12): 1164-1167.
[15] Shepelyakovskii K Z, Postnikov Y Y, Kal'ner V D, et al. (1971) Hardenability of steels for precision castings. J. Metal Science and Heat Treatment, 13: 713-716.
[16] Cheng, L.J., Luan, Y., Gu, Q. (2017) New version of GB/T 6394-2017 "Determination of estimating the average grain size of metal" standard analysis. J. Metallurgical Standardization and Quality, 55(3):8.
[17] Liu, X., Cao, J., Chen, W., et al. (2022) Post-Rolling Cooling Phase Transformation and Microstructure of High-Strength Anti-Seismic Rebars with Different Solute Nb and Austenite Microstructure. J. Metals, 12(10): 1734-1749.
[18] Nikravesh, M., Naderi, M. Akbari G. H. (2012) Influence of hot plastic deformation and cooling rate on martensite and bainite start temperatures in 22MnB5 steel. J. Materials Science and Engineering A, 540: 24-29.
[19] Cui, Z.Q., Liu, B.X. (2004). Metallurgy and heat treatment principles. M. Harbin Institute of Technology Press.
[20] Li, Y.Y., Hu, L., Wang, L., Tong, Y. (2018) Development of high-performance prestressed steel strand wire SWRH72BCr-1. J. Metal Heat Treatment, 43(09):31-36.
[21] Li, J., Huang, J., Jia, T. (2020) Effect of initial microstructure on austenite transformation and grain growth behavior of 20CrMnTi steel during the pseudo-carburizing process. J. Metal Heat Treatment, 45(06): 157-163.
[22] Cao, Y.G., Li, Z.D., Ding, C.C., et al. (2020) Effect of niobium addition on the hardenability of high-speed railway axle steel J. Iron and Steel, 55(06):107-112.

# Experimental study on shock isolation of metal rubber-disc spring composite shock absorber under different shock excitation

**Guoqiang Liu[1], Xuebao Wang[1], Zhirong Guo[1,2] and Yiqun Zhang[1]**

[1]Naval Petty Officer Academy, Bengbu City, Anhui Province, 233000, China

[2]Corresponding author's e-mail: 28267453@qq.com.

**Abstract:** Due to the instability and aiming accuracy of the onboard laser emission system cannot meet the requirements of its work tasks, it is necessary to carry out vibration isolation optimization design for the laser emission system. This paper designs a metal rubber-disc spring composite shock absorber and conducts relevant experimental research on the shock isolation performance of the metal rubber-disc spring composite shock absorber. The shock isolation performance tests of the shock absorber under different shock excitation conditions with pendulum heights of 300 mm, 600 mm, 900 mm, and 1200 mm were carried out respectively. Through the experiments, it was found that when the pendulum height is 300 mm, the shock isolation effect of the metal rubber-disc spring composite shock absorber is the best, and the shock isolation rate is above 95%.

## 1. Introduction

In modern air defense missions, laser launch systems have advantages such as efficient cost ratio and graded killing, which can effectively enhance mission execution capabilities [1, 2]. In the face of the complex and changing terrain environment of the island reef, higher requirements are placed on the stability and aiming accuracy of the operation of the onboard laser launch system. Therefore, it is necessary to optimize the vibration isolation design of the laser launch system. The passive vibration isolation method is mainly used for system equipment vibration isolation. Chen et al. [3] studied the influence of subsystem parameters on the vibration isolation characteristics of a double-layer vibration isolation system. Wang et al. [4] designed a new vibration isolation system with inertial capacity and negative stiffness components based on vibration theory, and conducted dynamic and static experimental studies on it. Some scholars [5, 6] have laid a physical foundation for shock isolation design by constructing physical structures with negative stiffness characteristics. Domestic scholars [7-11] have taken a different approach to achieve optimal design of the vibration isolation system through the manufacture of composite low yield points and multi-substructure of the power pack, effectively reducing the vibration response. Ikago et al. [12] proposed a type of Tuned Viscous Mass Damper (TVMD) isolator, and optimized its parameters using the traditional fixed point method [13] to apply the obtained optimal isolator to the vibration reduction research of buildings under seismic excitation. Whittaker et al. [14] and Tsai et al. [15] first proposed displacement bending energy dissipation dampers with X-shaped and triangular shapes; Tirca et al. [16] proposed a steel damper with in-plane force form and conducted performance analysis on mid to high-rise structures equipped with this damper, proving that this damper has good energy dissipation and seismic reduction capabilities. In addition, optimization algorithms have also been widely applied to the design of vibration isolation systems and achieved good application results.

The on-board ATP system plays an important role in detecting, locking, and targeting, so it is necessary to carry out vibration isolation design for the ATP system to reduce its vibration response and meet the optimization design requirements of the equipment. Therefore, this article designs a metal rubber disc spring composite isolator. In order to test the impact isolation performance of the metal rubber disc spring composite isolator, experimental studies were conducted on impact isolation under different impact excitations, and relevant experimental results were obtained.

## 2. Design of metal rubber disc spring composite shock absorber

In order to effectively reduce the shock response of the onboard laser launch system, a new metal rubber-disc spring composite shock absorber was designed in this paper. The shock absorber design mainly includes the selection of disc spring parameters and the determination of metal rubber density. (1) In the selection of disc spring parameters, based on engineering experience and relevant calculations, the outer diameter D=30, inner diameter d=9.5, thickness t=0.55, and free height h=1.15 were selected. The parts were processed according to the determined geometric parameters of the disc spring. (2) The selection of geometric parameters in the metal rubber mainly includes the determination of the density of metal rubber. (3) The external structure of the metal rubber-disc spring composite shock absorber. The structure of the shock absorber was processed according to the design parameters of the disc spring and the relative density of the metal rubber, so as to obtain the structure of the shock absorber. Figure 1 shows the prototype of the designed metal rubber-disc spring composite shock absorber. Number 1 in the figure represents metal rubber; Number 2 represents the disc spring.

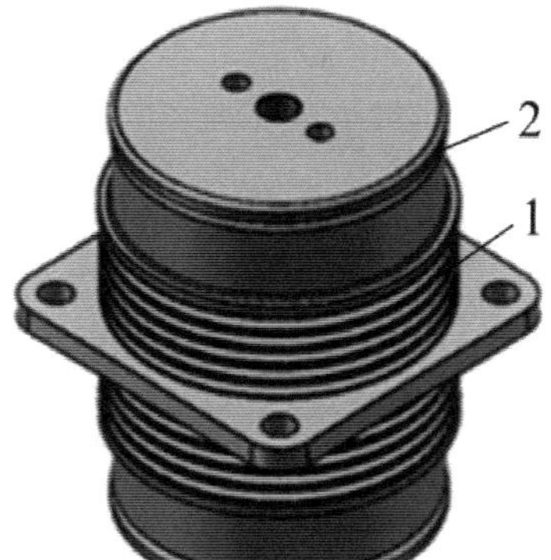


**Figure 1.** The principle profile of the metal rubber-disc spring composite isolator.

## 3. Impact isolation performance test

This section uses a pendulum impact test to determine the impact isolation performance of a metal rubber-disc spring composite shock absorber and compares it with the impact resistance performance of a GSHG-050 steel wire rope vibration isolator to analyze the advantages and disadvantages of the impact performance of the metal rubber-disc spring composite shock absorber.

### *3.1 Test protocol*

*3.1.1 Test system.* The impact test equipment for the metal rubber-disc spring composite shock absorber is shown in Figure 2. It mainly consists of a drop (pendulum) hammer light impact machine, two B&K4302 acceleration sensors, a Su Shi, and a Dong Ling vibration impact control system. During the test, the acceleration impact response values of the impact table and the simulated load are measured by the B&K4384 sensor. The Dong Ling vibration impact control system adjusts the height value of the drop hammer. When the height of the drop hammer reaches the set value, the operation switch is activated, causing the drop hammer to impact the impact table, thereby impacting the metal rubber-disc spring composite shock absorber.

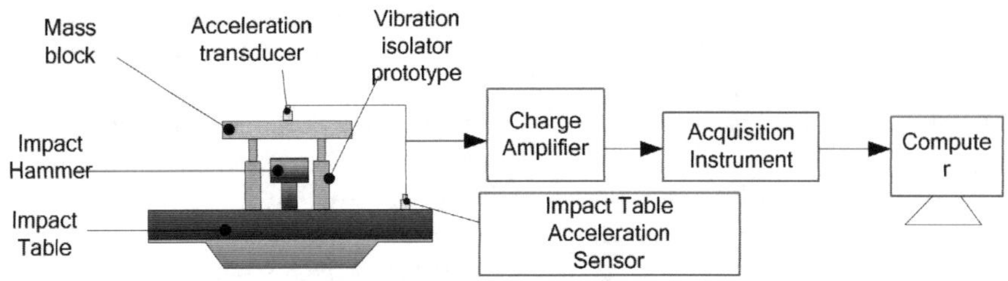


**Figure 2.** Flow chart of impact test equipment.

*3.1.2 Test conditions.* In order to conduct a more in-depth study on the impact isolation performance of the metal rubber-disc spring composite shock absorber, a vibration isolation manufacturer conducted a study on this type of shock absorber and the standard steel wire rope GSHG-050. According to the relevant test requirements of GJB150.18-86, we develop the following test conditions. This experiment adopts a vertical impact direction and a pendulum angle of 90 degrees. The impact isolation performance of the metal rubber disc spring composite isolator is studied using different impact excitations represented by the drop hammer heights of 0.3 m, 0.6 m, 0.9 m, and 1.2 m. See in table 1.

**Table 1.** Shock testing parameters.

| Charge amplifier mode | sensitivity of charge amplifier | Acquisition instrument model | Sensitivity | Sensor model | Sensor sensitivity | Measuring position and direction |
|---|---|---|---|---|---|---|
| SUSHI RX100 | 3.36 mv/g | Econ AVANT MI-7018 | 3.16 mv/g | B&K 4302V | 9.347 pc/g | Vertical direction |
| | | | | | 9.089 pc/g | Vertical load |

## 4. Analysis of test results

Taking the impact test of the composite shock absorber under four working conditions as an example, the impact performance of the composite shock absorber was explored. The relative density of metal rubber was selected, and the composite shock absorber with a load mass of 10 kg and the GSHG-050 steel wire rope vibration isolator with the same load mass were used. Under different impact strengths, namely different drop heights, 300 mm, 600 mm, 900 mm, and 1200 mm, and under the same impact direction and pendulum angle, the impact response results of the metal rubber-disc spring composite shock absorber are shown in Figure 3.

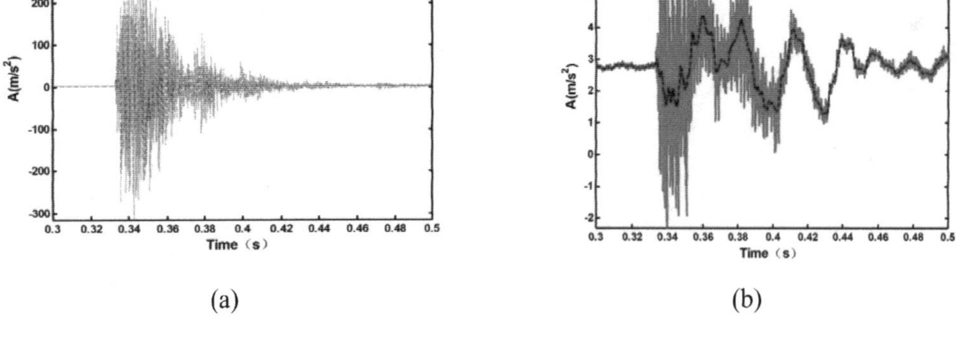


(a) (b)

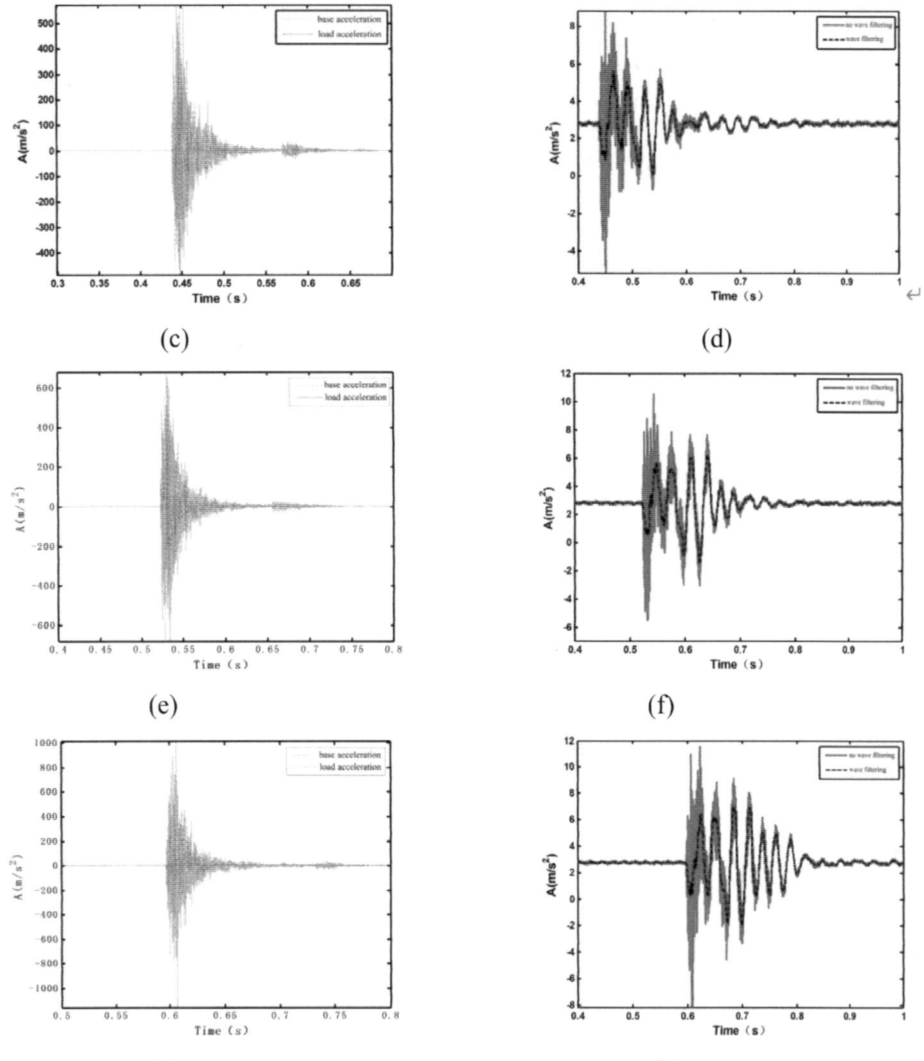


(c) (d)

(e) (f)

(g) (h)

**Figure 3.** Acceleration response of the load and the acceleration response of the shock absorber, where (a), (c), (e), and (g) are the acceleration responses of the load with a pendulum height of 300 mm, 600 mm, 900 mm, and 1200 mm, respectively; (b), (d), (f) and (h) are the acceleration responses of the shock absorber with a pendulum height of 300 mm, 600 mm, 900 mm, and 1200 mm, respectively.

As can be seen from the figure:

(1) For a certain pendulum height and a certain impact excitation, the load is much smaller than the base, the first waveform peak of the load acceleration response curve is greatly reduced, and the time response becomes longer, indicating that the metal rubber disc spring composite shock absorber can improve the impact resistance performance of the laser emission system by reducing the peak of the impact response and extending the time response.

(2) From Figures 3(a), (c), (e), and (g), it can be observed that under the four drop height conditions, there is a plateau period in the first peak of the absolute acceleration response curve of the load, which is due to the relatively small impact excitation response of the platform. From Figure 3(a), it can be seen that under the excitation of a drop height of 300 mm, there is a plateau period of approximately 2.85 m/s$^2$ before the first trough of the acceleration response of the load, which immediately decays to

1.84 m/s$^2$. Under the excitation of drop heights of 300 mm and 600 mm, the first trough curve of the acceleration response curve of the load exhibits a continuous “V” shape.

(3) The load occurs at the first trough curve, which is caused by two factors. First, during the vertical drop hammer impact, the force of the metal rubber and the Belleville spring in the composite shock absorber during the elongation process is about twice as large as its own gravity during the compression process. Secondly, the composite shock absorber elongation process is in the trough stage when subjected to double sine shock action. During this process, the impact action is low-frequency and strong, and its pulse width is wider compared to the sine wave. The elongation deformation produced by the composite shock absorber is longer than the compression deformation. On the other hand, the greater the deformation of the metal rubber-Belleville spring composite shock absorber is, the stronger the load force of the composite shock absorber is. Therefore, the peak and trough values of the acceleration time-domain curve increase continuously for a certain period of time. When the deformation reaches the limit of material properties, the peak and trough values of the acceleration response curve no longer increase.

(4) From Figure 3(a), it can be found that there is a small plateau period when the acceleration response of the load reaches the first peak, and the acceleration immediately increases and climbs to form a spike during this small plateau period. The main reason for this phenomenon is that the composite shock absorber increases the deformation range during the compression process, and the axial force of the composite shock absorber immediately increases within a certain period of time. Similarly, under the impact excitation of the falling hammer at three distances, the first valley curve of the acceleration response curve of the load presents a continuous “V” shape.

(5) From Figures 3(c), (e), and (g), it can be found that when the height of the pendulum is 600 mm, 900 mm, and 1200 mm, the platform period slowly disappears when the acceleration response of the load reaches the first peak. Especially at 1200 mm, there is almost almost no platform period. The reason for this phenomenon is the excessive impact load, and the composite shock isolator exhibits a hard stiffness characteristic. However, the first trough in the acceleration response of the load still presents a “V” shape. The shock isolation rate is an important indicator to test the performance of the composite vibration isolator, and the larger the shock isolation rate value is, the better the shock resistance performance of the composite vibration isolator will be. Its definition is:

$$\eta=\frac{A_{load}}{A_{base}}\times 100\% \tag{1}$$

Where $A_{load}$ represents the absolute acceleration response amplitude of the load, $A_{base}$ represents the absolute acceleration of the base, and $\eta$ represents the response amplitude.

According to Formula (1), the impact isolation rate at different pendulum heights was calculated. It was found that when the pendulum height was 300 mm, the impact isolation effect of the metal rubber-disc spring composite shock absorber was the best, with an impact isolation rate of over 95%.

(6) From Figure 3 (b), (d), (f), and (h), it is found that the stable values of impact isolation device A under impact excitation with pendulum heights of 300 mm, 600 mm, 900 mm, and 1200 mm are 0.60 s, 0.67 s, and 0.72 s, respectively. Under impact excitation with pendulum heights of 300 mm, 600 mm, and 900 mm, the impact isolation device can quickly return to a stable state. However, the performance of larger impact excitation with pendulum heights of 1200 mm is poor, and further optimization can be carried out in subsequent device design.

## 5. Conclusion

Faced with the reality that the vibration isolation performance of steel wire rope vibration isolators cannot meet the high stability requirements and strict aiming accuracy requirements of laser launch systems, this article mainly conducted experimental research on the shock isolation performance of metal rubber-disc spring composite shock isolators. The following conclusions were drawn: Based on the transient strong shock working environment of laser launch systems, the drop hammer test was selected to test the shock isolation performance of this type of composite shock isolator. Through comparative tests of different shock excitation parameters, it was concluded that when the pendulum height is 300 mm, the shock isolation effect of the metal rubber-disc spring composite shock isolator is the best, with a shock isolation rate of over 95%.

**References**

[1] AMY H. 2019 The Looming Swarm [EB/OL]//Air & Space Forces Magazine. https:// www.airandspaceforces.com/article/the-looming-swarm/.
[2] LUO L, TAN B T. 2022 Research on operational application of ship borne laser weapon. Laser & Infrared., 52(7): 1058-1063.
[3] Chen J, Ding J, Yan B, et al. 2018 Effect of subsystem parameters on vibration isolation characteristics of two-stage vibration isolation system. Journal of Traffic and Transportation Engineering., 18(3):114-128.
[4] Wang Kai, Wu L T, Liu F. 2005 The optimization design of a novel inerter-based isolation system with an element of negative stiffness. Journal of Vibration Engineering., 26(6):531-534.)
[5] Zou D L, Liu G Y, Rao Z S, et al. 2021 A device capable of customizing nonlinear forces for vibration energy harvesting, vibration isolation and nonlinear energy sink. Mechanical Systems and Signal Processing., 147:107-101.
[6] Hao Y, Shen Y J, Wang J F, et al. 2021 A piecewise negative stiffness mechanism and its application in dynamic vibration absorber. International Journal of Mechanical System Dynamics., 1(2):173–181.
[7] He H X, Wang X B, Zhang X F. 2018 Seismic performance of compound low yield point damper based on multi-target topology optimization. Journal of Vibration and Shock., 37(8):158-166.
[8] Yu K F, Zhang J R, Liu X B, et al. 2023 An optimization method for vibration isolation of a multi-substructure two-stage vibration isolation system of power pack. Journal of Vibration and Shock., 42(16):209-217.
[9] Li L, Xuan M, Jia H G, et al. 2017 Design and optimization of vibration isolation system for photoelectric stabilized platform. Computer Simulation., 34(3):77-81.
[10] Xi Z L, Peng Y M, Yang M L. 2020 Multi-objective Optimization of a Certain Type Electric Bus's Mount System . Noise and Vibration Control. , 40(1):104-110, 131.
[11] Wu Y J, Xu C Q, Chen J, et al. 2021 Sensitivity analysis and optimization design of parameters of vibration isolation for power pack. Journal of Central South University(Science and Technology)., 52(11):3872-3884.
[12] Ikago K, Saito K, Inoue N. 2012 Seismic control of single-degree-of-freedom structure using tuned viscous mass damper. Earthquake Engineering & Structural Dynamics., 41(3): 453-474.
[13] Den Hartog J P. 1956 Mechanical Vibrations. New York., McGraw-Hill.
[14] Whittaker A S,Bergeron V V,Thompson C I,et al.1991 Seismic testing of steel plate energy dissipation devices.Earthquake Spectra.,7(4):563-604.
[15] TSAIK K C,CHEN H W,HONG C P,et al.1993 Design of steel triangular plate energy absorbers for seismic resistant construction.Earthquake Spectra.,9(3):505-528.
[16] Tirca L D,Foti D,Diafe Rio M.2003 Response of middle rise steel frames with and without passive dampers to near field ground motions. Engineering Structures.,25(2):169-179.

# Low-cost porous silicon material based on fenton process assisted chemical etching and its application as anode for lithium-ion battery

**Junpeng Li[1,2], Yihao Wen[1,2] and Zi Wang[1,3]**

[1]National Key Laboratory of Advanced Micro and Nano Manufacture Technology, Shanghai Jiao Tong University, Shanghai, 200240, China
[2]Department of Micro/Nano Electronics, School of Electronic Information and Electrical Engineering, Shanghai Jiao Tong University, Shanghai, 200240, China

[3] Corresponding author's e-mail: ziwang@sjtu.edu.cn.

**Abstract.** Nanostructured silicon-based materials are considered as one of the most promising alternatives for anode electrodes of lithium-ion batteries. Nevertheless, the relatively high cost and complexity of most present processes make them difficult to scale up. In this manuscript, we would like to introduce a porous anode prepared by Fenton reaction-assisted chemical etching of low-cost ferrosilicon in the mixed solution of hydrogen fluoride and hydrogen peroxide. The experimental results show that hierarchical structures containing both mesopores and micropores can be formed by this convenient room-temperature etching process and its electrochemical performance as the anode is comparable to that of commercial product. Furthermore, the low cost of ferrosilicon ($1200~1500/ton) compared with the silicon powder of industrial grade (~$3000/ton) traditionally used in Li-ion battery anode preparation also renders this method potential and feasible for large-scale production of Si-based anode materials.

## 1. Introduction

With the increasing demand for energy density of lithium-ion batteries, silicon-based composites have been used as anodes. The reversible intercalation process of lithium ions in the silicon frameworks determines the available capacity of the anode. It is well known that silicon (Si) will change from crystalline to amorphous phase during the process of intercalation, forming $Li_{15}Si_4$ compound with a corresponding specific capacity of 3579 mAh/g, which is one of the best options to partially replace the graphite anode currently used [1]. However, the formation of Si-Li alloy by intercalation of lithium ions in a silicon framework will lead to anisotropic volume expansion of up to 300%. The subsequent extraction of lithium-ion and repeated charge-discharge cycles will lead to the pulverization of the anode and the destruction of the solid electrolyte interface (SEI). At present, a variety of processes have been developed to solve the problem of anisotropic volume expansion, one of which is the porous silicon-based materials [2].

In the past 60 years, dozens of processes for producing porous silicon have been developed. From the perspective of raw materials, these processes can be divided into two categories: bottom-up and top-down [3]. The former uses silicon-containing compounds as raw materials and often uses reduction reactions at high temperatures to prepare porous silicon. These processes are easy to scale, but often involve high-energy consumption processes, and their competitiveness will be limited in the context of sustainable development [4]. In contrast, many top-down processes can be carried out under mild conditions. This kind of process starts with the Si wafer or particles and mainly forms a porous structure through etching, including electrochemical etching, stain etching, metal-assisted chemical etching (MACE), etc [5]. However, these etching methods were originally designed for the field of microelectronics and are more suitable for silicon wafers. For example, anisotropic porous arrays formed by electrochemical etching are used for the preparation of photonic crystal devices [6]. On the other hand, MACE is a highly controllable and adaptable top-down chemical process, which can be applied to almost all forms of silicon materials (including wafers, powders and particles). However, the use of silver nitrate and other precious metal salts involved in MACE will inevitably increase the cost of large-scale production [7]. Considering the scale of the lithium-ion battery industry is so huge and still growing rapidly, it seems necessary to develop an electroless etching process using low-cost materials to produce porous silicon powder products.

Chemical corrosion of silicon in an acidic fluoride solution is one of the most important processes for the formation of porous silicon. This process is mainly based on the hole injection by oxidant into the silicon surface in solution and the subsequent reaction with fluorine ion and electron losing Si to form soluble compounds [8]. The oxidant not only needs to meet the thermodynamic requirement that the redox potential is higher than that of Si, but also has a great influence on the kinetics of this reaction. Compared with the most commonly used nitric acid and hydrogen peroxide, the hydroxyl radical has a higher redox potential (2.8 V), which is expected to be used in the chemical etching process of silicon. The Fenton reaction between iron-based materials and hydrogen peroxide in an acidic aqueous solution is one of the most commonly used methods to produce hydroxyl radicals [9]. The redox potential of hydroxyl radical is second only to fluorine, which can rapidly degrade a variety of organic pollutants, and is widely used in environmental engineering [10]. However, to our best knowledge, there have been no reports on the use of the Fenton reaction to generate hydroxyl radicals for the chemical etching of silicon.

Compared to commercial silicon powder which is commonly used in LIB anode fabrication, ferrosilicon, an alloy of silicon and iron, is more affordable and readily available. In this manuscript, we use ferrosilicon as the raw material, and the mixed solution of hydrofluoric acid and hydrogen peroxide as the etchant. The hydroxyl radicals generated by the Fenton reaction between iron and hydrogen peroxide are expected to promote the hole injection process, resulting in the preparation of hierarchical porous silicon-based materials with both mesopores and micropores. Additionally, the low cost of ferrosilicon (\$1200~1500/ton) compared with the silicon powder of industrial grade (~\$3000/ton) traditionally used in Li-ion battery anode preparation also renders this method potential and feasible for large-scale production of Si-based anode materials.

## 2. Experimental

5 g of the pristine ferrosilicon (FeSi) powder (Grade FeSi75Al1.5-A, 500 mesh) is first added into 200 mL HCl (2 M), and the mixture is stirred under ambient conditions for 60 min in order to reduce the metal content for further reaction. After washing with deionized water to remove excess acid, the HCl-treated Fe-Si powder (de-FeSi) is added into a mixed solution of HF (4.8 M) and $H_2O_2$ (5.0 M) for preparation of porous structure. The chemical etching is proceeded at 10°C with agitation in a mechanically refrigerated cold bath for 12 hours. Finally, the product (de-FeSi-P) is collected by centrifugation after washing with deionized water and anhydrous ethanol several times, and dried at 60°C under vacuum for 24 hours. Two control samples are prepared by direct etching of commercial Si powder (500 mesh) and pristine FeSi using $HF/H_2O_2$ solution, designated as Si-P and FeSi-P, respectively.

Scanning electron micrographs are collected under a ZEISS Sigma 300, with an accelerating voltage of 5 kV and InLens observation mode. The pore size distributions are evaluated by nitrogen adsorption (77 K) with a Quantachrome Autosorb-IQ3 system. The crystal structures of the prepared materials are characterized by a D8 ADVANCE X-ray Powder diffractometer (XRD). The elemental analysis is performed on an iCAP 7600 ICP-AES.

Electrochemical measurements are conducted at 25°C using CR2025 coin cells assembled in an argon glove box. The working electrode is prepared as a mixture of 70% active substance, 20% CMC, and 10% Super P, which is loaded with a mass of approximately 1.0 mg/cm$^2$. A lithium tablet was utilized as both the counter and reference electrode, and the Celgard 2400 membrane served as the separator. The electrolyte is by solving 1 M $LiPF_6$ in 1:1 ethylene carbonate (EC)/diethyl carbonate (DEC) with 10 wt % FEC additives. A LAND test system is utilized to evaluate the electrochemical performance of the assembled half-cell with a 0.01-2.0 V cut-off voltage. A CHI660E is used to obtain electrochemical impedance spectroscopy (EIS) and cyclic voltammograms (CVs) data.

## 3. Results and discussion

Figure 1 shows our synthetic procedure for the preparation of ferrosilicon-based hierarchical porous materials. Ferrosilicon is an iron-silicon alloy made by smelting coke, steel chips and silica as raw materials, with a high iron content and other metals such as aluminium (Al), calcium (Ca) and manganese (Mn). Therefore, before etching, the pretreatment by HCl is required to reduce the metal content to obtain dealloyed ferrosilicon powder. The results of the elemental analysis show that after dealloying by HCl treatment, the iron content is reduced from 23% to 15%. For the other elements, the contents of Al, Ca and Mn are reduced from 1.1%, 0.9% and 0.6% to 0.5%, 0.2% and 0.2%, respectively. The dealloyed Fe-Si powder is then etched in a mixture of HF and $H_2O_2$. In this acidic environment, ferrous ions can be continuously supplied by the residual iron, which later reacts with $H_2O_2$ to form hydroxyl radicals (OH·) [11].

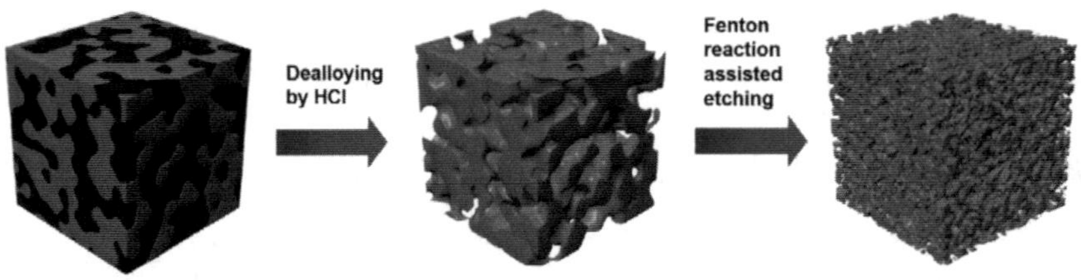


**Figure 1.** Fabrication process of porous framework based on Fenton reaction assisted etching of dealloyed ferrosilicon.

Figure 2 shows the scanning electron microscopy images of each prepared sample. As shown by Figure 2a and Figure 2b, the surface of pristine ferrosilicon particles is fairly smooth and their size is widely ranging from 50 microns to 100 nm. After dealloying with HCl, some columnar and shallow pored are found on the particle surface, which is the result of the partial removal of iron-rich phases (Figure 2c). Upon subsequent etching with HF/ $H_2O_2$, the morphology of ferrosilicon particles has transformed into a sponge-like porous structure (Figure 2d). As shown by the BJH pore size distribution data in Figure 3, in addition to the macropores ranging from 100-300 nm found by SEM, microporous structures with a pore size ranging from 1.5-4 nm can be formed in this etching process. The morphology of the control sample prepared from pure Si particles (Si-P) is shown in Figure 2e for further investigation of the effect of OH· generated by the Fenton reaction in HF/$H_2O_2$ solution. Similar macroporous frameworks are observed for Si-P. However, a very different pore size distribution without any micropores is given by nitrogen adsorption as shown in Figure 3a. The microporous structures found in de-FeSi-P should therefore be attributed to the presence of iron components, namely the hole injection by OH· generated by the Fenton reaction. Furthermore, another control sample FeSi-P is prepared by direct etching of pristine ferrosilicon powder in HF/$H_2O_2$ solution without HCl treatment, in order to check the necessity of the dealloying process. The irregular morphology of FeSi-P (Figure 2f) shows that an excessive metal content will lead to the failure of porous structure formation. This indicates the fine-tuning of metal content during the dealloying process is necessary for obtaining optimized porous structures in the subsequent etching step, which will be studied in detail in our future research.

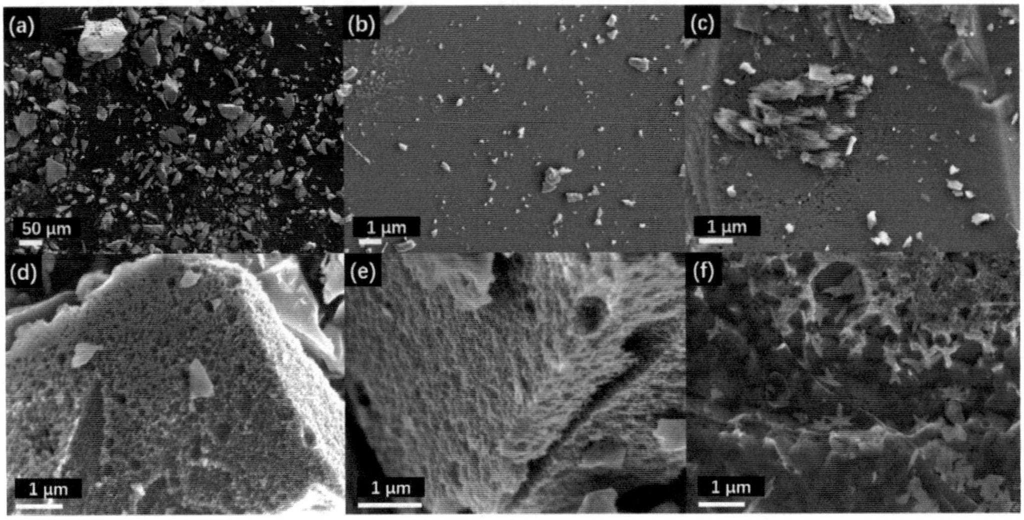


**Figure 2.** SEM images of samples: (a) pristine Fe-Si powder, (b) local magnification of (a), (c) de-FeSi, (d) de-FeSi-P, (e) Si-P, (f) FeSi-P.

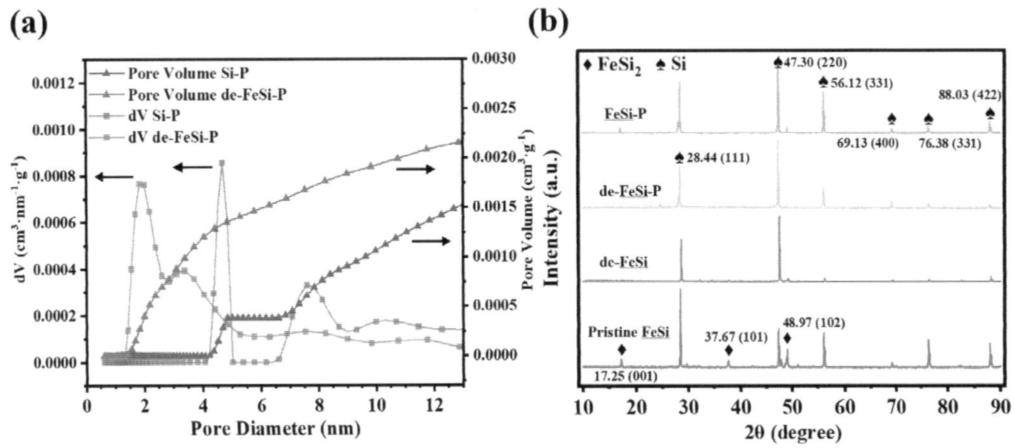


**Figure 3.** (a) Pore size distribution curves of de-FeSi-P and Si-P, (b) XRD patterns of pristine FeSi powder, de-FeSi, de-FeSi-P and FeSi-P.

Figure 3b shows the X-ray diffraction data of pristine FeSi, de-FeSi, de-FeSi-P and FeSi-P. The peaks located at 2θ values of 17.25, 37.67 and 48.97° are ascribed to the (001), (101) and (102) planes of crystalline $FeSi_2$, respectively [12]. These peaks are weakened but still observable in de-FeSi and de-FeSi-P owning to the dealloying process. On the other hand, the peaks located at 2θ values of 28.44, 47.30, 56.12, 69.13, 76.38 and 88.03° in all the samples are assigned to the (111), (220), (331), (400), (331) and (422) planes of silicon crystals, respectively [13]. According to these XRD data, the main crystal structures are not affected by the etching process, which is highly favoured by scaled production.

In order to investigate the applicability of de-FeSi-P for lithium-ion battery anode, a CR2025 half-cell is assembled and the first four cyclic voltammetry curves at a scan rate of 0.1 mV/s from 0.01V to 1.0 V (versus Li/Li+) are shown in Figure 4a. Two anodic peaks are clearly observed at 0.35 V and 0.53 V for all four cycles, respectively, which can be assigned to the transition of $Li_xSi$ to Li during the Li extraction process [14]. The increasing intensity of these two peaks indicates the gradual activation process during the cycles and the existence of crystalline silicon phase in de-FeSi-P, which is consistence with XRD data in Figure 3b. The cathodic peak observed at 0.15 V during the charging process can be assigned to the formation of the amorphous $Li_xSi$ phase during the Li insertion process [15]. Figure 4b shows the EIS curves of Si-P and de-FeSi-P for the comparison of the transportation properties of electrodes. The $R_{ct}$ of de-FeSi-P in the equivalent circuit model, corresponding to the charge-transfer resistance, is calculated as 82 Ω, much lower than that of Si-P (311 Ω). Inferred from the above experimental results, the residual $FeSi_2$ phase in de-FeSi-P can improve the electric performance without interfering with the insertion and extraction of Li ions, which is, therefore, a high potential candidate for LIB anode.

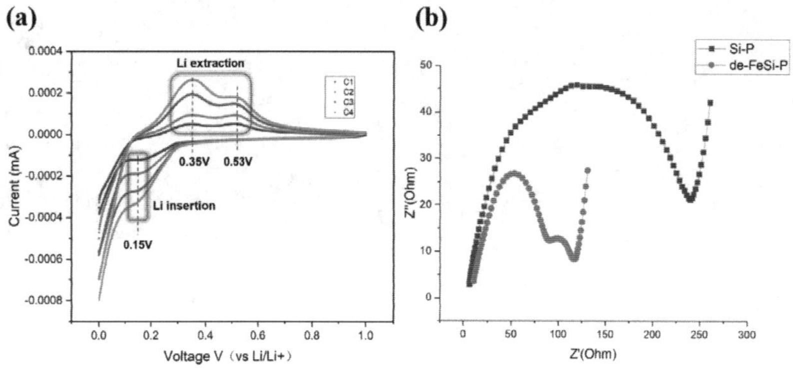


**Figure 4.** (a) CV curves for de-FeSi-P at 0.1 mV/s (b) EIS curves for Si-P and de-FeSi-P electrodes.

## 4. Conclusion

In conclusion, a low-cost porous material is successfully prepared by Fenton reaction-assisted chemical etching of partially dealloyed ferrosilicon powder in $HF/H_2O_2$. The electrochemical results show that this ferrosilicon-based hierarchical structure is capable of smooth lithium-ion insertion and extraction as a demonstration of the concept and feasibility for LIB application. Compared with the porous anode based on traditionally used commercial Si powder, this material has fine porous frameworks with pore size mainly ranging from 2 nm to 4 nm and a much lower charge-transfer resistance owing to the high conductivity of the partially retained $FeSi_2$ phase. Additionally, the low cost of ferrosilicon ($1200~1500/ton) compared with commercial silicon powder (~$3000/ton) traditionally used for anode preparation also renders this method attractive for large-scale production of Si-based anode materials.

## References

[1] Kim N, Kim Y, Sung J and Cho, J 2023 *Nat. Energy* 8 921.
[2] Song A, Zhang W, Guo H, Dong L, Jin T, Shen C and Xie K 2023 *Adv. Energy Mater.* 13 2301464.
[3] Chen Z, Jiang H, Zhang X, Cheng F, Wu M and Zhang H 2023 *Adv. Func. Mater.* 33 2301109.
[4] Chen S, Chen Z, Xu X, Cao C, Xia M and Luo Y 2018 *Small* 14 1704410.
[5] Zhang F, Wan L, Chen J, Li X and Yan X 2018 *Electrochim. Acta* 280 86.
[6] Maza D C, Garcia D S, Deriziotis I, Rodriguez A and Llorca J 2019 *J. Electrochem. Soc.* 166 B1010.
[7] Sreejith K P, Sharma A K, Basu P K and Kottantharayil A 2022 *Sol. Energ. Mat. Sol. C.* 238 111531.
[8] Gondek C, Lippold M, Röver I, Bohmhammel K and Kroke E 2014 *J. Phys. Chem. C*, 118 2044.
[9] He J, Yang X, Men B and D. Wang, J 2016 *J. Environ. Sci.* 39 97.
[10] Yu T, Chen H, Hu T, Feng J, Xing W, Tang L and Tang W 2024 *Appl. Catal. B* 342 123401.
[11] He J, Yang X, Men B and D. Wang, J 2016 *J. Environ. Sci.* 39 97.
[12] Li Y, Xiao P, Xiong J, Luo J, Zhou P, Xie X, Li Y and Chen M 2023 *J. Alloys Compd.* 945 169331.
[13] Chen M, Li B, Liu X, Zhou L, Yao L, Zai J, Qian X and Yu X 2018 *J. Mater. Chem. A* 6 3022
[14] Li P, Hwang J Y and Sun Y K 2019 *ACS Nano* 13 2624.
[15] Gan C, Ye X, Zhang S, Chen J, Wen W, Liu Y, Peng D, Tang L and Luo X 2021 *J. Energy Storage* 41 102930.

# Design of a digital twin system for micro-lens array embossing manufacturing

**Zhengjie Yang[1,3], Liping Ma[2,3,4], Jianyong Zhang[3], Wendi Dong[3], Qingmeng Zhang[3] and Rumeng Chen[3]**

[1]School of Mechanical and Materials Engineering, North China University of Technology, Beijing 100144, China
[2]Institute of Automation, CAS, Beijing 100190, China
[3]Binzhou Institute of Technology, Weiqiao-UCAS Science and Technology Park, Binzhou, Shandong Province 256606, China

[4]Corresponding author's e-mail: liping.ma@ia.ac.cn

**Abstract.** Micro-lens arrays, due to their unique structure and outstanding optical performance, have been widely applied in advanced domains such as display technology, optical imaging, and laser processing. Among the manufacturing techniques, embossing molding is considered an efficient and precise method for fabricating micro-lens arrays. However, given the intricate structural characteristics of these arrays, precise control over the process becomes paramount. This study introduces a digital twin system for micro-lens array embossing manufacturing, grounded on a process knowledge base. The system relies on the rich process knowledge within the backend database combined with real-time sensor data, providing a powerful drive for the digital twin, thereby ensuring the precision and stability of the embossing manufacturing process. Additionally, the digital twin technology not only offers real-time monitoring and feedback for the production process of the micro-lens arrays but also aids researchers in optimizing process parameters, further enhancing product quality and consistency.

## 1. Introduction

Micro-lens arrays, due to their unique structure at the microscopic level and superior optical performance, have become key components in the modern optoelectronic domain, being widely applied in advanced technological fields like display technology [1], optical imaging [2], and laser processing [3]. The fabrication of such microscopic components, given their high precision requirements, makes the selection and optimization of the production process a pivotal factor limiting their large-scale application.

Embossing molding, as an efficient method to produce micro-lens arrays, has been extensively researched and employed [4-6]. This method, by accurately controlling key parameters such as material, temperature, pressure, and time, can yield micro-lens arrays with very high precision. However, ensuring the stability and reliability of this process in large-scale production is crucial. Traditional manufacturing monitoring often relies on post-production quality checks, which not only increases production costs but also makes it difficult to intervene timely in case of anomalies during the production process.

In recent years, digital twin technology, as a novel approach capable of digitally simulating physical entities, has demonstrated its vast potential across various application domains [7]. Merging physical modeling, sensor technology, data analysis, and simulation techniques, digital twin provides a revolutionary tool for modern manufacturing in terms of production monitoring and optimization. For the embossing manufacture of micro-lens arrays, utilizing digital twin technology enables real-time monitoring of the production process, facilitating the prompt identification and rectification of deviations, thus ensuring the high quality and consistency of products.

This paper aims to explore the integration of digital twin technology with the embossing manufacturing process of micro-lens arrays, proposing a novel and efficient method for monitoring and optimizing the production of micro-lens arrays.

## 2. System introduction

The digital twin system framework for the micro-lens array embossing manufacturing center is illustrated in Figure 1, consisting of the physical system module of the embossing machine, the virtual twin system module, the backend knowledge database module, and the communication module.

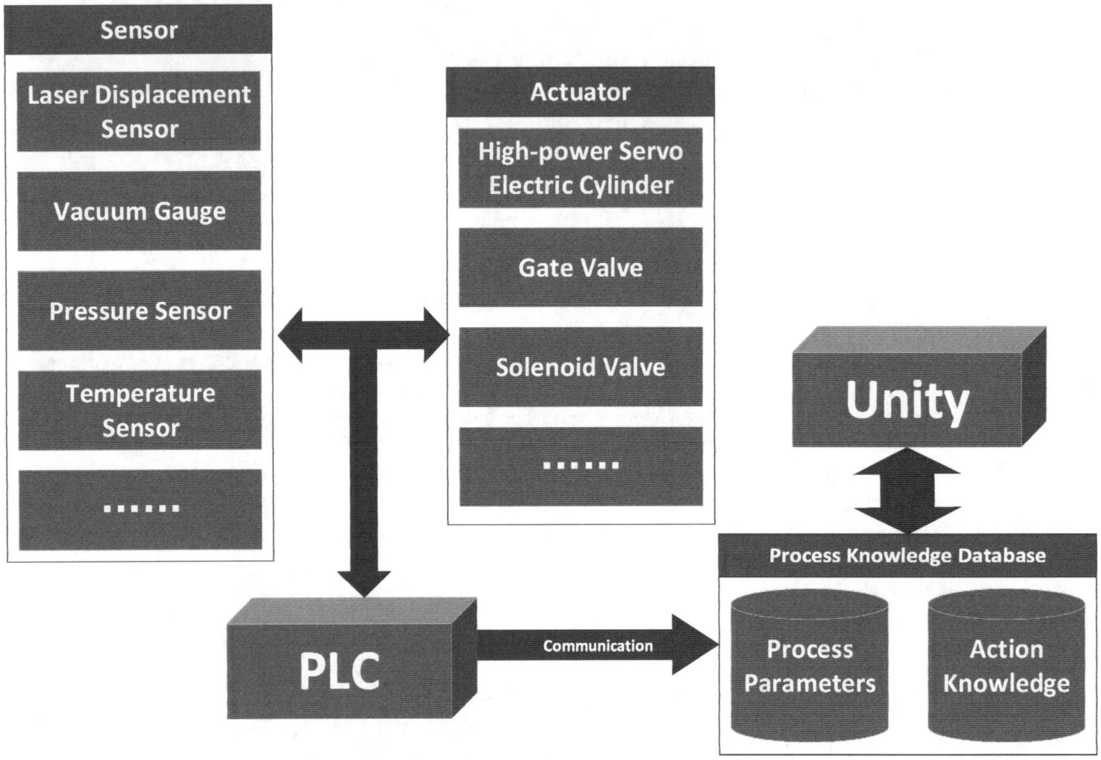


**Figure 1.** Digital twin system framework.

Firstly, the physical entity module of the embossing machine comprises the motion mechanism of the embossing machine, the heating module, and various sensors. Secondly, the virtual twin system module includes the twin model of the embossing machine and a monitoring interface. The backend process knowledge database for the embossing machine is built based on SQLServer [8], storing assembly process knowledge. Finally, the communication module employs the Modbus TCP protocol to achieve bidirectional communication between the virtual and physical modules of the embossing machine.

### *2.1. Physical system module of the embossing machine*

Figure 2 illustrates the physical system structure of the embossing machine. To meet the full-process requirements of the micro-lens embossing technique, this system mainly consists of actuating components and sensors.

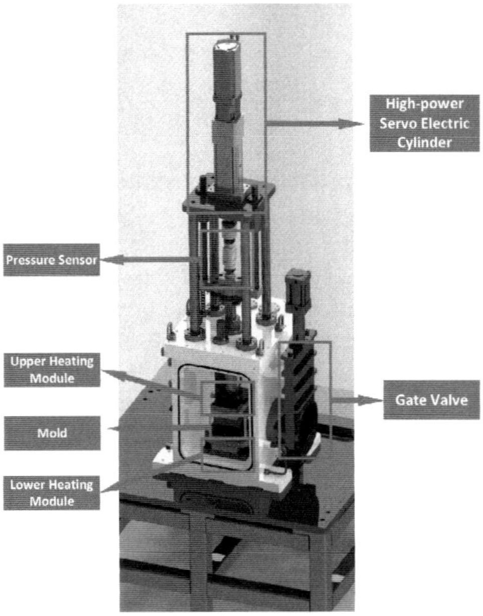


**Figure 2.** Physical system structure of the embossing machine.

The molding actuator achieves the molding of preforms through a high-power servo electric cylinder. Both the end of the servo cylinder and the mold placement platform are equipped with heating modules, ensuring that the preform is heated uniformly from both directions, enhancing heat conduction efficiency and allowing the material to quickly reach the desired temperature and state. To ensure the accuracy of pressure during the molding process, the servo cylinder is fitted with a precise pressure sensor for real-time monitoring and feedback. Furthermore, a vacuum environment is necessary during the molding process to prevent bubbles that affect the quality of micro-lenses, hence a complete vacuum system is employed, ensuring the sealing of the molding environment and precise control of the vacuum level.

### *2.2. Virtual twin system module*

In the construction of the digital twin, accurately simulating the motion relationships of the virtual model is of paramount importance. To achieve this, it is first necessary to deeply adjust and optimize the hierarchical structure of the 3D CAD model.

The output shaft of the servo actuator is a key dynamic component in the system. Together with the serial components, they form a whole with relative motion, and their motion relationships should be strictly consistent. To this end, the output shaft of the servo actuator and its related components are defined as an independent sub-assembly, ensuring that they are presented as a unified motion module in the overall model.

The gate valve is a critical device for ensuring the sealing integrity of the embossing system, and its actions directly affect the system's sealing efficiency. By simulating the gate valve, its opening and closing actions can be translated into the upward and downward movement of the gate, achieving a realistic simulation effect in a virtual environment.

To achieve high-quality visualization effects in the digital twin system, the Unity engine was chosen as the rendering platform [9]. For the original CAD model, to better integrate with the Unity engine, the 3DSMAX software was used to convert it into an fbx format. This format not only retains the hierarchical structure and animation information of the model but also ensures efficient rendering effects in Unity.

Through the above steps, the digital twin system of the embossing machine was successfully constructed, ensuring its performance in the virtual environment remains highly consistent with the actual physical system.

### *2.3. Knowledge base module*

In the digital twin system, the backend knowledge base is a crucial component, primarily encompassing action knowledge and process knowledge. These insights not only offer an in-depth understanding of the equipment for the entire system but also lay the foundation for precise control and real-time monitoring.

Action knowledge primarily covers the actions and operations the equipment carries out according to the process. It serves as a digital description of the movements of the equipment's executing mechanisms, offering a reference for judging any abnormal actions. On the other hand, process knowledge is central to the equipment manufacturing process. Integrating various process parameters of the device, it offers a reference for controlling the quantities in the process.

The core idea behind building the process knowledge base is to comprehensively describe the equipment's state at every moment. To this end, the knowledge base was constructed following the principle of "process step — execution time — action — temperature — vacuum level — pressure — status." This detailed categorization ensures every detail of the production process is captured and recorded:

- The process step identifies every step or stage of the manufacturing process.
- Execution time records the start and end times for each process step.
- Action details the specific operations the equipment performs in a particular process step.
- Temperature, vacuum level, and pressure are the key process parameters. All parameters in the knowledge base make up the entire process, influencing the quality and performance of the product.
- Status provides the specific conditions of the equipment at a given moment, such as whether it's running normally, needs maintenance, etc.

After integrating this detailed data, it is organized in tabular form and stored in the SQL Server database. This not only facilitates real-time monitoring and control but also ensures data security and integrity. This structured knowledge base backs the digital twin system robustly, guaranteeing the accuracy and stability of equipment operations.

### *2.4. Communication module*

As illustrated in Figure 3, the communication architecture of the digital twin system for the molding machine is layered and organized, which can be divided into two main parts.

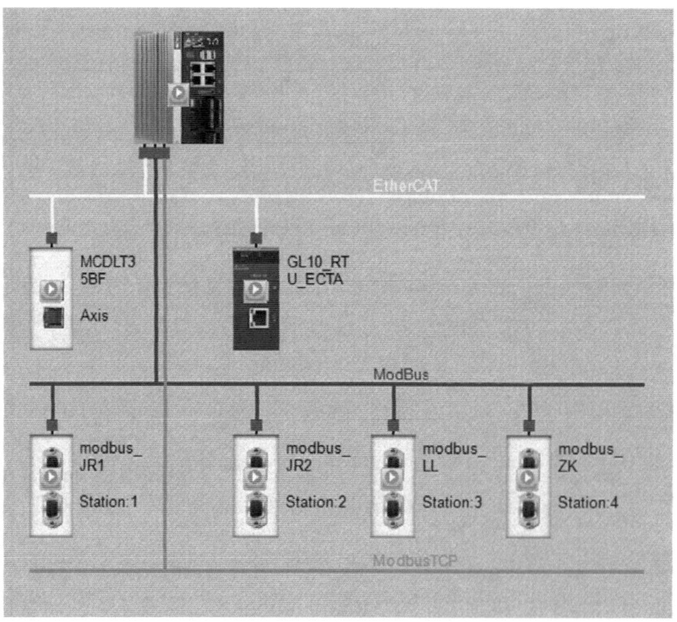


**Figure 3.** Communication architecture of the digital twin system for the molding machine.

The first part involves the communication between the actuator and sensors using the high-speed EtherCAT bus with the PLC, achieving real-time control commands and position feedback, providing low latency and high reliability in communication. Key sensors, such as pressure sensors, laser displacement sensors, and vacuum gauges, transmit various physical parameters to the PLC via the 485 communication protocol. The PLC then analyzes and processes this information, forming a unified data stream for the supervisory computer.

The second part pertains to the communication of integrated information in the PLC to the digital twin system supervisory computer using the Modbus TCP communication protocol [10]. To ensure stable communication, detailed parameter configurations were performed on the PLC. The supervisory computer has a specialized communication interface program that can read data from the PLC and perform type conversion and analysis to adapt to the Unity environment. Simultaneously, a timer ensures periodic data reading from the PLC and updates the device model in the supervisory computer, guaranteeing the real-time accuracy and precision of the digital twin system.

In summary, this communication approach offers an efficient, stable, and precise data transmission and processing solution for the digital twin system of the molding machine, ensuring the authenticity and real-time nature of the digital twin.

## 3. Knowledge-based driven digital twin system

To ensure that the data in the digital twin system precisely matches the data from the actual production site and, at the same time, integrates with the backend process knowledge base, a digital twin detection strategy was adopted.

Firstly, frame-by-frame detection is conducted within the twin system. The system continuously inspects every frame of data, comparing in real-time whether the motion of the physical entity matches the simulation of the digital twin. The objective of this phase is to accurately capture any potential minor discrepancies or deviations.

Next, parameters that deviate from the process knowledge base data undergo error correction. Once the system identifies any inconsistencies during the frame-by-frame detection, it immediately initiates an error correction mechanism. This mechanism adjusts the digital twin model based on data from the actual production site, ensuring synchronization and consistency between the two.

Lastly, during the production process control, ensuring the proper sequential operation of the execution mechanisms is crucial for the success of the entire process flow. To achieve this precise synchronized simulation and control, the concept of a state machine is employed. The state machine plays a pivotal role throughout the process flow, capable of transitioning states based on real-time sensor data, continuously tracking, and simulating every process step. Moreover, the process knowledge base, being the system's core, not only retains the current process sequence but also provides the necessary state parameters for the digital twin environment. This means that when the state of the physical production line changes, the digital twin system can simulate and compare in real time, ensuring production efficiency and quality stability.

## 4. Experiments and conclusions

Based on the existing mechanical structure of the molding machine and in conjunction with the methods proposed and described in this paper, a digital twin system of the microlens array molding machine was successfully constructed, as shown in Figure 4.

This digital twin system is not just a simple virtual model, but a highly integrated system that incorporates an advanced process knowledge database. Through this system, the manufacturing process of the microlens array can be monitored in real-time and effectively. Such real-time monitoring not only greatly improves production efficiency but also significantly reduces the error rate, providing a strong guarantee for ensuring product quality and performance.

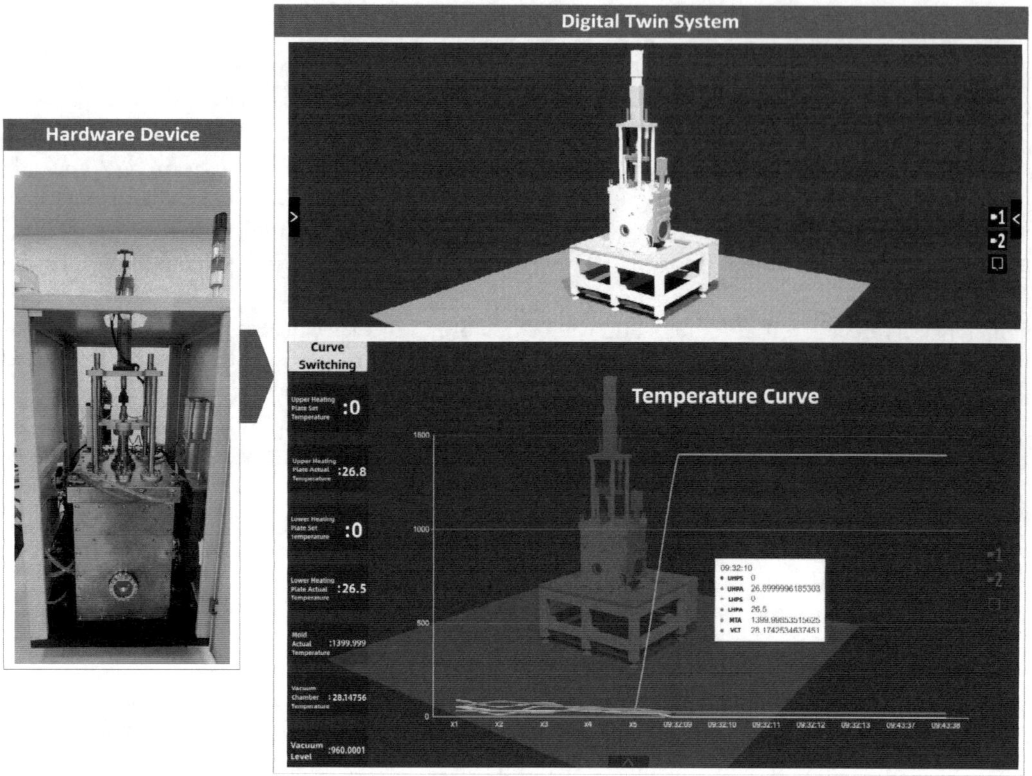


**Figure 4.** Digital twin system of the molding machine.

**Acknowledgments**
This work is supported by the National Defense Basic Research Program's key project with project funding (JCKY2020208B027).

## References

[1] Yan, S., Liu, Y., Wang, R., Zhang, H. and Liu, Z. (2018), P-9.10: Design and Fabrication of Micro-lens Array Based on Micro-LED Projector. SID Symposium Digest of Technical Papers, 49: 688-691. https://doi.org/10.1002/sdtp.12816.

[2] Ki-Chul Kwon, Young-Tae Lim, Chang-Won Shin, Munkh-Uchral Erdenebat, Jae-Moon Hwang, and Nam Kim, "Enhanced depth-of-field of an integral imaging microscope using a bifocal holographic optical element-micro lens array," Opt. Lett. 42, 3209-3212 (2017).

[3] Jinliang Han, Jun Zhang, Xiaonan Shan, Yawei Zhang, Hangyu Peng, Li Qin, Yannan Tan, Lijun Wang, Conversion of a Gaussian-distributed circular beam to a flat-top-distributed square beam in laser shock processing based on a micro-lens array structure, Optik, Volume 274, 2023,170525, ISSN 0030-4026, https://doi.org/10.1016/j.ijleo.2023.170525.

[4] Kangsen Li, Xinfang Huang, Qiang Chen, Gang Xu, Zhiwen Xie, Yuanyuan Wan, Feng Gong, Flexible fabrication of optical glass micro-lens array by using contactless hot embossing process, Journal of Manufacturing Processes, Volume 57, 2020, Pages 469-476, ISSN 1526-6125, https://doi.org/10.1016/j.jmapro.2020.07.019.

[5] Gao, P.; Liang, Z.; Wang, X.; Zhou, T.; Xie, J.; Li, S.; Shen, W. Fabrication of a Micro-Lens Array Mold by Micro Ball End-Milling and Its Hot Embossing. Micromachines 2018, 9, 96. https://doi.org/10.3390/mi9030096.

[6] Weng, Y.-J. Development of Belt-Type Microstructure Array Flexible Mold and Asymmetric Hot Roller Embossing Process Technology. Coatings 2019, 9, 274. https://doi.org/10.3390/coatings9040274.

[7] Botín-Sanabria, D.M.; Mihaita, A.-S.; Peimbert-García, R.E.; Ramírez-Moreno, M.A.; Ramírez-Mendoza, R.A.; Lozoya-Santos, J.d.J. Digital Twin Technology Challenges and Applications: A Comprehensive Review. Remote Sens. 2022, 14, 1335. https://doi.org/10.3390/rs14061335.

[8] Rafael Almeida, Pedro Furtado, and Jorge Bernardino. 2015. Performance Evaluation MySQL InnoDB and Microsoft SQL Server 2012 for Decision Support Environments. In Proceedings of the Eighth International C* Conference on Computer Science & Software Engineering (C3S2E '15). Association for Computing Machinery, New York, NY, USA, 56–62. https://doi.org/10.1145/2790798.2790808.

[9] Wang, Ziran (2021). Digital Twin Simulation of Connected and Automated Vehicles with the Unity Game Engine. TechRxiv. Preprint. https://doi.org/10.36227/techrxiv.14582943.v1.

[10] F. Katulić, D. Sumina, I. Erceg and S. Groš, "Enhancing Modbus/TCP-Based Industrial Automation and Control Systems Cybersecurity Using a Misuse-Based Intrusion Detection System," 2022 International Symposium on Power Electronics, Electrical Drives, Automation and Motion (SPEEDAM), Sorrento, Italy, 2022, pp. 964-969, doi: 10.1109/SPEEDAM53979.2022.9842239.

# Investigation on the improvement of microwave dielectric properties of plasma-prepared fluoropolymer-coated $CaTiO_3$-filled PTFE composites

**Xiaolei Zheng[1], Qun Wang[1,2], Xiaoqiang Li[1], Chenjie Shen[1], and Changkuan Zheng[1]**

[1]Faculty of Materials and Manufacturing, Bejing University of Technology, Beijing, 10024, PR China

[2] Corresponding author's e-mail: wangq@bjut.edu.cn

**Abstract.** High-performance polymer matrix composites preparation has been urgently hampered by poor interfacial adhesion between two phases. In this article, a thorough analysis of the microstructure and overall properties of composites made of polytetrafluoroethylene filled with fluorocarbon polymer film-coated $CaTiO_3$ was conducted. The similar fluorocarbon molecular structure of $CaTiO_3$ surface and PTFE provides a bridge between the two, significantly improving and promoting the microstructure and interface adhesion of the composite. The comprehensive properties of modified composites are obviously superior to those of untreated composites. Composites filled with 30 wt% $CaTiO_3$ have an excellent dielectric constant at 5.53, acceptable dielectric loss at 0.0047, an outstanding heat resistance index of 281.1°C, and relatively low hygroscopicity of 0.10%. At the same time, the mechanical properties are maintained at a high application level, which makes them promising for microwave devices.

## 1. Introduction

Nowadays, in the high-speed development of the information age, electronic devices dominated by 5G communication, satellite navigation, automotive radar, etc., show the development trend of high-frequency and high-speed information processing, high integration, and high function; the demand for microwave dielectric materials is increasingly urgent [1]. Polymer/ceramic microwave dielectric composites show promise because they combine the benefits of ceramic and polymer materials, have great dielectric tunability, are easy to process, and are more suitable for industrial production [2]. Polytetrafluoroethylene (PTFE) has a significant advantage over other polymers in microwave dielectric materials due to its extremely low dielectric loss, stable dielectric constant, strong chemical stability, excellent thermal stability, and mechanical properties [3,4]. However, a major challenge in ceramic-filled composites is the incompatible interface between the organic polymer and the inorganic ceramic filler, where the considerable disparity in surface characteristics between the two causes not only uneven filler dispersion but also the formation of pores at the interface of the distinct phases, deteriorating the properties of the composites.

The current effective method is to functionalize the surface of the ceramic filler, and many scholars have tried numerous surface functionalization reagents for PTFE-based composites. It has been proved that surface treatment with $C_{14}H_{19}F_{13}O_3Si$ (F8261) can significantly improve the surface properties of ceramics, making them exhibit strong hydrophobicity. The modified ceramic surface, because the molecular structure (carbon-fluorine bonding) is similar to that of PTFE, can be a better bridge to establish with the PTFE and promote the formation of a compact structure [5-7]. However, these treatment methods are mostly chemical methods, which have the defects of complicated operation, harsh reaction conditions, low efficiency, and producing a large amount of waste organic solvents. On the contrary, low-temperature plasma surface treatment technology has a short action time, simple and controllable operation, and high efficiency. It only occurs in the surface layer, has almost no effect on the performance of the body, and belongs to the dry treatment, with no pollution. Thus, it shows the advantages and potentials that traditional methods do not have. However, plasma technology is currently largely applied in other fields, such as polymer surface treatment, and the surface modification of micro-nano ceramic particles is less studied [8-10]. To this end, in earlier work, we fabricated a low surface energy fluorocarbon polymer-coated core-shell $CaTiO_3$ using $CH_4/CF_4$ plasma. It is expected to be used for PTFE composites to strengthen the interface bonding between the two phases and improve the comprehensive properties of the composites.

This study aims to improve the microstructure and balance the overall performance of composites by resolving the issue of poor interfacial compatibility of polymer-based microwave dielectric materials in an efficient and environmentally friendly method. Fluorocarbon polymer-coated core-shell $CaTiO_3$ was used as filler to fabricate the PTFE-based microwave dielectric composites through cold pressing and sintering. The microstructure and dielectric characteristics of the composite were examined in relation to filler content and surface modification. Moreover, the mechanical, hygroscopic, and thermal characteristics of the composites were studied from an application standpoint. The composite has the benefits of being easy to prepare and ecologically friendly, exhibiting excellent properties, and having a high application value in the field of microwave media.

## 2. Experimental

### *2.1. Sample preparation*

PTFE/$CaTiO_3$ composites were prepared by cold press molding and hot press sintering. $CaTiO_3$ powder (Shanghai Dianyang Industrial Co., Ltd, China) was pre-treated by a homemade microwave plasma device using $CH_4/CF_4$ gas to deposit a fluorocarbon polymer film on its surface to enhance its interaction with PTFE. Firstly, different contents (5 wt.%-50 wt.%) of $CaTiO_3$ were mixed with PTFE emulsion (60 wt.%, DISP30, Shanghai JinFu Chemical Technology Co., Ltd, China) while adding the appropriate amount of emulsion breaker in a high-speed disperser (JFS-750, Hangzhou Qi Wei Instrument Co., Ltd, China) for 1 h. The resulting mixture was then dried for 12 hours at 120°C in a vacuum oven and further cold-pressed into the required size and thickness. Finally, hot press sintering was carried out at 370°C and 10 MPa for 2 h using hot press equipment (SY-6210B, Dongguan ShiYan Precision Instrument Co., Ltd, China) to obtain a series of samples required for the test.

### *2.2. Characterization*

The cross-section morphology of the $CaTiO_3$/PTFE composite was observed under a scanning electron microscope (SEM, FEI Quanta 600FEG). The vector network analyzer (VAN, Agilent Vector X) was used to test the composites' microwave dielectric characteristics according to the rectangular waveguide method. TA Instruments Q50 was employed for thermogravimetric analysis (TG). TA Instruments Q100 was used for differential scanning calorimetry (DSC). To erase thermal history, the temperature was first raised quickly to 370°C and isothermalised for 3 minutes. It is followed by cooling to 40°C with a rate of 10°C/min to obtain a crystallization curve. Finally, the melting curve was obtained by heating to 370°C with rate of 10°C/min. The density and porosity of the composites were qualified using Archimedes' principle, and the hygroscopicity of the composites was tested according to IPC-TM-650 2.6.2. Bending tests were carried out using a universal materials machine (CMT4304) at room temperature, according to IPC-TM-650 2.4.4B.

## 3. Results and discussion

### *3.1. Microstructure*

SEM was used to examine the micromorphology of $CaTiO_3$/PTFE composites with different modified $CaTiO_3$ content, as shown in Figure 1. As a comparison, SEM of unmodified $CaTiO_3$/PTFE composites at 5 wt % content is shown in Figure 1 (a). Ca elemental scans of $CaTiO_3$/PTFE composites before and after modification at 5 wt % filler content were also performed to better analyze the variations in filler distribution. Results show that the modified $CaTiO_3$ has better connectivity with PTFE and a more uniform distribution of ceramic particles in the matrix. On the contrary, it was found in Figures 1 (a) and 1 (h) that the particles were obviously agglomerated, showing a porous microstructure. This is due to the surface energy difference between PTFE and $CaTiO_3$ which makes the two poor compatibility. After fluoride modification in the $CaTiO_3$ surface coating with a similar fluorocarbon structure with PTFE, a polymer layer can be established between the two "bridges", significantly improving and promoting the composite microstructure and interfacial adhesion. As the $CaTiO_3$ content increases, it can be observed that the ceramic particles start to become dense, and some agglomeration occurs, leading to the creation of voids between the separated particles. At the same time, the low mobility of PTFE also promotes the formation of pores [11]. In short, the modified $CaTiO_3$/PTFE composite has a relatively complete and compact internal structure, which will be conducive to improving the comprehensive properties of the composite.

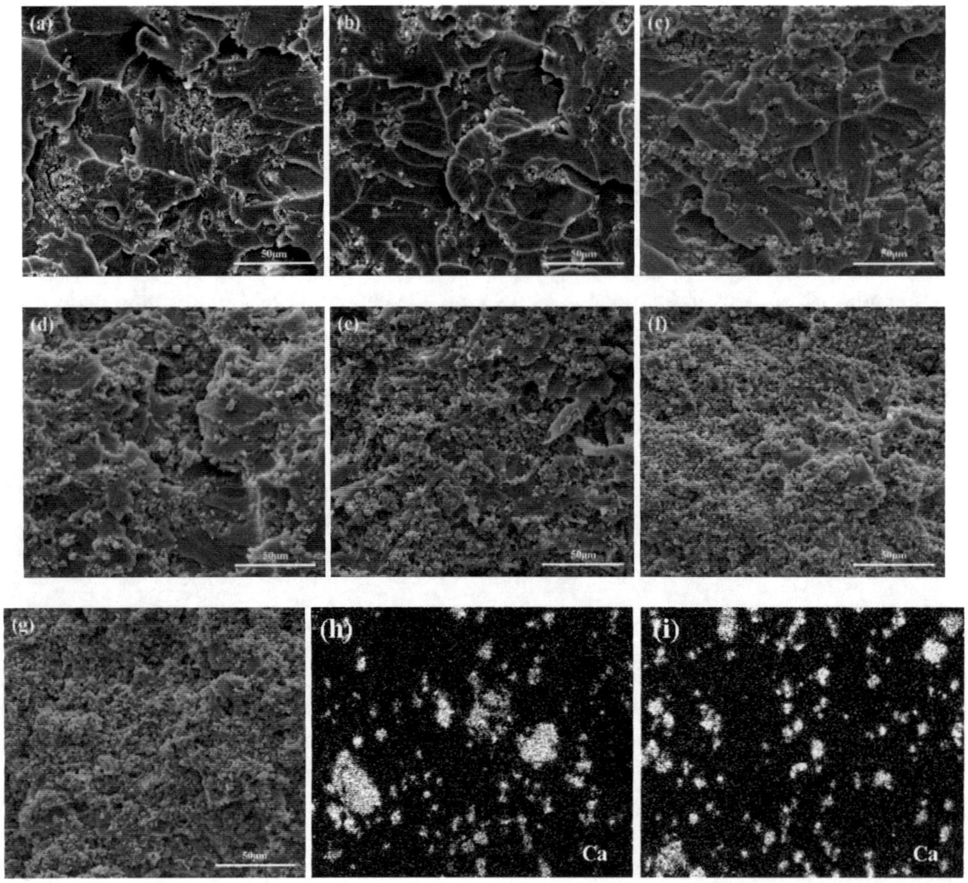


**Figure 1.** SEM images of $CaTiO_3$/PTFE composites in cross section: (a) 5% unmodified and (b) 5%, (c) 10%, (d) 20%, (e) 30%, (f) 40%, and (g) 50% modified $CaTiO_3$ filled composites, (h) and (i) Ca element mapping of (a) and (b).

### *3.2. Dielectric characteristics*

Figure 2 represents the dielectric characteristics of modified and unmodified $CatiO_3$-filled PTFE-based composites (about 10 GHz) as a function of $CaTiO_3$ content. It can be shown that as the $CaTiO_3$ content increases, so does the dielectric parameters of PTFE-based composites. The increase in dielectric constant is primarily due to the fact that $CaTiO_3$ has a higher dielectric constant than PTFE. The dielectric loss depends not just on the dielectric properties of each component of the composites. Furthermore, since the dielectric relaxation at such high frequencies is negligible or nonexistent, the effect of defects in the composites on the dielectric loss is very significant. It is noteworthy that the dielectric constants of modified $CaTiO_3$/PTFE composites are significantly higher than those of unmodified samples. In contrast, the dielectric loss exhibits the opposite trend, especially at high filler contents where the difference between the two is more pronounced. The optimization of the dielectric properties of the composite is mainly attributed to the improved interface connection between $CaTiO_3$ and PTFE after plasma fluorination and the compact microstructure of the composite. When the $CaTiO_3$ content is 30%, $CaTiO_3$/PTFE composites reach a dielectric constant of 5.53, with a dielectric loss of less than $6\times10^{-3}$, which is kept at 0.0047. The results show that the modified $CaTiO_3$/PTFE composites have excellent dielectric properties, which can be used for high-frequency microwave substrates and electronic packaging materials.

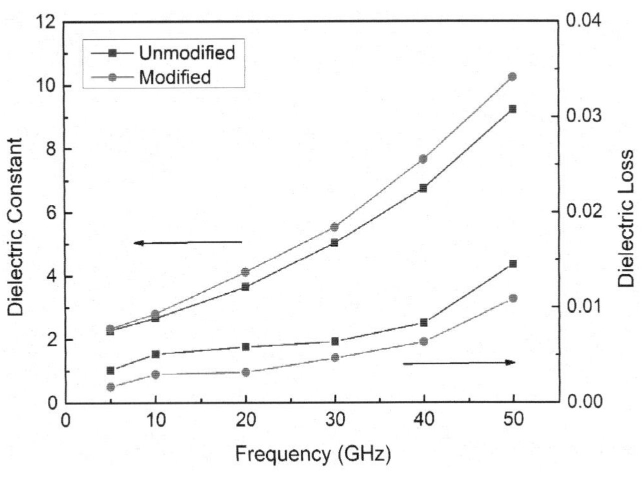


**Figure 2.** Variation of the dielectric characteristics of modified and unmodified $CaTiO_3$/PTFE composites with $CaTiO_3$ content.

### *3.3. Thermal properties*

Resonance-type relaxation polarization determines the dielectric characteristics of composites at high frequencies, such as electron cloud distortions, atomic polarisations, small changes in bond lengths and angles, or small vibrations of the lattice, whereas the crystal morphology has an important influence on the resonance-type relaxation [12,13]. As shown in Figure 3, the crystallization behavior of $CaTiO_3$/PTFE composites with different $CaTiO_3$ content was studied via DSC, and the statistical data on melting and crystallization were obtained. Where the degree of crystallinity ($\chi_c$) of the PTFE composite can be calculated by:

$$\chi_c = \frac{\Delta H_m}{\alpha \times \Delta H_m^0} \times 100\% \qquad (1)$$

Where $\Delta H_m$ Is the experimentally measured enthalpy of melting, $\Delta H_m^0$ Is the enthalpy of melting for 100% crystallization of PTFE (69 J/g) [14], and α is the mass fraction of PTFE. It can be seen that the melting points ($T_m$) of all the composites do not differ much from each other and are similar to the stable crystalline melting point of neat PTFE, implying that the catio$_3$ content has little influence on the melting point of catio$_3$/PTFE composites. The crystallization peak temperature ($T_c$) of the composites shifted obviously towards high temperature with the increase of catio$_3$ content. For instance, the $T_c$ of neat PTFE is 309.7°C. When the catio$_3$ content is 50 wt%, the $T_c$ of the composite increases by 5.3°C and reaches 315°C. This suggests that the addition of catio$_3$ promotes the crystallization of PTFE and plays the role of heterogeneous nucleation. However, the crystallinity showed a different trend. The crystallinity of catio$_3$/PTFE composites remains around 48% at low filler content and decreases obviously when it exceeds 10 wt.%. This is because catio$_3$ plays the role of heterogeneous nucleation at low content. At the same time, the excessive addition of catio$_3$ will reduce the relative content of PTFE components, and it will impede the movement of molecular chains, which will make grain growth difficult [15]. The interaction of these factors makes the crystallinity of catio$_3$/PTFE composites exhibits the above trend. It is noteworthy that the dielectric constant of the catio$_3$/PTFE composite rises as the catio$_3$ content does, indicating that the filler content is more important in improving the composite's dielectric properties than crystallinity in the high-filling system.

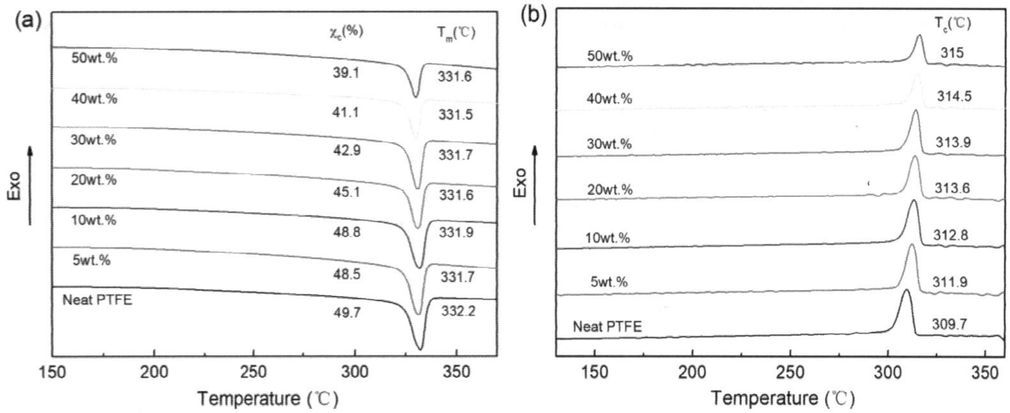


**Figure 3.** DSC (a) melting and (b) crystallisation curves of $CaTiO_3$/PTFE composites.

Thermal stability is a key aspect influencing the application of composites. Figure 4 (a) shows the thermal degradation behavior of $CaTiO_3$/PTFE composites for different content of $CaTiO_3$. It can be seen that the curve trend of all samples is the same, indicating that the addition of $CaTiO_3$ to the PTFE matrix did not affect the decomposition mechanism of PTFE itself. The residual mass of the sample is basically the same as the content of $CaTiO_3$ filler, indicating that $CaTiO_3$ is uniformly dispersed in PTFE. As demonstrated in Figure 4 (b), the thermal stability of the $CaTiO_3$/PTFE composite was assessed using the Heat resistance index (THRI) [16]. The steady increase in the thermal degradation temperature of $CaTiO_3$/PTFE composites with increasing $CaTiO_3$ content shows that $CaTiO_3$ improves composites' thermal stability. This is attributed to the fact that $CaTiO_3$ has a higher specific heat capacity compared to PTFE, which can absorb heat better, causing the PTFE molecular chains to degrade at higher temperatures. When filled with 30 wt.% $CaTiO_3$, the decomposition temperatures of the composites reached 551°C and 588.7 °C at weight loss of 5 wt.% and 30 wt.%, respectively, with a THRI value of 281.1°C, which showed excellent heat resistance.

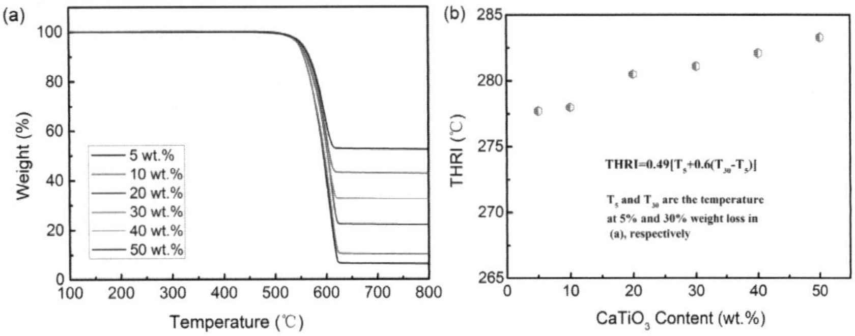


**Figure 4.** (a) TG curves and (b) THRI of $CaTiO_3$/PTFE composites.

*3.4. Moisture absorption properties*

Moisture absorption properties are critical to achieving stable and reliable applications of composites in a variety of environments. As water has a high loss, the dielectric properties deteriorate when water absorption increases. A composite's water absorption is primarily influenced by the water absorption characteristics of the component parts and the structural porosity. Figure 5 demonstrates the change of $CaTiO_3$/PTFE composite density, porosity, and water absorption with ceramic content. The theoretical density of $CaTiO_3$/PTFE composites can be computed using the mixing law :

$$\rho_{theory} = \sum_n \rho_n V_n \tag{2}$$

$$V_n = \frac{\frac{m_n}{\rho_n}}{\sum_n \frac{m_n}{\rho_n}} \tag{3}$$

Where $\rho_n$, $V_n$, and $m_n$ denote the density, volume fraction, and weight of each component in the composite, respectively. The volume percentage of the interior pores determines the porosity of the composite:

$$P = \left(1 - \frac{\rho}{\rho_{theory}}\right) \times 100\% \tag{4}$$

Where $\rho$ and $\rho_{theory}$ stand for the experimental and theoretical densities, respectively.

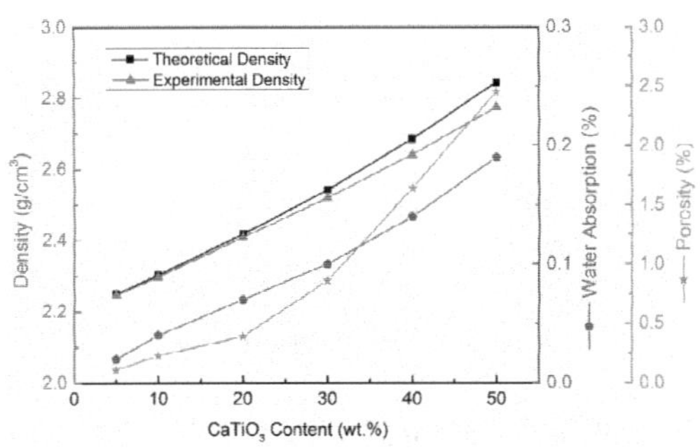


**Figure 5.** Density, water absorption and porosity of $CaTiO_3$/PTFE composites in relation to $CaTiO_3$ content.

 2720 (2024) 012045 doi:10.1088/1742-6596/2720/1/012045 

As seen in Figure 5, the density of the composites increases essentially linearly as the filler amount increases. This is attributed to the fact that $CaTiO_3$ ceramics have a higher density than PTFE. In addition, especially at lower ceramic fillings, the experimental data are in good agreement with the theoretical densities. However, the experimental densities of $CaTiO_3$/PTFE composites at high filler contents show an increased deviation from the theoretical values. The reason for this is that increasing the filler content causes agglomeration of $CaTiO_3$ fillers in the matrix and an increase in pores. The existence of pores not only reduces the density of $CaTiO_3$/PTFE composites but also has a negative impact on the composites' water absorption. In Figure 5, it is evident that the trends of water absorption and porosity of the composites are similar and that both rise with an increase in filler content. In particular, at high filler contents, the ceramic filler agglomeration is intensified, which further increases the porosity. As a result, when the $CaTiO_3$ ceramic filler content is low (5 wt%), the porosity and water absorption are low as well, at 0.11% and 0.02%, respectively. As the $CaTiO_3$ concentration is 30 wt.%, the $CaTiO_3$/PTFE composite has porosity and water absorption of 0.86% and 0.10%, respectively, meeting the requirements for engineering applications. The porosity of the composite rapidly increases as the filler content is raised to 50 wt.%. The water absorption of the composite still remained within an acceptable range, reaching 0.18%, due to the enhanced interfacial compatibility of the modified composites.

### *3.5. Bending strength*

Mechanical properties are also an important indicator of the practical value of composites. Figure 6 shows the typical force-displacement curves of $CaTiO_3$/PTFE composites with $CaTiO_3$ content. The corresponding bending strength and maximum bending force data are listed in Table 1. It is clear that the bending strength of the composite steadily decreases as the $CaTiO_3$ content increases. As the filler content increased from 5 wt.% to 50 wt.%, the bending strength decreased by only 22% from 18.98 MPa to 14.80 MPa. The interface between the two phases has a strong influence on the mechanical properties of the composite. In general, the large surface difference between the ceramic filler and the substrate tends to form a large number of interfacial defects and pores in the matrix. Thus, it weakens the interaction between the filler and matrix, resulting in a decrease in the composite's mechanical characteristics. Surface fluorination modification, which effectively promotes the bridging between filler and PTFE, improves the interfacial bonding state to make the composite material structure relatively dense and lowers the composite's mechanical properties loss.

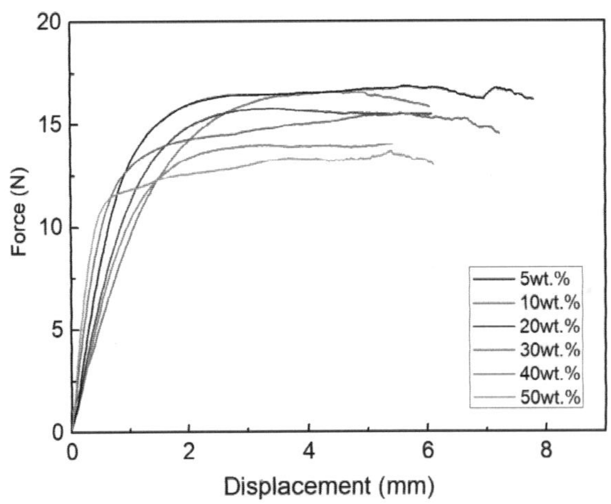


**Figure 6.** Force displacement curves of $CaTiO_3$/PTFE composites with different contents.

**Table 1.** The bending properties of $CaTiO_3$/PTFE composites.

| $CaTiO_3$ content (wt.%) | Bending strength (MPa) | Maximum bending force (N) |
|---|---|---|
| 5 | 18.98 | 16.87 |
| 10 | 17.57 | 16.59 |
| 20 | 16.86 | 15.75 |
| 30 | 16.53 | 15.50 |
| 40 | 15.47 | 14.02 |
| 50 | 14.80 | 13.68 |

## 4. Conclusion

In this paper, microwave dielectric composites were prepared using fluorocarbon polymer-coated $CaTiO_3$-filled PTFE. The study systematically examined how the filler content affected the composites' microstructure, dielectric, thermal stability, hygroscopic, and mechanical properties. The results show that the introduction of PTFE-like fluorocarbon molecular structure on the surface of the filler improves the compatibility between the filler and PTFE, which makes the composites exhibit excellent dielectric properties, good heat resistance, very low hygroscopicity, and excellent mechanical properties. In contrast to the unmodified composites, the 30 wt% filled modified $CaTiO_3$/PTFE composites showed a 10% increase in dielectric constant to 5.53 and a 27% reduction in dielectric loss to 0.0047. In parallel, the composites' heat resistance dramatically increased, with a heat resistance index of 281.1°C, a water absorption rate of only 0.10%, and a bending strength maintained at 15.5 MPa. Hence, the obtained modified $CaTiO_3$/PTFE composites have excellent potential for use as high-frequency microwave substrates and electronic packaging materials.

**Acknowledgments**

The authors appreciate the funding from the National Key Research Programme (Grant No 2016YFB1200602-37).

## References

[1] Yu, Y.Y., Hou, D.J., Zhou, J.J., Shen, J., Zhang, P.C., Chen, W., & Zhou, J. (2023) Improved dispersion and interfacial interaction of $SiO_2$@ polydopamine fillers in polytetrafluoroethylene composites for reduced thermal expansion and suppressed dielectric deterioration. Ceramics International, 49(13): 21492-21501.

[2] Yang, Z., Qing, Z., Li, E., Tang, B., & Yuan, Y. (2023) High microwave dielectric constant and improved thermal stability of functionalized $Zn_{0.15}Nb_{0.3}Ti_{0.55}O_2$-filled polyolefin composites. Journal of Alloys and Compounds, 957: 170417.

[3] Li, Y., Zhou, J., Shen, J., Li, Q., Qi, Y., & Chen, W. (2022) Ultra-low permittivity HSM/PTFE composites for high-frequency microwave circuit application. Journal of Materials Science: Materials in Electronics, 33(13): 10096-10103.

[4] Abiodun, S., Krishnamoorti, R., & Bhowmick, A.K. (2023) Polytetrafluoroethylene Nanocomposites with Engineered Boron Nitride Nanobarbs for Thermally Conductive and Electrically Insulating Microelectronics and Microwave Devices. ACS Applied Nano Materials, 6(5): 3781-3796.

[5] Ren, J., Yang, P., Peng, Z., & Fu, X. (2021) Novel $Al_2Mo_3O_{12}$-PTFE composites for microwave dielectric substrates. Ceramics International, 47(15): 20867-20874.

[6] Wang, H., Zhou, F., Guo, J., Zhang, Y., Yang, H., & Zhang, Q. (2020) Surface-modified $Zn_{0.5}Ti_{0.5}NbO_4$ particles filled polytetrafluoroethylene composite with extremely low dielectric loss and stable temperature dependence. Journal of Advanced Ceramics, 9: 726-738.

[7] Luo, F., Tang, B., Fang, Z., Yuan, Y., Li, H., & Zhang, S. (2019) Effects of coupling agent on dielectric properties of PTFE based and $Li_2Mg_3TiO_6$ filled composites. Ceramics International, 45(16): 20458-20464.

[8] Luque-Agudo, V., Hierro-Oliva, M., Gallardo-Moreno, A.M., & González-Martín, M.L. (2021) Effect of plasma treatment on the surface properties of polylactic acid films. Polymer Testing, 96: 107097.

[9] Primc, G. (2020) Recent advances in surface activation of polytetrafluoroethylene (PTFE) by gaseous plasma treatments. Polymers, 12(10): 2295.

[10] Kehrer, M., Rottensteiner, A., Hartl, W., Duchoslav, J., Thomas, S., & Stifter, D. (2020) Cold atmospheric pressure plasma treatment for adhesion improvement on polypropylene surfaces. Surface and Coatings Technology, 403: 126389.

[11] Tomer, V., Manias, E., & Randall, C.A. (2011) High field properties and energy storage in nanocomposite dielectrics of poly (vinylidene fluoride-hexafluoropropylene). Journal of Applied Physics, 110(4): 044107.

[12] Xiao, D.P., Tian, Q. (2001) The analysis of the polarization mechanism and dielectric properties of dielectric medium. College Physics, 20(9): 44-46.

[13] Hoffman, J.D., Williams, G., & Passaglia, E. (1966) Analysis of the α, β, and γ relaxations in polychlorotrifluoroethylene and polyethylene: Dielectric and mechanical properties. Journal of Polymer Science Part C: Polymer Symposia, 14(1): 173-235.

[14] Kim, E.S., & Jeon, C.J. (2011) Dependence of dielectric properties on the crystallinity of ceramic/polymer composites. IEEE transactions on ultrasonics, ferroelectrics, and frequency control, 58(9): 1939-1946.

[15] Zheng, X., & Wen, B. (2019) Practical PBT/PC/GNP composites with anisotropic thermal conductivity. RSC advances, 9(62): 36316-36323.

[16] Gu, J., Du, J., Dang, J., Geng, W., Hu, S., & Zhang, Q. (2014) Thermal conductivities, mechanical and thermal properties of graphite nanoplatelets/polyphenylene sulfide composites. RSC Advances, 4(42): 22101-22105.

# Determination of fan height based on FDM nozzle temperature field analysis—an example of babbitt alloys

**Jianxiu Liu [1, 2, 3], Tianchang Zhou [1], Yi Li [1], and Pengcheng Zhang [1, 2]**

[1] School of Mechanical and Electrical Engineering, Zhengzhou University of Light Industry, Henan 450003, China
[2] Department of Engineering, Yellow River University of Science and Technology, Henan 450061, China

[3] Corresponding author's e-mails: jianxiuliu@126.com.

**Abstract:** In this paper, the phenomenon of nozzle clogging or drooling droplets in the process of FDM metal 3D printing of tin-based Babbitt alloys was evaluated by simulation to assess the effects of different heights of cooling fans on the printer's nozzle cooling module, and the optimal height of the fans was determined. Based on the thermal analysis results, a thermodynamic coupling analysis was carried out to analyze the thermodynamic changes of the nozzle at a given temperature. Finally, the printing trials of tin-based Babbitt alloy were carried out using self-developed forming equipment to verify the accuracy of the theory. The results show that the print quality can be effectively improved by controlling the heat dissipation module of the printer nozzle. The printed material exhibited a uniform and fine microstructure in the 3D printing trials of tin-based Babbitt alloy.

## 1. Introduction

FDM (Fused Deposition Modeling) metal 3D printing technology is an additive manufacturing technology based on the discrete stacking principle, where the digital model of the object is used to build a three-dimensional entity by stacking material layer by layer. This technology is efficient, accurate, and capable of creating complex objects, making it a valuable tool in modern manufacturing. It can use molten materials for continuous superposition and stacking to rapidly manufacture complex shapes with great design freedom and realize structural shapes that are difficult to complete by traditional processes. Its substantial performance has been recognized in various fields, including mechanical manufacturing, medical, aerospace, military, and automotive industries [1][2]. 3D printing technology is conducive to promoting industrial production to produce digital, networked, intelligent direction, transforming and upgrading the traditional manufacturing industry, and has huge development prospects [3][4].

Babbitt alloy holds a significant position in the material of axle shaft tiles, with a wide range of applications in large-scale industrial machinery. Due to the rapid iteration of the current industrial machinery and equipment, the existing bearing performance has been unable to meet the ever-increasing performance requirements. Therefore, exploring new Babbitt alloy preparation technology has become an urgent problem to solve. Tin-based Babbitt alloys are commonly prepared by centrifugal casting, which can cause defects such as tissue segregation and coarse organization due to the influence of centrifugal force and non-equilibrium crystallization [5][6][7][8]. They are prone to fatigue damage, abrasion, pitting, and other failures during the operation of bearings.

In this paper, tin-based Babbitt alloy (SnSb11Cu6) is prepared by FDM technology, and firstly, the finite element simulation method is used to analyze the heat dissipation module of the FDM printhead using Fluent in ANSYS workbench. The simulation of the temperature field with the addition of fans of varying heights was performed to ascertain the optimal location for installing the heat dissipation fan [9]. Based on this, a thermal coupling analysis of the nozzle device was also carried out. Finally, self-developed forming equipment was used to test the printing of tin-based Babbitt alloys.

## 2. Nozzle temperature field analysis

### *2.1. Mathematical model of nozzle temperature distribution*

The 3D printing process is primarily characterized by heat conduction and convection. The heating rod is chiefly responsible for generating heat transfer as the heat source. Convective heat transfer is further classified into forced convection heat transfer and natural convection heat transfer. In this paper, the wind pressure is selected as the design standard. The temperature of the heat sink is affected by the wind pressure and the position of the fan. Under the premise of ensuring the installation distance and the installation size, selecting the fan with the largest airflow can accelerate the airflow and reduce the temperature of the heat sink. The area not affected by the fan is subject to the natural convection of the air to dissipate heat. The following is a mathematical description of the design's temperature field changes during printing.

(1) Mathematical description of thermal conductivity processes

The general form of the three-dimensional unsteady thermal conductivity differential equation in the right-angle coordinate system is:

$$\rho C_p \frac{\partial T}{\partial t} = \frac{\partial}{\partial y}\left(\lambda \frac{\partial T}{\partial x}\right) + \frac{\partial}{\partial y}\left(\lambda \frac{\partial T}{\partial y}\right) + \frac{\partial}{\partial z}\left(\lambda \frac{\partial T}{\partial z}\right) + s \tag{1}$$

Where $\rho$, $C_p$, $\lambda$, $S$, and $t$ denote the density, specific heat capacity, thermal conductivity, and heat generated by the endothermic source per unit volume per unit time and time of the object [10].

In the process of 3D printing, the printer starts to print when the temperature of the nozzle reaches the set value, and the heating process of the heating rod is not studied, so the temperature field calculated in this paper is the steady state temperature field when the nozzle temperature reaches the extrusion temperature. There is only a heating rod as an energy source in the system and no internal heat source. In the calculation, the physical properties of the object are assumed to be constant physical properties. The differential equation for the thermal conductivity of the 3D printing process with constant physical properties, steady state, and no internal heat source is given by:

$$\frac{\partial^2 T}{\partial x^2} + \frac{\partial^2 T}{\partial y^2} + \frac{\partial^2 T}{\partial z^2} = 0 \tag{2}$$

(2) Mathematical description of the convective heat transfer problem

Assuming that air is an incompressible fluid, the differential equation for the incompressible, no internal heat source, steady state, and constant physical properties three-dimensional convection heat transfer problem is given by:

Mass conservation equation:

$$\frac{\partial u}{\partial x} + \frac{\partial v}{\partial y} + \frac{\partial w}{\partial z} = 0 \tag{3}$$

Conservation of momentum equation:

$$\begin{cases} \rho\left(u\frac{\partial u}{\partial x}+v\frac{\partial u}{\partial y}+w\frac{\partial u}{\partial z}\right)=F_x-\frac{\partial p}{\partial x}+\mu\left(\frac{\partial^2 u}{\partial x^2}+\frac{\partial^2 u}{\partial y^2}+\frac{\partial^2 u}{\partial z^2}\right) \\ \rho\left(u\frac{\partial u}{\partial x}+v\frac{\partial u}{\partial y}+w\frac{\partial u}{\partial z}\right)=F_y-\frac{\partial p}{\partial y}+\mu\left(\frac{\partial^2 v}{\partial x^2}+\frac{\partial^2 v}{\partial y^2}+\frac{\partial^2 v}{\partial z^2}\right) \\ \rho\left(u\frac{\partial u}{\partial x}+v\frac{\partial u}{\partial y}+w\frac{\partial u}{\partial z}\right)=F_z-\frac{\partial p}{\partial z}+\mu\left(\frac{\partial^2 w}{\partial x^2}+\frac{\partial^2 w}{\partial y^2}+\frac{\partial^2 w}{\partial z^2}\right) \end{cases} \quad (4)$$

Energy conservation equation:

$$u\frac{\partial T}{\partial x}+v\frac{\partial T}{\partial y}+w\frac{\partial T}{\partial z}=\frac{\lambda}{\rho C_p}\left(\frac{\partial^2 T}{\partial x^2}+\frac{\partial^2 T}{\partial y^2}+\frac{\partial^2 T}{\partial z^2}\right) \quad (5)$$

Where $u$, $v$, and $w$ denote the components of velocity in the $x$, $y$, and $z$ directions, respectively. $F_x$, $F_y$, and $F_z$, denote the components of volumetric force in the $x$, $y$, and $z$ directions, respectively. $\rho$ denotes the density of the air, $C_p$ denotes the specific heat capacity, $\lambda$ indicates the thermal conductivity, $\mu$ denotes the kinetic viscosity, and p denotes the pressure.

### *2.2. Numerical simulation of nozzle temperature distribution*

The calculations ignore small structures such as threads and small holes to save computational resources. At the same time, air domains and fan inlets are created to represent the areas of airflow and cooling fans. The calculation model for simulating the heat sink block's temperature field is depicted in Figure 1, with Air Domain 1 indicating the airflow area and Air Domain 2 as the schematic range of the cooling fan's action area, which will be different in actual calculations due to the influence of gravity, object obstruction, airflow, and other factors. The selected fan is DC12V (radius is 20 mm, air pressure is 25.9 Pa), and the air domain calculated size is 80 mm × 60 mm × 90 mm. On the other hand, this paper selects the printing silk material for SnSb11Cu6, the printing nozzle material using aluminum alloy, steel, and its related material parameters, which are shown in Table 1.

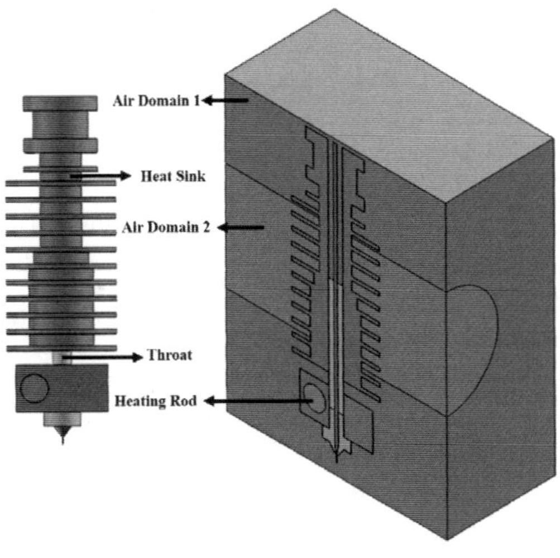


**Figure 1.** Numerical modelling of heating and cooling modules.

**Table 1.** Material performance parameters.

| Materials | Air | Aluminium alloy | Steel |
|---|---|---|---|
| Density $\rho$ (kg · m$^{-3}$) | 1.225 | 2770 | 7350 |
| Specific heat capacity of melting $C$ (J · kg$^{-1}$ · K$^{-1}$) | 1006.43 | 875 | 480 |
| Thermal conductivity $\lambda$ (W · m$^{-1}$ · k$^{-1}$) | 0.0242 | 175 | 202.4 |
| Dynamic viscosity $\mu$ (kg·m$^{-1}$·s$^{-1}$) | 1.7894×10$^{-5}$ | | |

The meshing tool Meshing in Fluent performs different meshing thicknesses for the computational model. In the division process of the fluid-solid coupling model, the solid domain selects a finer mesh to better calculate the temperature transfer in the nozzle, the first layer of the wall mesh is only conductive heat transfer in the normal direction during the flow process, and a laminar transition mesh region is set up in the air-fluid domain and the solid domain of the nozzle, as shown in Figure 2. The relevant boundary conditions are set as shown in Table 2.

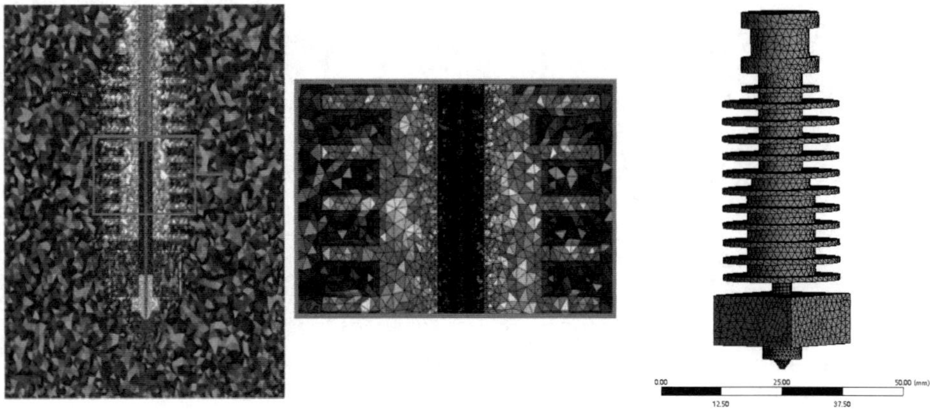


**Figure 2.** Nozzle mesh division.

**Table 2.** Nozzle boundary conditions.

| Boundary condition | Application location | Initial conditions |
|---|---|---|
| Speed Entrance | Silk entrance | 0.005 m/s, 20°C |
| Pressure inlet | fan | 25.9 Pa |
| Pressure outlet | air outlet | 0.1 MPa, 20°C |
| Coupling | Fluid and solid domains | Coupling |
| Thermostatic region | Heating rod area | 360°C |

In addition, 100 temperature observation points were set equidistant from the silk as the central axis in the simulation model to obtain the temperature values of the inside of the silk and the air near the nozzle. The main body of the nozzle is divided into five parts, as shown in Figure 3. $L_1$ (0~30 mm) is the region where the silks inside the heat sink are not in contact with the throat. $L_2$ (30~50 mm) is the region where the silks inside the heat sink are in connection with the heat sink in the throat. $L_3$ (50~60 mm) is the region where the exposed throat and the heating compartment are connected to the heat sink in the throat. $L_4$ (60~70 mm) is the region where the exposed nozzle and the heating compartment are connected to the nozzle region. $L_5$ (70~72 mm) is for the extrusion nozzle outside the silk, and outside the nozzle is the air region.

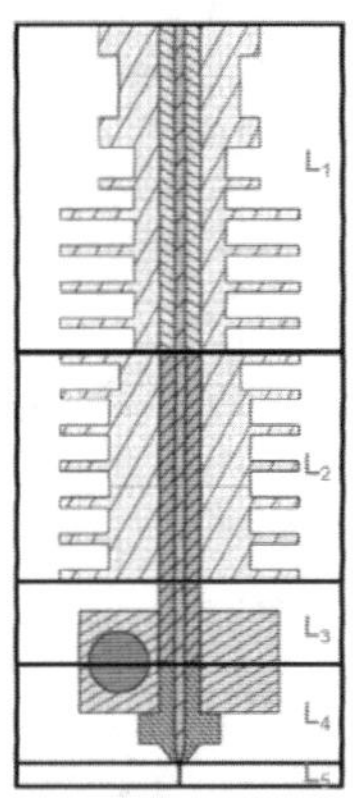


**Figure 3.** Nozzle structure diagram.

### *2.3. Analysis of calculation results*

The velocity cloud of the heat sink module is shown in Figure 4. Based on the cloud diagrams (a) and (c), we need to analyze how the installation position of the cooling fan affects the printer printhead temperature. To do this, we will use CFD-Post to obtain the heat flow and surface area of the model heating compartment and heat sink. We will then use the empirical formula to calculate the convective heat transfer coefficients hs and hb for the heating compartment and heat sink respectively. By quantitatively analyzing these coefficients, we can better understand how the installation position of the cooling fan affects the temperature of the printer printhead. L is the distance from the center of the cooling fan to the top of the heat sink. When L = 35 mm, hs = 38.165 (W/m$^2$·°C), and hb = 205.61 (W/m$^2$·°C). When L = 45 mm, hs = 18.362 (W/m$^2$·°C) and hb = 212 (W/m$^2$·°C). According to the calculation results, when the installation height is 35 mm, the temperature loss of the heating compartment is less, which is conducive to the temperature aggregation of the heating compartment. The temperature loss of the heatsink area is faster than that of silk feeding, at which time the cooling fan is utilized with the best effect.

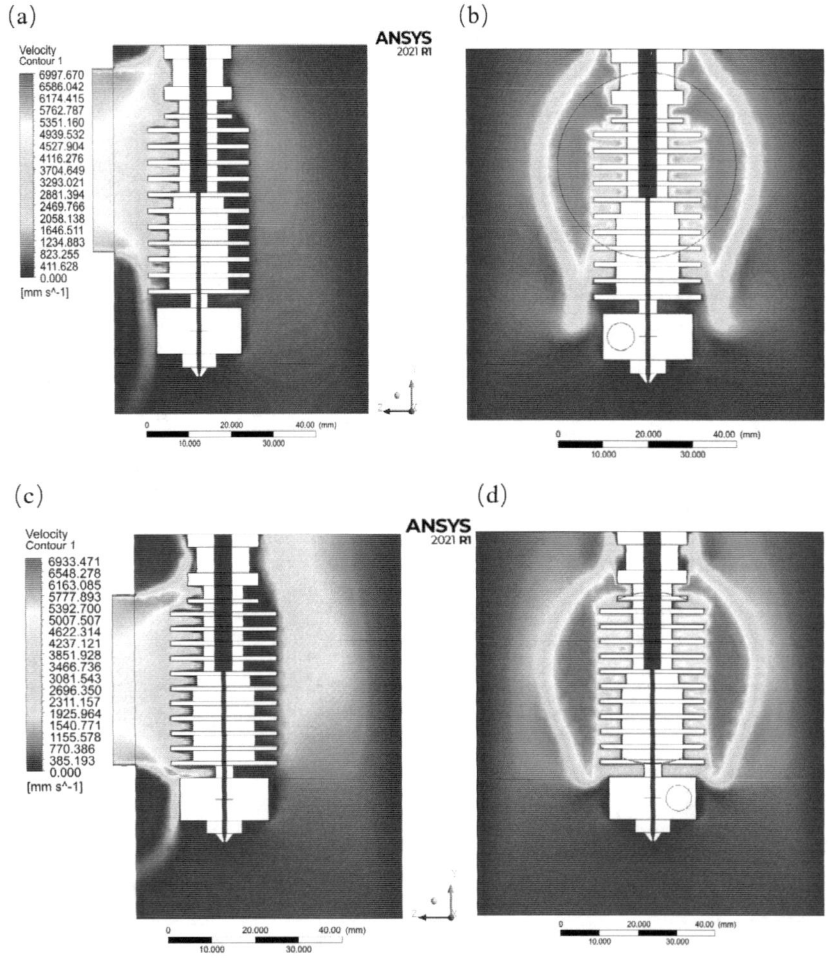


**Figure 4.** Fan airflow diagrams for different mounting positions. (a) X-axis view at L=35 mm; (b) Z-axis view at L=35 mm; (c) X-axis view at L=45 mm; (d) Z-axis view at L=45 mm.

In the printing process, the silk emerges in three stages. At room temperature, the state is solid. When the temperature rises to 161°C, the silk appears embrittlement. Premature material embrittlement caused by the increase in silk feeding resistance is not conducive to extrusion. When the temperature rises to 355°C, the material changes to a molten viscous flow state. When studying the heating and heat dissipation process, it is crucial to focus on the temperature of the silk at the nozzle and the temperature of the silk in the heat dissipation block. The primary goal is to ensure that the temperature at the nozzle remains constant to enable the material at the nozzle to become a printable molten state. This helps avoid clogging of the nozzle due to low temperature or droplets due to high temperature. Secondly, the critical point of silk softening should be lowered so that the silk softens as late as possible to ensure that the silk can be extruded smoothly. Figure 5 shows the temperature cloud of the nozzle structure when the fan mounting position is 35 mm, and the constant temperature of the heating rod is 360°C. As can be seen in Figures 5 and 6, the temperature of $L_2$ decreases due to the heat dissipation effect of the heat sink. The temperature of $L_3$ increases rapidly due to the insufficient heat dissipation capability of the heat sink. The temperature of $L_4$ increases due to the heating of the specified temperature of the heating compartment wire heating to a molten state. The temperature of $L_5$ decreases due to the extrusion of the wire out of the nozzle, which results in a decrease in the temperature of the wire in the air, and the temperature of $L_5$ decreases sharply after the temperature gradually tends to room temperature. The temperature of the silk at the nozzle is measured at a steady state when the temperature of the heat source is different. Among them, it can be seen from Figure 6 that the temperature of the silk at the nozzle is 356.95°C when the temperature of the heat source is 360°C, and the silk is in a molten state and can be cooled to a solid state quickly after printing and extruding. And see in figure 7.

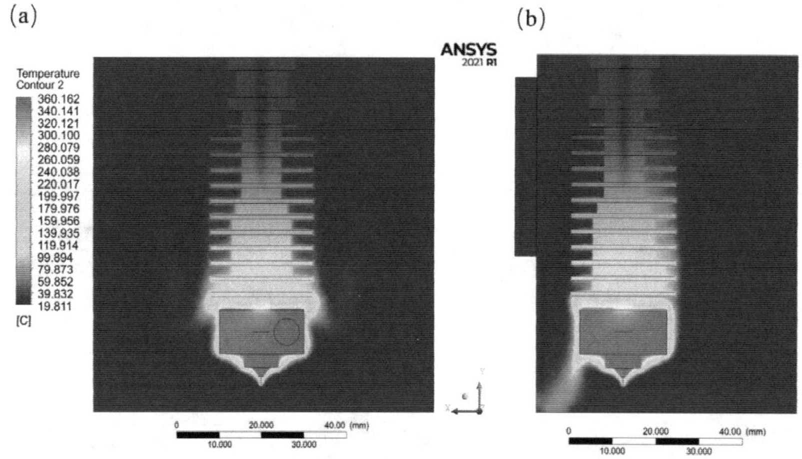


**Figure 5.** Temperature cloud diagram. (a) Temperature cloud in the Z-axis direction for L=35 mm; (b) Temperature cloud in the X-axis direction for L=35 mm.

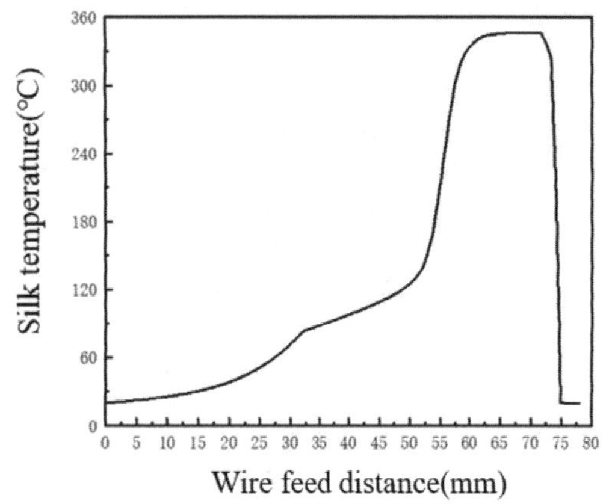


**Figure 6.** The temperature at each point is in axial direction of the silk.

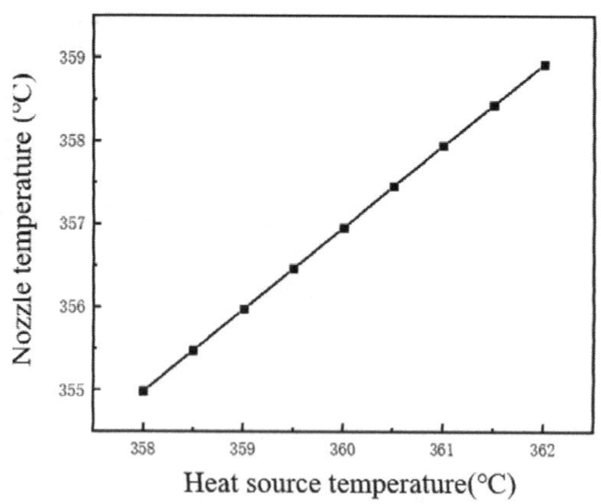


**Figure 7.** Effect of heat source temperature on nozzle temperature.

After the thermal analysis, the steady state temperature data model of the nozzle unit at L=35 mm obtained in Figure 5 was transferred to Workbench for thermal coupling analysis. As depicted in Figure 8, the region where the heat sink ring is fixed at the upper end and where the heating rod is installed requires attention. Stress-strain is most apparent. The maximum equivalent force is 227.58 MPa, and the maximum equivalent strain is 0.002663 mm/mm. The upper end of the heat sink block is fixed on the support frame through the ring and bolt. The inner side of the heat sink in contact with the ring is subjected to the greatest stress strain, followed by a particular stress-strain on the contact surface of the parts of the heating rod and the heating compartment. Still, the overall structure of the stress-strain is small, far below the yield limit of the material, in line with the requirements of the design of the material. Figure 8(c) shows that the heating block area is subject to the most serious thermal deformation. The maximum thermal strain is 0.00525 mm/mm, and the deformation amount meets the design requirements. Figure 9 shows the maximum thermal deformation at the nozzle is 0.0018 mm, the deformation is minimal, and the impact on the silk extrusion printing is almost negligible to meet the actual requirements.

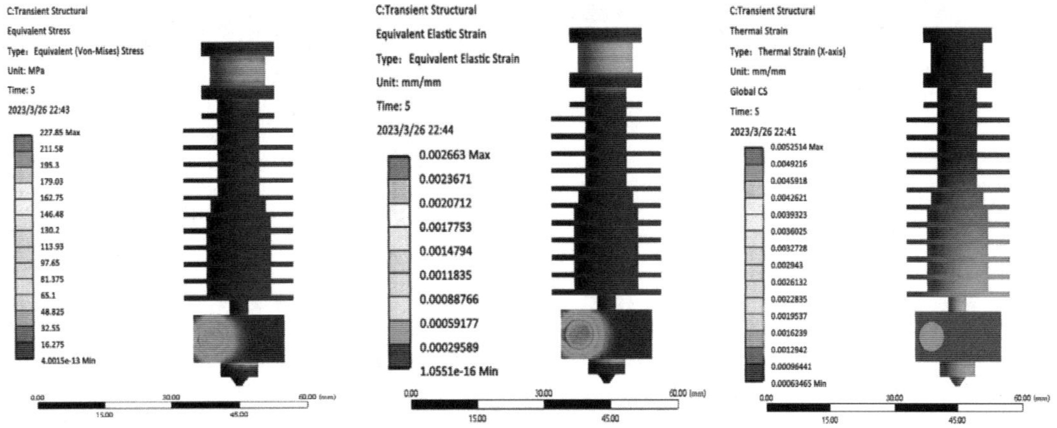


**Figure 8.** Stress-strain cloud of nozzle structure. (a) equivalent force map of nozzle structure; (b) equivalent strain map of nozzle structure; (c) thermal strain cloud of the nozzle structure.

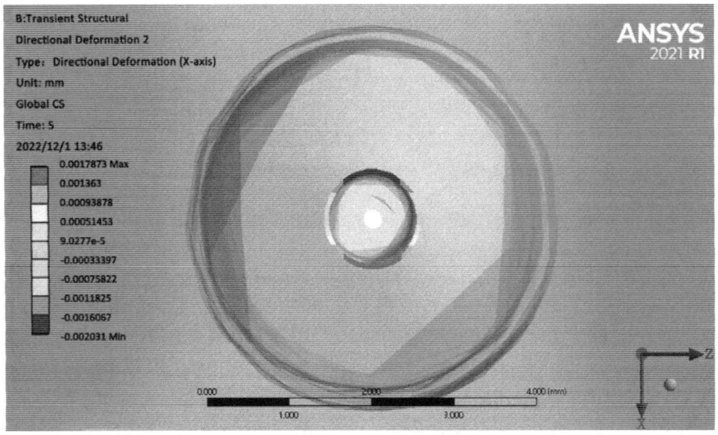


**Figure 9.** Cloud view of nozzle directional deformation.

## 3. FDM metal 3D printing forming test

In this paper, based on the above simulation results of the temperature field of the molten metal 3D direct writing process, the 3D printing forming test of tin-based Babbitt alloy (SnSb11Cu6) was carried out by using self-developed forming equipment. The test process was carried out under a nozzle diameter of 0.2 m, silk feeding speed of 9 mm/s, scanning speed of 9 mm/s, molten metal temperature of 360°C, and substrate temperature of 60°C. From Figure 10, it can be seen that the length × width × height of the molded parts and thin-walled parts are 10.0 mm×10.0 mm×10 mm, with a total of 45 layers.

The microstructure of the 3D printed, prepared tin-based Babbitt alloy was observed using a Phenom XL SEM scanning electron microscope. It is evident from Figure 11 that the 3D-printed prepared Babbitt alloy has a fine and uniform microstructure.

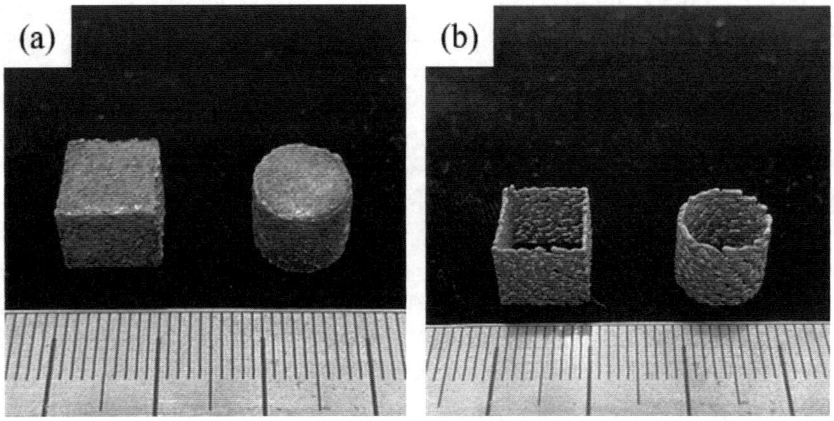


**Figure 10.** 3D printing of tin-based babbitt alloys. (a) tin-based babbitt alloy molding; (b) tin-based babbitt alloy thin-walled parts.

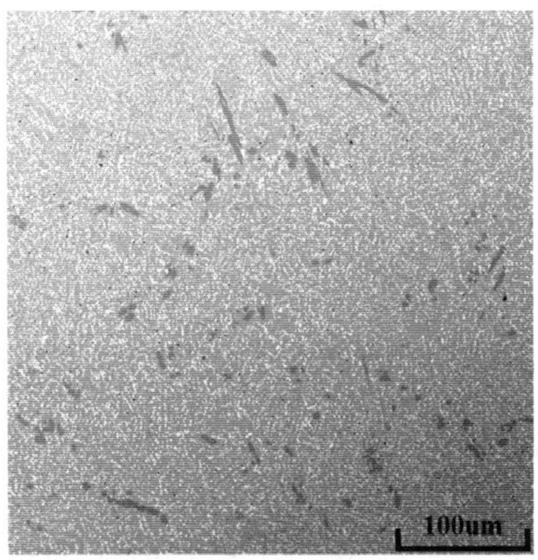


**Figure 11.** Microstructure of tin-based Babbitt alloys.

## 4. Conclusion

In this paper, the temperature field simulation of the heat dissipation module of the printhead is first verified to analyze the effect of different fan positions and the thermodynamics of the printhead. Then, the FDM printer printhead structure is optimized. In addition, based on the nozzle temperature field simulation results, 3D printing experiments of tin-based Babbitt alloys were carried out using self-developed forming equipment. The corresponding conclusions are as follows:

(1) After optimization of the printer printhead structure, when the fan height is 35 mm, there is less temperature loss in the heating compartment, which is conducive to temperature aggregation in the heating compartment. The temperature loss in the heatsink area is faster, thus improving the silk-guiding performance of the silk-feeding system.

(2) At a heat source temperature of 360°C, the silk is molten and cools quickly to a solid state after printing and extruding, greatly improving nozzle clogging and premature softening.

This paper provides a theoretical and practical basis for further improving the printing accuracy of 3D printers and a reference for the optimized design of printhead structures.

**Acknowledgments**
This research is a phase outcome of the National Natural Science Foundation of China project "Design and regulation of microstructure of Sn-Sb-Cu alloy based on liquid metal 3D printing and molding" (U1904175).

**Reference**

[1] Buchbinder, D., Schleifenbaum, H., Heidrich, S., Meiners, W., & Bültmann, J. J. P. P. (2011). High power selective laser melting (HP SLM) of aluminum parts. Physics Procedia, 12, 271-278. [CrossRef].

[2] Bourell, D., Kruth, J. P., Leu, M., Levy, G., Rosen, D., Beese, A. M., & Clare, A. (2017). Materials for additive manufacturing. CIRP annals, 66(2), 659-681. [CrossRef].

[3] Hao, W., Liu, Y., Zhou, H., Chen, H., & Fang, D. (2018). Preparation and characterization of 3D printed continuous carbon fiber reinforced thermosetting composites. Polymer Testing, 65, 29-34. [CrossRef].

[4] Goyanes, A., Wang, J., Buanz, A., Martínez-Pacheco, R., Telford, R., Gaisford, S., & Basit, A. W. (2015). 3D printing of medicines: engineering novel oral devices with unique design and drug release characteristics. Molecular pharmaceutics, 12(11), 4077-4084. [CrossRef].

[5] Azad, M. A., Olawuni, D., Kimbell, G., Badruddoza, A. Z. M., Hossain, M. S., & Sultana, T. (2020). Polymers for extrusion-based 3D printing of pharmaceuticals: A holistic materials–process perspective. Pharmaceutics, 12(2), 124. [CrossRef].

[6] Leszczynska-Madej, B., & Madej, M. (2013). Effect of the heat treatment on the microstructure and properties of tin Babbitt. kovove Mater, 51, 101-110. [CrossRef].

[7] Wang, Y., Qian, Z., Li, X. Y., & Tandon, K. N. (1997). Sliding wear properties of TiAl alloys with/without TiN coatings. Surface and Coatings Technology, 91(1-2), 37-42. [CrossRef].

[8] Gajmal, S. S., & Raut, D. N. (2022). An investigation on wear behaviour of ASTM B23 tin-based Babbitt alloy developed through microwave-assisted casting. International Journal of Metalcasting, 1-19. [CrossRef].

[9] Lee, S. W., Cho, S. J., & Kim, W. (2022). Development of thermo-fluid simulation technique for extruder and chamber of FDM-type 3D printer for printing high-melting-point materials. Microsystem Technologies, 28(10), 2225-2234. [CrossRef].

[10] Roberts, I. A., Wang, C. J., Esterlein, R., Stanford, M., & Mynors, D. J. (2009). A three-dimensional finite element analysis of the temperature field during laser melting of metal powders in additive layer manufacturing. International Journal of Machine Tools and Manufacture, 49(12-13), 916-923. [CrossRef].

# Numerical simulation of air flow and distribution in the blast system of blast furnace

**Wenxuan Xu[1,2,3], Gele Qing[1,2], Yanglong Li[1,2], Lijun Zhang[1,2], Meng Yu[1,2] and Fengqin Wang[1,2]**

[1]Research Institute of Technology, Shougang Group Co., Ltd., No. 69, Yangzhuang Road, Shijingshan District, Beijing, 100043, P. R. China
[2]Beijing Key Laboratory of Green Recyclable Process for Iron and Steel Production Technology, No. 69, Yangzhuang Road, Shijingshan District, Beijing, 100043, P.R. China

[3]Corresponding author's e-mail: xuwenxuan_ustb@126.com

**Abstract.** A three-dimensional geometric model of the blast system of the 5500 m$^3$ blast furnace was established in this study. The velocity distribution of the blast system was calculated by OpenFOAM, and the velocity distribution at cross sections of the round pipe and the branch pipe was analysed. The main conclusions are as follows: (1) When the hot air enters the branch pipe, the air flow speed gradually increases and reaches its maximum value (approximately 266.7 m/s) at the tuyere outlet. (2) When the hot air enters the round pipe from the main pipe, there is momentum loss. When the hot air further reaches the opposite side of the junction between the round pipe and the main (Tuyere 1#), the speed of the hot air in the round pipe decreases to the minimum. (3) The maximum air flow average speed of the tuyere outlet is located at Tuyeres 20# and 21# at the junction of the main pipe and the round pipe, which can reach 264.5 m/s. The minimum air flow average velocity of the tuyere outlet is located at Tuyeres 19# and 22#, approximately 258.5 m/s.

## 1. Introduction

The blast system of the blast furnace has great significance for the efficient and stable operation of the blast furnace, and it mainly consists of a main pipe, a round pipe, a branch pipe, and a tuyere. With the development of large-scale blast furnaces, the number of tuyeres in the blast system is gradually increasing. The number of tuyeres in a 5500 m$^3$ blast furnace can reach 40. The main function of the blast system of the blast furnace is to blow hot air around 1200°C into the interior of the blast furnace to provide heat and oxygen. As one of the four major operating systems of the blast furnace, the blast system mainly affects the depth and height of the raceway in the furnace, and directly determines the initial distribution of gas flow in the furnace. Moreover, with the development of large-scale blast furnaces, the circumferential uniform distribution of gas flow is also crucial [1-4]. Therefore, it is necessary to study the uniformity of air flow distribution at each tuyere of the blast system of the blast furnace.

At present, adjusting the diameter and length of the tuyere is the main method to improve the uniformity of air distribution of the tuyere. Due to the characteristics of high speed and high temperature in the blast system of a blast furnace, the method to study the flow and distribution of hot blast mainly depends on numerical simulation. Liu et al. [5] used the CFD method to simulate the air volume at each tuyere of the 2580 $m^3$ blast furnace in Ansteel. The results showed that under the same tuyere diameter conditions, the air volume distribution at the blast furnace tuyere was not uniform. Liu et al. [6] studied the dual hot air main pipe structure in the blast system of a 3200 $m^3$ blast furnace and calculated that under the condition of all tuyere diameters of 120 mm, the air volume distribution at each tuyere in the circumferential direction was uneven, with a change in air volume of about 4.4%. Ni et al. [7] took the blast system of a 4000 $m^3$ blast furnace as the research object and used CFD numerical simulation method to analyse the influence of tuyere layout, tuyere diameter, and total air volume on the pressure and velocity distribution at the tuyere of the blast system of the blast furnace. The results indicate that the maximum velocity of tuyere is generally located at the two branch pipes closest to the main pipe, and the overall tuyere air volume distribution is uneven.

This article takes the blast system of a 5500 $m^3$ blast furnace as the research object, uses OpenFOAM to calculate the hot air flow field of the blast system, and analyses the current situation of uneven air distribution at each tuyere of the blast system.

## 2. Model and boundary conditions

### *2.1. Geometry model*

Figure 1 shows the three-dimensional geometric model of the blast system of the 5500 $m^3$ blast furnace, which mainly consists of a hot blast main pipe, a hot blast round pipe, a hot blast branch pipe, and 40 tuyeres. The hot blast generated by the hot blast stove enters the hot blast round pipe from the hot blast main pipe, is distributed to each hot blast branch pipe in a certain proportion, and finally enters the interior of the blast furnace from the tuyere. In order to reduce the influence of the mesh type on the calculation results, all mesh elements are hexahedrons, as shown in Figure 2.

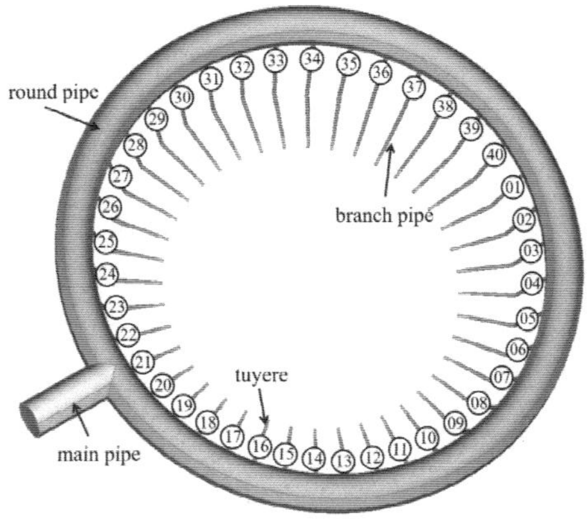


**Figure 1.** 3D geometric model of the blast system of the blast furnace.

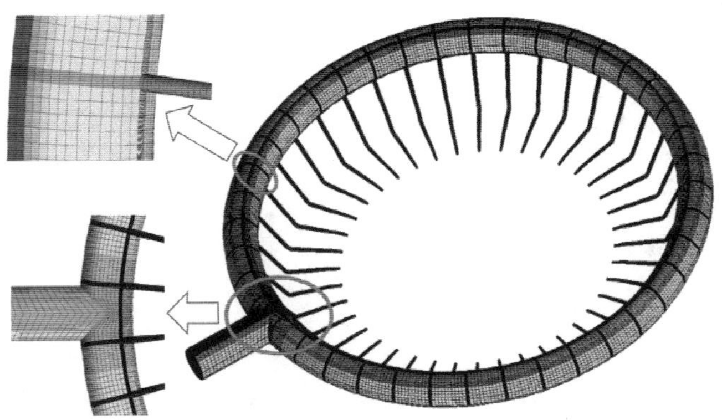

**Figure 2.** The hexahedral mesh of the blast system.

### *2.2. Conservation equations and boundary conditions*

Numerical simulations have been performed to examine the velocity profile of the blast system of the blast furnace. The 3D steady-state, incompressible governing equations for the conservation of mass (Equation (1)) and momentum (Equation (2)) are solved using an open-source available software OpenFOAM.

$$\rho \nabla(\vec{v}) = 0 \tag{1}$$

$$\rho(\nabla \vec{v})\vec{v} = -\nabla p + \mu \nabla^2 \vec{v} \tag{2}$$

Where $\rho$ and $\mu$ are the mass density (1.27 kg $m^{-3}$) and the dynamic viscosity ($1.78\times10^{-5}$ Pa s), respectively, of air. In this blast system, the length and diameter of the tuyere are 650 mm and 130 mm, respectively. The inlet of the hot blast main pipe is set as a volumetric flow inlet, 8400 $Nm^3$/min, and each tuyere outlet is set as a pressure outlet, 460000 Pa.

## 3. Result analysis and discussion

Figure 3 shows the contour of velocity distribution in the blast system of a blast furnace. It can be obtained that the gas flow velocity gradually increases when the hot air enters the branch pipe from the round pipe due to the reduction of the pipe diameter, and the hot air velocity reaches its maximum value at the tuyere outlet.

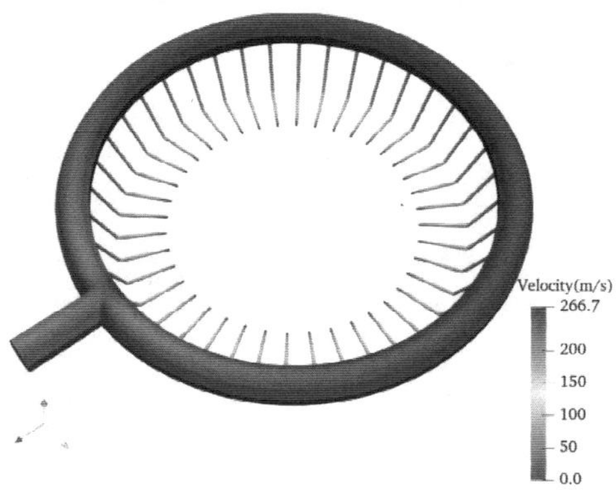


**Figure 3.** The contour of velocity distribution in the blast system of the blast furnace.

Figure 4 shows the contour of velocity distribution at the cross-section of 1# tuyere. It can be obtained that the air velocity in the round pipe is smaller, which is similar to the velocity of the air inlet. When the hot air enters the branch pipe, the airspeed gradually increases and reaches its maximum value (approximately 266.7 m/s) at the tuyere outlet.

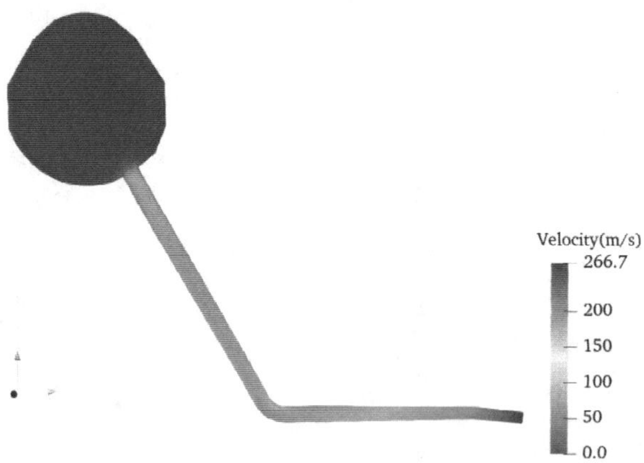


**Figure 4.** The contour of velocity distribution at the cross-section of 1# tuyere.

Figure 5 shows the velocity vector of air at the cross-section of 1# tuyere. It can be obtained that in the round pipe, the hot air mainly moves along the axis of the round pipe in a circumferential direction. At the junction of the branch pipe and the round pipe, the air flow begins to change direction, moves along the axis of the branch pipe, and finally flows out from the tuyere outlet.

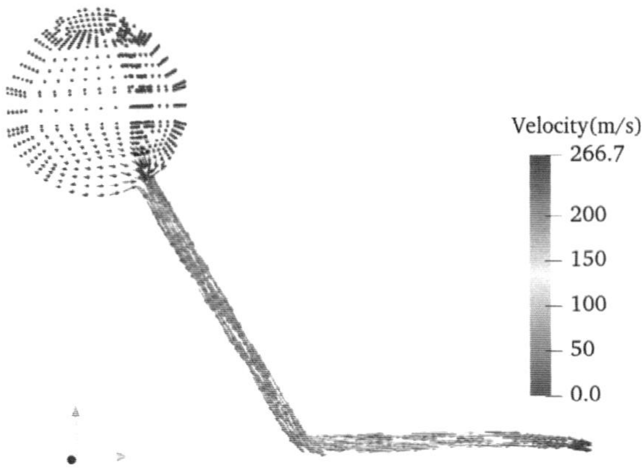


**Figure 5.** The velocity vector of air at the cross-section of 1# tuyere.

Figure 6 shows the contour of velocity in the horizontal cross-section of the blast system. It can be obtained that there is momentum loss after the hot air enters the round pipe from the main pipe, which is due to the angle between the main pipe and the round pipe being close to 90 °. The speed of the hot air decreases to the minimum in the round pipe when the hot air further reaches the opposite side of the junction between the main pipe and the round pipe, and the reason is that the hot air in both directions meets here, causing the air flow speed to decrease to the minimum.

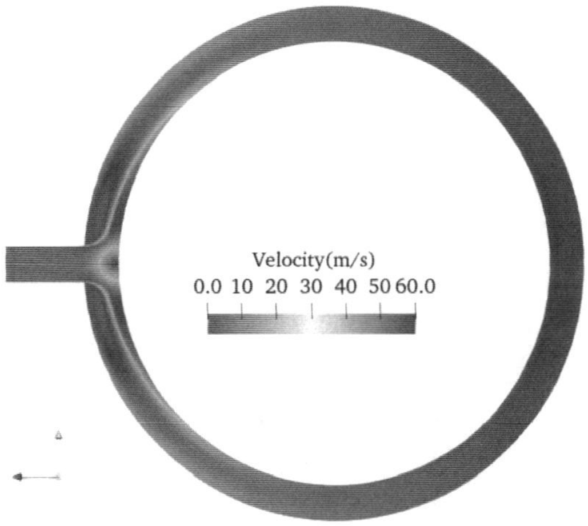


**Figure 6.** The contour of velocity in the horizontal cross-section of the blast system.

Figure 7 shows the velocity vector distribution in the horizontal cross-section of the blast system. It can be obtained that the flow direction changes when the hot air enters the round pipe from the main pipe, the speed near the outer side of the round pipe is lower, and the speed near the inner side of the round pipe is higher at the junction of the main pipe and the round pipe. Afterward, the hot air moves along the axis of the round pipe until it meets at the opposite side of the junction of the main pipe and the round pipe, and the speed of the hot air decreases to the minimum.

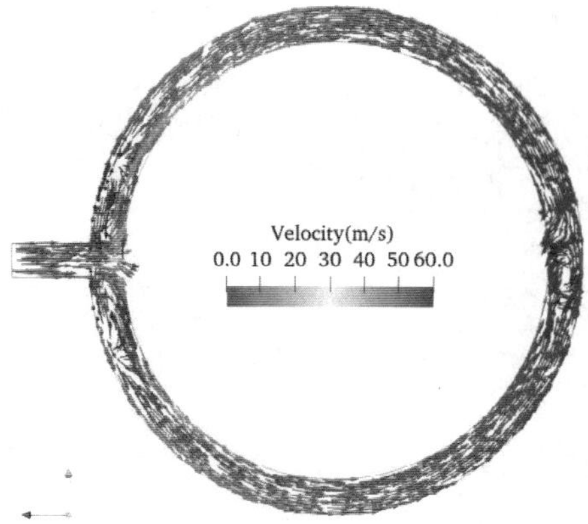


**Figure 7.** The velocity vector distribution in the horizontal cross-section of the blast system.

Figure 8 shows the average velocity distribution of each tuyere. It can be obtained that the maximum average speed of the air outlet is located at the 20# and 21# tuyeres at the junction of the main pipe and the round pipe, which can reach 264.5 m/s. The minimum average velocity of the air outlet is located at the 19# and 22# tuyeres, about 258.5 m/s, which is 6 m/s different from the maximum value.

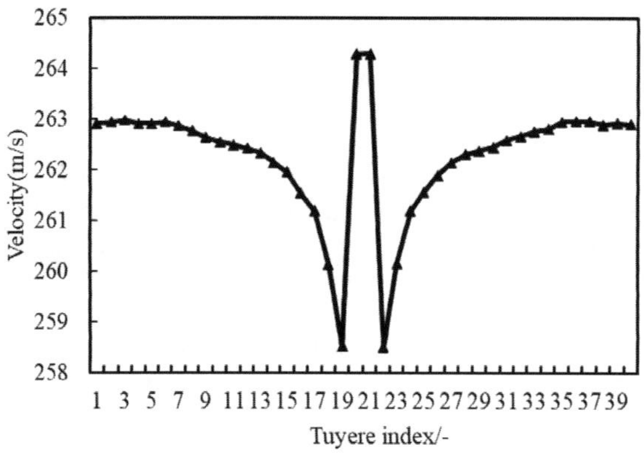


**Figure 8.** The average velocity distribution of each tuyere.

## 4. Conclusions

A three-dimensional geometric model of the blast system of the 5500 $m^3$ blast furnace was established in this study, the velocity distribution of the blast system was calculated by OpenFOAM, and the velocity distribution at cross sections of the round pipe and the branch pipe was analysed. The main conclusions are as follows:

(1) When the hot air enters the branch pipe, the air flow speed gradually increases and reaches its maximum value (approximately 266.7 m/s) at the tuyere outlet.

(2) When the hot air enters the round pipe from the main pipe, there is momentum loss. When the hot air further reaches the opposite side of the junction between the round pipe and the main (Tuyere 1#), the speed of the hot air in the round pipe decreases to the minimum.

(3) The maximum air flow average speed of the tuyere outlet is located at Tuyeres 20# and 21# at the junction of the main pipe and the round pipe, which can reach 264.5 m/s. The minimum air flow average velocity of the tuyere outlet is located at Tuyeres 19# and 22#, approximately 258.5 m/s.

**Acknowledgments**

The authors are grateful to the ironmaking department for providing resources for this work. The authors are also grateful to the management of Shougang Group Co., Ltd. for granting permission to publish the work.

**References**

[1] Radyuk A G, Titlyanov A E, Sidorova T Yu. (2020) Effect of slurry coating on the resistance of thermal insulation insert in blast furnace air tuyere. Metallurgist. 63: 1153-1159.
[2] Chattejee R, Nag S, Kundu S. (2019) Erosion behaviour of blast furnace tuyere. ISIJ International. 59: 1732-1734.
[3] Polinov A A, Pavlov A V, Onorin O P, et al. (2018) Blast distribution over the air tuyeres of a blast furnace. Metallurgist. 62: 418-424.
[4] Lyalyuk V P, Tarakanow A K, Kassim D A, et al. (2018) Improvement of the uniformity of blast distribution over the circumference of blast furnace hearth. Metallurgist. 62: 119-124.
[5] Liu H S, Shi B H, Liu B N, et al. (2022) Numerical simulation of the air volume distribution uniformity in blast furnace tuyere. Journal of Materials and Metallurgy. 21: 255-260.
[6] Liu W P, Jin Y, Dong X S. (2017) Simulation on blast volume distribution of blast furnace tuyeres with double hot blast main structure. The Chinese Journal of Process Engineering: 17: 1148-1155.
[7] Ni A, Li C Z, Zhang W, et al. (2022) Numerical simulation research on tuyere air supply characteristics of large blast furnace. Journal of Iron and Steel Research. 34: 411-420.

# First-principles study on interfacial reaction of SiCO/LTaO for all-solid-state lithium-ion batteries

**Changgui Xu[1], Zhirong Chen[1] and Ningbo Liao[1,2]**

[1]College of Mechanical & Electrical Engineering Wenzhou University, Wenzhou, 325035, P.R. China

[2]Corresponding author's e-mail : nliao@wzu.edu.cn

**Abstract:** Silicon oxycarbide is an excellent anode material for all solid batteries, characterized by excellent stability and some oxidation resistance. The structure of the SiCO/LTaO system has been investigated using first principles. We have investigated the electron density, DOS, and energy band structure at the SiCO/LiTaO interface. The calculations note that this particular configuration is expected to be applied to such as lithium-ion batteries and supercapacitors. This work will provide some theoretical support for the difficult problem of lowering impedance in the preparation of all-solid-state lithium-ion batteries.

## 1. Introduction

Silicon oxycarbide is a silicon-based polymer-derived ceramic characterized by excellent stability and some oxidation resistance [1-3]. The presence of silicon as a neutral mixed bond tetrahedron and amorphous carbon nanographene in the SiCO structure is considered to be the reason for the higher lithium storage. However, the prepared all-solid-state battery still suffers from high interfacial impedance as well as lithium dendrites during cycling, and the interface of the electrode-electrolyte is a key element, which is related to space charge. LTaO is considered to be a very promising candidate due to its high chemical stability, good mechanical properties, and small volume change during lithium embedding/detachment. Several research groups have also investigated LTaO as an ASSB material. However, the electrochemical mechanisms on the atomic scale have not been fully investigated so far. In the present work, we have studied the DOS, electron density as well and the energy band structure of the SiCO/LTaO hybrid system by first principles. The computational results show that LTaO material is expected to be a high-quality buffer layer material for solid lithium batteries with high impedance. Provides theoretical support for the commercialization of all solid batteries.

## 2. Computational method

All calculations in this paper are performed using the first principles (DFT) method [4] and are implemented using the CASTEP module [5]. PBE-GGA (generalized gradient approximation) is chosen as the gradient correction. The energy cutoff of 500 eV and a 3×3×2 k-point lattice. A vacuum layer with a thickness of 15 Å is constructed.

The system is constructed from supercells SiCO and $LiTaO_3$, and to eliminate the problems posed by lattice mismatches, 2×2 and 3×3 supercell operations are performed on SiCO and $LiTaO_3$, respectively. Geometric optimization is used to relax the SICO/LTaO interface until the force per atom is less than $10^{-2}$ eV/Å, and the energy convergence is less than $10^{-5}$ eV.

## 3. Discussion

The geometry of the SiCO/LTaO heterogeneous interfacial system has been completely relaxed as shown in Figure. 1.

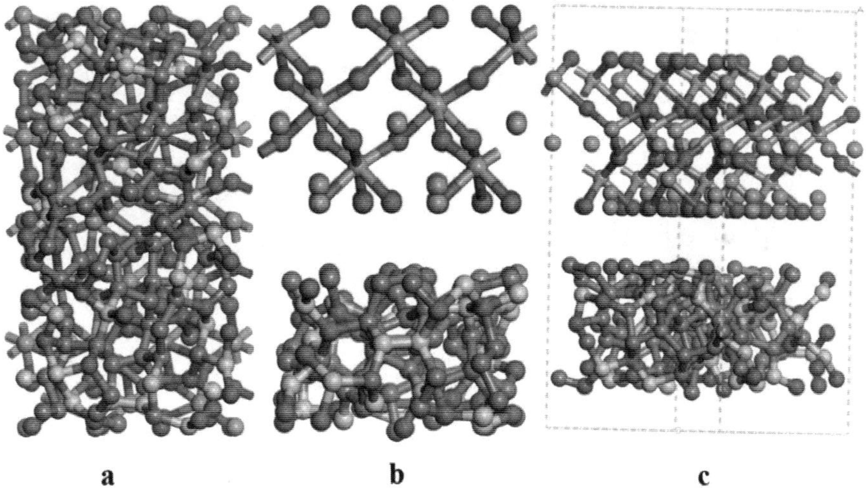


**Figure 1.** Heterogeneous interfaces for the SiCO/LTO systems: (a) up view of SiCO/LTaO; (b) side view of SiCO/LTaO; (c) A single supercell for the SiCO/LTaO system.

As can be seen in Figure. 2(a), after 40 iterations of geometry optimization, the structure remains stable and its equilibrium energy is -127250.08 eV. The maximum ion force, maximum ion displacement, and total energy convergence set for this study are 0.1E-04/atom, 0.27000E-01, and 0.400E-02Å, respectively. The calculated results are shown in Figure. 2(b) as $1.578336E^{-6}$ eV/atom, $2.9501534E^{-2}$eV/Å, and $6.02645E^{-4}$ Å. All the values are lower than our preset values, which indicates that our calculated results are in line with the computational accuracy.

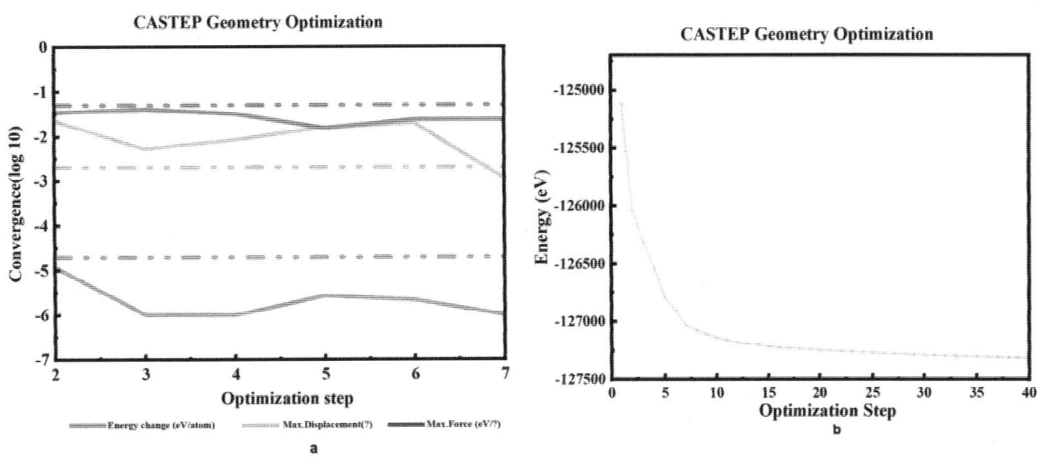


**Figure 2.** (a) Geometric optimization, (b) energy optimization convergence convergences.

The energy band structure of the SiCO/LTaO system is shown in Figure. 3, and the band gaps of the two materials constituting the system are 0.291 eV and 0.064 eV, respectively, which are optimized from the original semiconducting properties to the metallic properties, proving that the charge insulating layer at the SiCO/LTao interface has disappeared.

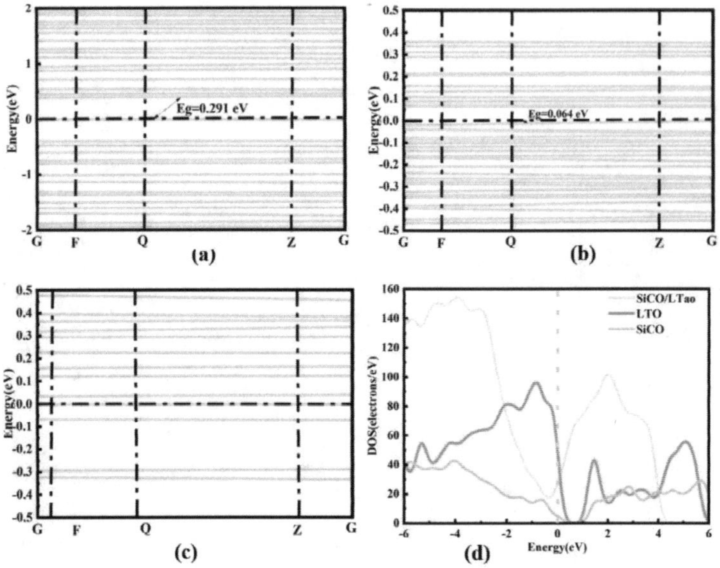


**Figure 3.** The band gap of (a) SiCO (b) LTaO (c) SiCO/LTaO (d) DOS of SiCO/LTaO, SiCO, LTaO.

The electron density maps before and after adsorption at the two interfaces are shown in Figure. 4, which noted that the middle charge layer has been conducted after the interface matching, which indicates that the SiCO/LTaO composite can improve the conductivity of the electrode.

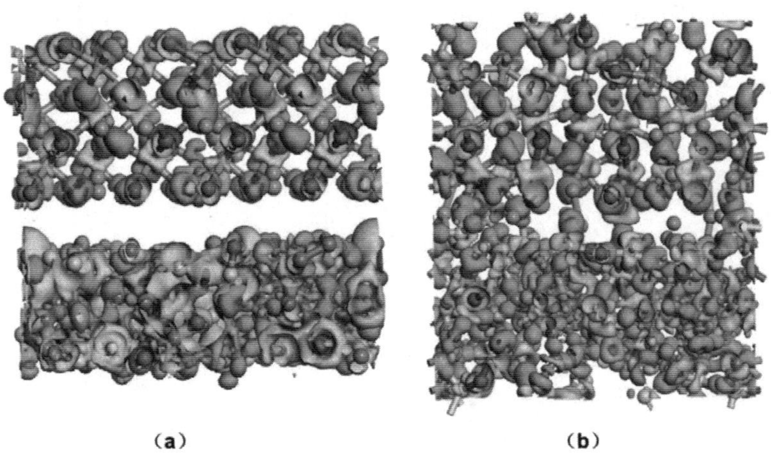


**Figure 4.** (a) Charge density map before interface matching (b) charge density map after interface matching.

In order to study the diffusion path of lithium atoms at the SICO/LTO interface, the activation energies of lithium atoms migrating to the SICO/LTO layer interface or within the electrode layer were calculated by the elastic band-climbing method, and two migration paths were selected in this paper. For convenience of illustration, we name the path of lithium atoms diffusing from the electrolyte LTO body to the SICO body as Path2, and the path of lithium atoms diffusing from the electrolyte body to the SICO/LTO interface as Path1, which is calculated as shown in Fig.5. The calculated energy barrier from path 1 is 0.48 eV,its value is small, which indicates that it is easier for lithium atoms to migrate to the interface of SICO and LTO, while the calculated migration barrier value from path 2 is 1.25 eV,its value is more than twice as much as that of the migration to path 1, which indicates that it is harder for lithium atoms to diffuse into the body of SiCO. Only a small amount of lithium atoms diffuse into the body, which is due to the fact that lithium atoms are adsorbed by the oxygen atoms on the surface of SICO when they migrate into the body of SiCO. As a result, a large number of lithium atoms migrate to the SICO/LTO interface, consistent with the above findings.

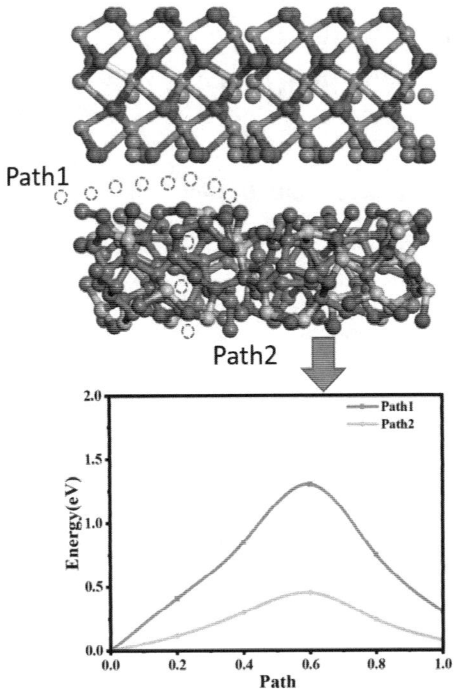


**Figure 5.** Two path lines of SICO/LTO and corresponding lithium ion jump energy barriers

## 4. Conclusions

We investigated the interfacial structure of SICO/LTaO by Density Functional theory. We have studied the electron density, DOS, and energy band structure at the SiCO/LiTaO interface. The calculations note that this particular configuration is expected to be applied to such as lithium-ion batteries and supercapacitors. This work will provide some theoretical support for the difficult problem of lowering impedance in the preparation of all-solid-state lithium-ion batteries.

## Acknowledgments

The authors would like to acknowledge the support of the Natural Science Foundation of Zhejiang province (LZ21E050001).

## References

[1] M. Arango-Ospina, F.T. Xie, I.G. Juan, R. Riedel, E. Ionescu, A.R. Boccaccini, Review: Silicon oxycarbide based materials for biomedical applications, Applied Materials Today, 18 (2020).
[2] L. David, R. Bhandavat, U. Barrera, G. Singh, Silicon oxycarbide glass-graphene composite paper electrode for long-cycle lithium-ion batteries, Nature Communications, 7 (2016).
[3] M. Faryad, F.A. Memon, F. Morichetti, M.I. Abro, G. Iseni, C. Somaschini, U. Aftab, A. Melloni, Synthesis, Characterization and Optical Constants of Silicon Oxycarbide, EPJ Web of Conferences, 139 (2017).
[4] Y.L. Yong, H.L. Cui, Q.X. Zhou, X.Y. Su, Y.M. Kuang, X.H. Li, C2N monolayer as NH3 and NO sensors: A DFT study, Applied Surface Science, 487 (2019) 488-495.
[5] M. Liangruksa, P. Sukpoonprom, A. Junkaew, W. Photaram, C. Siriwong, Gas sensing properties of palladium-modified zinc oxide nanofilms: A DFT study, Applied Surface Science, 544 (2021).

# Photoluminescence properties of $Sr_3Y_2(BO_3)_4$: $Eu^{3+}$ phosphor codoped with $Gd^{3+}$ and $La^{3+}$ under ultraviolet excitation

**Linyan Jia[1,2], Zichun Cong[1], Liping Zhang[1]**

[1]College of Chemistry and Chemical Engineering, Mudanjiang Normal University, 157011, China

[2] Corresponding author's e-mail: jialinyan1971@163.com

**Abstract**: $Sr_3Y_2$ $(BO_3)_4$: $Eu^{3+}$ phosphor codoped with $Gd^{3+}$ and $La^{3+}$ was generated using the solid state method, and XRD and fluorescence spectrum researched the structure and luminescence properties. The strongest excitation peak of $Sr_3Y_2$ $(BO_3)_4$:$Eu^{3+}$ is in the near ultraviolet region of 394 nm, which comes from the $^7F_0 \rightarrow {}^5L_6$ transition of $Eu^{3+}$. With the stimulation of 394 nm ultraviolet-light, the strongest emission of $Sr_3Y_2$ $(BO_3)_4$:$Eu^{3+}$ phosphor is a red light with a wavelength of 618 nm. It is produced by the transition of $^5D_0 \rightarrow {}^7F_2$ of $Eu^{3+}$. The optimum preparation temperature is 1000°C, and the doping amount of $Eu^{3+}$ is 5% (mol). The doping of $Gd^{3+}$ and $La^{3+}$ greatly enhanced the luminescent intensity of $Sr_3Y_2$ $(BO_3)_4$:$Eu^{3+}$ phosphor, and the doping of 20% (mol) was suitable.

## 1. Introduction

White LED has attracted great attention for its excellence in high efficiency, low energy consumption, and environmental protection [1,2]. Using a near-ultraviolet InGaN chip (380-410 nm) to excite three primary colours phosphor to achieve white LED has become one of the international research hotspots. At present, the performance of the red powder in terms of stability, luminous efficiency, and color rendering properties needs to be further improved. There are many types of oxysalt that can be used as a phosphor matrix, including aluminate, borate, vanadate, tungstate, molybdate, etc. Among them, the borate matrix has been generally considered by researchers because of its characteristics of high luminescence efficiency, low preparation temperature, and good chemical stability [3]. Li et al. [4] prepared the series of phosphors $Eu^{3+}$ and $Dy^{3+}$ doped alkaline earth borate. Li et al. [5] investigated the preparation of $Eu^{3+}$ doped $LiBaB_9O_{15}$ phosphor and the energy transfer process by high temperature solid-phase method. $Ce^{3+}$, $Tb^{3+}$ co-doped $Sr_2B_2O_5$ matrix phosphor was prepared by Sun et al. [6]. All the research on phosphor structure and luminescence performance proved that the alkaline earth borate system can be explored in white LED applications.

This paper applies a high-temperature solid phase method to generate $Sr_3Y_2(BO_3)_4$:$Eu^{3+}$ phosphor and digs into the influence of codoped $Gd^{3+}$ and $La^{3+}$ on the structure and luminescent properties.

## 2. Experimental section

The stoichiometric amounts of $Eu_2O_3$, $Y_2O_3$, $La_2O_3$, $SrCO_3$, and $H_3BO_3$ (20% excess of $H_3BO_3$ to make up for the loss in the high temperature calcining process) were fully mixed and ground in a quartz mortar, then placed in an alumina crucible, and calcined in a high temperature oven at a given environment for 5 h.

With the utilization of Cu Kα radiation of XRD (Rigaku D/max-2200), the structure of the phosphor was depicted. The fluorescence spectrophotometer (Hitachi F-7000) measured the luminescent properties under UV excitation.

## 3. Results and discussion

### *3.1 Luminescent properties of $Sr_3Y_2(BO_3)_4$: $Eu^{3+}$ phosphor*

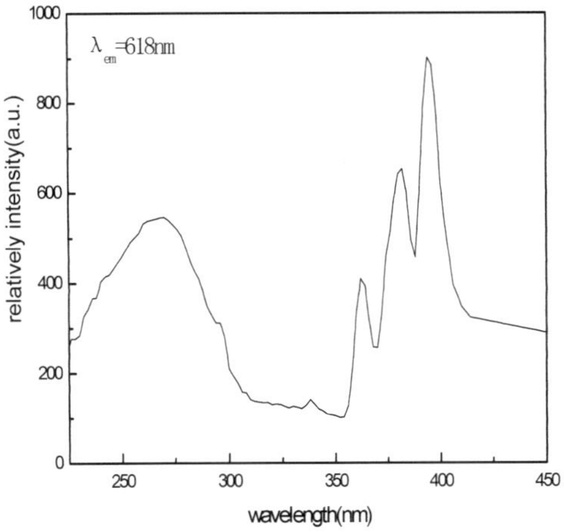


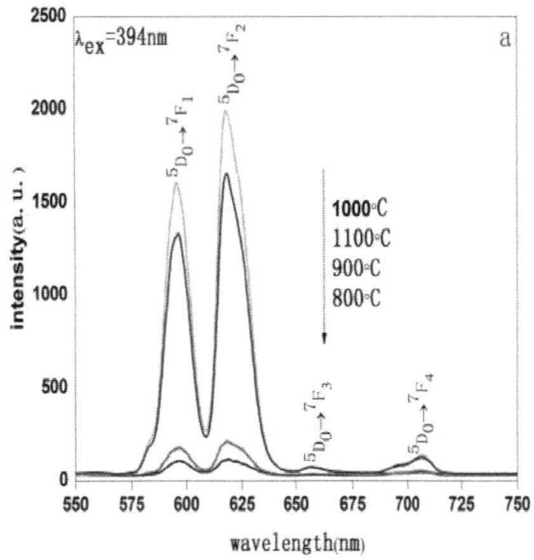


**Figure 1.** The excitation spectra of $Sr_3Y_2(BO_3)_4$: $Eu^{3+}$ phosphors.

**Figure 2.** The emission spectra of $Sr_3Y_2(BO_3)_4$: $Eu^{3+}$ phosphors.

Figure 1 portrays the excitation spectra of $Sr_3Y_2(BO_3)_4$:$Eu^{3+}$ generated at a calcination temperature of 1000°C and $Eu^{3+}$ doping of 5% (mol), measured at a monitoring wavelength of 618 nm. It portrays the formation of an excitation spectrum containing a broadband around 270 nm and several narrow peaks between 360-400 nm. The transition of the 2p electron of $O^{2-}$ in the matrix to the f-orbitals of $Eu^{3+}$ generated the broadband around 270 nm. $Eu^{3+}$ is strongly coupled to the lattice in the process, which results in the form of broadband peaks in the excitation spectra. The rest of the peaks originate from the f→f transition of $Eu^{3+}$. The highest excitation peak at 394 nm originates from $Eu^{3+}$ of $^7F_0 \rightarrow {}^5L_6$ transition. Therefore, $Sr_3Y_2$ $(BO_3)_4$:$Eu^{3+}$ can be applied as a luminescence material for near-ultraviolet-excited LEDs.

Figure 2 demonstrates the emission spectra of phosphor prepared at different calcining temperatures from 800°C to 1100°C. The phosphors generated under different calcining temperatures have two strong emission peaks at 596 nm and 618 nm. The 596 nm emission peak is the magnetic dipole transition of $^5D_0 \rightarrow {}^7F_1$ of $Eu^{3+}$, and the stronger one at 618 nm is the electric dipole transition of $^5D_0 \rightarrow {}^7F_2$ of $Eu^{3+}$. It indicates that the doping of $Eu^{3+}$ ions mainly occupies the asymmetric center lattice position in the structure. The other main emission peaks, 656 nm, and 709 nm, belong to the energy level transition $^5D_0 \rightarrow {}^7F_3$ and $^5D_0 \rightarrow {}^7F_4$ of $Eu^{3+}$ ions. From the intensity of the peaks, it is revealed that the luminescent intensity of the prepared phosphor enhances with the rising of the calcining temperature between 800°C and 1000°C and reaches the maximum at 1000°C. The luminescence intensity of the material starts to decrease when it continues to rise to 1100°C, which is caused by the temperature quenching effect.

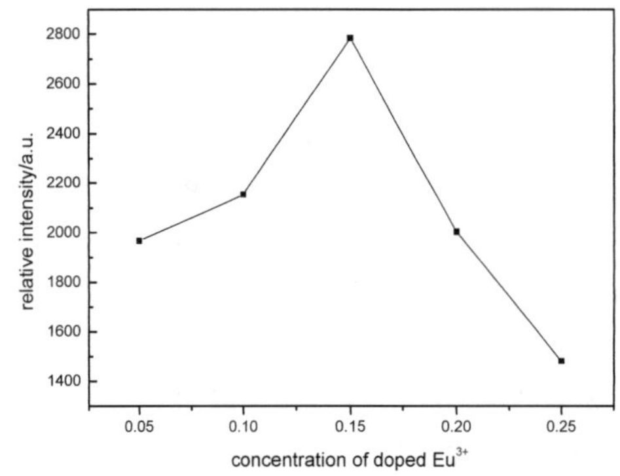


**Figure 3.** The emission intensity of $Sr_3Y_{2-x}(BO_3)_4$:$xEu^{3+}$ with different $Eu^{3+}$ doping molar ratios.

Figure 3 shows the relationship between luminescence intensity and $Eu^{3+}$ ion doping amount of $Sr_3Y_{2-x}(BO_3)_4$:$xEu^{3+}$ ($x \leq 0.25$) phosphors stimulated by UV light (394 nm). It is observed that from Figure 3, when the $Eu^{3+}$ doping amount is less than 15%, the luminescent intensity of $Sr_3Y_{2-x}(BO_3)_4$:$xEu^{3+}$ phosphor increasingly enhances with the gradual rise in the doping amount of $Eu^{3+}$. The intensities reached a maximum when it was 15% (mol), and then increased the europium-doped amount of luminescence intensity instead of weakening, which is caused by the non-radiative ionic transition, that is, there is the effect of concentration quenching.

*3.2 Luminescence properties of $Gd^{3+}$, $La^{3+}$ co-doped $Sr_3Y_2(BO_3)_4$:$Eu^{3+}$ phosphors*

The XRD pictures of $Sr_3Y_{2-y}M_y(BO_3)_4$:$Eu^{3+}$ (M=Gd, La) phosphors generated with a calcining temperature at 1000°C for 5 h with 5% (mol) $Eu^{3+}$ doping and the standard patterns of $Sr_3Y_2(BO_3)_4$ crystals are depicted in Figure 4.

What can be observed from Figure 4(a) is the affection of $Gd^{3+}$ doping amount on the structure of $Sr_3Y_{1.95-y}Gd_y(BO_3)_4$:$0.05Eu^{3+}$ phosphor. When the gadolinium content y value is less than or equal to 0.6, the position of $Sr_3Y_{1.95-y}Gd_y(BO_3)_4$:$0.05Eu^{3+}$ diffraction peak is the same as that of $Sr_3Y_2(BO_3)_4$ standard card PDF No.54-1120, indicating that the small doping amount of $Gd^{3+}$ will not affect the crystal structure of $Sr_3Y_2(BO_3)_4$. This is because the three ions of $Gd^{3+}$, $Eu^{3+}$, and $Y^{3+}$ have a similar ionic radius, and the entry of $Gd^{3+}$ ions and $Eu^{3+}$ ions into $Sr_3Y_2(BO_3)_4$ crystals to replace part of the $Y^{3+}$ will not change its crystal structure.

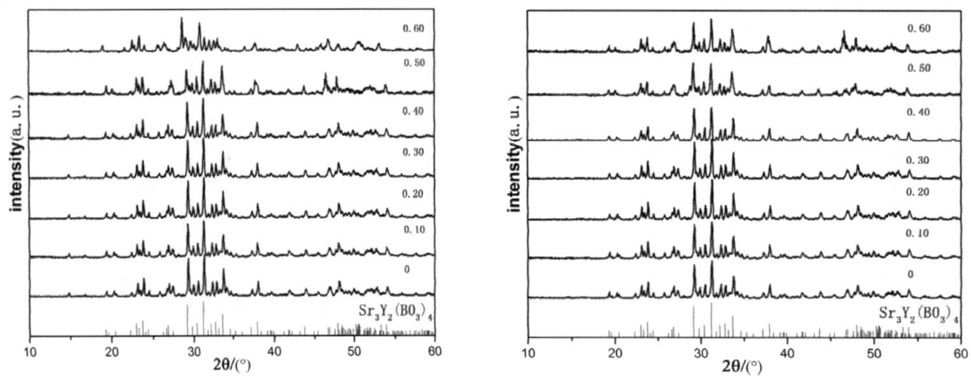


(a) $Sr_3Y_{2-y}Gd_y(BO_3)_4$:$Eu^{3+}$ (b) $Sr_3Y_{2-y}La_y(BO_3)_4$:$Eu^{3+}$

**Figure 4.** XRD patterns of $Sr_3Y_{2-y}M_y(BO_3)_4$:$Eu^{3+}$(M=Gd, La) phosphors.

Figure 4(b) displays the XRD patterns of phosphor $Sr_3Y_{2-y}La_y(BO_3)_4$:$Eu^{3+}$ prepared by $La^{3+}$ ion doping. It can be seen that the relative strength and location of the diffraction peaks are the same as those of the standard card when the amount of $La^{3+}$ doping is $\leq$ 25%, which indicates that the small doping amount of $La^{3+}$ cannot change the crystal structure of the substrate. When the $La^{3+}$ doping amount is greater than 30% (mol), the position of each diffraction peak is changed to different degrees, and the diffraction peaks are broadened.

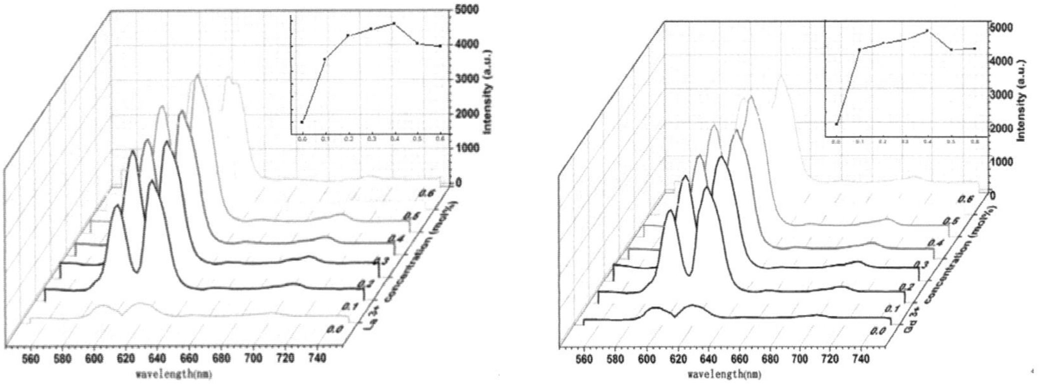


(a) $Sr_3Y_{2-y}Gd_y(BO_3)_4:Eu^{3+}$ (b) $Sr_3Y_{2-y}La_y(BO_3)_4:Eu^{3+}$

**Figure 5.** The emission spectra of $Sr_3Y_{2-y}M_y(BO_3)_4:Eu^{3+}$(M=Gd, La) phosphor.

The emission spectra of $Sr_3Y_{2-y}Gd_y(BO_3)_4:Eu^{3+}$ and $Sr_3Y_{2-y}Lay(BO_3)_4:Eu^{3+}$ phosphors are indicated in Figure 5. It demonstrates that the luminescence intensity of $Sr_3Y_{2-y}M_y(BO_3)_4:Eu^{3+}$ (M=Gd, La) phosphor is significantly higher than that of phosphor $Sr_3Y_2(BO_3)_4:Eu^{3+}$, the doping of $Gd^{3+}$ and $La^{3+}$ improves the luminescence intensity of the $Sr_3Y_2(BO_3)_4:Eu^{3+}$ phosphor greatly, but the position of the emission peaks is not changed. In $Sr_3Y_{2-y}M_y(BO_3)_4:Eu^{3+}$ (M=Gd, La) phosphor, $Gd^{3+}$ and $La^{3+}$ ions acted as sensitizers to improve their luminescence efficiency. The luminescence intensity of phosphor $Sr_3Y_{1.95-y}Gd_y(BO_3)_4:0.05Eu^{3+}$ showed a tendency of enhancement initially and then diminution with the increase in the number of $Gd^{3+}$ ions. When the $Gd^{3+}$ concentration is less than or equal to 20% (mol), the luminescence intensity is enhanced with the enlargement of the $Gd^{3+}$ doping amount, and the luminescence strength reaches the maximum value when the $Gd^{3+}$ doping amount is 20% (mol). The luminescence intensity increased by 40% under the optimal preparation conditions. With the rise of $Gd^{3+}$ ion concentration, the luminous intensity decreases correspondingly. It can be explained that because of the growth of $Gd^{3+}$ ion concentration, the distance between the $Gd^{3+}$ ion and $Eu^{3+}$ ion in the lattice decreases, resulting in the increase of non-radiative transfer of $Eu^{3+}$ ion and the weakening of luminous intensity.

The same can be discovered. The luminous intensity of doped $La^{3+}$ ions is obviously better than that of undoped phosphors, and the optimal lanthanum content is also 20% (mol). Under the best preparation conditions, doped $La^{3+}$ ions can increase the luminous intensity by 42%.

Table 1 shows the CIE color coordinates of phosphors prepared from $Sr_3Y_{1.45}Gd_{0.4}(BO_3)_4:0.15Eu^{3+}$ and $Sr_3Y_{1.45}La_{0.4}(BO_3)_4:0.15Eu^{3+}$ calcined at 1000°C for 5 h. It is observed that it is close to NTSC red (0.67, 0.33), indicating that the prepared phosphor meets the red phosphor standard for white LEDs.

**Table 1.** CIE (Chrominance coordinates) of phosphor.

| NO. | Phosphor | CIE (x, y) |
|---|---|---|
| a | NTSC | (0.67, 0.33) |
| b | $Sr_3Y_{1.45}Gd_{0.4}(BO_3)_4:0.15Eu^{3+}$ | (0.648, 0.351) |
| c | $Sr_3Y_{1.45}La_{0.4}(BO_3)_4:0.15Eu^{3+}$ | (0.649, 0.350) |

The data listed in Table 1 depicts the CIE color coordinate diagram, as shown in Figure 6. The coordinate points are close to the red light area, indicating that the prepared phosphor powder has good color purity.

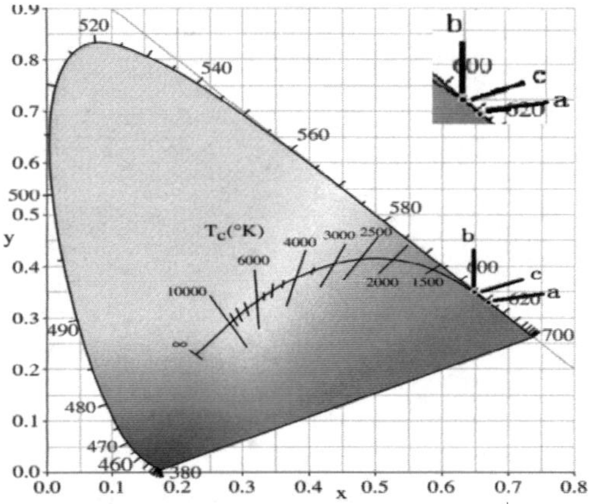


**Figure 6.** Chromaticity coordinates of phosphors.

## 4. Conclusion

This research has generated $Eu^{3+}$ and $Gd^{3+}$, $La^{3+}$ co-doped $Sr_3Y_2(BO_3)_4$ series phosphors using the solid phase method and investigated luminescence properties. The results show that the strongest emission peaks of the phosphors are the electric dipole transition of $Eu^{3+}$ ions $^5D_0 \rightarrow {}^7F_2$ under near-ultraviolet excitation at 394 nm. Doping a small amount of $Eu^{3+}$, $Gd^{3+}$, and $La^{3+}$ will not change the substrate structure of $Sr_3Y_2$ $(BO_3)_4$. The doping of $Gd^{3+}$ and $La^{3+}$ increases the luminous intensity of phosphors $Sr_3Y_2$ $(BO_3)_4$:$Eu^{3+}$ by about 40%, and the color coordinates have a very close distance from the standard red coordinates specified by the NTSC standard. $Sr_3Y_{2-y}Gd_y(BO_3)_4$:$Eu^{3+}$ and $Sr_3Y_{2-y}La_y(BO_3)_4$:$Eu^{3+}$ series of red phosphors have the potential application prospect in the white light LED.

**Acknowledgments**

This project is supported by the fund of Mudanjiang Normal University (YB2021004).

**References**

[1] Lakde J., Mehare C. M., Pandey K. K., et al. (2021) Recent development of $Eu^{3+}$-doped phosphor for white LED application: a review. Journal of Physics: Conference Series. 1913: 1-8. DOI 10.1088/1742-6596/1913/1/012029

[2] Mehare M. D., Mehare C. M., Swart H. C., et al. (2023) Recent development in color tunable phosphors: A review. Progress in Materials Science, 133: 1-52. https: //doi. org /10.1016/j. pmatsci.2022.101067

[3] Tawalare, P. K. (2022). Luminescent inorganic mixed borate phosphors materials for lighting. Luminescence, 37(8): 1226-1245. https://doi.org/10.1002/bio.4301

[4] Li P, Yang Z, Pang L, et al. (2008) Luminescent characteristics of $Ba_3Y_2(BO_3)_4$:$Eu^{3+}$ phosphor for white LED. Journal of Rare Earths, 26(1): 44-47. Doi: 10.1016/S1002-0721(08)60034-4.

[5] Li T., Li P.L., Wang Z.J., et al. (2015) Structure, luminescence properties and energy transfer of $Tb^{3+}$–$Eu^{3+}$ codoped $LiBaB_9O_{15}$ phosphors. The Royal Society of Chemistry 44: 16840–16846. https://doi.org/10.1039/C5DT01849H.

[6] Sun J.Y., Lai J.L, Zhu J.C., et al. (2012) Luminescence properties and energy transfer investigations of $Sr_2B_2O_5$:$Ce^{3+}$, $Tb^{3+}$ phosphors, Ceramics International. 38: 5341–5345. https://doi.org/10.1016/j.ceramint.2012.03.040.

# Numerical simulation of spinning forming of A356 aluminum alloy hub using dual spinning rollers

**Hao Zhong [1], Shengguan Qu[1, 3], Wang Long [2], Rixin Deng [2], and Wenxue Zhou [2]**

[1] South China University of Technology, Guangzhou, Guangdong Province, 510640 China
[2] Liuzhou Arays Technology Co., Ltd., Liuzhou, Guangxi Province, 545000 China

[3] Corresponding author's e-mail: qusg@scut.edu.cn

**Abstract**: A finite element model of one-pass forming of an A356 aluminum alloy hub using dual spinning rollers was established, and the effectiveness of the finite element model was verified through experiment. The established finite element model was used to study the influence of the feed rate, the spindle speed, the axial distance of the two spinning rollers, and the friction coefficient on the forming quality of the hub. The result shows that a good forming quality could be obtained when the feed rate is 4 mm/s, the spindle speed is 220 rpm, the axial distance between the two spinning rollers is 3 mm or 7 mm, and the friction coefficient is 0.2. The simulation results can provide a reference for actual spinning production processes.

## 1. Introduction

In recent years, with the urgent need to save resources and protect the environment, automotive lightweight has become an important guiding direction for the development of the automotive parts industry. As one of the key components determining the driving safety of automobiles, aluminum alloy hubs are constantly improving and innovating their production and manufacturing technology while the automotive industry is developing rapidly [1]. For traditional aluminum alloy hub manufacturing enterprises, how to achieve the lightest wheel hub quality while ensuring product performance is a challenging issue.

The production and preparation processes of aluminum alloy hubs mainly include the casting method, forging method, and casting-spinning forming method. Among them, the spinning formed hubs have become an important development direction for hub production due to their advantages such as lightweight and good mechanical properties [2]. However, the spinning production of wheel hubs is a large plastic deformation process of metal, which is prone to defects such as wrinkling, uneven wall thickness, and unqualified dimensions, resulting in material waste and reducing the production efficiency of enterprises. Therefore, studying the quality control of spinning forming is of great significance for reducing energy consumption and improving production efficiency. At present, researchers mainly use numerical simulation and experimental methods to study the spinning process of hubs. For example, Chen et al. [3] used the Simufact platform to establish the spinning forming model of the aluminum alloy wheel hub, studied the influence of process parameters on the roundness of the wheel hub and the amount of spring back, and found that the feed rate and the thinning rate have the greatest influence on the forming quality of the wheel hub; Bi et al. [4-5] used the finite element software DEFORM 3D to improve the spinning forming process of aluminum alloy wheels and steel wheels and investigated the influence laws of different feed rates and spindle speeds on their forming process. Roy and Majer [6] established a thermal coupling model for hub spinning forming and predicted the final geometric shape of the results. The model predictions were consistent with the experimental results under small deformation conditions. Hwang et al. [7] studied the effects of spindle load, feed rate, radial thinning rate, and the geometric shape of the spinning roller on product accuracy through experiments and simulations, and obtained the optimal process parameters for each pass. Cha et al. [8] studied the mechanism of material backflow defects during the hub spinning process through simulation and experiments, and the results showed that the spindle load is the main factor causing material backflow.

The above research on hub spinning forming primarily concentrates on multi-pass spinning forming. There is limited research on the one-pass forming of hubs using dual rollers. Using two spinning rollers for one pass spinning has a faster production speed and a greater wall thickness reduction rate, which means it will achieve higher production efficiency and a better strengthening effect. At the same time, it will generate greater spinning pressure and more unstable factors, which affect the quality of spinning forming. Therefore, examining the one-pass spinning technique of aluminum alloy hubs with dual spinning rollers is of great importance for improving production efficiency and product quality. This study employs finite element analysis software to simulate the one-pass spinning process of an aluminum alloy hub using two spinning rollers and investigates the effects of spindle speed, feed rate, axial distance between the two rollers, and friction coefficient on the forming quality of the hub. The results identify optimal process parameters, which can guide actual hub-spinning production processes.

## 2. Establishment and verification of finite element model

The schematic diagram of the spinning forming process of the aluminum alloy hub is shown in Figure 1. At the beginning of forming, the workpiece is first compressed onto the mandrel by applying a certain amount of pressure through the tail stock, and then the mandrel rotates at a constant angular velocity $\omega$ with the workpiece and the tail stock. During the forming process, the two spinning rollers maintain a certain axial distance $l_y$ on the $y$ axis, and perform three degrees of freedom movements in the $xy$ plane according to the preset motion trajectory, including linear motion in the $x$ direction and $y$ direction and rotational motion around the $y$ axis under the frictional contact with the hub. Among them, the feed rate $v_y$ of the two spinning rollers in the $y$ axis direction is the same. The spinning rollers keep in contact with the workpiece as it is fed in a straight line, applying a constant force to press and deform it until it finally takes the shape of a hub.

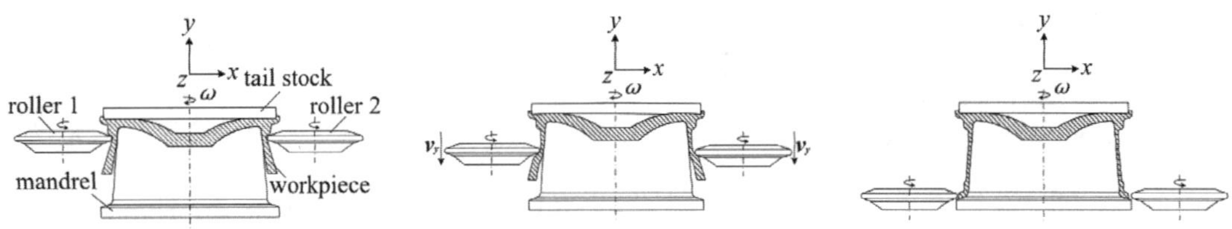


a) Start of spinning forming b) In spinning forming c) End of spinning forming

**Figure 1.** Schematic diagram of hub spinning forming.

Based on the diagram of the principle of hub spinning forming provided above, a one-pass spinning forming finite element model of aluminum alloy hub using dual spinning rollers is established in Figure 2. The tail stock, workpiece, and mandrel are fixed while the two spinning rollers rotate around the workpiece, feeding the movement in a straight line.

The aluminum alloy hubs used in production are made from cast A356 aluminum alloy, with a spinning forming temperature of 360°C. Table 1 and Figure 3 display the material parameters adopted in the finite element model and the simplified stress-strain curve obtained from the high-temperature hot pressing experiment at 360°C.

**Table 1.** Material properties employed in the finite element model.

| ρ (kg/m$^3$) | E (GPa) | Poisson's ratio |
|---|---|---|
| 2, 750 | 75 | 0.3 |

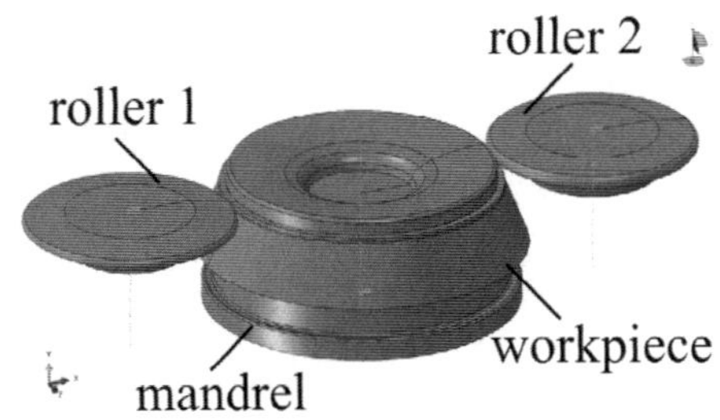


**Figure 2.** Finite element model for hub spinning forming.

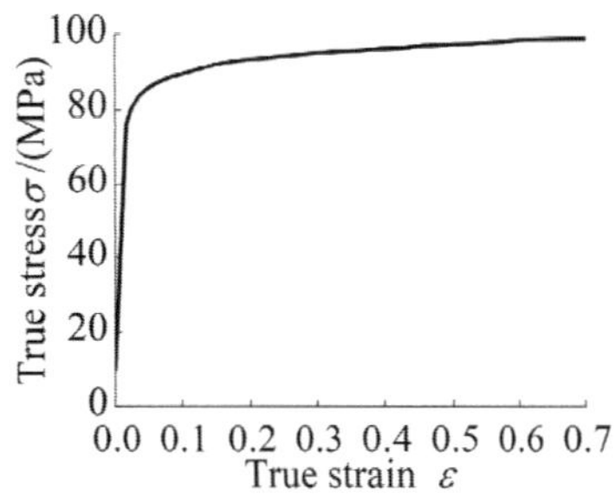


**Figure 3.** True stress-true strain curve of A356 at 360°C.

To verify the accuracy of the established finite element model, the actual spinning manufacturing process parameters are substituted into the finite element model, and the final spinning forming result of the aluminum alloy hub is obtained as shown in Figure 4a, from which it can be seen that the rim has better final formability. Figure 4b illustrates the comparison between the cross-sectional features of the rim section produced by both experimentation and simulation under the same operating conditions. The results indicate that a satisfactory level of agreement exists between the two methods with respect to the rim profile. The rim thickness measurements were taken at varying distances from the start point as presented in Figure 5a. A comparison between simulation and experimental results is illustrated in Figure 5b, demonstrating that the relative error of the rim wall thickness is below 10%. This indicates the established finite element analysis model is dependable and enables further simulation studies into aluminum alloy hub spinning.

a) Finite element simulation results b) Comparison of rim sections

**Figure 4.** Finite element simulation results and comparison of experimental and simulation results.

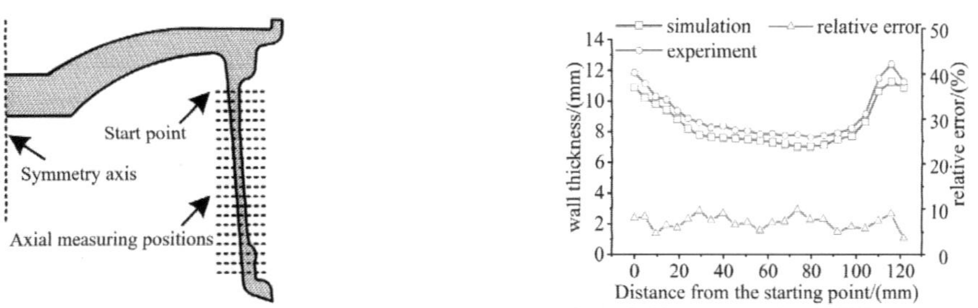


a) Measuring positions of the rim b) Comparison of simulated and experimental results

**Figure 5.** Positions for measuring rim wall thickness and comparison of rim wall thickness.

## 3. Results and discussion

After establishing a dependable finite element analysis model, we conducted four spinning simulation tests in accordance with the process parameters specified in Table 2. The purpose of these tests was to analyze the effect of process parameters such as feed rate, spindle speed, axial distance between the two rollers, and friction coefficient on the forming quality of the formed rims.

**Table 2.** Spinning process parameters in the finite element model.

| Simulation number | Feed rate $v_y$ (mm/s) | Spindle speed $\omega$ (rpm) | Axial distance $l_y$ (mm) | Friction coefficient |
|---|---|---|---|---|
| 1 | 3/4/5/6/7 | 220 | 3 | 0.15 |
| 2 | 5 | 180/200/220/240/260 | 3 | 0.15 |
| 3 | 5 | 220 | 3/4/5/6/7 | 0.15 |
| 4 | 5 | 220 | 3 | 0.1/0.15/0.20/0.25/0.30 |

### *3.1 Quality assessment criteria for spin forming*

The values of the wall thickness $d_i$ ($i$=1-3) with 50 samples and inner diameter $r_i$ ($i$=1-3) with 50 samples were measured at the same axial measurement position and circumferential direction of the rim after simulated forming, as illustrated in Figure 6. The simulation's dimensional accuracy was evaluated by calculating the average deviation of the wall thickness at the axial position from the design dimensions $\overline{\Delta d}$ and the average deviation of the inner diameter at the axial position from the design dimensions $\overline{\Delta r}$. Additionally, the homogeneity of the rim forming in the circumferential direction was assessed by calculating the standard deviation of the wall thickness $\sigma_d$ and the inner diameter $\sigma_r$.

**Figure 6.** Schematic of post-simulation measurement.

### *3.2 Impact of feed rate on rim forming quality*

The impact of varying feed rates of the spinning rollers on the quality of the rim is shown in Figures 7a, 7b, 7c, and 7d. Figure 7a illustrates that the rim's wall thickness deviation is minimal in the midpoint and highest at the starting point when using different feed rates. This trend can be linked to the finite element analysis model's simplification. The model in this study employs the beginning end face as a fixed boundary condition with little deformation, resulting in the highest wall thickness deviation value. Furthermore, the diagram illustrates an increase in wall thickness deviation of the rim between 0-60 mm from the starting point, corresponding with the increase in feed rates. However, in the second half of the rim, there is a decrease in wall thickness deviation with the rise of feed rate.

Figure 7b illustrates the correlation between the rim's inner diameter deviation and the feed rate. The results indicate that the deviation in the middle section of the rim is the most substantial for varying feed rates, while it is the smallest at both ends. This phenomenon is known as inner diameter expansion and should be avoided during the spinning process [9]. Meanwhile, it can be seen from the figure that the deviation of the inner diameter of the rim increases with the increase of the feed rate.

Figures 7c and 7d present how the standard deviation of wall thickness and inner diameter varies with the feed rate. The wall thickness uniformity in the center of the rim is the most optimal, while the inner diameter uniformity is the least favorable. Both the wall thickness and inner diameter uniformity increase as the feed rate increases, suggesting that lower feed rates produce more uniform forming.

In summary, the lower the feed rate of the spinning roller is, the better the dimensional accuracy and forming uniformity of the rim is, but a too-low feed rate will seriously reduce the production efficiency. Combined with the above figure, the 4 mm/s feed rate of the spinning roller has both better forming quality and production efficiency.

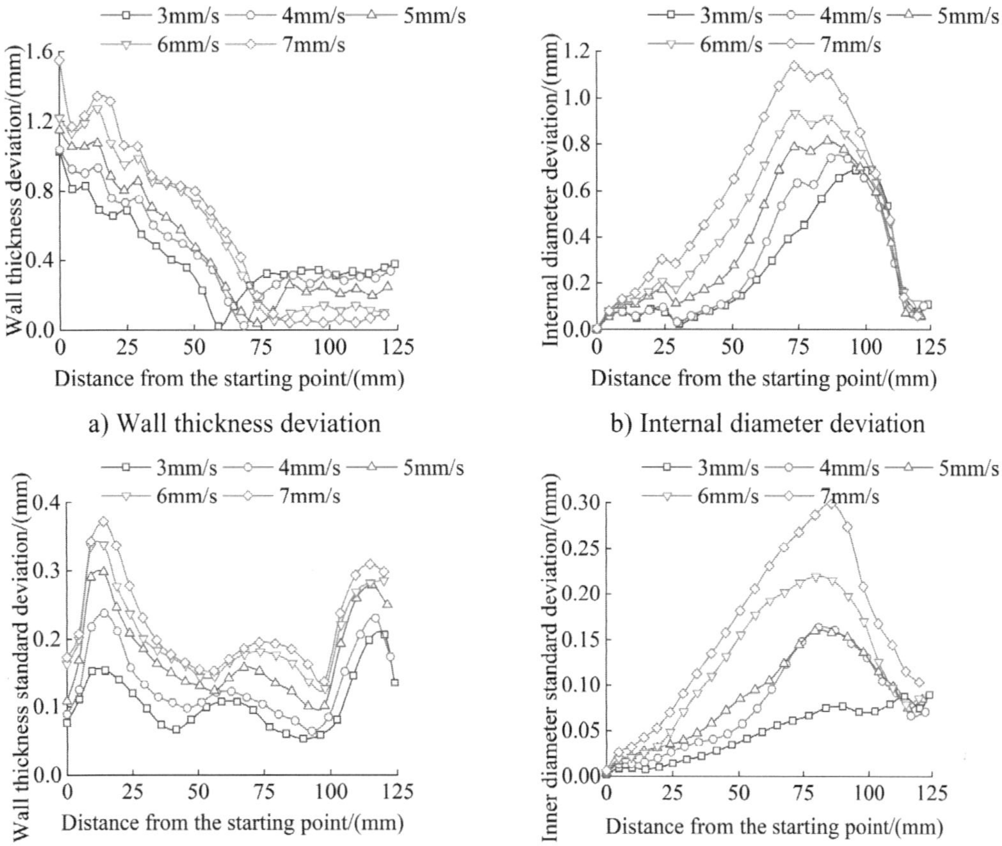


a) Wall thickness deviation b) Internal diameter deviation

c) Wall thickness standard deviation d) Inner diameter standard deviation

**Figure 7.** Influence of feed rate on rim forming quality.

*3.3 Effect of spindle speed on rim forming quality*

Figures 8a, 8b, 8c, and 8d show the effect of spindle speed on the forming quality of the rim. From Figure 8a, the wall thickness deviation of the rim decreases with the increase of spindle speed in the rim part between 0 to 60 mm from the starting point and increases with the rise of spindle speed in the second half, indicating that the spindle speed can be appropriately increased in the first half of the spinning forming process and can be decreased to reduce the wall thickness deviation of the rim in the second half.

The variation of the wall thickness standard deviation of the rim with spindle speed is shown in Figure 8c, from which the wall thickness standard deviation of the rim decreases as the spindle speed increases, so the wall thickness uniformity of the rim is better at higher spindle speeds.

The variation of internal diameter deviation and internal diameter standard deviation with spindle speed is shown in Figures 8b and 8d. Both internal diameter deviation and internal diameter standard deviation rise with increasing spindle speed, which indicates that the accuracy of internal diameter dimensions of the rim is improved with decreasing spindle speed. The same pattern has been found in the paper [10], because the centrifugal force on the material also increases by increasing the spindle speed, thus gradually deflecting it from the mandrel.

In summary, by selecting a spindle speed of 220 rpm, the rim can achieve both better wall thickness accuracy and internal diameter accuracy.

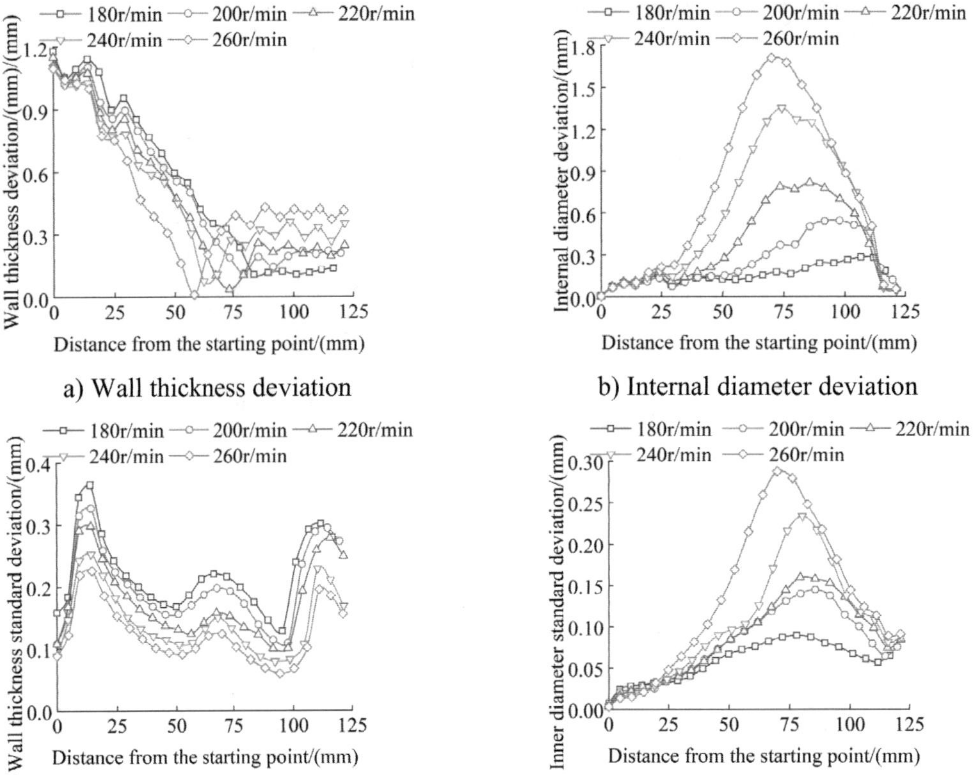


a) Wall thickness deviation b) Internal diameter deviation

c) Wall thickness standard deviation d) Inner diameter standard deviation

**Figure 8.** Influence of spindle speed on rim forming quality.

### *3.4 Effect of axial distance on rim forming quality*

The influence of axial distance on rim forming quality is shown in Figures 9a, 9b, 9c, and 9d. From Figures 9a and 9c, the deviation and the standard deviation of the wall thickness of the rim do not change significantly with the axial distance. From Figures 9b and 9d, the deviation and the standard deviation of the inner diameter of the rim increase with the rise of the axial distance from 3 mm to 6 mm, while the difference between the values of the axial distance of 7 mm and 3 mm is very small. Therefore, a better forming quality can be obtained by selecting 3 mm or 7 mm for the axial distance.

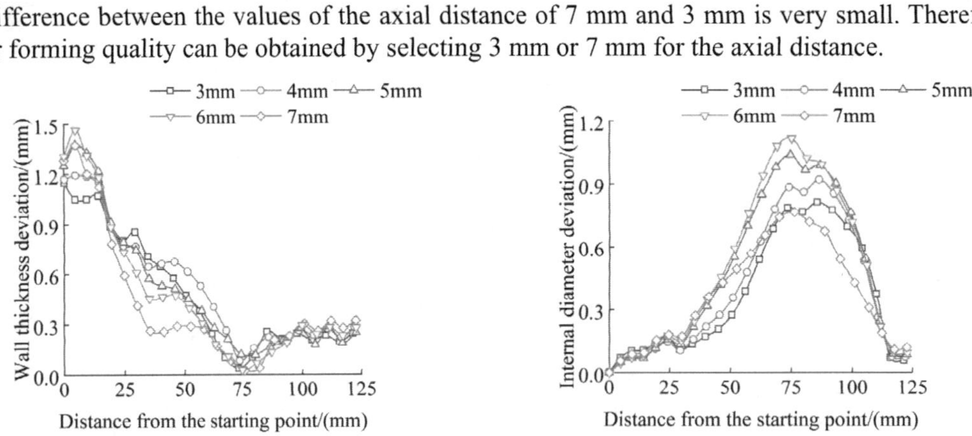


a) Wall thickness deviation b) Internal diameter deviation

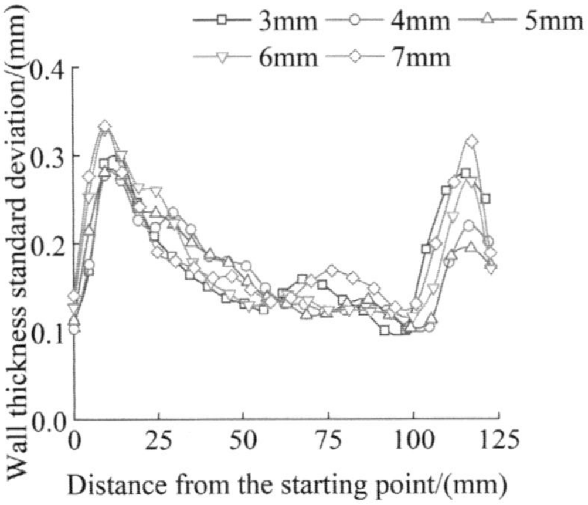


c) Wall thickness standard deviation

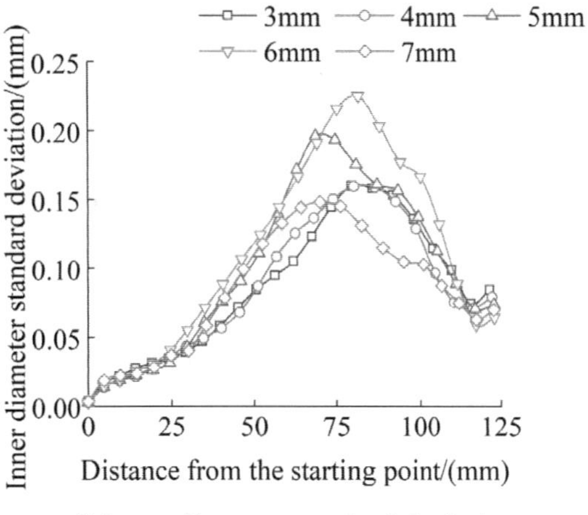


d) Inner diameter standard deviation

**Figure 9.** Influence of axial distance on rim forming quality.

### *3.5 Effect of friction coefficient on rim forming quality*

The impact of different friction coefficients on rim forming quality is shown in Figures 10a, 10b, 10c, and 10d. From Figure 10a, the wall thickness deviation of the rim increases with the rise of friction coefficient in the rim part between 0 to 60 mm from the starting point and decreases with the rise of friction coefficient in the second half, but there's very little difference between them. The wall thickness standard deviation, as shown in Figure 10c, increases with the increase of friction coefficient. Therefore, it can be concluded that to obtain higher wall thickness accuracy, the friction coefficient needs to be reduced appropriately.

It is obvious from Figure 10b and Figure 10d that the inner diameter deviation and the inner diameter standard deviation of the rim decrease as the friction coefficient increases, indicating that the larger the friction coefficient is, the higher the dimensional accuracy of the inner diameter of the rim is. This is because a high coefficient of friction allows the inner surface of the rim to be closely fitted to the mandrel during the spinning process. It is advisable to choose a coefficient of friction of 0.2. A too-high coefficient will reduce the dimensional accuracy of the wall thickness, so the coefficient should be neither too high nor too low.

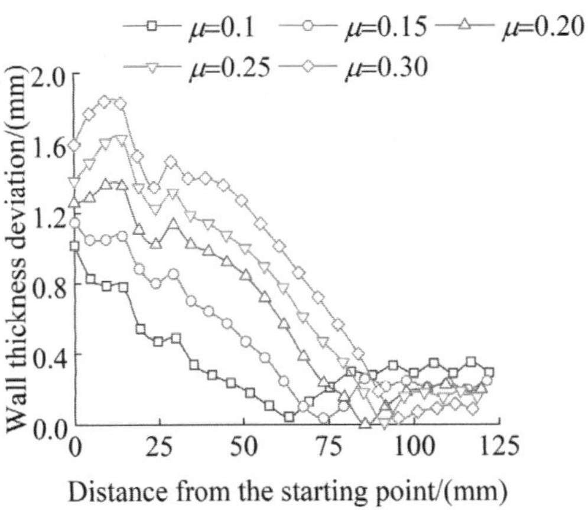


a) Wall thickness deviation

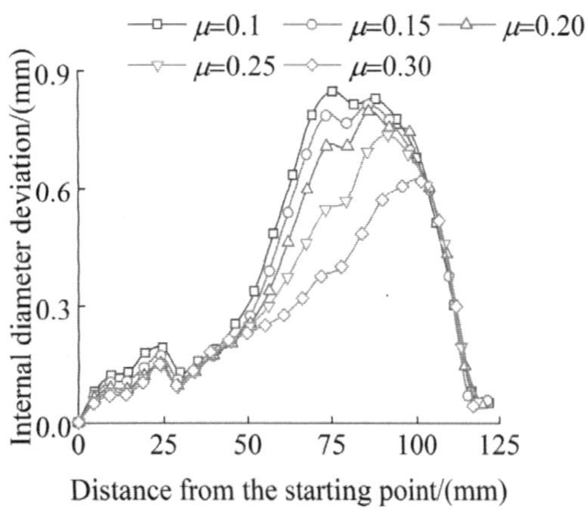


b) Internal diameter deviation

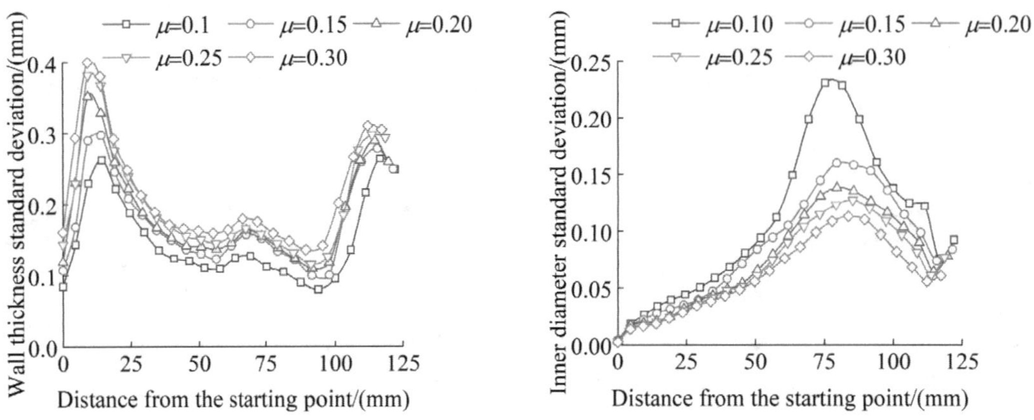


c) Wall thickness standard deviation d) Inner diameter standard deviation

**Figure 10.** Influence of friction coefficients on rim forming quality.

## 4. Conclusions

A finite element model of one-pass forming of an A356 aluminum alloy hub using dual spinning rollers was established, and the accuracy of the model was verified by experiment. A simulation experiment was conducted by using the established finite element model to investigate the effects of feed rate, spindle speed, axial distance between the two rollers, and friction coefficient on rim forming quality. The following conclusions were obtained:

(1) The feed rate of 4 mm/s can simultaneously achieve better forming accuracy and production efficiency. However, to get a higher production speed, the feed rate must be increased, which reduces dimensional accuracy.

(2) A higher spindle speed will improve the forming accuracy of the rim wall thickness, but it will cause serious expansion of the inner diameter of the rim, so the forming quality of the rim is best when the spindle speed is 220 rpm.

(3) The forming quality of the rim is better at the axial distance of 3 mm and 7 mm, and the forming accuracy of the inner diameter of the rim decreases with increasing axial distance in the range of 3 mm to 6 mm.

(4) The forming quality of the rim is better when the friction coefficient is 0.2. A larger coefficient of friction will reduce its wall thickness forming accuracy, while a smaller coefficient of friction will reduce its inner diameter forming accuracy.

## Acknowledgments

This work was supported by the Liuzhou Science and Technology Plan Project (No. 2021AAD0103) and the Zhuhai Industrial Core Research Project (No. 2220004002348).

## References

[1] Zhao H. W., Zhang R. B., and Bin Z. Y. (2018). A Review of Automotive Lightweight Technology. In: Proceedings of the 2nd International Conference on Mechanics, Electronics, Control and Automation Engineering (MECAE), Qingdao, pp. 30-31.

[2] Zhang M. X., Song H. W., Yan W. H., Yue F. L., Yang B. C., Chen S. F., and Zhang S. H. (2023). Properties and life prediction study on casting and spinning aluminum alloy wheel fatigue. Fatigue & Fracture of Engineering Materials & Structures, 46: 3, 446-3, 457.

[3] Chen S. Y., Shu X. D., Lu Y. J., and Zheng J. B. (2023). Effect of Spinning Process Parameters on Roundness of Aluminum Alloy Wheel Hub Surface and Analysis of Springback Characteristics. In: Proceedings of the ASME 2022 International Mechanical Engineering Congress and Exposition. Columbus.

[4] Bi D. S., Yang G., Chu L., Zhang J., and Wang Z. H. (2011). Numerical Simulation on Spinning Forming Process of Automotive Wheel Rim. Materials Science Forum, 704-705: 1, 458- 1, 464.
[5] Bi D. S., Yang G., Chu L., and Zhang J. (2011). Numerical Simulation Analysis of Spinning Forming Process of Tubeless Steel Wheel on Technical Parameters. Advanced Materials Research, 189-193: 2, 322-2, 328.
[6] Roy M. J. and Majer D. M. (2015). Analysis and modeling of a rotary forming process for cast aluminum alloy A356. Journal of Materials Processing Technology, 226: 188-204.
[7] Hwang S. Y., Kim N., and Lee C. S. (2015). Numerical Investigation of the Effect of Process Parameters during Aluminum Wheel Flow-Forming. Journal of Mechanical Engineering, 61: 471-476.
[8] Cha W. G., Hwang S. Y., Kim N., and Lee C. S. (2014). Analysis of the mechanism of backflow defect of the aluminum wheel flow forming. International Journal of Precision Engineering and Manufacturing, 15: 1, 075-1, 080.
[9] Yu Z. Q., Meng Y., Zhou Y., and Zhao Y. X. (2023). Study on multi-roller flow forming process of aluminum alloy cylinders with cross ribs. The International Journal of Advanced Manufacturing Technology, 125: 3, 489-3, 501.
[10] Ebrahimi M., Tabei K. H., Naseri R., and Djavanroodi F. (2018). Effect of flow-forming parameters on surface quality, geometrical precision, and mechanical properties of titanium tube. Proceedings of the Institution of Mechanical Engineers, Part E: Journal of Process Mechanical Engineering, 232: 702-708.

# Medium-range order structure analysis of metallic glasses based on reverse Monte Carlo modeling

**Yuan Liu[1], Jingrun Luo[1], Xin Huang [1,2] and Jiaxing Li[1]**

[1]Institute of Systems Engineering, China Academy of Engineering Physics, Mianyang, 621999, PR China

[2] Corresponding author's e-mail: huangx@caep.cn

**Abstract:** The elucidation of amorphous alloy characteristics presents a formidable challenge due to the absence of crystalline arrangements and enigmatic imperfections. This study aimed to tackle this challenge through a comprehensive investigation utilizing synchrotron X-ray diffraction (XRD), extended X-ray absorption fine structure spectroscopy (EXAFS), *ab initio* molecular dynamics (AIMD) simulations, and reverse Monte Carlo (RMC) modeling. The primary objective was to delve into the medium range order (MRO) structure of metallic glasses (MGs) containing $Cu_XZr_{100-X}$, wherein X represents values of 50, 56, 60, and 64. The results obtained from this research demonstrate that an increase in the Cu atom concentration within Cu-Zr MGs leads to the formation of complete icosahedral clusters and high-coordination polyhedral connections, thereby bolstering the stability and heterogeneity of local structures. Examination of the stiff full icosahedral bones shows that defects exist in the network connections on the MRO, resulting in a lack of connectivity in the positive icosahedron. Conversely, the density of defects decreases with an increase in Cu content. The observation suggests that the observed low MRO defects in $Cu_{64}Zr_{36}$ MGs may be accountable for their excellent glass-forming ability and kinetic stability.

## 1. Introduction

Microscopic defects are ubiquitous in various materials and significantly impact their macroscopic mechanical behavior [1]. As a result, a considerable number of microstructure-based studies have been conducted to comprehend and characterize the intrinsic material defects and reveal their response mechanisms under external loading [2]. Recently, Voronoi analysis has been introduced in metallic glasses (MGs) to provide better insights into the short-order range microscopic arrangement of such disordered structures [2, 3]. On this basis, the Voronoi index [2] and the local five-fold symmetry (LFFS) [4] parameter, among others, have been proposed to establish a relationship between MG structure and properties. Previous molecular dynamics simulations [5] have unveiled that MGs possess defective forms like voids and MRO boundaries at scales of 1-2 nm, which are closely involved in porosity and shear transformation zone events [6]. However, pure MD simulation cannot entirely reflect objective facts due to significant differences in time and space scales compared to the real world. To remove this limitation, the RMC method was introduced, combining AIMD and synchrotron XRD to remodel the three-dimensional atomic structure of the Cu-Zr series [7].

However, the partial pair distribution function (ppdf) in AIMD used as a constraint in the RMC modeling has some limitations. While density flooding theory-based AIMD can represent some main characteristics of glass in the objective world [8], it still differs from the real cooling rate of the glass in the time domain, making this modeling method incomplete. To overcome this challenge, the K-edge spectrum of Cu/Zr obtained by the EXAFS experiment was integrated as a supplementary constraint in the RMC simulation process. The combination of AIMD, EXAFS, and synchrotron XRD data through RMC simulations leading to the acquisition of an MGs atomic structure model with diffraction characteristics is very similar to that of the actual sample. Subsequently, we applied our developed technique, as outlined in reference [9], to elucidate the atomic structure of MGs and provide robust evidence regarding the presence of medium-range order (MRO) defects. Additionally, a comparative analysis was conducted between the previous 3D atomic models [7] to discern the disparities and investigate their underlying origins.

## 2. Method

### 2.1. *Experiment*

The four alloys (including $Cu_{50}Zr_{50}$, $Cu_{56}Zr_{44}$, $Cu_{60}Zr_{40}$ $Cu_{64}Zr_{36}$) were synthetically prepared in the laboratory, followed by the acquisition of high-resolution synchrotron XRD datasets at an energy of 17.9 keV (λ = 0.693 Å). The complete data processing workflow was executed employing PDFgetX3 software [10]. To obtain Cu and Zr K-edge EXAFS spectra, a transmission mode setup was utilized with subsequent data processing performed using Athena software [11]. The AIMD calculation involves all 256 atoms per sample model and is performed on a supercomputing platform using CP2K [12] software. The specific preparation process, synchrotron radiation experimental, and AIMD calculation procedures are detailed in previous work [7, 9].

### 2.2. *RMC simulation*

Reverse Monte Carlo (RMC) computational simulation methodologies facilitate the generation of three-dimensional structural models that align with experimental data, primarily derived from diffraction techniques. Here, $S^{Exp}(Q)$ from synchrotron, XRD, Cu/Zr K-edges from EXAFS, and three ppdfs from AIMD (data from reference [7]) are employed as constraints within the RMC simulation. It is widely acknowledged that incorporating both long-range ordered structure information from synchrotron X-ray diffraction and short-range ordered structure information dataset from EXAFS in the RMC simulation can enhance the reliability of the final structure. In the RMC simulation, each sample contains 20000 atoms with dimensions of 7×7×7 $nm^3$. The initial simulated boxes were randomly placed according to the material-chemical composition ratio. The whole RMC process was performed using RMC_POT software [13]. The simulation process was closely monitored utilizing an error factor, denoted as $\chi^2$, which serves as a metric for quantifying the disparity between the experimental and simulated data.

$$\chi^2 = \sum_i^m (S^{Cal}(Q_i) - S^{Exp}(Q_i))^2 / \sigma(Q_i)^2 \tag{1}$$

In Equation (1), $S^{Exp}(Q_i)$ and $S^{Cal}(Q_i)$ represent the experimental and simulation data, respectively. The variable $i$ pertains to the intersection point between the two datasets concerning $Q$, with $m$ denoting the count of such intersections. Furthermore, $\sigma(Q_i)$ signifies the experimental error associated with the ith data point. It is important to highlight that during the fitting of EXAFS data, only $S^{Exp}(Q_i)$ necessitates replacement with the corresponding K-edge spectral data. Throughout the simulation procedure, a state of dynamic equilibrium is attained as $\chi^2$ progressively decreases, indicating a favorable agreement between the simulation model and the experimental data.

## 3. Results and discussion

### 3.1. *Reverse monte carlo model*

The structural characteristics of four MGs with the composition $Cu_XZr_{100-X}$ (X= 50, 56, 60, 64) were systematically analyzed by integrating six unique data sets for each constituent material. These encompass three separate partial pair distribution functions (PDFs) generated from AIMD, structure factors derived from synchrotron XRD, and two distinct sets of K-edge spectra (Cu and Zr) acquired through EXAFS experiments. Figure 1 depicts the fit outcomes for all six constituents, indicating a strong concurrence between the computed results and the experimental measurements, which provides direct evidence that the RMC program produces a dependable and trustworthy configuration. In contrast to the previous fitting outcomes [7] that did not incorporate EXAFS data, our current K-edge spectra fitting results for Cu and Zr exhibit a remarkable consistency with the experimental measurements. Despite some discrepancies observed in the fitting process to ppdfs, as illustrated by the slightly deviated fitting outcome in Figure 1(c), the inclusion of EXAFS data has significantly enhanced the tractability of our model. Furthermore, it is important to acknowledge the phase correction applied to the K-edge spectra obtained from EXAFS measurements.

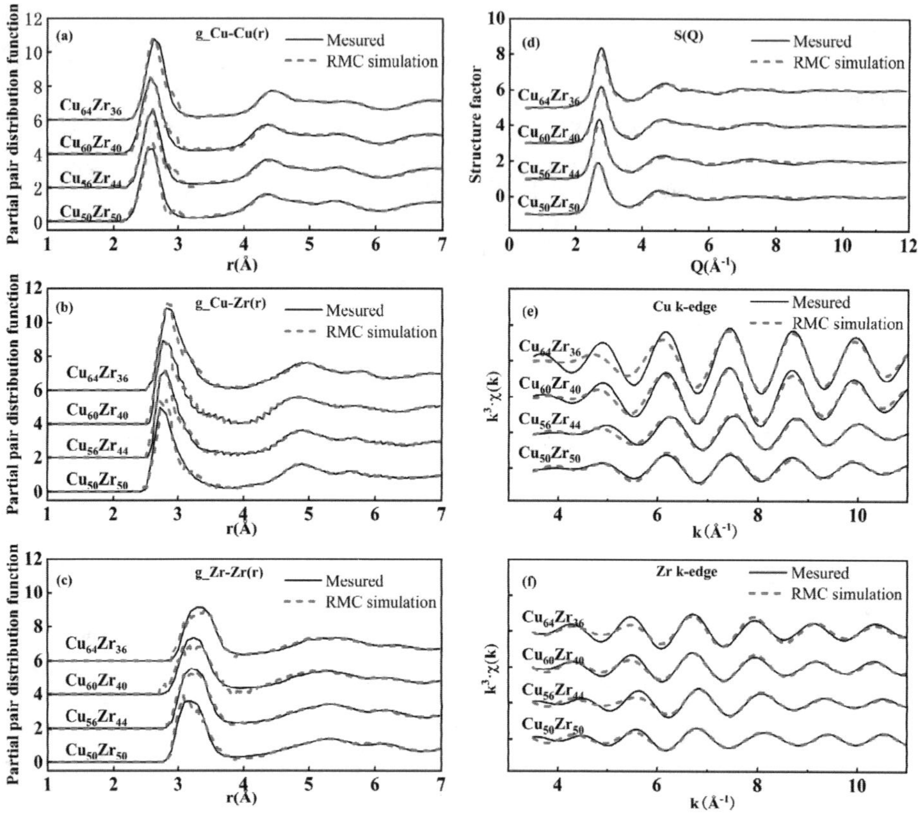


**Figure 1.** RMC Calculation results. Each subfigure (a-f) corresponds to one of the four MGs investigated, namely $Cu_XZr_{100-X}$ (X = 50, 56, 60, 64).

*3.2 Short-range order structure*

We conducted Voronoi tessellation analysis [14] to assess the short-range order of the four MGs samples investigated in this study. The statistical analysis of the twelve most prevalent cluster atoms is presented in Figure 2, which demonstrates that icosahedral-like clusters are the most abundant, particularly as the Cu atom content increases. Moreover, we quantified the 3 (including<0, 0, 12, 0>, <0, 1, 10, 2>, <0, 2, 8, 2>) occurrence of cluster atoms exhibiting high levels of fifth-fold symmetry and the highest content, which account for 0.1497%, 0.18812%, 0.25137%, and 0.25941% of the composition in four alloys MGs, respectively as shown in Table 1. These findings suggest that the solid-like region with high resistance to loading [2] expands as the Cu content increases, indicating that $Cu_{64}Zr_{36}$ has a more stable structure than the other three compositions. Furthermore, we compared the RMC model before and after incorporating K-edge spectra obtained from EXAFS experiments, as detailed in Table 1. The findings indicate a minor reduction in the quantity of high-content icosahedral-like clusters while preserving the overall trend after the inclusion of K-edge spectra.

Furthermore, through our analysis, we have recognized that the incorporation of K-edge spectra into the simulation leads to the development of greater numbers of icosahedral-like clusters with high coordination numbers, such as <0, 1, 10, 4>, <0, 1, 10, 5>, and so on, which could account for the decrease in statistical rate. This observation supports the idea that the RMC model improves the packing distribution of certain elements with high coordination numbers, such as Zr, by introducing new fitting parameters.

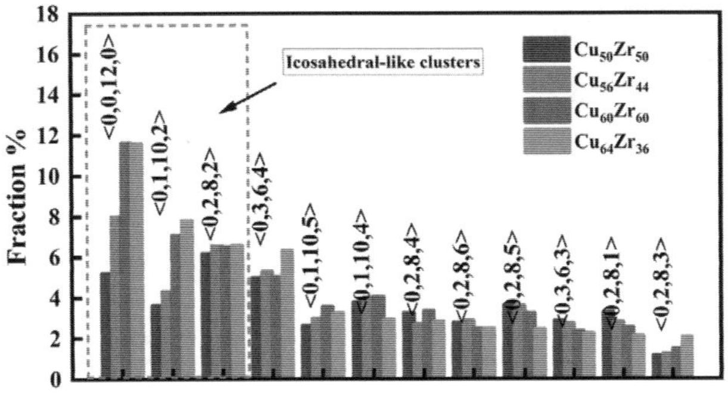


**Figure 2.** Frequency distribution of Voronoi polyhedron of Cu-Zr MGs.

**Table 1.** Statistics of high-content icosahedral-like clusters of Cu-Zr MGs from different RMC model.

| Parameters of the RMC model | ppdfs+S($Q$) [7] | ppdf+S($Q$)+K-edge |
|---|---|---|
| $Cu_{50}Zr_{50}$ | 0.15276 | 0.14970 |
| $Cu_{56}Zr_{44}$ | 0.19010 | 0.18812 |
| $Cu_{60}Zr_{40}$ | 0.24944 | 0.25137 |
| $Cu_{64}Zr_{36}$ | 0.26261 | 0.25941 |

In the subsequent analysis, LFFS parameters were employed to visualize the local structural patterns of the four distinct components, as depicted in Figure 3. Our observations indicate that the distribution of LFFS is uneven, with the full icosahedron, <0, 0, 12, 0>, exhibiting a red color and an LFFS value of 1. Moreover, an increase in the Cu content was found to correlate with a broader range of red coloration, indicating a higher abundance of dominant class icosahedra as depicted in Figure 1 and Table 1.

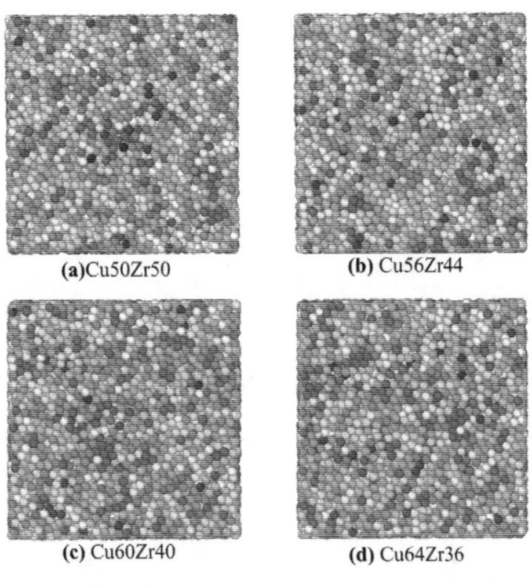


**Figure 3.** The distribution of LFFS in distinct MGs components is illustrated, with the red indicating atoms exhibiting a high LFFS and the blue representing a low LFFS.

*3.3 Medium range order structure*

Extensive reports [15] have highlighted that the MRO characteristics of MGs tend to provide greater insight into their dynamics, thermal properties, and plastic deformation. In our previous investigations, we employed a customized methodology [9] to extract the core structures of complete icosahedral clusters. Subsequently, we focused on characterizing the structural defects associated with MRO in our subsequent studies [7]. Specifically, we identified a type of missing degree of full icosahedral connectivity represented by weakly connected regions, which play a critical role in glass failures [5, 9].

Similarly, in our new RMC model, the full icosahedral bones were extracted using a self-developed procedure as shown in Figure 4. An evident observation from our analysis is that the $Cu_{50}Zr_{50}$ sample with low Cu content contains numerous super large cluster groups. With an increasing Cu content, a super bone progressively emerges, encompassing a significant portion of the atoms. The presence of MRO defects within this structure is manifested through weakly connected regions between freely formed complete icosahedral clusters and the prominent bones of the configuration. Moreover, a specific defect structure was extracted, as illustrated in Figure 5(b). Additionally, Figure 5(a) showcases the MRO defect structure obtained from the RMC model without the incorporation of K-edge spectrum data [7]. A comparison of Figures 5(a) and (b) highlights the presence of a similar boundary layer, which represents the MRO defect in both RMC models. Furthermore, this defect was also identified in our model through molecular dynamics simulations [5], suggesting that this MRO defect is prevalent across various three-dimensional atomic models.

Our investigation encompassed a comprehensive statistical analysis of the atomic structure characteristics for each component, as detailed in Table 2. Notably, a significant rise in the number of full icosahedra is observed as the Cu content swiftly escalates from 50% to 60%, indicating an expansion of the entire full icosahedral cluster of atoms and the gradual formation of an extensive bone network. Although no further increase in full icosahedra is observed between Cu content values of 60-64%, the solid-like regions (groups of full icosahedral clusters) tend to expand, suggesting the emergence of potential MRO defect regions. However, there is an overall increase in the maximum atomic number of bones, suggesting enhanced structural stability as Cu atoms are introduced within a specific range. This observation provides partial insight into the favorable kinetic properties and strong glass-forming ability exhibited by the $Cu_{64}Zr_{36}$ system, which is consistent with our earlier research outcomes [7].

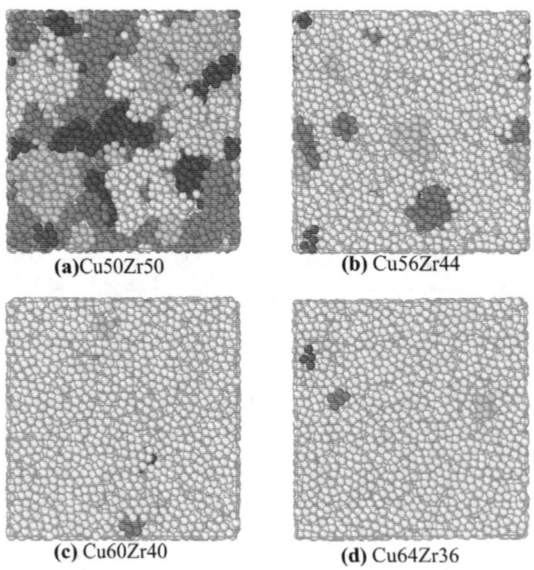


**Figure 4.** The cube bones of Cu-Zr MGs are depicted in Figures (a–d) correspondingly. Clusters that are interconnected have been assigned the same color for visualization purposes.

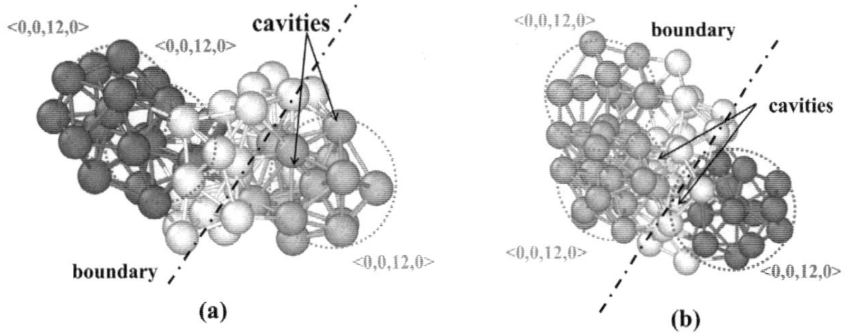


**Figure 5.** MRO defects. (a) RMC model without EXAFS data [7] (b) RMC model includes EXAFS data.

**Table 2.** Statistical analysis was conducted to examine the parameters of full icosahedral bones in Cu-Zr MGs.

| Sample Components | $Cu_{50}Zr_{50}$ | $Cu_{56}Zr_{44}$ | $Cu_{60}Zr_{40}$ | $Cu_{64}Zr_{36}$ |
|---|---|---|---|---|
| Number of <0, 0, 12, 0> cluster | 1022 | 1569 | 2282 | 2276 |
| Number of solid-like regions | 122 | 51 | 14 | 15 |
| Number of atoms of <0, 0, 12, 0> cluster | 8220 | 10793 | 13209 | 13643 |
| Number of atoms of bone | 1518 | 9589 | 12983 | 13403 |

## 4. Conclusion

In summary, a novel three-dimensional atomic model was constructed through the integration of AIMD, synchrotron XRD, and EXAFS data using RMC simulations. The remarkable agreement observed between the simulated and experimental data provides substantial support for the credibility of this newly proposed atomic structure. Subsequently, we conducted a comprehensive analysis of the characteristic properties of both the previous and current models employing Voronoi tessellation, LFFS, and an assessment of MRO defects. The specific conclusions derived from this analysis are enumerated below:

(1) The inclusion of EXAFS data in the RMC modeling process offers increased structural information and enhances the feasibility of the proposed atomic model.

(2) Our investigation into MRO defects revealed that these defects are a common intrinsic feature across various models.

(3) It can be found that the full icosahedral atomic bone expands with increasing Cu content over a specific range, resulting in a more stable structure. This observation provides insight into why $Cu_{64}Zr_{36}$ MGs exhibit superior amorphous formation ability and strong stable kinetic properties.

## Acknowledgments

This work is financially supported by the National Natural Science Foundation of China under grant number 11772313.

## References

[1] Huang X, et al. (2014) Ductile-to-brittle transition in spallation of metallic glasses. J. J Appl Phys, 116(14): 445-69.

[2] Sheng H W, et al. (2006) Atomic packing and short-to-medium-range order in metallic glasses. J. Nature, 439(7075): 419-25.

[3] Sun Y L, Shen J. (2009) Icosahedral ordering in $Cu_{60}Zr_{40}$ metallic glass: Molecular dynamics simulations. J. Journal of Non-Crystalline Solids, 355(31-33): 1557-60.

[4] Hu Y C, et al. (2015) Five-fold symmetry as indicator of dynamic arrest in metallic glass-forming liquids. J. Nat Commun, 6: 8310.
[5] Huang X, et al. (2016) Intrinsic structural defects on medium range in metallic glasses. J. Intermetallics, 75: 36-41.
[6] Wang X D, et al. (2008) Atomic structure of binary Cu [sub 64.5] Zr [sub 35.5] bulk metallic glass. J. Appl Phys Lett, 92(1).
[7] Liu Y, et al. (2023) Investigation of Medium Range Order Defects in $Cu_xZr_{100-x}$ (x = 50, 56, 60, 64) Metallic Glasses Using Reverse Monte Carlo Modeling. J. Metals, 13(1).
[8] Yu Q, et al. (2016) Atomic packing in Fe-based metallic glasses. J. Acta Mater, 102: 116-24.
[9] Liu Y, et al. (2023) Identification of medium range order defects and their critical effect on spallation of $Cu_{64}Zr_{36}$ metallic glass. J. J Alloys Compd, 932.
[10] Juhás P, et al. (2013) PDFgetX3: a rapid and highly automatable program for processing powder diffraction data into total scattering pair distribution functions. J. J Appl Crystallogr, 46(2): 560-6.
[11] Ravel B, Newville M. (2005) ATHENA, ARTEMIS, HEPHAESTUS: data analysis for X-ray absorption spectroscopy using IFEFFIT. J. J Synchrotron Radiat, 12(Pt 4): 537-41.
[12] Kuhne T D, et al. (2020) CP2K: An electronic structure and molecular dynamics software package - Quickstep: Efficient and accurate electronic structure calculations. J. J Chem Phys, 152(19): 194103.
[13] Gereben O, et al. (2007) A new version of the RMC++ Reverse Monte Carlo programme, aimed at investigating the structure of covalent glasses. J. OPTOELECTRONICS AND ADVANCED MATERIALS-RAPID COMMUNICATIONS, 9(10): 3021-7.
[14] Voronoi G. (1908) Nouvelles applications des paramètres continus à la théorie des formes quadratiques. Premier mémoire. Sur quelques propriétés des formes quadratiques positives parfaites. J. Journal für die reine und angewandte Mathematik (Crelles Journal), 1908(133): 97-102.
[15] Shen Y, Perepezko J H. (2017) Al-based amorphous alloys: Glass-forming ability, crystallization behavior and effects of minor alloying additions. J. J Alloys Compd, 707: 3-11.

# Continuous carbon fiber 3D printing partitioning method based on stress distribution

**Wen Qiu[1], Fan Zhang[1,2], Yuegang Tan[1], Jun Zhang[1] and Yiwen Tu[1]**

[1]Mechanical and electrical engineering, Wuhan University of Technology, Wuhan, 430070, China

[2]Corresponding author's e-mail: fanzhang@whut.edu.cn

**Abstract.** 3D printing of continuous carbon fiber composites presents a need for the directivity of fiber distribution. For parts with actual working conditions, the stress distribution of the same part under the same stress condition is different, but the existing partitioning algorithm mainly fills the path of the subregion according to the shape of the part, and does not combine the working condition of the part. To solve this problem, a partitioning algorithm based on stress contour density is proposed. Based on the results of finite element analysis, the stress contour of the parts is obtained, and the stress concentration area is partitioned by calculating the stress contour density. The stress concentration area is filled with a more suitable path to improve the performance of the parts.

## 1. Introduction

The application of 3D printing technology combined with continuous carbon fiber composite materials in the manufacturing field can reduce the complexity of the carbon fiber molding process, improve molding efficiency, give play to molding advantages, and have broad application prospects [1-2]. In the actual use of composite material sheets in automobiles, the mechanical opening is required to facilitate connection and maintenance, etc., but the opening will destroy the continuity of fibers, produce stress concentration, and reduce the mechanical properties of parts [3-4]. Due to the directivity and continuity of continuous carbon fiber composite materials, the combination of the path planning of parts and the actual working conditions can effectively improve the performance of parts. For the continuous carbon fiber 3D printing path research and the improvement of part performance, domestic and foreign scholars have done a lot of research.

Zhai et al. [5] used a generalized Voronoi diagram and duality operation to divide molecular regions for porous structures, and filled the subregions with helical curves to achieve path continuity in porous regions and ensure mechanical properties around holes. Wang et al. [6] developed a stress vector tracking algorithm through topology optimization to generate a print path along the load transmission path, showing better load-bearing performance. Li et al. [7] used topological optimization to analyze the load transfer path and generate the print path, which reduced the stress concentration but changed the original shape of the part. Li et al. [8] used grayscale images to delimit molecular regions, and the sub-regions generated print paths aligned with the main direction field to improve the tensile strength of the parts. The stress concentration area of the part is determined and targeted reinforcement is carried out, but the partition boundary of the method is inaccurate and irregular, which will affect the print quality during the process of carbon fiber 3D printing. Kamil et al.[9] Steuben et al. used Von Mises stress fields to generate horizontal set filling paths, but these paths were unevenly spaced due to the irregular shape of the stress fields.[12] Yusuke et al. used genetic algorithm to adjust the placement of fibers and optimize the orientation of bent carbon fibers, and the optimized fiber position showed excellent fracture improvement.[13]

Determined the interval of layers to be reinforced through the finite element calculation results, reflecting the selective reinforcement of carbon fiber between different layers, but no selective reinforcement was carried out within the same layer. Combined with the finite element analysis, domestic and foreign scholars have studied and planned the 3D printing path of continuous carbon fiber composite materials according to the stress distribution results in the finite element calculation, and mainly enhanced the performance of parts by generating paths through load transfer. Chen et al. [10] extracted the equivalent curve of the stress field to generate the path of fiber reinforcement and improve the mechanical properties of the parts. However, due to the irregular shape of the stress field, the shape and spacing of these paths are not uniform.

It can be seen that the existing path planning method still has the following problems: (1) the zoning path planning is mainly divided by the shape characteristics of the parts, and the zoning path planning is not carried out according to the actual working conditions of the parts, which cannot give full play to the anisotropy characteristics of continuous carbon fiber composites; (2) Due to the irregularity of the stress field, the shape of the area after zoning according to the stress field boundary is irregular, which will affect the molding effect of the parts during the 3D printing of carbon fiber composite materials.

In this paper, a zonal path planning algorithm based on stress contour lines is proposed. The stress contour lines are obtained through finite element analysis of the specific use environment of the parts, and the stress concentration areas are determined by calculating the density of the stress contour lines. On this basis, different paths are selected for different areas for targeted enhancement.

## 2. Based on stress contour partitioning algorithm

The traditional partition method based on the shape of the part is not specific to the working condition, and the irregular shape of the partition boundary based on the finite element analysis will affect the print quality. In order to solve the problems of the above partitioning methods, a partitioning algorithm based on stress isoline is proposed in this paper. Partitioning is to process STL surfaces after processing the finite element analysis data. The partitioning algorithm inputs STL files with model surface information and the finite element analysis result data, outputs the regional boundary information of the partition after algorithm calculation and processing, and segments the STL surfaces according to the boundary information to obtain regions with different stress concentration degrees. The main flow of the algorithm is shown in Figure 1

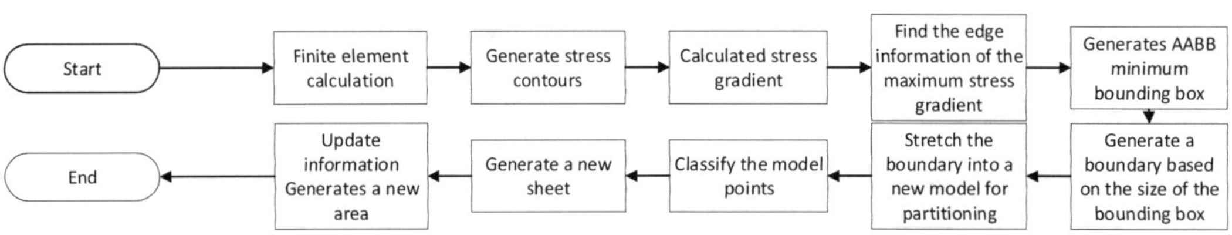


**Figure 1.** Flow chart of stress contour partitioning algorithm.

### *2.1 Method of obtaining stress contour*

In the process of identifying the stress concentration area and non-concentration area of the model, the stress distribution in the specific model cannot be obtained only from the simple distribution of isolines. Therefore, the partition calculation should be carried out by combining the relevant data of the model and the stress isolines, and the specific process of generating isolines is shown in Figure 2:

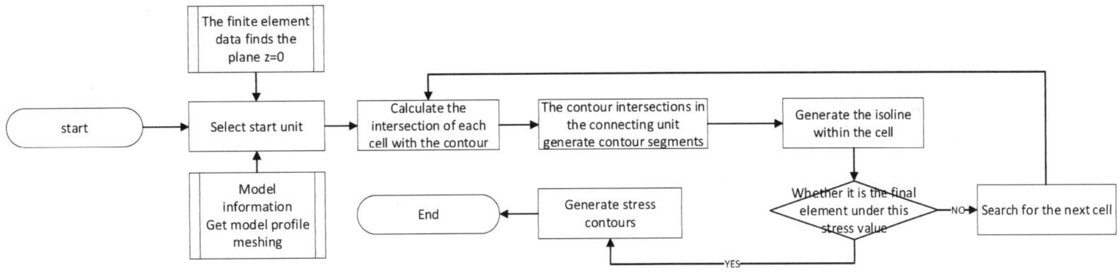


**Figure 2.** Contour generation process.

For the grid cell four vertices$(X_0, Y_0), (X_0, Y_1), (X_1, Y_0), (X_1, Y_1)$, four vertices corresponding stress values respectively $F_{00}, F_{01}, F_{10}, F_{11}$, forming an isoline with a stress value of $F_t$ in the element. The calculation of the intersection point between grid elements and isolines is mainly to find the intersection point between edge lines and isolines of each element. Assuming that the function changes linearly within the element, the method of vertex judgment and edge interpolation can be used to calculate the intersection point. The specific steps are as follows:

(1) The grid point is divided into two states, IN and OUT, indicating that the point is inside or outside the contour line. If $F_{ij} \leq F_t$, then the vertex $(X_i, Y_j)$ is IN, denoted as $-$. If $F_{ij} \geq F_t$, then the vertex $(X_i, Y_j)$ is OUT, denoted as $+$.

(2) If all four vertices of the cell are $+$ or $-$, then the grid cell has no intersection with the contour of $F_t$. Otherwise, for cell points where the two vertices are $+$ and $-$, linear interpolation calculates the intersection of the contours on this edge. If the two vertices $(X_0, Y_0)$ is $-$, $(X_0, Y_1)$ is $+$, then the intersection point is

$$\begin{cases} X_t = X_0 \\ Y_t = Y_0 + \dfrac{F_t - F_{00}}{F_{01} - F_{00}}(Y_1 - Y_0) \end{cases} \tag{1}$$

(3) After calculating the intersection point between the contour line and the grid element in each cell, the contour line segment in the cell is formed by using the intersection point.

(4) The contour lines in the unit are connected to form the contour lines in the unit.

(5) The resulting contours are labeled.

*2.2 Calculation of density of stress contour*

The stress contour can clearly describe the stress variation in the model in the finite element calculation, but it does not obtain the coordinate range, which makes it impossible to accurately partition in the printing. Therefore, the calculated stress contour gradient information in this section is used to fit the intensity of stress contour change, and the density of stress contour. The Sobel algorithm is widely used in image edge detection [11].

The predetermined threshold T=0 is set, and each pixel of the gradient image is compared with the predetermined threshold T. If the gradient exceeds the threshold, the pixels are regarded as edges. Otherwise, they are non-edge points, and all pixels are regarded as edge points. The gradient values of all pixels are compared to obtain the edge point with the largest gradient value, that is, the boundary of the area with dense stress isolines, and the edge information of the area with concentrated stress, that is, the edge isoline information.

The contour lines in the finite element analysis were obtained by the grid sequence method in 1.1. After the finite element analysis, the contour lines have irregular shapes such as sharp corners. The printing quality of continuous carbon fiber will be affected if the region is divided by the contour lines as the boundary. The region calculated by contour density is used to generate an AABB minimum bounding box composed of regular shapes with the information of the points on the contour boundary of the region, and the center circle surrounded by the stress concentration area is generated with the information of the AABB minimum bounding box.

**3. Simulation experiment**

In this paper, the composite plate with holes is taken as a typical test sample for calculation, analysis and experimental verification. The length of the perforated plate model is 100 mm, the width is 40 mm, the diameter of the central circular hole is 12 mm, and the thickness is 4 mm. The left end of the perforated plate is fixed, and the displacement W is applied to the right end. The stress partitioning results of the parts are obtained through the partitioning calculation in Section 1, and the partitioning path planning for different areas is carried out according to this, as shown in Figure 3 and Figure 4.The contour line of stress distribution is shown in Figure 5, and the partition range calculated according to the contour line is shown in Figure 6.

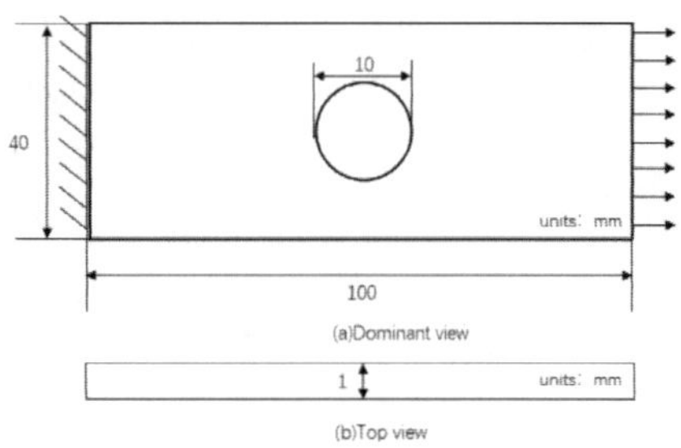


**Figure 3.** Composite plate with holes.

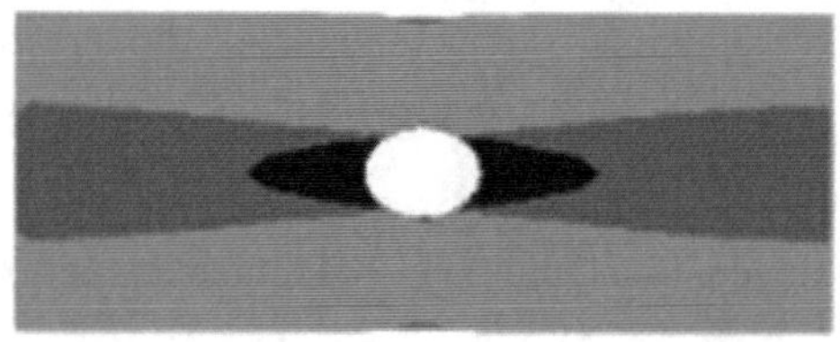

**Figure 4.** Finite element analysis cloud image.

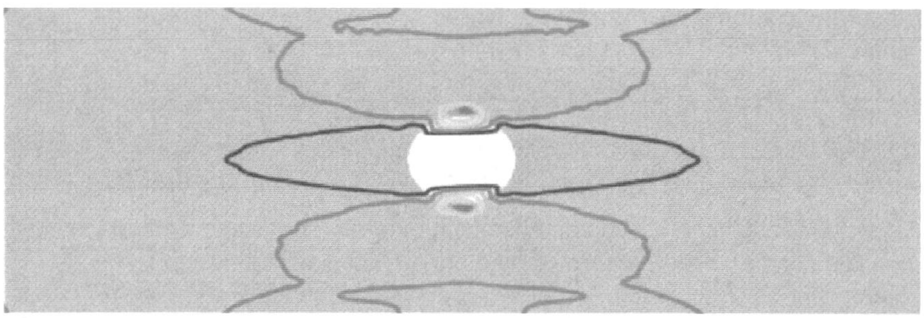

**Figure 5**. Contour calculation diagram.

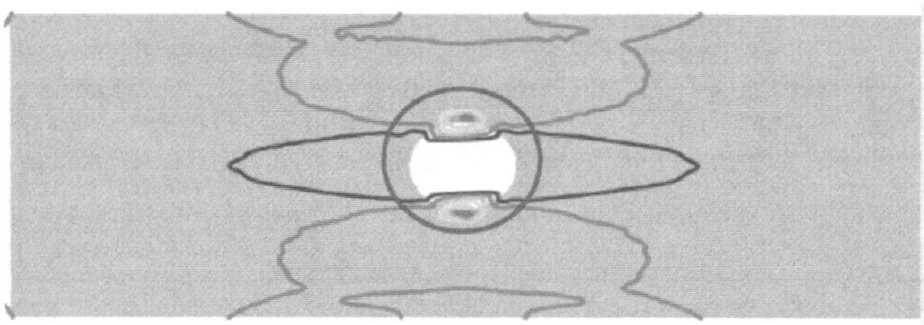

**Figure 6.** Partition result.

*3.1 Path planning simulation*

The simulation experiment of traditional path planning was carried out on the perforated plate. Figure 7 shows the linear scanning and filling path. It can be seen that the traditional path planning only has one circle of continuous path for the planning of the stress concentration area of the perforated plate. Under the working condition of the formed part under tension, the hole part will break quickly, affecting the use of the formed part.

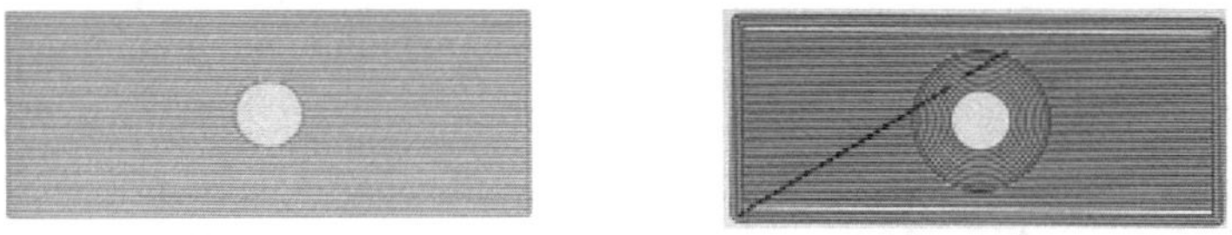

(a) Traditional path planning method (b) Partition path planning method

**Figure 7**. Comparison diagram of path simulation planning.

Compared with traditional path planning, path planning is carried out for different regions according to the results of zoning. For the model with holes, a continuous path is formed around the hole in the stress concentration area to ensure performance, and a straight filling path is selected for the non-stress concentration area when the force direction of the part is satisfied.

### *3.2 Print test*

After the carbon fiber composite material printing experiment was carried out on the above paths through the double-nozzle printer, the molded parts realized different path planning in different areas of the same model, realized the printing of different paths in the same layer, and realized the combination with the actual working conditions of the parts, as shown in Figure 8.

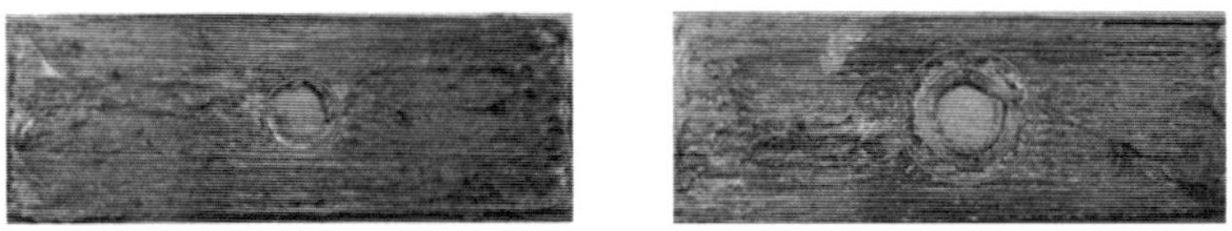

(a) Straight path print diagram (b) Partition print

**Figure 8.** Partition printing and straight line printing results.

## 4. Conclusion

By processing the finite element analysis data of the parts, the stress isoline was obtained. The density of the stress isoline was calculated according to the partition to realize the analysis of the actual working conditions of the parts. The distinction between the concentrated and non-concentrated areas was realized, and the path planning was carried out according to different areas. The path planning is realized according to the actual working conditions of the parts.

## References

[1] Li Guangji, Liu Xinling. Review of research status of automotive lightweight technology [J]. Materials Science and Technology, 2020, vol. 28 (5): 47-61.
[2] SONG Yanli, Yang Long, Guo Wei, Hua Lin. Key Technologies of Carbon Fiber Composites for Automotive Lightweight Applications [J]. Materials Review, 2016, 30(17):16-25+50.
[3] PENG Xing-Shuang. Process Optimization of 3D printing fiber reinforced Composites and Study on energy absorption characteristics of automobile honeycomb [D]. Liaoning: Dalian University of Technology, 2020.
[4] Qin Guofeng, Qin Ruijian, Mi Peiwen, Li Ming. Optimization design of pore shape and layup of composite plates with holes [J]. Journal of Traffic and Transportation Engineering, 2022, vol. 22 (4): 223-231.
[5] Xiaoya Zhai; Falai Chen. Path Planning of a Type of Porous Structures for Additive Manufacturing [J]. Computer-Aided Design, 2019, Vol. 115: 218-230.
[6] Wang, Ting; Li, Nanya; Link, Guido; Jelonnek, John; Fleischer, Jürgen; Dittus, Jörg; Kupzik, Daniel. Load-dependent path planning method for 3D printing of continuous fiber reinforced plastics. [J]. Composites: Part A, Applied Science & Manufacturing, 2021, Vol. 140: 106181.

[7] Nanya Li; Guido Link; Ting Wang; Vasileios Ramopoulos; Dominik Neumaier; Julia Hofele; Mario Walter; John Jelonnek. Path-designed 3D printing for topological optimized continuous carbon fibre reinforced composite structures [J]. Composites Part B: Engineering, 2020, Vol. 182: 107612.
[8] Yingguang Li; Ke Xu; Xu Liu; Mengyuan Yang; James Gao; Paul Maropoulos (1). Stress-oriented 3D printing path optimization based on image processing algorithms for reinforced load-bearing parts [J]. CIRP Annals, 2021, Vol. 70(1): 195-198.
[9] Kamil Opalach; Joanna Porter-Sobieraj; Przemysław Zdroik. Stacking optimization of 3D printed continuous fiber layer designs [J]. Advances in Engineering Software, 2022, Vol. 164.
[10] Xiangjia Chen; Guoxin Fang; Wei-Hsin Liao; Charlie C.L. Wang. Field-Based Toolpath Generation for 3D Printing Continuous Fibre Reinforced Thermoplastic Composites [J]. Additive Manufacturing, 2022, Vol. 49: 102470.
[11] Chen Hao, Huang Xun, ZHAO Zhiming. Research on Apple image edge extraction algorithm based on improved Sobel operator [J]. Tianjin Agricultural Sciences, 2019, 26(09):51-55.
[12] Steuben, John C.;Iliopoulos, Athanasios P.;Michopoulos, John G..Implicit slicing for functionally tailored additive manufacturing.[J].Computer-Aided Design,2016,Vol.77: 107-119.
[13] Yusuke Yamanaka;Akira Todoroki;Masahito Ueda.Fiber Line Optimization in Single Ply for 3D Printed Composites[J].Open Journal of Composite Materials,2016,Vol.6(4): 121-124,126-128,130-131.

# Estimation of the residue capacity of lithium iron phosphate battery based on the internal resistance obtained from charging voltage drop

**Jingyun Yin[1,3], Jingmei Yu[2], Wenfeng Zhang[1], Qinhe Zhou[1] and Canjun Huang[1]**

[1]Guangdong Polytechnic of Water Resources and Electric Engineering, Guangzhou, Guangdong, 510925, China
[2]Beijing CELLLONG Tech Limited Liability Company, Beijing, 100055, China

[3]Corresponding author's e-mail: yinjingyun@gdsdxy.cn

**Abstract:** This study takes the 80 Ah lithium iron phosphate (LFP) prismatic battery that is from the vehicle and is in the middle or end of life as the research target, and the voltage-drop resistance (VDR), which is calculated through the voltage drop at the charging end, is used to for residue capacity estimation. The relationship between the VDR and the residue capacity of the battery is systematically studied, including factors such as charging cut-off voltage, charging current, charging end time, and charging current change mechanism. The results show that the residue capacity is a quadratic function relationship with the charging cut-off voltage and VDR. The VDR is affected by the charging cut-off voltage, charging cut-off current, and charging end time. There is no significant difference in the influence of VDR obtained by the charging modes of step charging and constant current charging. By testing the VDR under different charging conditions and different resting times, and standardizing the VDR, the residue capacity estimation error of 97% of the vehicle's batteries is within ±5%, which meets the application requirements for residue capacity estimation.

## 1. Introduction

Residual capacity is one of the important parameters for the lithium-ion battery. With the aging of lithium-ion batteries, the residue capacity is gradually decreasing, affecting the range and safety of electric vehicles [1-2]. Therefore, the estimation of the residue capacity of the whole vehicle has always been the focus of attention of electric vehicle enterprises. Lithium iron phosphate battery (LFP) is widely used in electric vehicles and energy storage due to its advantages of good safety performance, long cycle life, low price, and low environmental pollution [3-5]. However, the voltage platform of the LFP battery is very flat, and its SOC (state of charge) estimation and residual capacity estimation are the focus and difficulty in the industry [6-7].

The internal resistance of the battery is a parameter to measure the difficulty of carrier movement in the electrode. Numerous experiments show that the internal resistance is closely related to the battery residual capacity [8-10]. Generally, the discharge capacity of a battery with a small internal resistance is higher, while the discharge capacity of a battery with a large internal resistance is lower. Especially in the middle or end-of-life stage, the increase of the internal resistance and the decline of the life of the battery are more significant [11]. The internal resistance of the battery is affected by the SOC, current, temperature, and aging degree. By establishing the equivalent model or electrochemical model of the internal resistance of the battery and obtaining the corresponding parameters, the prediction relationship between the internal resistance and the capacity can be obtained [12]. Compared with the capacity estimation method based on the electrochemical model, the performance fitting estimation method based on experimental or vehicle operation data can grasp the internal relationship between battery capacity, battery health status, and key parameters without an in-depth understanding of battery geometric structure, material system and internal mechanism of capacity decay, and use the relationship to estimate the evolution trend of battery residual capacity.

In this study, the 80 Ah LFP prismatic battery from the vehicle that has been run for more than hundreds of thousands of kilometers is taken as the research subject, and the VDR at the charging end is proposed to estimate the residue capacity. The quantitative relationship between the VDR and the residue capacity of the battery is systematically studied. The factors considered include the charging cut-off voltage, charging current, charging end time, and step charging mechanism. The results show that the residual capacity is a quadratic function of the charge cut-off voltage and the VDR. The VDR is affected by the charge cut-off voltage, the charge cut-off current, and the end time of the charge. The variable current charging mechanism in the charging process has little effect on the VDR. By testing the VDR under different charging conditions and different standing times and carrying out standardization treatment on the VDR, the estimation deviation of the residual capacity of 97% of the battery monomers of the whole vehicle is within ±5%, which meets the application requirement of the estimation of the residual capacity.

## 2. Experimental process

### *2.1. Battery sample and capacity calibration*

To avoid the single working condition of the sample, there are 20 batteries in the experiment, which come from buses in three different cities in China, including Xiamen (7), Beijing (7), and Chongqing (6). After long-term operation of the bus, the battery capacity has different degrees of attenuation. These batteries are produced by a domestic power battery company, all of which are LFP prismatic batteries with a rated capacity of 80 Ah. $1/3C_0$ current is used at both the charge and discharge process at the ambient temperature of 25°C ± 2°C for capacity calibration, and $C_0$ is taken as 80 A. The battery charging and discharging instrument is the Sunway XW10005C16, each device has 16 standard channels, the limit current of a single channel is 100 A, the limit voltage is 5 V, the current detection accuracy is 0.001 A, and the voltage detection precision is 1 mV. The sampling interval is 1 s. The constant capacity results are shown in Table 1. The residue capacity interval of 20 batteries is [61.6 Ah, 71.4 Ah], and the corresponding capacity retention rate is [79%, 89%], indicating that the batteries are at the middle or end of life, and also the interval state that needs special attention in practical application.

**Table 1**. The Capacity Distribution of 20 Batteries Used in the Experiment.

| Capacity (Ah) | [60-62) | [62-64) | [64-66) | [66-68) | [68-70) | [70-72) |
|---|---|---|---|---|---|---|
| Sample number | 2 | 2 | 4 | 4 | 4 | 4 |

### *2.2. Sample charge and discharge test*

2.2.1. Galvanostatic charge test. To simulate the different charging states of the whole vehicle during use, the samples after capacity calibration shall be charged according to the matrix listed in Table 2. In order to prevent test fluctuation, each data point is tested twice. To detect whether the capacity is attenuated, the sample capacity is re-calibrated after each set of current tests. According to Table 2, 20 cells are tested for constant current charging at different currents, and the data of the standing process are obtained.

**Table 2.** Battery Constant Current Charging Test Matrix.

| Charge Current (A) | Charge Cut-off Voltage (V) | | | | | | | |
|---|---|---|---|---|---|---|---|---|
| | 3.38 | 3.40 | 3.42 | 3.44 | 3.46 | 3.50 | 3.54 | 3.65 |
| 7.5 | | | | | | | | |
| 15 | | | | | | | | |
| 26.67 | | | | | | | | |
| 40 | | | | | | | | |
| 50 | | | | | | | | |

2.2.2. Step charging test. In order to compare the difference of the VDR under the two different charging modes of step charging and constant current charging, one battery from each of the six different residue capacity sections in Table 1 (6 batteries in total) is taken for the comparison test. In the step charging mode, the battery is charged to 80%, 90%, and 95% SOC respectively with a current of 40 A. Then the battery is charged to 100% SOC (3. 65 V) with a current of 7.5 A. The data of the charging, discharging, and standing processes are recorded, and the internal resistance of voltage drop under 100% SOC is calculated. In the constant current charging mode, the battery is charged to 100% SOC (3.65 V) at a constant current of 7.5 A. The charge-discharge and standing data are also recorded, and the VDR at 100% SOC is calculated.

## 3. Analysis and discussion

### *3.1. Characteristics of charging curve and resting curve of LFP battery*

Figure 1 shows the charging curves and the change of the voltage at the end of charging of the same sample at different charging currents and different charging cut-off voltages. It can be seen that the VDR at the end of the charge of the same sample is consistent with the previous internal resistance model, and the VDR is related to the current, the charge cut-off voltage, and standing time, so the control variable method is used to study the corresponding relationship between the VDR and the residue capacity under different charge systems, and ultimately achieves the purpose of estimating the residue capacity through the VDR.

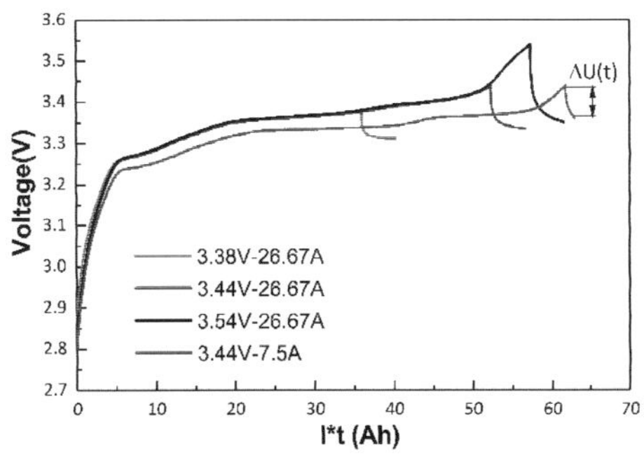


**Figure 1.** Static characteristics of charging terminal voltage with different charging cut-off voltages and charging currents.

### *3.2. Relationship between charge cut-off voltage, VDR, and residual capacity*

The internal resistance of lithium-ion batteries is closely related to SOC, especially at the end of charging, the change of internal resistance is more obvious. Figure 2 fits the relationship between the VDR and residue capacity at different charging cut-off voltages (the charging current is 1/3 C 26.67 A), and it can be seen that the VDR and the residue capacity are highly correlated. To obtain the functional relationship among the VDR, the charging cut-off voltage, and the residual capacity, the curve fitting toolbox in Matlab$^{TM}$ is used for fitting analysis, and it is found that the relationship among the VDR, charge cut-off voltage, and residual capacity meets Equation (1):

$$\text{Re}\,sidue\ capacity = 73.9 - 143.6x + 0.0955y - 1.799x^2 + 39.13xy \quad (1)$$

Where $x$ is the VDR and $y$ is the charge cut-off voltage, indicating that the residue capacity can be estimated if the VDR and charge cut-off voltage are known.

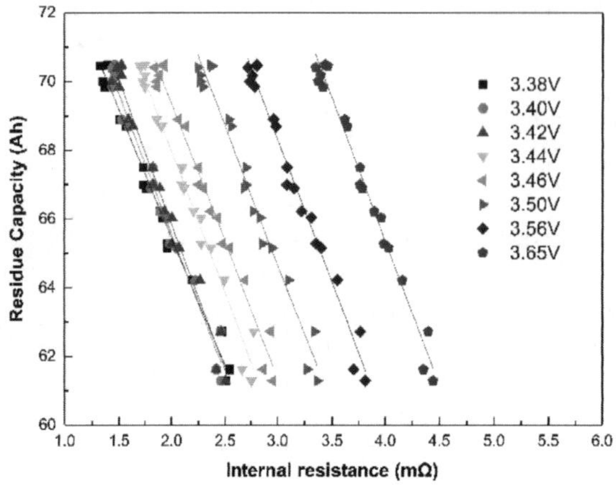


**Figure 2.** VDR-residue capacity fitting relationship under different charging cut-off voltages.

### *3.3. Effect of charging current and standing time after charging on VDR*

Considering that the charging current and the standing time after charging vary greatly in the actual application of the vehicle, $R_{(I=26.67\,A,\ t=40\,s)}$ is defined as the standard state internal resistance. By studying the relationship between the VDR under other different conditions and the standard state voltage drop internal resistance $R_{(I=26.67\,A,\ t=40\,s)}$, the VDR under different conditions is transformed into the standard state internal resistance $R_{(I=26.67\,A,\ t=40\,s)}$, facilitating the estimation of the residue capacity. Figure 3a is the relationship between the internal resistance $R_{(I=7.5\,A,\ 15\,A,\ 40\,A,\ 50\,A,\ t=40\,s)}$ of the voltage drop at the charging end and the internal resistance $R_{(I=26.67\,A,\ t=40\,s)}$ of the voltage drop at the standard state of the same sample under the same cut-off voltage, the same standing time and different charging currents, which is obtained by using the Matlab$^{TM}$ Curve fitting toolbox. It can be seen that there is a linear relationship between the non-standard current internal resistance and the standard current internal resistance, which can be fitted by the linear expression of $R_{(I)} = a * R_{(26.67\,A)} + B$, that is, the non-standard current internal resistance can be converted into the standard current internal resistance. Figure 3b shows that the fitting coefficient a is linearly and positively correlated with the current, while the value of the coefficient B is negatively correlated with the current. Therefore, the conversion Equation (2) between the internal resistance of the voltage drop at different currents and the VDR at the current of 26.67 A is obtained:

$$R_{(26.67A,\ 40s)} = (0.0299 \times I + 0.2209) \times R_{(I,\ 40s)} + 0.0003 \times I^2 - 0.0507 \times I + 1.0922 \quad (2)$$

The VDR under different charging currents can be converted into the voltage drop internal resistance under $1/3C_0$ standard charging by using the relation in Equation (2), thereby facilitating the horizontal comparison of the VDR.

Due to the influence of the electrochemical polarization and concentration polarization, the VDR increases exponentially with time. Therefore, the relationship between the VDR at the end of charging and the standing time of each sample under the same charging system is further studied, and the results are shown in Figure 3c-d. It shows that when the standing time is between 20 and 100 s, the VDR of all samples is linearly related to the voltage drop internal resistance when the standing time is 40 s, and the internal resistance conversion Equation (3) under different standing times is obtained by further fitting the correlation coefficient:

$$R_{(40s)} = (-0.4331 \times \ln(t) + 2.662) \times R_{(t)} + 0.4914 \times \ln(t) - 1.8058 \quad (3)$$

To prevent the fluctuation of voltage acquisition, Equation (3) can be used to estimate the residue capacity by calculating the average VDR for a period of time, reducing the interference of voltage noise.

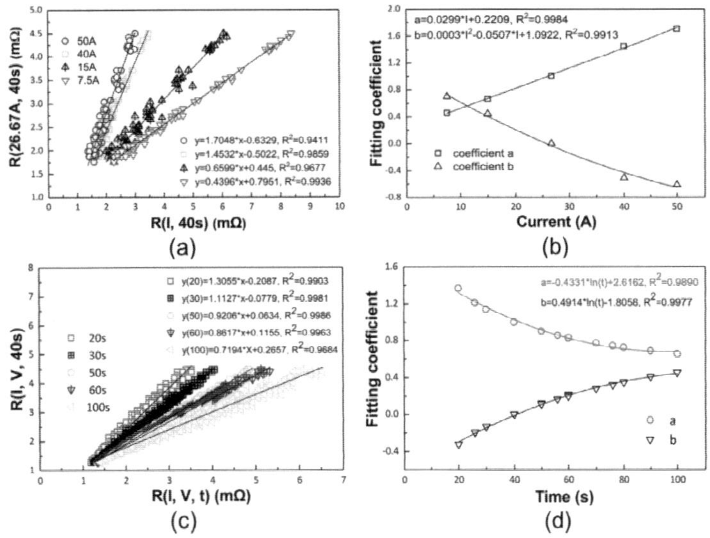


**Figure 3.** (a) The conversion relationship of VDR with different charging currents; (b) The relationship between the parameters in R(I)-R(26.67 A) fitting equation and current I; (c) The VDR transformation relationship at different times; (d) The relationship between the parameters in R(t)-R(40 s) fitting equation and time t.

### *3.4. Impact of the step charging*

To give consideration to both fast charging and life management of electrical vehicles, the step charging mode is generally adopted. At present, the large current is used to charge to 80% -95% SOC, and then the current is reduced to the end of constant current charging [13]. Because the VDR at the charging end is related to the charging state of the battery, whether the current change in the charging process will affect the internal resistance of the voltage drop needs further verification. Figure 4 compares the difference of the VDR under the two different charging modes of step charging and constant current charging. The first group of step charging is to charge the battery to 80% SOC with a current of 40 A, and then to 100% SOC (3.65 V) with a current of 7.5 A. The remaining two groups of step charging are to charge the battery to 90% SOC and 95% SOC respectively with a current of 40 A, and then charge the battery to 100% SOC (3. 65 V) with a current of 7.5 A. The battery is charged to 100% SOC (3. 65 V) with a standard charging current of 7.5 A. The results show that there is no significant difference between the VDR under step charging mode and the constant current, which means that even if the whole vehicle adopts the step charging mode, the current at the end of charging can be used as the estimation data of the VDR.

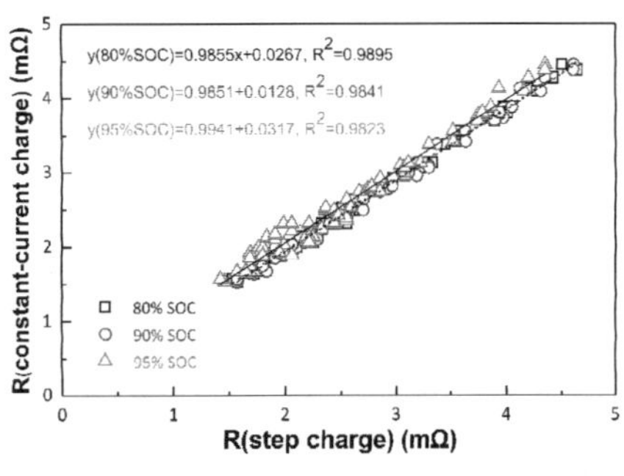


**Figure 4.** Comparison of VDR under variable current and constant current test conditions.

### *3.5. Verification based on vehicle operation data*

To verify the reliability of this method, 100 LFP batteries are disassembled from the battery pack of the whole vehicle, the VDR is tested and the residual capacity is estimated according to the above method, as shown in Figure 5. The column height in Figure 5b represents the proportion of battery cells in different error intervals, and the dotted line data represents the cumulative growth of cells from -6% error to 5% error interval. The results show that the deviation between the real residual capacity and the estimated residual capacity of 97% of the batteries is within ± 5%, and the error of the residual capacity of most of the batteries is between -2% and + 5%, which meet the application requirements of residual capacity estimation. The method is simple to operate and can be used for estimating and correcting the residual capacity of the BMS battery at the vehicle end.

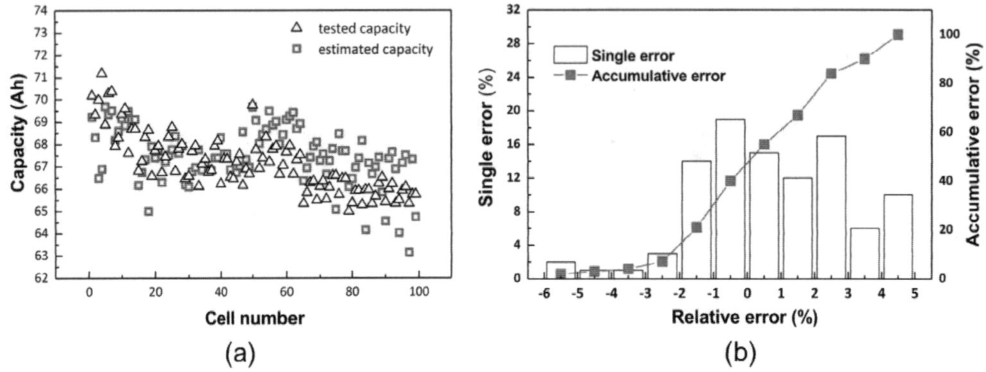


**Figure 5.** (a) Comparison of actual residue capacity and estimated residue capacity of batteries from electrical vehicle; (b) Capacity estimation error statistics.

## 4. Conclusion

In this study, the 80 Ah LFP prismatic battery in the middle or end of life is taken as the research subject, the residual capacity is estimated by the VDR, and the quantitative relationship between the internal resistance of the battery and the residual capacity is systematically studied. At the same time, the influence of the charge cut-off voltage, the charge current, the charge end time, and the charge current change mechanism on the relationship between the VDR and the residue capacity is investigated. At room temperature, the charging cut-off voltage, charging cut-off current, and charging end time affect the VDR, and there is no significant difference between the VDR obtained by the step charging mode and the constant current charging mode. The method realizes the standardized treatment of the VDR by testing the VDR under different charging conditions and standing times and finally completes the capacity estimation of the whole vehicle under different use conditions, and the method can realize that the error of more than 97% of samples is within ±5%.

## Acknowledgments

This work was financially supported by the Young Innovative Talent Project of Guangdong Provincial Universities (Natural Science) (Grant Nos. 2022KQNCX200), Special Projects of Key Fields in Guangdong Province (high-end equipment manufacturing) (Grant Nos. 2023ZDZX3068), Young Scientific Research Projects of Guangdong Polytechnic of Water Resources and Electric Engineering (Grant Nos. XK202213).

## References

[1] Yuliya Preger, Loraine Torres-Castro, Taina Rauhala, and Judith Jeevarajan (2022). Perspective-On the Safety of Aged Lithium-Ion Batteries. J. Electrochem. Soc., 169: 030507. http://doi.org/10.1149/1945-7111/ac53cc.

[2] Pierre Kuntz, Loïc Lonardoni, Sylvie Genies, Olivier Raccurt and Philippe Azaïs (2023). Evolution of Safety Behavior of High-Power and High-Energy Commercial Li-Ion Cells after Electric Vehicle Aging. Batteries, 9 (8): 427. https://doi.org/10.3390/batteries9080427.

[3] Rotem Marom, S. Francis Amalraj, Nicole Leifer, David Jacoba and Doron Aurbacha (2011). A review of advanced and practical lithium battery materials. J. Mater. Chem., 21: 9938-9954. https://doi.org/10.1039/C0JM04225K.

[4] Li-Xia Yuan, Zhao-Hui Wang, Wu-Xing Zhang, Xian-Luo Hu, Ji-Tao Chen, Yun-Hui Huang, and John B. Goodenough (2011). Development and challenges of $LiFePO_4$ cathode material for lithium-ion batteries. Energy Environ. Sci., 4: 269-284. https://doi.org/10.1039/C0EE00029A.

[5] Arumugam Manthiram (2020). A reflection on lithium-ion battery cathode chemistry. Nat Commun., 11: 1550. https://doi.org/10.1038/s41467-020-15355-0.

[6] Maitane Berecibar, Maitane Garmendia, Iñigo Gandiaga, Jon Crego and Igor Villarreal (2016). State of health estimation algorithm of $LiFePO_4$ battery packs based on differential voltage curves for battery management system application. Energy, 784-796. https://doi.org/10.1016/j.energy.2016.02.163.

[7] Samppa Jenu, Ari Hentunen, Jari Haavisto and Mikko Pihlatie (2022). State of health estimation of cycle aged large format lithium-ion cells based on partial charging. Journal of Energy Storage, 46: 103855. https://doi.org/10.1016/j.est.2021.103855.

[8] Stephen J. Harris, David J. Harris, and Chen Li (2017). Failure statistics for commercial lithium-ion batteries: A study of 24 pouch cells. Journal of Power Sources, 342: 589-597. https://doi.org/10.1016/j.jpowsour.2016.12.083.

[9] Lin Chen, Zhiqiang Lü, Weilong Lin, Junzi Li and Haihong Pan (2018). A new state-of-health estimation method for lithium-ion batteries through the intrinsic relationship between ohmic internal resistance and capacity. Measurement, 116: 586-595. https://doi.org/10.1016/j.measurement.2017.11.016.

[10] Rasheed Ibraheem, Calum Strange and Gonçalo dos Reis (2023). Capacity and Internal Resistance of lithium-ion batteries: Full degradation curve prediction from Voltage response at constant Current at discharge. Journal of Power Sources, 556: 232477. https://doi.org/10.1016/j.jpowsour.2022.232477.

[11] Xiaowei Ma, Jessie E. Harlow, Jing Li1, Lin Ma, David S. Hall, Samuel Buteau, Matthew Genovese, Marc Cormier, and J. R. Dahn (2019). Hindering Rollover Failure of $Li[Ni_{0.5}Mn_{0.3}Co_{0.2}]O_2$/Graphite Pouch Cells during Long-Term Cycling. J. Electrochem. Soc., 166: A711. http://doi.org/10.1149/2.0801904jes.

[12] Alexandros Nikolian, Joris Jaguemont, Joris de Hoog, Shovon Goutam, Noshin Omar, Peter Van Den Bossche and Joeri Van Mierlo (2018). Complete cell-level lithium-ion electrical ECM model for different chemistries (NMC, LFP, LTO) and temperatures (-5°C to 45°C) - Optimized modeling techniques. International Journal of Electrical Power & Energy Systems, 98: 133-146. https://doi.org/10.1016/j.ijepes.2017.11.031.

[13] S. Hemavathi and A. Shinisha (2022). A study on trends and developments in electric vehicle charging technologies. Journal of Energy Storage, 52: 105013. https://doi.org/10.1016/j.est.2022.105013.

# The effect of tightening torque on the strength of composite-metal hybrid joints

**Jingchao Wei[1,2], Su Cao[1], Hui Zhang[1], Xiangming Chen[1], and Yu Yang[1]**

[1]National Key Laboratory of Strength and Structural Integrity, Aircraft Strength Research Institute of China, Shanxi, Xian, 710065, China

[2]Corresponding author's e-mail: weijingchao66@aliyun.com

**Abstract.** In the manufacturing and assembly process of aircraft composite-metal hybrid structures, tightening torque is one of the important factors that control assembly quality and affect connection strength. This paper experimentally studies the mechanical properties of composite-metal single-bolt single-lap joints under different tightening torques, reveals the impact mechanism of different tightening torques on the manufacturing and assembly of composite-metal hybrid structures, and also establishes a finite element analysis model of hybrid connection structure considering the influence of drilling and assembly processes. The results show that the ultimate load of the composite-metal connection structure increased with the increase of tightening torque. When the tightening torques were 3 N·m, 6 N·m, and 9 N·m, the ultimate failure loads of the composite-metal hybrid joints were 12.9 kN, 13.8 kN, and 14.6 kN respectively.

## 1. Introduction

Composite materials are widely used in the aerospace field due to their advantages such as high specific strength, specific modulus, good fatigue performance, good corrosion resistance, strong structural designability, and integrated molding. As the proportion of composite materials used in aircraft structures increases, composite and metal hybrid connection structures will inevitably be used in aircraft structure design. In aircraft structures, hybrid structures usually use mechanical joints, and the nonlinear changes in structural strength caused during the manufacturing and assembly process will complicate the strength prediction of hybrid structures [1].

Ireman et al. [2] generated a design allowable value chart based on the method of applying computers to predict the extrusion strength value under non-uniaxial extrusion bypass load. The stress distribution in the thickness direction of the laminate near the nail holes was obtained by using a three-dimensional finite element model. Since interlaminar stress was three-dimensional (3D) and always occurred near the free boundary of composite materials, a 3D finite element model was required to accurately design the stress distribution of bolted joints.

Based on the finite element model, the authors [3] can predict the relative displacement between the two sides of the specimen, but the experimental measurement results of the displacement were quite different from the FEM results. In addition, the strain energy measured through experiments can intuitively reflect secondary bending. Failure load occurs when the strain-load curve becomes nonlinear. Differences in experimental testing may be due to deviations caused by non-linearity and friction effects between the two plates of the sample. The result showed that the finite element model was in good agreement with the experimental strains. Another conclusion was also drawn that the 3D finite element model is an effective method to study the thickness direction performance of laminate joints.

Sun et al. [4] studied the effect of clamping area on connection stiffness and failure load for single-nail double-shear laminate (T800G/3900-2 carbon/epoxy) bolted connection structures, and compared and analyzed three different connection forms. The results showed that the failure load of bolted connections with and without washers was higher than that of pin connections without laterally restrained pin connections. In general, the failure load of the connector increased with the increase of the clamping area. Park HJ [5] conducted a study on five layups of mechanically connected (pin-connected and bolted) carbon/epoxy (HT145/RS122) laminates. When the surface layer of the laminate was a 90° layer, it helped to increase the laminated extrusion strength; if the surface layer was a 0° layer, it reduced both the final extrusion strength and the laminated extrusion strength.

McCarthy MA et al. [6] conducted a large number of experiments to study the effect of gap amounts (0%, 1%, 2%, 3%) on the stiffness and extrusion of single-nail (including convex and countersunk) single-lap composite connection structures. Wei JC et al. [7] studied the effects of different interference fit sizes on bearing strength and fatigue life through finite element analysis and experiments.

Gray et al. [8] proposed a global modeling method for composite bolt connections in large-scale modeling of aircraft and other applications. Liu FR et al. [9] also studied the mechanical properties of single-nail single-lap and single-nail double-lap joints under the secondary bending effect. They found that because the contact area of single-shear joints and laminates decreases faster than double-shear joints during load application, therefore single-shear joints will fail faster than double-shear joints.

Therefore, this paper carries out experimental research on typical hybrid connection structures considering different tightening torques during assembly, and based on the comparison between simulation results and experimental results, the correction and calibration of the finite element simulation model can be achieved, which can be used for composite-metal single nail and single shear hybrids. These works can provide the experimental basis and technical support for the design of composite-metal single-bolt and single-shear hybrid connection structures.

## 2. Specimens and experimental procedures

The tensile test of the composite-metal hybrid structure adopts the composite-aluminum alloy single bolt single lap connection method and designs different tightening torques as test variables to conduct static strength tests to obtain the hybrid structure connection strength. The specimens are shown in Figure 1.

**Figure 1.** Composite-metal single-bolt single-lap connection specimens.

According to the specimen condition adjustment requirements in the ASTM D 5961 standard, the specimen was placed under laboratory-standard environmental conditions for 48 hours. Environmental chambers, humidifiers, and other equipment are used to maintain the laboratory temperature at laboratory standard environmental conditions (temperature: 23±2°C; relative humidity: 50±10%).

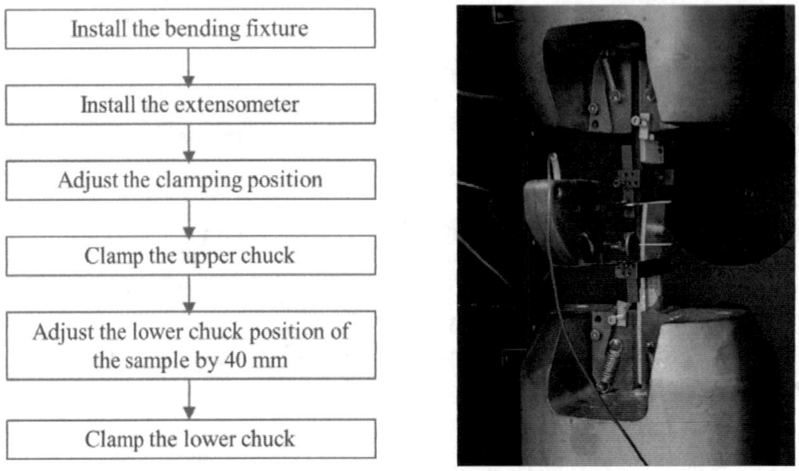


(a) The program of experimental procedures (b) Installation status diagram

**Figure 2.** Static strength test of composite-metal hybrid single-bolt single-shear connection.

A static strength tensile test without the influence of tightening torque is carried out under dry environmental conditions at room temperature. The test methods, procedures, measurements, and data are carried out according to the ASTM D 5961 standard. First, the specimen was installed on the bending clamp to eliminate the secondary bending effect during the stretching process, and an extensometer was used to measure the deformation of the bearing hole. Then they assembled the specimen and clamped it into the chuck of the testing machine, adjusted the clamping position to ensure that the specimen was parallel to the loading block of the lower chuck of the testing machine, tightened the upper chuck, and placed one side of the sample close to the lower chuck on the limit strip to ensure axial loading of the specimen. Finally, the position of the lower chuck was adjusted to align with the 40 mm long dash line of the sample and clamped. The test status after installation is shown in Figure 2.

## 3. Results and discussion

The failure morphology of composite-metal single-lap structure specimen with different tightening torques is shown in Figure 3. The failure mode of the bearing hole was mainly a shearing failure of the composite laminate. The failure location was mainly around the bolt holes of the composite laminate. The cracks in the laminate extended along the 45° direction. This was because during the tensile process, the specimen was subjected to shear force, and the contact area between the bolt and the hole was small, resulting in a serious stress concentration phenomenon. At the same time, the failure strength of aluminum alloy was much greater than that of composite materials, so the composite laminate first reached the failure strength and broke. Therefore, due to the anisotropy of the composite laminate and the setting of the plying direction, the cracks generated when the laminate fails and damage extended along the 45° direction.

**Figure 3.** The failure mode of specimens.

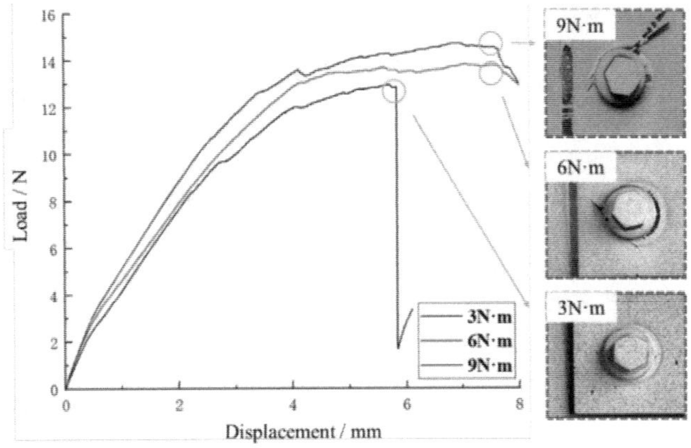


**Figure 4.** Load-displacement curves of single-shear joints under different tightening torques.

The static tensile load-displacement curves of the composite metal single-lap joint under different tightening torques are shown in Figure 4. During the test, the 3 N·m moment group sample suffered premature bolt failure during the loading process, causing the curve to suddenly drop the load. This could be due to the small preload force applied to the sample and the nut failing to fully compress the washer. This caused the bolt to undergo bending deformation during the stretching process, reaching its limit bearing load in advance and causing fracture failure. In addition, the 6 N·m and 9 N·m moment group specimens had larger pre-tightening force and the bolts did not undergo severe bending deformation. Therefore, their load-bearing capacity and tensile displacement were larger, and their main failure mode was the shearing failure of composite laminates.

The curves were two inflection points in Figure 4. First, before the first inflection point, the curve was in the linear growth stage, and the tensile load increased rapidly with the increase of displacement. This was due to the composite material around the connecting hole was extruded and deformed under the combined effect of pre-tightening and bolted connection. The increase in material density led to an increase in stiffness in the area around the hole. In addition, the tangential friction between the connecting plates of two different materials also shared part of the external load, so the slope of the curve was larger before the first inflection point. As the tightening torque increased, the slope of the curve also rose slightly at this stage.

Secondly, between the second inflection point and the first inflection point, the slope of the curve became smaller and the load growth rate slowed down.

This was due to the forced extrusion of the materials around the holes of the laminate, and the compressive stress in the bearing area gradually increased, causing composite small-scale local damage to the material, and material properties deteriorated, resulting in a gradual decrease in structural stiffness, as well as slight sliding between the plates. The external load between the upper and lower plates also gradually decreased, so the slope of the load curve decreased significantly. At the same time, as the tensile displacement increased, the tensile load also gradually increased. This was due to the different characteristics of CFRP laminate materials and traditional metal materials. Due to the multi-phase structural material characteristics of composite materials, the damage to the laminate was a gradual degradation process, and the failure of the weak point would cause the load to be redistributed. Therefore, a wider area around the hole would help disperse the external load and improve the structural bearing capacity.

Finally, as the damage area expanded, it ultimately affected the structural bearing capacity until it reached the limit strength of the structure, which was the second inflection point in the curve. This was because the specimen had been seriously damaged during the loading process. As the damage to the specimen continued to expand, the load would drop sharply until the loading of the testing machine was terminated. In addition, the structure took a long time and had a large displacement from damage degradation to finally reaching the ultimate bearing capacity. On the one hand, the elastic deformation of the composite-metal single-lap joint was relatively large. On the other hand, under the action of the secondary bending moment, the contact area between the hole and the bolt had experienced a process of full contact around the hole-local contact-full contact again, and the generated local displacement would be relatively large.

The laminate was meshed into an eight-node linear hexahedral reduction integral unit (C3D8R), in which the mesh around the bolt-hole was refined. Since the secondary element was not very accurate in simulating contact problems, a linear element was used. However, when subjected to bending loads, linear fully integrated elements were prone to shear self-locking problems due to the rigidity of the elements, which can be solved by linear reduced integration elements such as C3D8R. At the same time, the hourglass problem was also introduced, which would lead to severe mesh distortion in the model, so hourglass control should be carried out to ensure the accuracy of the results. For simplifying the model and reducing the contact pairs, one end of the hexagonal head bolt was simplified to a circular end with the bolt envelope circle, and the eight-node linear reduction integration unit (C3D8R) was still used. The mesh around the bolt hole was refined, and 1152 units were arranged along the hole circumference, with a total of 18432 elements and 9696 bolt units. A meshing diagram of the hybrid joint model is shown in Figure 5.

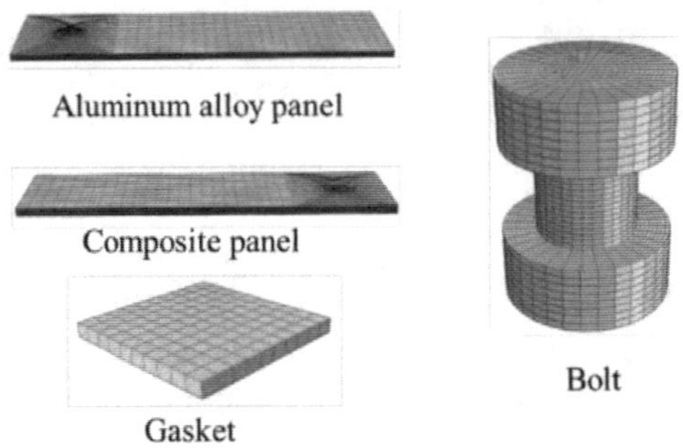


**Figure 5.** Finite element model diagram of a composite-metal hybrid connection structure.

As shown in Figure 6, the tightening torque caused tensile stress in the thickness direction of the bolt-hole, and the composite material under the washer produced compressive stress, and with the increase of tightening torque, the stress peak also increased. The compression effect could effectively improve the load capacity of the structure. In the loading process, the compression effect could enhance the friction between the contact surfaces, and the friction force could carry a large part of the load, at the same time the damage was mainly concentrated in the bolt-hole contact area, Therefore, the diversion of friction to the load can effectively delay the damage initiation time of the material.

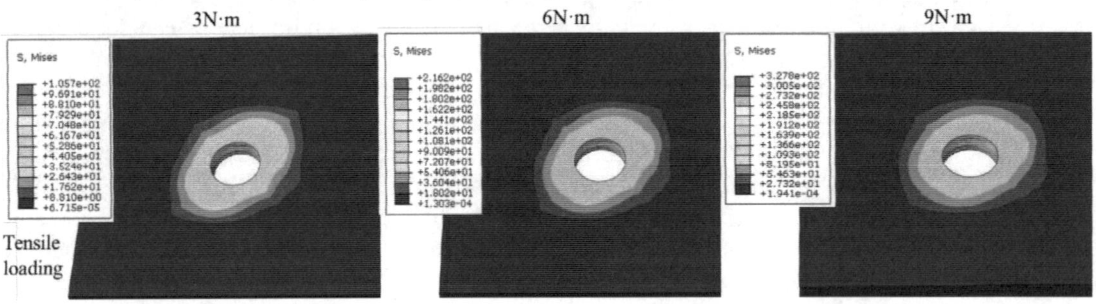


**Figure 6.** Simulation of the ultimate loads at different tightening torques.

**Table 1.** Comparison of test and simulation results under different tightening torques.

| Specimen number | Tightening torques / N·m | Test results / kN | Simulation results / kN |
|---|---|---|---|
| NJL-1 | 3 | 12.9 | 16.5 |
| NJL-2 | 6 | 13.8 | 16.6 |
| NJL-3 | 9 | 14.6 | 16.7 |

To further analyze and study the effect of different tightening torques on the load-bearing capacity and tensile strength of composite/aluminum alloy single-shear structures, a finite element analysis model of hybrid structures considering different tightening torques was established. The relative interference between the bolts and the hole circumference in the model was 0%, and the tightening torque is set to 3 N·m, 6 N·m, and 9 N·m according to the test. Combining the test and finite element analysis results (as shown in Table 1), when the tightening torque was 3 N·m, the experimental value was 12.9 kN and the simulation value was 16.5 kN; when the tightening torque was 6 N·m, the experimental value was 13.8 kN and the simulation value was 16.6 kN; when the tightening torque was 9 N·m, the experimental value was 14.6 kN and the simulation value is 16.7 kN. According to the simulation results (as shown in Figure 7), it could be seen that the numerical simulation limit load was larger than the test limit load. This was because the simulation process was based on numerical simulation under completely ideal conditions, and its structural damage was smaller than the actual assembly process. Therefore, the numerical simulation results of the hybrid connection structure were high.

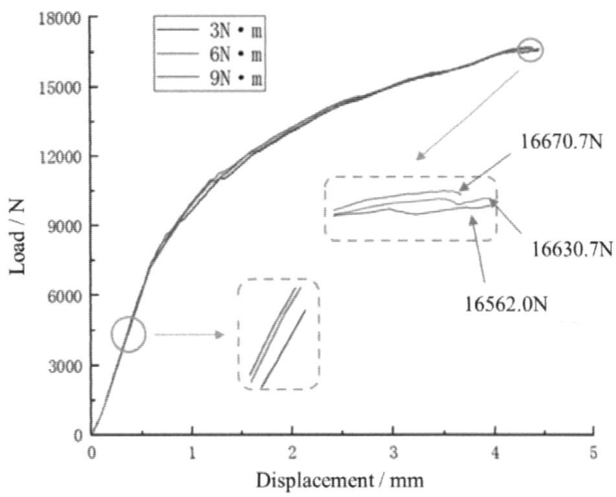


**Figure 7.** Numerical simulation results under different tightening torques.

Figure 7 shows the load-displacement curve of the static strength test under numerical simulation. Comparative analysis with the test showed that in the growth stage of the load curve, a higher tightening torque corresponded to a larger curve growth slope. That is, under the same tensile displacement, the greater the tightening torque of the specimen, the greater the tensile load it bears. On the one hand, the clamping force made the upper and lower plates of the joint contact each other, and the tangential friction force on the contact surface between the plates effectively shared part of the tensile load. The clamping force effectively inhibited the initiation and proliferation of cracks in the area around the hole, improving the structural load-bearing capacity. On the other hand, the tightening torque provided "lateral support" to the connection structure, which reduced the impact of secondary bending effects and improved the structural stiffness in the area around the hole. However, the effect provided by the tightening torque did not increase infinitely. The strength of the composite material along the thickness direction was mainly provided by the resin matrix, which was much smaller than that of the carbon fiber material. Therefore, an excessive tightening torque would crush the composite material before it was subjected to tensile load, causing damage and failure and reducing the bearing capacity of the specimen.

## 4. Conclusion

To obtain the effect of assembly factors on the static strength of hybrid structures, experiments, and finite element simulation analysis were conducted to consider the influencing factors of different tightening moments. The ultimate load of the composite/aluminum alloy connection structure increased with the increase of tightening torque, and the maximum failure loads were 12.9 kN, 13.8 kN, and 14.6

kN respectively. To a certain extent, an increase in tightening torque would increase the limit bearing load of the structure, but if the tightening torque was too large, the specimen would be crushed and the structural bearing capacity would decrease.

**References**

[1] Wallner, C. and Almeida, S.F.M.D. (2023) Fracture criteria and surrogate models for bearing strength of mechanically fastened joints on hybrid composite laminates. Journal of Composite Materials, p.00219983231215484. DOI: https://doi.org/10.1177/00219983231215484.

[2] Ireman T, Nyman T, Hellbom K (1993) On design methods for bolted joints in composite aircraft structures. Composite Structures, 25(1-4): 567-78. DOI: https://doi.org/10.1016/0263-8223(93)90205-5.

[3] Ireman T (1998) Three-dimensional stress analysis of bolted single-lap composite joints. Composite Structures, 43(3): 195-216. DOI: https://doi.org/10.1016/S0263-8223(98)00103-2.

[4] Sun H T, Qing X, Chang F K (2002) The response of composite joints with bolt-clamping loads part II: model verification. J Composite Material, 36(1): 69-91. DOI: https://doi.org/10.1177/0021998302036001302.

[5] Park H J (2001) Effect of stacking sequence and clamping force on the bearing strengths of mechanically fastened joints in composite laminates. Composite Structure, 53(2): 213-21. DOI: https://doi.org/10.1016/S0263-8223(01)00005-8.

[6] McCarthy M A, Lawlor V P, Stanley W F, McCarthy CT (2002) Bolt-hole clearance effects and strength criteria in single bolt, single lap, composite joints. Compos Science and Technology, 62(10-11): 1415-31. DOI: https://doi.org/10.1016/S0266-3538(02)00088-X.

[7] Wei J C, Jiao G Q, Jia P R, Huang T (2013) The effect of interference fit size on the fatigue life of bolted joints in composite laminates. Composites Part B: Engineering, 53: 62-68. DOI: https://doi.org/10.1016/j.compositesb.2013.04.048.

[8] Gray P J, McCarthy C T (2011) A highly efficient user-defined finite element for load distribution analysis of large-scale bolted composite structures. Composites Science and Technology, 12(71): 1517-27. DOI: https://doi.org/10.1016/j.compscitech.2011.06.011.

[9] Liu F R, Lu X H, Zhao L B, Zhang J Y, Hu N, Xu J F (2018) An interpretation of the load distributions in highly torqued single-lap composite bolted joints with bolt-hole clearances. Composites Part B: Engineering, 138: 194-205. DOI: https://doi.org/10.1016/j.compositesb.2017.11.027.

Journal of Physics: Conference Series **2720** (2024) 012055 doi:10.1088/1742-6596/2720/1/012055

# Research electrical properties of Cu doped $FeTe_2$ prepared by solid state reaction

**Linghu Zeng [1], Yunhao Shi [1], Guangji Yang [1], Panpan Hu [1], Lang Zhang [2] and Bingke Qin [1,3]**

[1] Liupanshui Normal University, Liupanshui, Guizhou 553004, China
[2] Dalian University, Dalian 116622, China

[3] Corresponding author's e-mail: qinbingke@126.com

**Abstract:** The present study successfully prepared $Cu_xFeTe_2$ (x=0.01, 0.05, 0.1, 0.15, and 0.2) compounds using a high-temperature solid state reaction sintering technique under preparation conditions ranging from 623 K to 823 K. The phase structure and microstructure of $Cu_xFeTe_2$ were determined using X-ray diffraction and scanning electron microscopy. Simultaneously, we studied the Seebeck coefficient and electrical resistivity of the sample. It was found that the samples prepared at 673 K had a purer phase, and the thermal power coefficient was better for samples with a Cu doping concentration below 0.1.

## 1. Introduction

With the rapid development of social productivity and changes in human lifestyles, environmental pollution, and energy scarcity have become increasingly serious issues. Thermoelectric conversion technology, as a new, efficient, and environmentally friendly energy conversion technology, has received wide attention domestically and abroad. Thermoelectric devices prepared using thermoelectric materials have significant advantages in energy conversion, such as being small in size, having no moving parts, no pollution, being lightweight, having long steady-state operating cycles, and being able to work under extreme conditions. This provides a feasible solution for direct energy conversion between heat and electricity [1]. Thermoelectric materials can use natural temperature differences and industrial waste heat to generate thermoelectric power. They can also use non-polluting natural energy sources and have comprehensive social benefits.

Thermoelectric materials are functional materials that can directly convert thermal energy into electrical energy through the movement of internal carriers [2-4]. However, different materials have different thermoelectric properties. We chose the pyrite-type compound $FeTe_2$, which belongs to the orthorhombic crystal system with high symmetry, with lattice parameters a=5.26Å, b=6.27Å, and c=3.87Å. The crystal structure of $FeTe_2$ has a high degree of symmetry and special properties in electricity, optics, magnetism, and superconductivity, and some studies have shown that it is also a potential thermoelectric material [5, 6].

As early as the 1980s, Alvi and Nazir [7] calculated a new type of band structure for ternary alloys such as NiMnSb and PtMnSb and called these compounds "half-metallic" magnetic materials. These materials are a new type of functional material. The traditional preparation methods for $FeTe_2$ are high-temperature and high-pressure, long-time high-temperature reactions, hydrothermal methods, and high-temperature solid state reaction methods. During the high-temperature solid state reaction sintering process, the low melting point and volatility of tellurium have an important impact on the preparation of samples, depending on the heating rate used [8]. Copper is relatively cheap compared to rare earth elements, and copper metal has low electrical resistivity and excellent mechanical processing properties. Therefore, in this study, we use high-temperature solid state reaction sintering to prepare Cu-doped $FeTe_2$ alloys and analyze and study their preparation methods and electrical properties under different heating rate conditions. This study investigated the thermoelectric properties of the alloy compound $Cu_xFeTe_2$, the effect of Cu doping on its thermoelectric properties, and provided technical references for the research of this system.

## 2. Experiment

The purity of the experimental raw material is 4n, with a particle size of about 300 mesh, Fe powder, Te powder, and Cu powder as starting materials, accurately weighed according to the chemical formula $Cu_xFeTe_2$ (x=0.01, 0.05, 0.1, 0.15, and 0.2). After weighing, the mixture is evenly mixed under inert gas protection and then pressed into cylindrical samples. The samples are synthesized in a vacuum furnace at different temperatures and holding conditions and removed after cooling to room temperature.

The vacuum furnace model used for solid state reaction sintering is HMZ-1700-20. The phase structure analysis of the sample was performed using an X-ray diffractometer (Cu Kα radiation, diffraction angle from 20° to 80°). At room temperature, the electrical resistivity of the samples was measured using an RTS-9 four-probe testing instrument. The Seebeck coefficient was measured using a self-made. It calibrated the Seebeck tester, maintaining a temperature difference of ~10°C at both ends of the sample to measure its thermoelectric potential, from which the Seebeck coefficient of the sample was calculated, with a measurement error of ±5%. The room temperature thermal conductivity of the samples was measured using an LFA475 laser thermal conductivity meter from NETZSCH in Germany. The ZT value of the samples was obtained through the formula $ZT = S^2\sigma T/\kappa$.

## 3. Results and discussions

The XRD patterns of samples $Cu_xFeTe_2$ with different Cu contents at a synthesis temperature of 400°C are shown in Figure 1. The diffraction peaks of the test spectrum match well with the $FeTe_2$-PDF#14-0419 card in the PDF database, indicating that the phase of synthesized samples is mainly composed of the marcasite-type compound $FeTe_2$. As the Cu content gradually increases, impurity peaks in the diffraction peaks also increase, and a certain amount of tellurium elemental is precipitated inside the sample. The melting points of copper, tellurium, and iron are quite different, and under prolonged heating and insulation conditions, tellurium is easily volatile. The sample is cooled and reprecipitated as an element when the temperature decreases. Generally, for preparing $Cu_xFeTe_2$ (x=0.01, 0.05, 0.1, 0.15, and 0.2) by solid-phase reaction, Cu doping should not be too high at 400°C, and less than 0.1 is better. Samples with Cu content above 0.1 will have pores and cavities inside after sintering and changing their properties.

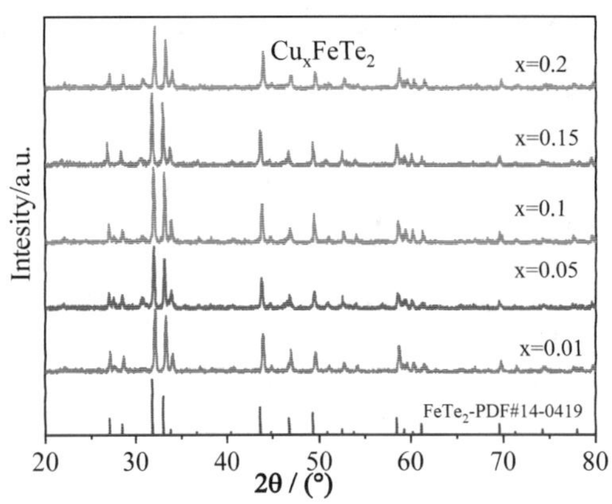


**Figure 1.** The XRD spectra of $Cu_xFeTe_2$ with different Cu contents.

**Figure 2.** SEM image of the cross-section of the $Cu_xFeTe_2$ sample.

The SEM image of the internal cross-section of the sample $Cu_xFeTe_2$ is shown in Figure 2, with a magnification of 5000 times. The synthesis temperature of the sample was 400°C. It can be seen from the SEM images that there are many pores inside the sample, which is caused by the solid state reaction. This phenomenon may be because the sample volatilizes and forms during the high-temperature synthesis process due to the low melting point of the Te element. It can be seen from Figure 2(a) that the crystallinity of the sample is poor, and the grain boundaries are blurred, which is consistent with the XRD measurement results. The SEM energy spectra in (b), (c), and (d) show that there is a certain segregation of Te, Fe, and Cu in the sample at this time, and there are many pores present.

The Cu-doped $FeTe_2$ sample prepared by the solid state reaction was tested for resistivity, and the resistivity-temperature curve is shown in Figure 3(a). The resistivity-temperature curves of $Cu_xFeTe_2$ with x values of 0.01, 0.05, 0.1, 0.15, and 0.2 were tested from 350°C to 550°C. As can be seen from Figure 3(a), the resistivity of the sample decreases with increasing synthesis temperature, and the decrease becomes gentle when the synthesis temperature is greater than 723 K. The sample with a Cu content of 0.05 has the lowest resistivity; its minimum resistivity is 1.74 mΩ·cm. The change is more obvious at temperatures below 723 K. The sample of Cu-doped $FeTe_2$ prepared by the solid state reaction was tested for the Seebeck coefficient, and the Seebeck coefficient-temperature curve was obtained.

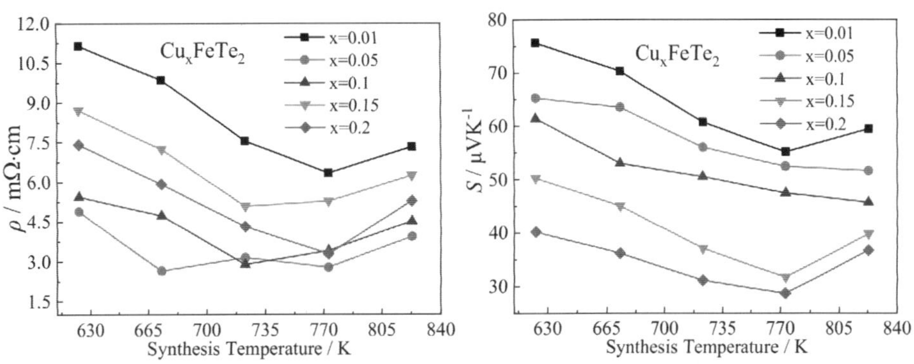


(a) (b)

**Figure 3.** Resistivity and seebeck coefficient versus temperature of $Cu_xFeTe_2$.

The functional relationship between the Seebeck coefficient and the temperature of sample $Cu_xFeTe_2$ is shown in Figure 3(b). It can be observed that the Seebeck values are all positive, indicating that the prepared thermoelectric material under these preparation conditions is a p-type semiconductor material. With increasing Cu doping, the curve becomes smoother as temperature increases. However, at higher temperatures, the sample undergoes greater changes, and the absolute value of Seebeck decreases with increasing temperature. The maximum Seebeck value occurs at Cu doping levels between 0.01 and 0.1, with the highest Seebeck coefficient of 75.66 $\mu VK^{-1}$ obtained.

According to the Seebeck coefficient and resistivity measured for $Cu_xFeTe_2$ (x=0.01, 0.05, 0.1, 0.15, 0.2), the power factor curve versus temperature was calculated $PF=S^2\sigma$ using the formula. Figure 4(a) shows the curve of power factor versus temperature for $Cu_xFeTe_2$ (x=0.01, 0.05, 0.1, 0.15, 0.2). From the figure, it can be observed that the power factor is the best at 400°C, and the largest power factor is obtained when Cu content is 0.05. However, the power factor decreases with the preparation temperature. The maximum power factor of 95.24 $\mu W/mK^2$ was obtained when the sample synthesis temperature was 823 K. Although the largest power factor was obtained at 823 K, according to the phase analysis of the sample, the optimum temperature for $FeTe_2$ is still 723 K. Figure 4(b) shows the calculated ZT values near room temperature for the samples. The figure shows that the maximum ZT value of 0.0275 is obtained when the doping amount of copper is 0.05 and the synthesis temperature of the sample is 673 K. In future research, optimizing its electrical properties can be achieved through element doping and changing the heating and holding time.

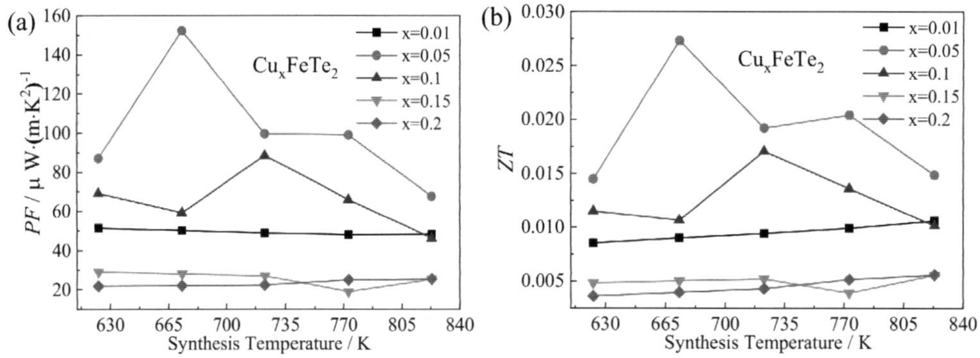


**Figure 4.** The variation of power factor and ZT of $Cu_xFeTe_2$ with synthesis temperature.

## 4. Conclusion

The effects of preparation and electrical properties of $Cu_xFeTe_2$ (x=0.01, 0.05, 0.1, 0.15, 0.2) were investigated by solid state reaction at different sintering temperatures and with different Cu contents. The following conclusions were obtained: The temperature range for preparing $Cu_xFeTe_2$ by solid state reaction was 350~550°C, with a heating time of 40 minutes and holding time within 60 minutes; the optimum temperature and Cu content were 400°C and 0.05, respectively. The minimum electrical resistivity of the sample was obtained at 1.74 mΩ·cm at a temperature of 400°C and Cu content of 0.05. The maximum Seebeck coefficient of 75.66 $\mu VK^{-1}$ was obtained at a temperature of 350°C and Cu content of 0.01. The experiment obtained the maximum power factor at 95.23 $\mu W/mK^2$.

**Acknowledgments**

This research was supported by the Scientific Research Cultivation Project (LPSSY2023KJZDPY02), College Students' Innovation Training Program (202110977039) and (S202210977050), Natural Science Foundation of China (52062031), Materials and Chemicals Direction Team (LPSSY2023XKTD06), Carbon Neutral Engineering Research Center of Guizhou colleges and universities in Coal Industry (Qian Jiao Ji [2023] No. 044), Guizhou Province first-class professional construction point (GZSylzy202103) and Youth Fund of Education Department of Guizhou Province (Qianjiaohe KY Zi [2022] No. 050), Liupanshui Normal University Fund Project (LPSSYLPY202230).

**References**

[1] An Y B, Park S J, Park O, et al. Electronic, Thermal, and Thermoelectric Properties of Ni-Doped $FeTe_2$ Polycrystalline Alloys [J]. Electronic Materials Letters, 2023, 19(3): 260-266.
[2] Li Guipeng, Song Guihong, Wang Nan, et al. Sputtering deposition and thermoelectric properties of Ag-doped β-$Cu_2Se$ thin films [J]. Surface Technology, 2021, 50(08): 218-226.
[3] Chibani H, Yahi H, Ouettar C. Tuning the magnetic properties of $FeTe_2$ monolayer doped by (TM: V, Mn, and Co) [J]. Journal of Magnetism and Magnetic Materials, 2022, 552: 169204.
[4] Park S J, Kwak H, Kim H S, et al. Evolution of electrical transport properties in $FeTe_2$-$CoTe_2$ solid solution system for optimum thermoelectric performance [J]. Journal of Alloys and Compounds, 2023: 170850.
[5] Zhang Z, Gao Y, Wu Y, et al. P-type doping of transition metal elements to optimize the thermoelectric properties of $CuGaTe_2$ [J]. Chemical Engineering Journal, 2022, 427: 131807.
[6] Qin Bingke, Ji Yonghua, Su Taichao, et al. Rapid synthesis of $FeTe_2$ compound and its electrical transport properties [J]. Rare Metal Materials and Engineering, 2019, 48(03): 931-935.
[7] Alvi K, Nazir S. Thermodynamics and robust n-type charge carrier density in Co-doped $FeTe_2$: strain strategy [J]. Physica Scripta, 2023, 98(6): 065915.
[8] Rahman A, Zhang D, Rehman M U, et al. Multiple magnetic phase transitions, electrical and optical properties of $FeTe_2$ single crystals [J]. Journal of Physics: Condensed Matter, 2019, 32(3): 035808.

# Process characteristics of Q890D joint using hybrid fiber laser-arc hybrid welding

**Yuhang Shen[1,2], Jiecai Feng[1] and Chuanwan Luo[1]**

[1]School of Mechanical Engineering and Automation, Shanghai University, Shanghai 200444, PR China

[2]Corresponding author's e-mail: 837935377@shu.edu.cn

**Abstract**: This article conducted a study on the process characteristics, structure, and properties of the Q890D joint. The results indicated that three-layer welding methods can obtain a joint free of pores and defects. The base metal was composed of a large amount of polygonal ferrite and dispersed granular carbides, while the weld metal consisted of lath martensite. The results also revealed that the average hardness of WM was 417.24 HV (20.7% higher than BM). The tensile performance of welded joints was better than that of BM. There were no obvious cracks on the surface of the bending specimens. Compared with BM, the impact strength and toughness of WM had been improved, since the impact toughness of HAZ was highly dispersed.

## 1. Introduction

Hitherto, high-strength steel has been widely used in the shipbuilding industry [1-3], and other engineering steel structures. However, when welding thick plate high strength steel, traditional welding efficiency is greatly reduced owing to the low penetration of arc welding. Recently, people have begun to apply laser-arc hybrid welding technology to industrial, with high energy density [4], high welding efficiency [5], deep penetration [6], and good welding quality [7].

Welding process research mainly focuses on the influence of various welding parameters on weld formation. The parameters of laser arc hybrid welding mainly include laser power, defocusing amount, welding speed, arc current, arc voltage, and wire feeding speed.

## 2. Test details

The materials used in this study were 12 mm thick Q890D. The heat treatment state was adjusted. The chemical composition of Q890D and wire are listed in Tables 1 and 2.

**Table 1.** Chemical element of Q890D.

| Element | C | Si | Mn | P | S | Cu | Ni | Mo | B | Cr | Nb | Ti |
|---|---|---|---|---|---|---|---|---|---|---|---|---|
| Percentage | 0.20 | 0.80 | 2.00 | 0.025 | 0.015 | 0.50 | 2.00 | 0.70 | 0.12 | 1.50 | 0.06 | 0.05 |

**Table 2.** Properties of ER83-G Wire.

| $\sigma_S$ (MPa) | $R_m$ (MPa) | Elongation (%) |
|---|---|---|
| 821 | 762 | 21 |

The metallurgical specimens were etched with 5% nitric acid alcohol solution and were observed by OLYMPUS BX-53 optical microscope. An HV-1000 hardness tester was used to measure the hardness of three layers. Tensile, bending, and impact performance tests were performed on an MTS electronic universal testing machine. The size and notch characteristics of tensile specimen, bending specimen and impact specimen are shown in Figure.1.

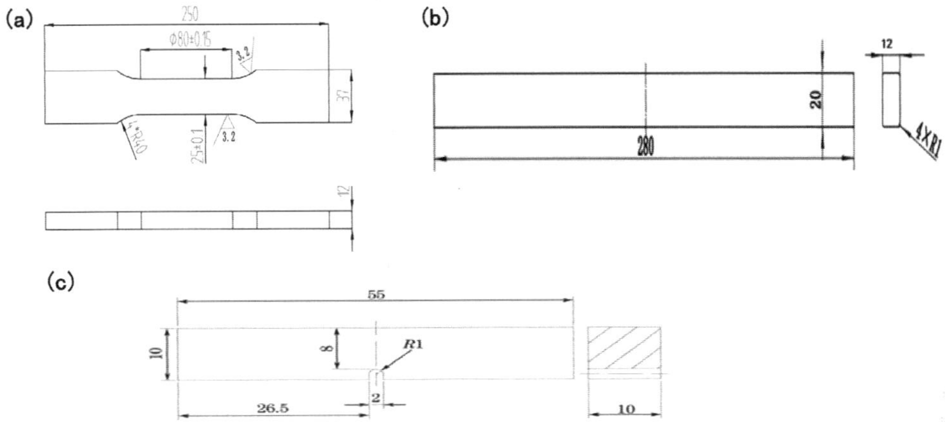


**Figure 1.** Dimensions of: (a) tensile specimen; (b) bend specimen; (c) impact specmen.

## 3. Welding joint qualification

### *3.1 Welding process*

Since the thickness of the steel selected in this welding is 12 mm, it is decided to use three welds to achieve material connection. When welding, a high-power laser is used in principle to obtain as much laser energy as possible on the surface of the workpiece and to achieve deep penetration.

The initial selection of test parameters for the current 205 A, welding speed 1.6 m/min. Figure 2 indicates that the laser power in 5000 W and 5400 W can achieve full penetration. The top of the weld is well-shaped without obvious defects, while the bottom of the weld is uniform residual height. At 5400 W, the top weld is rougher and has obvious spatter. At the same time, the residual height of the bottom weld is not uniform.

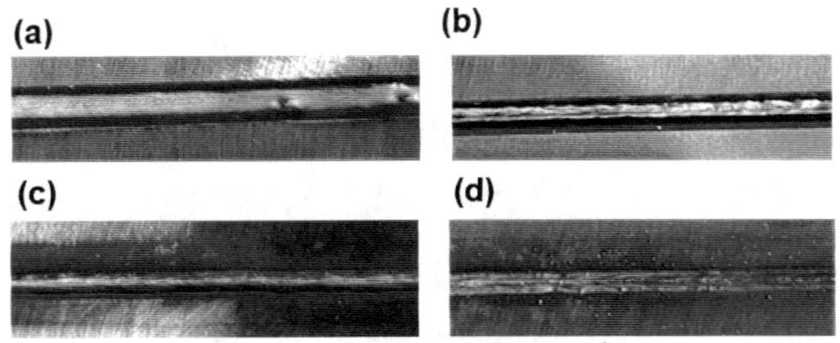


**Figure 2.** Appearance of weld: (a) top of the weld (5000 w); (b) laser power (5400 w); (c) back of the bottom layer (5000 w); (d) back of the bottom layer (5400 W).

It is easily seen from Figure 3 that the weld can achieve full penetration at all currents. When the arc current is low, a large spatter is produced and the weld is discontinuous. As the current increases, the arc ionization increases, and the preheating effect on the steel is increasing, which helps to improve the absorption of laser energy on the surface of the steel. Meanwhile, the increase in current makes the weld width wider. When the arc current is 205 A, the weld has the best morphology, and a continuous and stable full-penetration weld can be obtained.

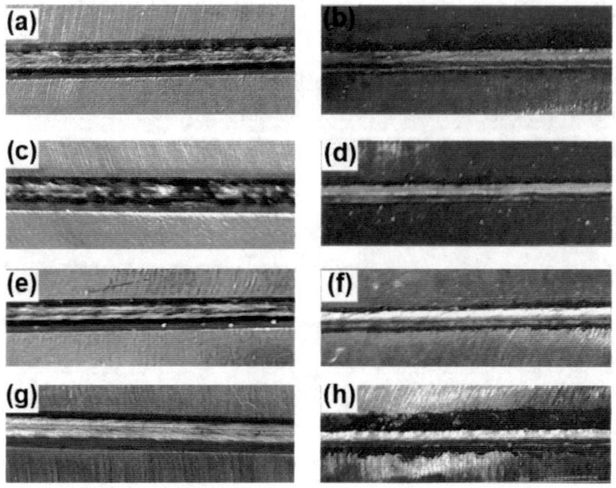

**Figure 3.** Weld morphology under different arc currents: (a) 182 A; (b) 182 A weld back; (c) 186 A; (d) 186 A weld back; (e) 189 A; (f) 189 A weld back; (g) 205 A; (h) 205 A weld back.

The filling layer has to use a higher arc current to melt more wire. Figure 4 demonstrates that low arc current (290 A) sidewalls are not completely fused. Excessive arc current will cause spatter, biting edge, and other welding defects. It can be seen that the most appropriate filling layer arc current should be 300 A.

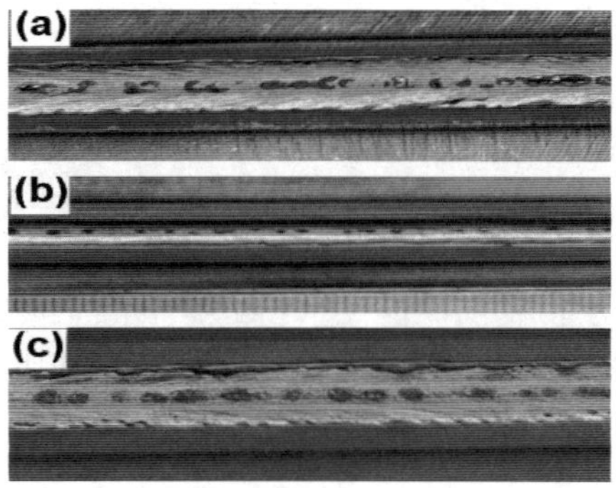

**Figure 4.** Weld morphology of filling layer under different arc currents: (a) 290 A; (b) 300 A; (c) 330 A.

The welding process of the cover layer was chosen to use similar process parameters as those of the filler layer, and a low-power laser of 800 W was selected to make the weld surface more continuous and stable. Under the 240 A arc current, the weld seam is uniform (Figure 5(c)).

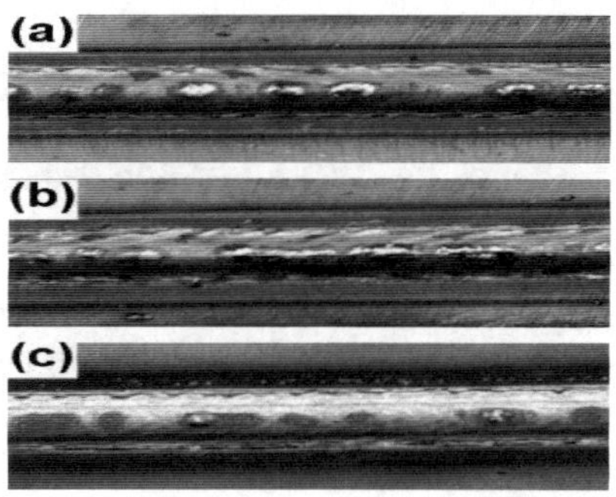

**Figure 5.** Weld morphology of filling layer under different arc currents: (a) 200 A; (b) 220 A; (c) 240 A.

Figure 6 shows that the front of the weld joint is uniform, without obvious undercuts or collapses. The back is transparent, and the excess height is small (Figure 6(a)). The X-ray non-destructive testing (Figure 6(d)) of the weld joint indicates that there are no internal cracks or air pores. The joint is in the shape of a wine cup. The bottom layer is thin and long, which presents the characteristics of laser welding. Meanwhile, the filling layer and cover layer are short and wide, which presents the characteristics of arc welding. It is worth noting that the high-power laser in the bottom layer results in a greater depth-width ratio than the other two layers. The widths of the bottom layer, filling layer, and cover layer are 1.240 mm, 6.196 mm, and 9.384 mm. The average penetration of the bottom layer is similar to the other two layers, while the width of the bottom layer is only 1/5, 1/8 of that of the other two layers.

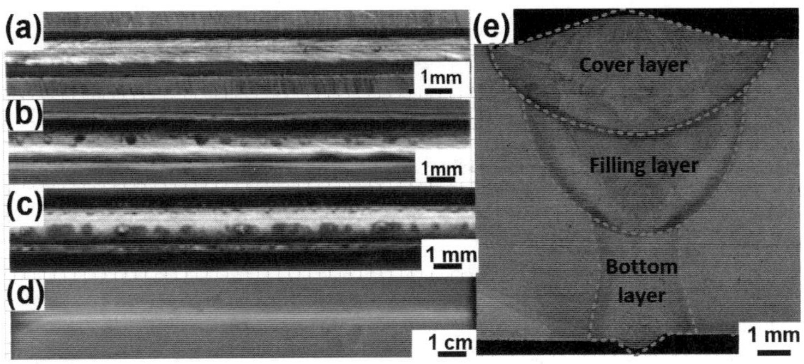


**Figure 6.** Morphology of welded joint: (a) bottom layer; (b) filling layer; (c) cover layer; (d) x-ray detection results; (e) cross-sectional morphology of welded joint.

*3.2 Microstructure*

Figure 7 shows the microstructure of welded joints. WM is composed of martensite. The partially transformed region, fine-grained region, and coarse-grained zone are composed of granular bainite and ferrite. Besides, the base metal consists of a large amount of massive ferrite and carbides.

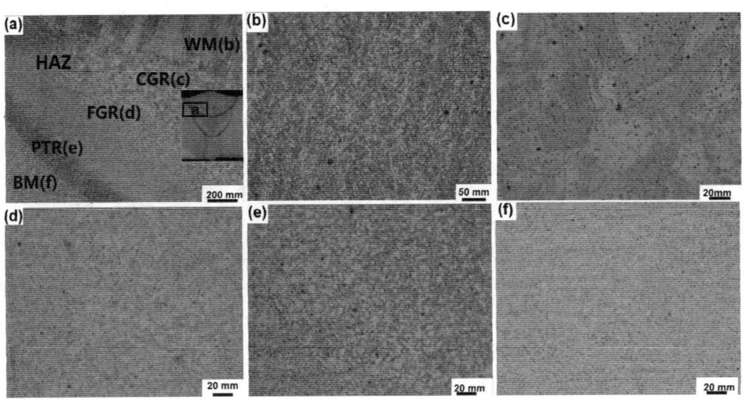


**Figure 7**. Microstructure of joint: (a) schematic diagram, (b) weld metal, (c) coarse grained zone, (d) fine grained zone, (e) partially transformed zone, (f) base metal.

*3.3 Mechanical property*

Figure 8 presents the variation region of hardness in the joint. The average harnesses of HAZ, WM, and BM are 399.1 HV, 415.3 HV, and 342.7 HV, and the hardness of WM has increased by 20% compared to the BM. The lateral hardness of the weld joint does not change much. The highest hardness appears in the HAZ of the cover layer, reaching 460 HV. The minimum hardness (296 HV) appears in the softening zone near the BM in the HAZ, and the hardness is about 20% lower than that of BM.

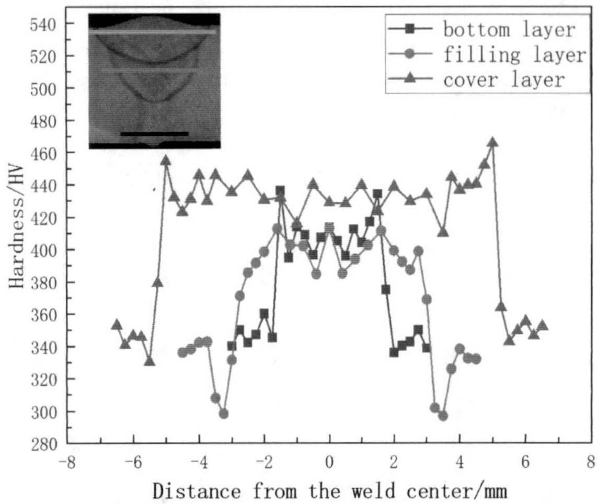


**Figure 8.** Hardness distribution curve.

It can be known from Figure 9 that the tensile strength is 993.67 MPa and the elongation is 22.5%. Besides, cracks sprout from the BM. The tensile strength has increased by nearly 200 MPa and the average elongation has increased by 1.5% than Q890D. Observation under high-power electron microscopy shows that the fracture surface is composed of pores and deep dimples, which indicates that the material has good toughness. In this experiment, the narrow width of the softening zone does not affect the overall tensile properties of the joint.

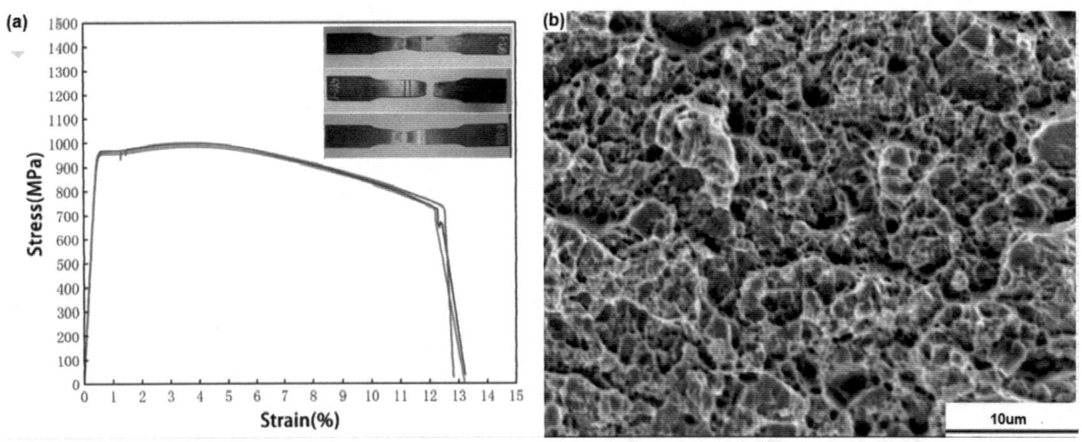


**Figure 9.** Tensile test results: (a) stress-strain curve; (b) microfracture morphology of tensile specimen.

Conduct front and back bending tests on welded joints. No surface cracks were found on both the front and back sides (Figure 10(a)(b)), indicating that the joint has good ductility and sufficient bending strength.

Table 3 shows the impact properties of WM and HAZ. The reason why the impact toughness of HAZ is higher than that of the weld is that HAZ contains a portion of ferrite and diffuse carbide composition, and its toughness is higher than that of martensite. The impact absorption energy and impact toughness of the weld are only one-third of those of HAZ, but the numerical distribution is relatively uniform, and the numerical dispersion of HAZ is large, which is also related to the greater heterogeneity of the HAZ. Figure 10(d)The impact fracture exhibits typical ductile fracture characteristics.

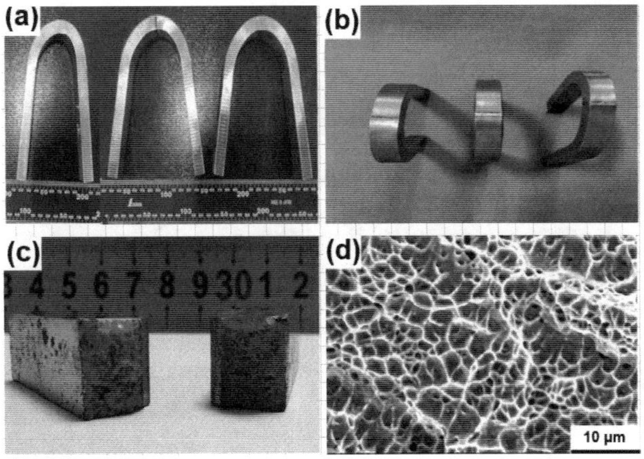


**Figure 10.** Bending and impact specimens: (a) macroscopic morphology of test pieces; (b) macroscopic morphology of bending specimens; (c) fracture morphology of impact specimens; (d) microscopic fracture morphology of impact specimens.

**Table 3.** Impact performance of weld metal and heat affected zone.

| Number | Notch position | Absorption energy (J) | Impact toughness (kJ/m$^2$) |
|---|---|---|---|
| 1 | BM | 63 | 0.762 |
| 2 | BM | 50 | 0.606 |
| 3 | BM | 49 | 0.585 |
| 4 | HAZ | 175 | 2.257 |
| 5 | HAZ | 147 | 1.887 |
| 6 | HAZ | 144 | 1.802 |

## 4. Conclusion

(1) The joint is in the shape of a wine cup. The bottom layer is thin and long, while the filling layer and cover layer are short and wide. The average penetration of the bottom layer was similar to that of the filler layer and cover layer.

(2) High-power lasers were suitable for bottom welding. Current affects the shape of the weld indication.

(3) The hardness of the weld metal was higher than BM. There was a softening zone with a width of about 1 mm at HAZ.

(4) The tensile specimens were fractured at BM. The weld has no obvious cracks on the bending specimens. The average values of -20°C impact toughness absorption work in WM and HAZ of the welded joints are 54.6 J and 155.4 J.

## Reference

[1] S.K. Sharma, S. Maheshwari (2016), A review on welding of high strength oil and gas pipeline steels, Journal of Natural Gas Science and Engineering., 38: 203-217.

[2] J.Z. Li, Q.J. Sun, K.X. Kang, Z.Y. Zheng, Y.B. Liu, J.C., Process stability and parameters optimization of narrow-gap laser vertical welding with hot wire for thick stainless steel in nuclear power plant, Optics & Laser Technology 123.

[3] Balakrishnan, J., Vasileiou, N. A., Francis, A. J., Smith, C. M., Roy, J. M. (2018), Residual stress distributions in arc, laser and electron-beam welds in 30 mm thick SA508 steel: A cross-process comparison, International Journal of Pressure Vessels and Piping.

[4] M. Sokolov, A. Salminen, M. Kuznetsov, I. Tsibulskiy (2011), Laser welding and weld hardness analysis of thick section S355 structural steel, Materials & Design 32(10): 5127-5131.
[5] Nakanishi, Motoki, Wang, Hongze, Kawahito, Yosuke (2017), Effects of welding speed on absorption rate in partial and full penetration welding of stainless steel with high brightness and high power laser, Journal of Materials Processing Technology.
[6] M. Zhang, G. Chen, Y. Zhou, S. Liao (2014), Optimization of deep penetration laser welding of thick stainless steel with a 10kW fiber laser, Materials & Design 53 (Jan.) 568-576.
[7] Y. Kawahito, M. Mizutani, S. Katayama (2009), High quality welding of stainless steel with 10 kW high power fibre laser, Science and Technology of Welding & Joining 14(4) p.288-294.

# Analysis of disc movement during the closing process of axial flow check valves

**Ting Wang[1,2], Na Li[1], Beibei Kong[1], Xiaomei Ni[1], and Xiangbao Li[1]**

[1]College of Intelligent Manufacturing and Control Engineering, Qilu Institute of Technology, Jinan, Shandong Province, 250200 China

[2]Corresponding author's E-mail:wt6805@163.com

**Abstract**: In a long-distance pipeline system, when the pump is stopped, the axial flow check valve will suddenly close under the pressure difference force of the medium, and the instantaneous valve disc will collide and impact the valve seat at a certain speed, which will have a certain impact on the structural strength and service life of the check valve. This article uses explicit dynamics to simulate the movement process of the instantaneous valve disc when it is completely closed and analyzes the velocity of the valve disc after the collision. Based on the spring damping model theory, the velocity of the valve disc after collision is derived. After error analysis, the reliability of numerical simulation has been verified, which has a certain guiding significance for the practical application and design improvement of axial flow check valves.

## 1. Introduction

Axial flow check valves are widely used in long-distance pipeline systems in large-scale petroleum refining, ethylene, and other petrochemical projects, as well as in pipeline systems with harsh operating conditions in large devices, due to their advantages of small flow resistance, stable operation, low noise, and small overall dimensions [1]. Axial flow check valves are key equipment in pipeline systems, whose performance, service life, and operational reliability can all affect the safety of the pipeline system. To determine the safety and reliability of check valves under various working conditions, a large number of experiments are required to verify. With the development of technology, numerical simulation technology has become a reliable method. For complex collision and impact processes, finite element software based on dynamic nonlinear theory is used and this process can be well analyzed and reproduced [2].

Han et al. [3] selected the RNG turbulence model and used CFD to dynamically simulate and analyze the opening process of a new type of reverse thrust check valve. It was found that the valve disc velocity was close to zero at the moment of complete opening and closing of the valve, indicating that the reverse thrust check valve would not experience impact. Zhang and Yu [4] used numerical analysis software Fluent and dynamic grid technology to analyze the oscillation phenomenon and reasons during the opening process of axial flow check valves, and studied the dynamic characteristics of check valve opening through structural improvement. To extend the service life of swing check valves and reduce energy loss, Zhang et al. [5] established a valve disc motion model combined with a structural dynamics model and used Fluent and Ansys/Ls-Dyna finite element software and analysis modules to simulate and analyze the valve disc motion and impact collision phenomenon during the valve closing process. Zhang et al. [6] obtained the variation law of the inlet medium pressure of the axial flow check valve from the full characteristic curve of the centrifugal pump and used CFD software simulation and UDF dynamic grid technology to dynamically simulate the two-dimensional flow field during the closing process of the axial flow check valve. Due to its short action time and high destructiveness, collisions are difficult for researchers to directly measure. In this paper, the explicit dynamic numerical simulation method is used to simulate the collision and impact process between the valve disc and valve seat when the axial flow check valve is completely closed and the pump suddenly stops. Based on the spring damping model theory, the velocity of the valve disc after collision is derived. After error analysis, the reliability of the numerical simulation is proved.

## 2. Basic concepts of impact motion

Collision is a special form of object motion, with a short time history but an extremely complex change process. The collision mechanism is still a hot topic in the theoretical field and many application fields. Due to the large collision force and acceleration generated by object collisions but limited velocity changes, the collision process is usually described macroscopically based on the difference in collision impulse and momentum.

The research in this article belongs to the problem of system dynamic response. For such problems, the dynamic response results of the system after a collision are related to the motion state of the system before the collision, the recovery coefficient during the collision contact process, the collision impulse, etc. [7].

## 3. Explicit dynamics

ANSYS has invested a lot of effort in explicit dynamics, with three explicit dynamics product modules, namely ANSYS LS-DYNA, ANSYS AUTODYN, and ANSYS Explicit Dynamics. LS-DYNA is a world-renowned explicit nonlinear dynamic analysis program that focuses on Lagrangian algorithm, display solving, structural solving, and nonlinear dynamic analysis. It has excellent parallel computing capabilities and excellent mesh adaptation technology [8].

## 4. Simulation and result analysis of ansys workbench

### *4.1 Establishment of a numerical model*

The inlet and outlet diameters of the valve body are both 252 mm, the total length of the check valve chamber is 622 mm, and the front and rear lengths of the valve seat are 156 mm and 462 mm, respectively. The valve disc mass is 4.08 kg, and the spring stiffness is 3,500 N/m. During the working process, the spring is always in a compressed state. When the valve is fully opened, the spring reaches a maximum compression of 86.8 mm, and when the valve is fully closed, the spring reaches a minimum compression of 18 mm. The medium is water. Under normal operating conditions, the medium enters from the left and exits from the right. When the pump is stopped, the medium flows back, and the high-pressure medium flows in from the right side and out from the left side.

We grid the model in ANSYS Workbench, and the explicit dynamic analysis diagram and grid diagram are shown in Figure 1.

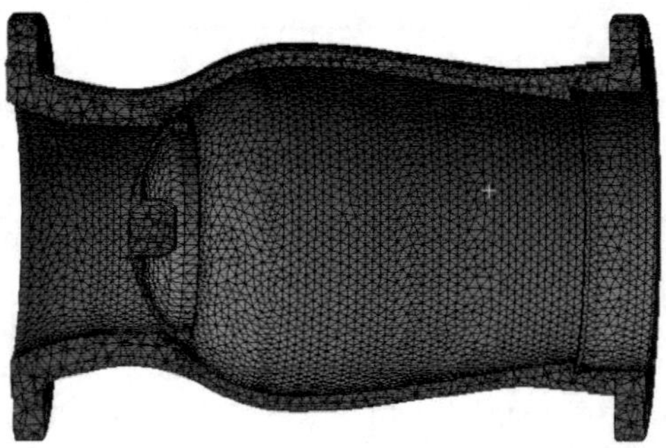

**Figure 1.** Explicit dynamic analysis diagram and grid diagram.

*4.2 Analysis settings*

We create a 3D geometric model by using SolidWorks and save it. x_t format file is imported into ANSYS Workbench for analysis and use. In engineering data, the performance parameters of the material were input, and to reduce the computational complexity of iterative solutions, the impact geometry model was simplified. The geometric model only contains two components, the valve body and the valve disc. The valve body material is WCB, the valve disc material is CF8M, the number of nodes is 22,322, and the number of elements is 93,480. Given an initial velocity of -3.624 m/s in the initial condition, we keep the remaining settings as default, apply force to the valve body, calculate the k file, and import the k file into the LS-DYNA solver for solving.

*4.3 Response result of valve disc speed*

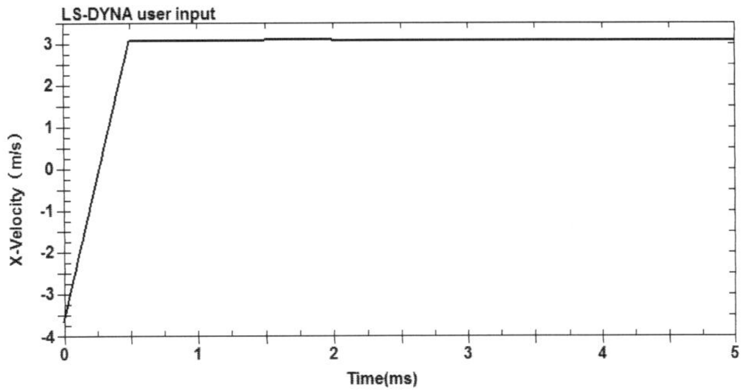


**Figure 2.** The change diagram of disc velocity at the moment of collision.

As shown in Figure 2, at 0.5 ms, the collision contact between the valve disc and the valve seat is completed. Due to the collision effect and energy loss during the collision process, the velocity of the valve disc changes from the initial velocity of -3.624 m/s (direction to the left) to 3.11 m/s (direction to the right). After 0.5 ms, the velocity remains at 3.11 m/s without any change, so the collision time is 0.5 ms.

## 5. Theoretical derivation of the velocity of the valve disc after collision

*5.1 Establishment of a collision mathematical model*

When the pump is stopped, the impact of the valve disc on the valve seat can be simplified as a spring-damping model of a mass impacting the elastic wall. The valve disc accelerating along the negative x direction eventually collides with the valve seat at a speed of 3.624 m/s, corresponding to the free mass colliding with the elastic wall at a positive speed, as shown in Figure 3. The stiffness coefficient and damping coefficient of the elastic wall are denoted as k and c, respectively. When t=0, it is the start time of the collision. During the collision period, the motion of the free mass satisfies the following initial value problem [9-10].

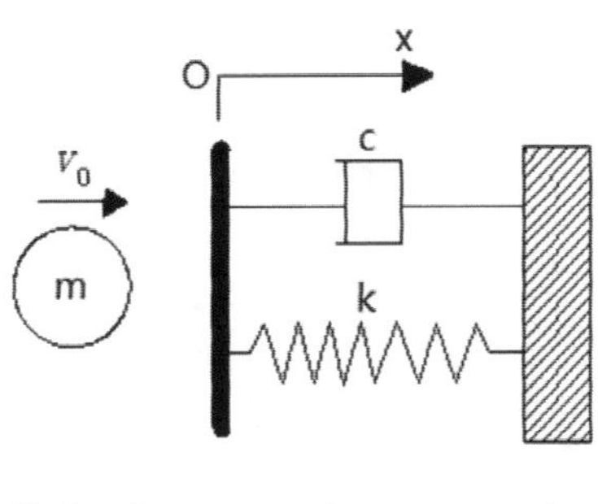


**Figure 3.** Collision between a free mass and an elastic wall.

$$\begin{cases} mx'' + cx' + kx = 0 \\ x(0) = 0, x'(0) = v_0 \end{cases} \quad (1)$$

During a collision like this, the motion of a moving object that undergoes repeated vibrations to reach its equilibrium position belongs to underdamped motion. For the underdamped case, the solution of the differential equation (i.e., Equation (1)) is:

$$x(t) = A\exp(-\xi\omega_n t)\sin\omega_d t \quad (2)$$

Where $\xi = c/(2m\omega_n) < 1$; $A = v_0/\omega_d$; $\omega_d = \sqrt{1-\xi^2}\,\omega_n$; $\omega_n = \sqrt{k/m}$, in which k is spring stiffness, c is damping coefficient, $v_0$ is the initial velocity of mass motion, m is the mass of a freely moving mass, and t is impact time.

*5.2 Optimization of k and c in Maple*

In practical engineering applications, it is difficult to determine the actual values of spring stiffness k and damping coefficient c, so it is necessary to obtain the values of k and c under the research conditions. The following is to optimize the accurate values of k and c in Maple.

After transforming the above parameters $\xi$, $A$, $\omega_d$ into functions related to m, k, and c, it can be concluded that:

$$A = 2mv_0 \frac{1}{\sqrt{4km - c^2}} \quad (3)$$

$$\xi = \frac{c}{2m \cdot \sqrt{k/m}} \quad (4)$$

$$\omega_d = \frac{\sqrt{4mk - c^2}}{2m} \quad (5)$$

Substituting Equations (3), (4), and (5) into Equation (2) yields:

$$x(t) = 2mv_0 \frac{1}{\sqrt{4km - c^2}} \sin\left(\frac{\sqrt{4mk - c^2}}{2m} t\right) \exp\left(-\frac{c}{2m} t\right) \quad (6)$$

We write a Maple operation program according to Equation (6). The relationship between the maximum impact displacement $x(t)$, damping c, and spring stiffness k obtained in Maple is shown in Figure 4. As the damping coefficient increases, the maximum impact displacement gradually decreases in a parabolic shape. As the spring stiffness increases, the maximum impact displacement decreases in a straight line.

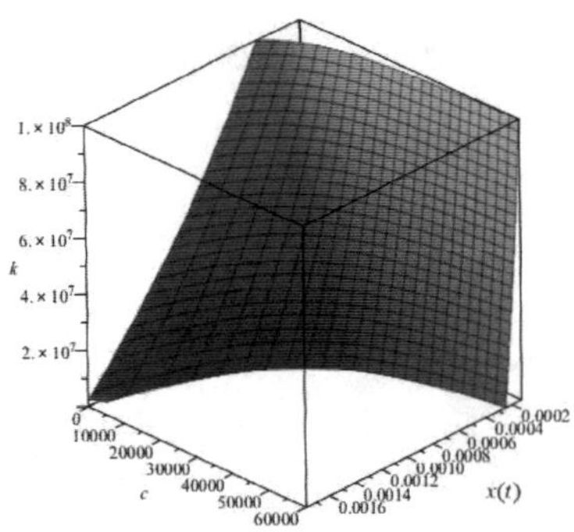


**Figure 4.** The three-dimensional relationship $x(t)$ of k and c.

The collision schematic diagram is shown in Figure 5. According to previous numerical simulations, the maximum strain values of valve body A and valve disc B at t=0.25 ms are $\varepsilon_A = 2.131\times10^{-3}$ and $\varepsilon_B = 8.3565\times10^{-4}$. Based on the curvature radius of the contact circle $r = 104\text{mm}$, the maximum impact displacement along the radius r direction of the vertical direct contact surface is obtained as $x(t) = x_1 + x_2 = r\cdot\varepsilon_A + r\cdot\varepsilon_B = 0.000309\text{m}$. Under this condition, the two-dimensional relationship between k and c is shown in Figure 6. As the damping coefficient continues to increase, the spring stiffness roughly increases in a straight line with a slope of 1.

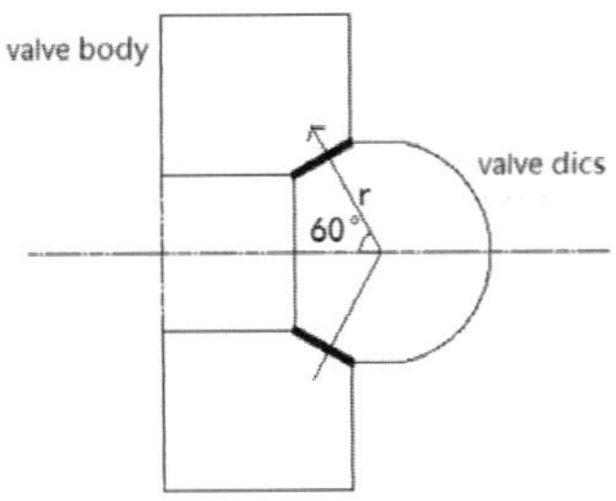


**Figure 5.** Collision diagram.

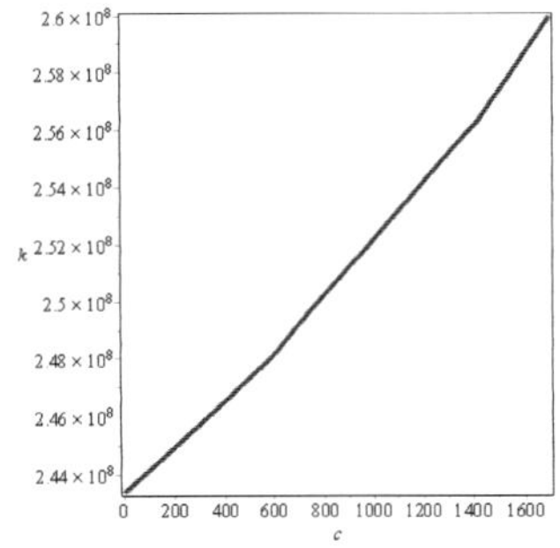


**Figure 6.** Relationship curve of k and c.

The calculation equation for spring stiffness is $k \approx 5Ga$, in which G is the shear elastic modulus. The shear elastic modulus of the valve body WCB is $G_1 \approx 84000\text{Mpa}$. The shear elastic modulus of the valve disc material CF8M is $G_2 \approx 80000$ MPa, and thus G=8, 200 MPa. Considering the collision system as a cube with side length, the spring stiffness is obtained as $k \approx 5Ga = 5\times82000\times622 = 2.55\times10^8\,\text{N/mm}$ [11]. When the spring stiffness is $2.55\times10^8\,\text{N/mm}$, we can obtain $c = 1400$ from Figure 6.

*5.3 Theoretical derivation process*

Taking the first derivative of Equation (2) yields:

$$x^{\cdot}(t) = A\exp(-\xi\omega_n t)(\omega_d \cos\omega_d t - \xi\omega_n \sin\omega_d t) \tag{6}$$

Taking the second derivative yields:

$$x^{\cdot\cdot}(t) = -A\exp(-\xi\omega_n t)[(1-2\xi^2)\omega_n \sin\omega_d t + 2\xi\omega_d\omega_n \cos\omega_d t] \tag{7}$$

We order $t = \Delta t$ and $x^{\cdot\cdot}(t) = 0$ and then obtain:

$$(1-2\xi^2)\omega_n \sin\omega_d\Delta t + 2\omega_d\omega_n \cos\omega_d\Delta t = 0 \tag{8}$$

The contact time is the minimum positive solution of Equation (8), which is:

$$\Delta t = \frac{\pi\text{-arctan}\lambda_1}{\omega_n\sqrt{1-\xi^2}} \qquad \lambda_1 = \frac{2\xi\sqrt{1-\xi^2}}{1-2\xi^2} \tag{9}$$

The velocity of a free mass before collision is known to be $v_0$. Through the relationship between velocity before and after collision, we obtain the velocity of the free mass at the end of the collision $v_1$.

$$\begin{aligned} v_1 &= x^{\cdot}(\Delta t) \\ &= A\exp(-\xi\omega_n\Delta t)(\omega_d \cos\omega_d\Delta t - \xi\omega_n \sin\omega_d\Delta t) \\ &= \frac{v_0}{1-2\xi^2}\exp[-\frac{\xi}{\sqrt{1-\xi^2}}(\pi - \arctan\lambda_1)]\cos(\pi - \arctan\lambda_1) \end{aligned} \tag{10}$$

We substitute the initial conditions $v_0 = -3.624\text{m/s}$, $k = 2.55\times10^8\text{N/mm}$, $c = 1400$, $m = 4.08\text{kg}$, and $\Delta t = 0.5\text{ms}$ into Equation (10) to obtain the final post-collision velocity as $v_1 = 2.985\text{m/s}$.

The equation for Newton's recovery coefficient is:

$$e = -\frac{v_1}{v_0} = \frac{v_0}{1-2\xi^2}\exp[-\frac{\xi}{\sqrt{1-\xi^2}}(\pi - \arctan\lambda_1)]\cos(\pi - \arctan\lambda_1) \tag{11}$$

We substitute $v_1$ into Equation (11) to obtain Newton's recovery coefficient of 0.825.

*5.4 Error analysis*

According to transient dynamic numerical simulation, the velocity of the valve disc after contact collision is 3.11 m/s, the contact collision velocity obtained from theoretical derivation is 2.985 m/s, and the relative error obtained is:

$$\delta_r = \frac{|x - x^*|}{x} = \frac{|v_1 - v_s|}{v_1} = \frac{|2.985 - 3.11|}{2.985} = 4.2\% \tag{12}$$

The errors generated above belong to model errors and truncation errors, which are generated during theoretical derivation and simplification of the model. In numerical simulation calculations, discretization of continuity equations results in truncation errors.

## 6. Conclusion

1) The velocity of the valve disc after collision is obtained through numerical simulation of display dynamics at 3.11 m/s;

2) In theoretical derivation, due to the difficulty in determining the damping coefficient c in the spring damping model, the impact dynamics equation was used to optimize the relationship curve between k and c in Maple, in which $k = 2.55\times10^8$ and $c$=1, 400.

3) According to theoretical derivation, the velocity of the valve disc after collision is 2.985 m/s, and the relative error compared with the numerical simulation value is 4.2%, proving the reliability of the numerical simulation.

In summary, the research in this article has certain guiding significance for the practical application of axial flow check valves under pump shutdown conditions.

## Acknowledgments

This work has received support from the school-level project of the Qilu Institute of Technology (Approval number: QIT22NN015). The author is very grateful for the valuable and insightful comments from the reviewers, which undoubtedly improved our manuscript.

## References

[1] Feng Ming. Numerical Simulation Study of Axial Flow Check Valves [J]. Mechanical Manufacturing, 2022, 60 (05): 56-59.

[2] Li Mengke. Analysis of the Opening and Closing Process and Drag Reduction Design of Axial Flow Check Valves [D]. Lanzhou University of Technology, 2020.

[3] Han Xu, Zheng Mingguang, and Yu Yiqi. Hydrodynamic characterization and optimization of containment check valve by numerical simulation [J]. Annals of Nuclear Energy, 2011, 38 (6): 1, 427-1, 437.

[4] Zhang Xiheng and Yu Rongfang. Analysis of dynamic characteristics and structural improvement of axial flow check valve disc opening [J]. Chemical Machinery, 2023, 50 (05): 649-654+684.6.

[5] Zhang Xiheng, Zhang Sunli, Bai Yazhou, et al. Transient closing process and collision response analysis of swing check valves [J]. Mechanical Strength, 2021, 43 (05): 1, 155-1, 161.

[6] Zhang Xiheng, Wang Yan, Wang Ting, et al. Dynamic analysis of transient closure process of axial flow check valve disc [J]. Journal of Gansu Science, 2019, 31 (03): 103-107+113.

[7] Ma Xiaoqing. Impact Dynamics [M]. Beijing: Beijing University of Technology Press. 1992.7.
[8] Pu Guangyi. Basic Tutorial and Example Explanation of ANSYS Workbench (Third Edition) [M]. Beijing: China Water Resources and Hydropower Press, 2019.
[9] Jin Dongping and Hu Haiyan. Collision Vibration and Control [M]. Beijing: Science and Technology Press, 2005: 30-46.
[10] Yu Tongxi and Qiu Xinming. Impact Dynamics [M]. Beijing Tsinghua University Press, 2011.
[11] K. L. Johnson. Contact Mechanics [M]. Beijing: Higher Education Press, 1992.

# Design of a superconducting UWB bandpass filter with tunable notch band

**Zhaojiang Shang[1,2], Jianyong Huang[1], Huibin Zhang[1], and Weijin Yang[1]**

[1]College of Sciences, Shanghai Institute of Technology, Shanghai 201418, China

[2]Corresponding author's e-mail: shangzj@sit.edu.cn

**Abstract**: By studying the principles of ultra-wideband (UWB) and band-stop, a novel band-pass filter with adjustable band-stop characteristics is proposed. The filter achieves 3.1-10.6 GHz UWB bandpass characteristics through an improved multi-mode resonator. An L-shape λ/2 notch resonator loaded with a 2-bit reconfigurable interdigital capacitor (IDC) array is coupled to the multiple-mode resonator and generates a tunable notch band to reject any undesired signal that may interfere with the UWB passband. By connecting the IDC to the ground or not, the capacitance of each IDC and the corresponding notch frequency can be adjusted. After design and optimization, the final size of the superconducting filter is 20.0 mm × 6.9 mm, and the test results and simulation results are highly consistent.

## 1. Introduction

Since 2002, when the U.S. FCC voted to allow the commercial use of the 3.1-10.6 GHz frequency band, ultra-wideband wireless communication technology has attracted considerable academic and industrial attention [1]. Compared to other communication systems, ultra-wideband wireless systems have many unique advantages, including higher data transmission speeds, low power consumption, and higher immunity to interference. To transmit and receive high-quality signals, as one of the key devices of ultra-wideband wireless systems, ultra-wideband filters with compact structure and superior performance have received more and more attention. Recently, various designs of UWB filters, including MMRs (multiple-mode resonators) [2], cascaded bandstop and bandpass filters [3], and electromagnetic bandgap structures [4], have been reported. However, in some cases, some interfering signals (e.g., Bluetooth and wireless LAN signals around 5 GHz) happen to be within the transmission spectrum of the ultrawideband system, and this interference may cause some interference to the performance of the ultrawideband system. To solve this problem, the existing research work is to introduce a blocking band with a fixed center frequency within the original passband to suppress the unwanted interference signals [5-6]. The frequency of the external interference signal is not fixed because it depends on factors such as external power supply, other interfering devices, and electromagnetic radiation. Therefore, it is important to study tunable band-stop ultra-wideband filters from both academic and engineering perspectives.

HTS films are known to have significantly lower surface resistivity than conventional materials in the microwave region. High-performance HTS filters with low in-band insertion loss, high band-edge steepness, and high out-of-band rejection characteristics can be fabricated by using high-temperature superconducting materials [7]. In this paper, an ultra-wideband superconducting filter with a tunable notch is successfully realized by using a λ/2 L-type reconfigurable resonator loaded with a reconfigurable 2-bit IDC array coupled with a multimode resonator.

## 2. L-shape 1/2-wavelength reconfigurable resonators

*2.1 Conventional coupled resonators*

Typical methods reported in the literature for introducing notch characteristics within the ultrawideband passband include coupled resonator, embedded open-circuit short-circuit line, and asymmetric input/output (I/O) coupling. The resistive band frequencies introduced by these methods are often constant and cannot effectively deal with the chaotic and complex electromagnetic interference problems in the outside world. The coupled resonator method introduces a resistive band with steeper band edges than the other methods (embedded open-circuit truncation method and asymmetric input/output coupling method). The basic principle is that when λ/2 resonators are coupled to a main transmission line, the response curves all produce a band resistance characteristic with a small bandwidth [8].

The coupled resonator method uses an externally coupled λ/2 resonator to determine the center frequency of the resistive band, which can be adjusted by changing the electrical length of the resonator. Inspired by this, it is believed that if the frequency of the externally coupled resonator is freely changed, the center frequency of the resistive band can also be freely adjusted. Previous literature has reported various shapes of coupled resonators, such as U-shape, spiral shape, and ring shape [9]. However, the frequencies of most of these coupled resonators are constant and cannot be adjusted, and there are few reports on frequency-variable coupled resonators.

*2.2 Theoretical computational analysis of L-shaped reconfigurable resonator*

In this paper, an L-type λ/2 reconfigurable resonator is proposed, as shown in Figure 1, which consists of an L-type superconducting resonator main structure and an end-loaded 2-bit fork-finger capacitor array. The equivalent circuit diagram of this tunable resonator is shown in Figure 2, where $C_1$, $C_2$, and L are the C-L-C equivalent circuit elements of the superconducting resonator, $C_{i1}$ and $C_{i2}$ are the capacitance values of the two superconducting fork-finger capacitors, and $C_{g1}$ and $C_{g2}$ are the capacitance values between the two IDCs and the ground plane. We ignore the small differences between the two parts of the IDCs interacting up and down. We assume that they have the same capacitance value to ground and that the switch in Figure 1 is a lossless ideal switch.

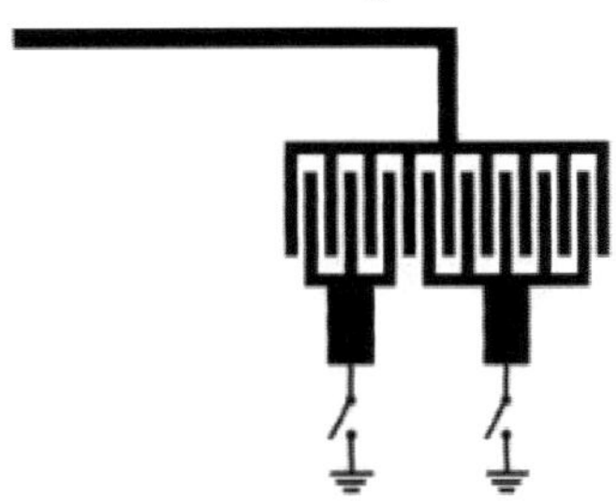

**Figure 1**. Configuration of the L-shaped λ/2 reconfigurable resonator.

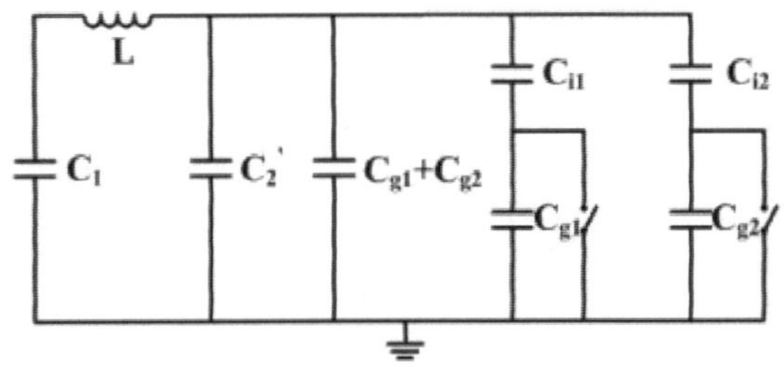


**Figure 2**. Equivalent circuit diagram.

It is assumed that the inductance of this L-type reconfigurable resonator remains unchanged at different grounding states, and the frequency tuning range can be expressed as follows:

$$\frac{\Delta f}{f_0} = \frac{-\Delta C}{2C} \tag{1}$$

where $f_0$ and C denote the resonance frequency and the equivalent lumped parameter capacitance of this resonator, and Δf and ΔC denote their variation ranges, respectively. When the IDC array is in different ground states, the loaded capacitance value of this resonator is different, and thus the resonance frequency is also different. In Figure 1, we refer to the "closed" state of the resonator as State "1" and the "open" state of the resonator as State "0". Then, the reconfigurable resonator has four states, (00), (01), (10), and (11), which represent different configurations of the two switches. By adjusting the body length of the resonator and the structure of the IDC array, the desired bandwidth frequency and tuning range can be realized.

## 3. Design of ultra-wideband superconducting filters with tunable band-stop characteristics

Based on an L-type λ/2 reconfigurable resonator, an ultra-wideband filter with tunable band-stop characteristics is designed by the coupled resonator method. The filter utilizes a multimode resonator with a U-shaped topology to achieve an ultra-wideband passband with appropriate external cross-finger coupling.

Figure 3 shows the topology of the final realized ultra-wideband filter with a tunable bandstop. The filter consists of an ultra-wideband bandpass filter and an L-type λ/2 resonator coupled with a 2-bit IDC array. The ultra-wideband bandpass filter is designed based on the U-shaped multimode resonator, the L-type λ/2 resonator is coupled to the middle part of the multimode resonator, and the coupling strength is controlled by varying the distance $g_1$ between them.

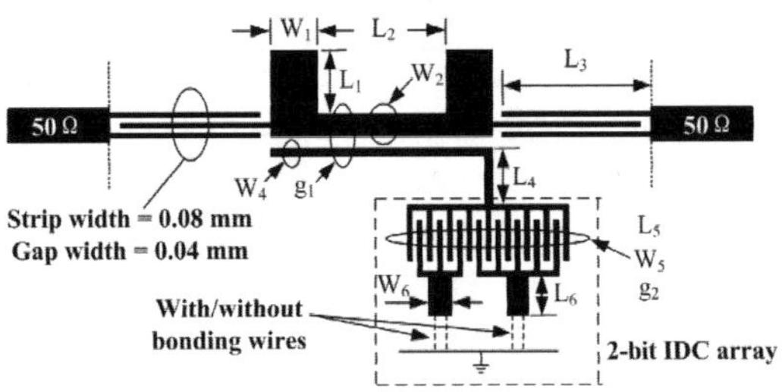


(a)

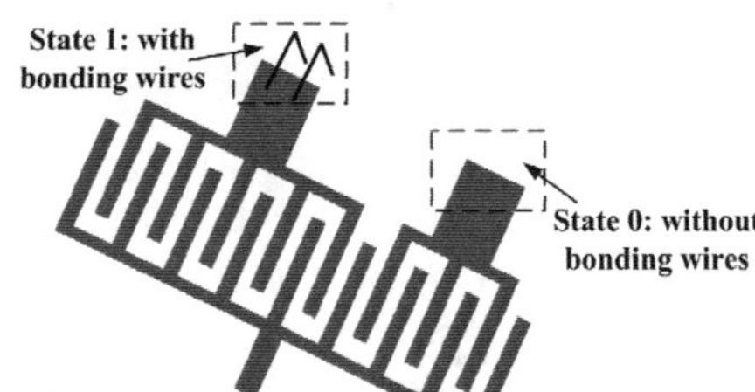


(b)

**Figure 3**. (a) The proposed UWB BPF (not to scale); (b) Using bonding wires to connect the IDC array to the ground (State 1) or disusing them (State 0).

One end of this L-shaped reconfigurable resonator is loaded with a 2-bit-IDC array, which is used to change the frequency of this resonator to produce a tunable resistive band within the ultrawideband passband. When a λ/2 resonator is coupled to a primary transmission line, a narrow-bandwidth resistive band characteristic is produced. As shown in Figure 2, a λ/2 L-type resonator is coupled to the center portion of the multimode resonator at a distance of $g_1$. The center frequency of the resistive band is determined by the L-type resonator, i.e., the wavelength of the L-type resonator is equal to the half electrical length corresponding to the center frequency of the resistive band. The suppression regime of the resistive band is determined by the coupling spacing $g_1$. The smaller the spacing is, the stronger the coupling is, and the higher the resistive band suppression is. The magnitude of the capacitance of each loaded fork finger capacitor can be varied by grounding the IDC with a solder wire (labeled as State "1") or without grounding it with a spot solder wire (labeled as State "0"), as shown in Figure 3 (b). Therefore, by selecting whether the solder wire is grounded or not, the resonant frequency of the L-type resonator can be controlled to change the center frequency of the stop band, thus realizing the design of an ultra-wideband filter with tunable bandstop characteristics. In practice, we use two spot weld lines for grounding to ensure the stability of grounding.

The simulation characteristics of the filter shown in Figure 3 (a) were completed through the electromagnetic field simulation software Sonnet. In the simulation process, the "ground" simulation is to simulate the actual machined AlSi spot welding wire connection by connecting the IDC to the metal box wall by using a microstrip wire with a width of 0.02 mm. Optimized by software, the final filter size selection is $L_1$ = 1.16 mm, $W_1$ = 0.86 mm, $L_2$ = 3.08 mm, $W_2$ = 0.38 mm, $L_3$ = 4.60 mm, $g_1$ = 0.16 mm, $W_4$ = 0.16 mm, $L_4$ = 2.00 mm, $L_5$ = 0.60 mm, $W_5$ = 0.04 mm, and $g_2$ = 0.04 mm. For the convenience of spot welding, the length $L_6$ and width $W_6$ of the microstrip block used for spot welding ground at the outer end of the IDC are 0.36 mm and 0.33 mm, respectively. The circuit size of the tunable band-stop wideband filter (excluding the feeder) is 15.1 mm × 4.8 mm, approximately 0.91 $\lambda_g$ × 0.29 $\lambda_g$, where $\lambda_g$ is the waveguide wavelength of the 50 Ω impedance microstrip line corresponding to the center frequency of the filter.

The simulation performance of the filter under different ground states is shown in Figure 4, and the specific parameters are shown in Table 1. The passband of UWB ranges from 3.2 GHz to 10.4 GHz, and the 3-dB relative bandwidth is 105.9%. The center frequency of the stopband is adjustable from 6.00 GHz to 6.90 GHz. In all ground states, the 3-dB bandwidth of the stopband is less than 4% and its rejection is greater than 20 dB, showing high selectivity and high rejection performance. In addition, by changing the structure of the loaded IDC array, more tunable states and a wider tuning range can be achieved. Over most of the passband, the reflection coefficient is higher than 15 dB.

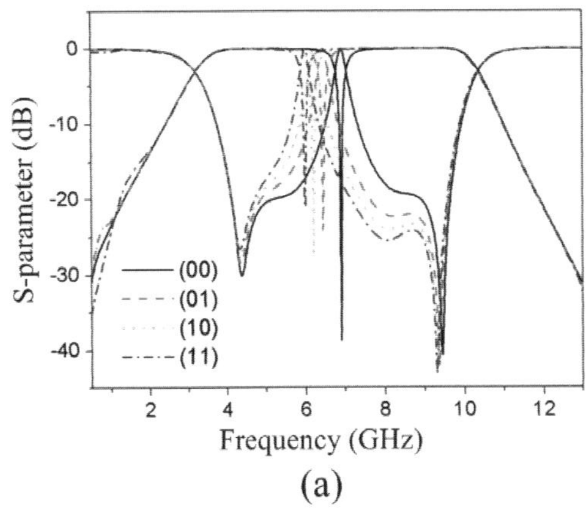


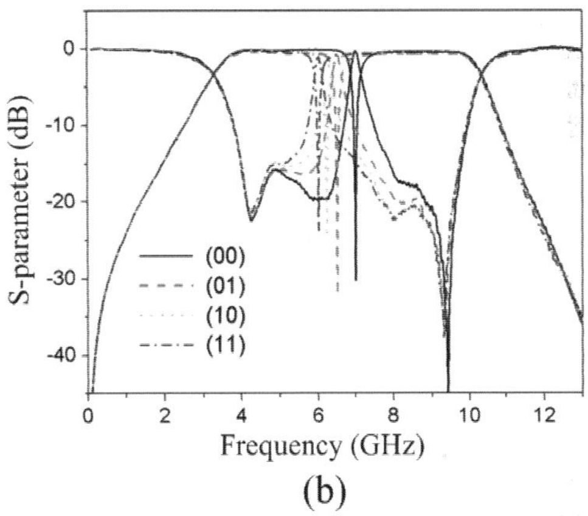


(a) Simulated response (b) Measured response

**Figure 4**. Frequency responses of the proposed UWB BPF in four states.

**Table 1**. Simulated and measured responses of the proposed filter in four states.

| State | Notch frequency (GHz) | | 3-dB BW (%) | | $S_{21}$ of notch band (dB) | |
|---|---|---|---|---|---|---|
| | Simulated | Measured | Simulated | Measured | Simulated | Measured |
| (00) | 6.90 | 7.01 | 3.9 | 3.9 | 38.69 | 29.42 |
| (01) | 6.44 | 6.51 | 3.7 | 3.4 | 24.13 | 31.43 |
| (10) | 6.20 | 6.24 | 3.7 | 3.4 | 27.14 | 23.79 |
| (11) | 6.00 | 6.02 | 3.8 | 3.3 | 21.04 | 22.94 |

## 4. Filter fabrication and test results

The filter was fabricated on a 20.0 mm × 6.9 mm MgO substrate. The substrate has a dielectric constant of 9.8 and a thickness of 0.5 mm, which is coated on both sides with the superconducting film of YBCO with a thickness of 600 nm and has a surface resistance in the μΩ range. One side of the superconducting film is processed into a filter circuit by etching and other processes, and the other side is used for the overall ground layup. Finally, the ultra-wideband superconducting filter is placed in a metal shielding box for encapsulation.

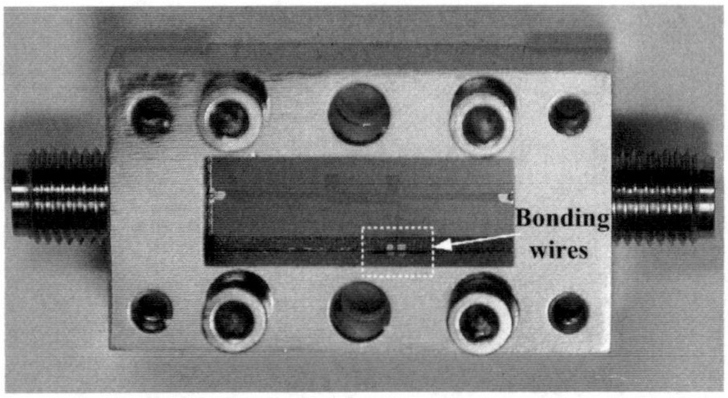


**Figure 5**. Photograph of the fabricated UWB BPF in State (11).

Figure 5 depicts the fabricated UWB BPF in State (11). The filter state was tuned by disusing or using the bonding wires. Using a Network Analyzer (Agilent N5230C Vector), performances of the proposed filter in various states were measured at 55 K (the input power is set to –25 dBm), as shown in Figure 4 (b) and Table 1. The experimental performances with no tuning are highly consistent with the simulation results. The measured 3-dB passband had a range of 3.3-10.4 GHz and its fractional bandwidth was 102.2%. The notch band shifted from 6.02 to 7.01 GHz within the UWB passband. It exhibits a high selectivity (<4% for 3-dB bandwidth) and a high rejection rate (greater than 22 dB) in all four states. Its return loss exceeded 15 dB in most cases. The differences between measured results and simulated results can be attributed to fabrication errors.

## 5. Conclusion

This paper presents and demonstrates the design, fabrication, and test process and results of an ultra-wideband high-temperature superconducting filter with tunable stopband characteristics. The tunable stopband is realized by coupling a $\lambda/2$ L-type resonator loaded with a reconfigurable 2-bit IDC array at one end of the UWB filter. Test results show excellent filter performance with high selectivity (3 dB bandwidth less than 4%) and high rejection ratio (more than 22 dB) for the tunable stopband.

**Acknowledgments**
This work was supported by the National Natural Science Foundation of China under Grant 61901270.

## References

[1] Federal Communications Commission (2002). Revision of Part 15 of the commission's rules regarding ultra-wideband transmission systems. Tech. Rep. ET-Docket 98-153, FCC02-48.

[2] Zhu L., Sun S., and Menzel W. (2005). Ultra-wideband (UWB) bandpass filters using multiple-mode resonators. J. IEEE Microwave and Wireless Components Letters, 15 (11): 796-798.

[3] Tang C. W. and Chen M. G. (2007). A Microstrip Ultra-Wideband Bandpass Filter with Cascaded Broadband Bandpass and Bandstop Filters. J. IEEE Transactions on Microwave Theory and Techniques, 55 (11): 2, 412-2, 418.

[4] Garcia J., Bonache J., and Martin F. (2006). Application of Electromagnetic Bandgaps to the Design of Ultra-Wideband Filters with Good Out-of-Band Performance. IEEE Transactions on Microwave Theory & Techniques, 54: 4, 136-4, 140.

[5] Ghazali A. N. and Sazid M. (2017). A Compact Broadside Coupled Notched Band UWB Filter with a Wide Upper Stopband. J. IETE Journal of Research, 63 (2): 145-150.

[6] Bakali H., Elftouh H., Farkhsi A., et al. (2020). Design of a Super Compact UWB Filter Based on Hybrid Technique with a Notch Band Using Open Circuited Stubs. J. Advanced Electromagnetics, 9 (3): 39-46.

[7] Lu X., Wei B., Xu Z., et al. (2015). Superconducting Ultra-Wideband (UWB) Bandpass Filter Design Based on Quintuple/Quadruple/ Triple-Mode Resonator. J. IEEE Transactions on Microwave Theory and Techniques, 63 (4): 1, 281-1, 293.

[8] Hong J. S. and Lancaster M. J. (2001). Microstrip filters for RF/microwave applications. New York, NY: John Wiley & Sons.
[9] Thomson N. and Hong J. S. (2007). Compact Ultra-Wideband Microstrip/Coplanar Waveguide Bandpass Filter. J. IEEE Microwave and Wireless Components Letters, 17 (3): 184-186.

# Dynamic compressive property of extruded VW75-Ti (wt.%) alloy under high strain rates

**Tong Zhang [1,2,3], Minglong Ma [1,2,4], Yongjun Li [1,2], Jiawei Yuan [1,2], Xinggang Li [1,2], Guoliang Shi [1,2], Zhaoqian Sun [1,2], and Kui Zhang [1,2]**

[1]State Key Laboratory of Nonferrous Metals and Processes, GRINM Group Corporation Limited (GRINM), Beijing 100088, China
[2]GRIMAT Engineering Institute Co., Ltd., Beijing 101407, China
[3]General Research Institute for Nonferrous Metals, Beijing 100088, China

[4]Corresponding author's e-mail: maminglong @grinm.com

**Abstract**: The dynamic compressive properties of extruded VW75 and VW75+Ti alloys were investigated by using a split Hopkinson pressure bar (SHPB) at 25°C under high strain rates of 1, 600-2, 700 $s^{-1}$ in the present work, and the microstructure evolution after deformation was inspected by OM, SEM, and EBSD. The results demonstrated that the maximum stress value rises with the escalating strain rate in the course of impact, and the true stress-true strain curves of the two alloys exhibit strain strengthening and positive strain effect. The addition of Ti particles refines alloy grains, increases fracture strength, and decreases the twinning ratio.

## 1. Introduction

Magnesium alloys offer the benefits of low density, high specific strength, and high specific stiffness, and are employed in the automotive industry, aerospace, rail transit, and other fields [1]. With the expansion of magnesium alloy application fields, the service environment of magnesium alloys has become increasingly complex, and magnesium alloy components are very vulnerable to high-speed impact during the use of automobile bodies, wheel hubs, armor plate missiles, etc. Studying the dynamic deformation behavior of magnesium alloys is essential [2].

Magnesium matrix composites, leveraging the benefits of both the magnesium matrix and reinforcement, have garnered significant attention. The reinforcement phases are mainly particles, fibers, and whiskers. Ceramic particles are in a brittle phase and have a weak deformation ability in coordination with the magnesium matrix. Fiber reinforcements have poor wettability with magnesium matrix, so it is usually necessary to coat the fiber surface. Affected by the surface van der Waals force, whisker reinforcement is not easily dispersed, and the whisker orientation is random, resulting in an increased cracking tendency of magnesium matrix composites [3]. Due to the inherent deficiency of ceramic particles, fibers, and whiskers as reinforcement, finding more suitable reinforcement is the focal point of current research. Some scholars have found that adding metal particle reinforcement with good plasticity and good wettability with magnesium matrix can enhance the comprehensive properties of the alloy. Typical metal particles include Ti particles, high entropy metal particles, Cu particles, etc. Ti has low density, a close-packed hexagonal structure with a magnesium matrix, and excellent wettability with magnesium melt [4]. At the same time, there is no brittle intermetallic compound formation between Ti and Mg, so it is very suitable to use as a strengthening phase of metal composites. Kurt and Kurnaz [5] found that the addition of titanium particles can significantly refine alloy grains and enhance the strength of magnesium alloys, but there are few investigations on the high-speed impact resistance of magnesium alloys reinforced by Ti particles. Therefore, this paper carried out a study on the high-speed impact resistance behavior of VW75 magnesium alloys reinforced by Ti particles, hoping to further expand the application range of magnesium matrix composites.

## 2. Experiment

VW75 and VW75+Ti (wt.%) alloys were chosen as the research subjects for this study. The VW75 magnesium alloy is melted and mixed with Ti particles at higher melting temperatures (between 993 K and 1,023 K). Mechanical stirring speeds up the dispersion of Ti particles and quickly cools and obtains the ingot of Φ120×350 mm. After homogenizing the as-cast VW75 and VW75+Ti alloys for 24 h at 793 K, isothermal extrusion was performed to obtain a size of 25 mm bars. During the process, the extrusion ratio and extrusion speed are 23 and 0.6 mm/s, respectively.

Φ8×6 mm cylindrical sample was chosen as the impact test sample along the direction of extrusion. Using a SHPB (Figure 1), high-speed impact tests were carried out on the extruded VW75 and VW75+Ti alloys at 25°C. Extruded VW75 and VW75+Ti alloys were taken through a series of dynamic deformation tests at 25°C with SHPB at high strain rates ranging from 1,600 $s^{-1}$ to 2,700 $s^{-1}$, with the direction of extrusion aligning to the direction of load-bearing. The microstructure and fracture morphology were evaluated by using SEM, OM, and EBSD. The EBSD sample electrolytic polishing electrolyte was 20% nitric acid-alcohol solution and electrolyzed for 30 s under the conditions of a DC power supply voltage of 20 V and -30°C.

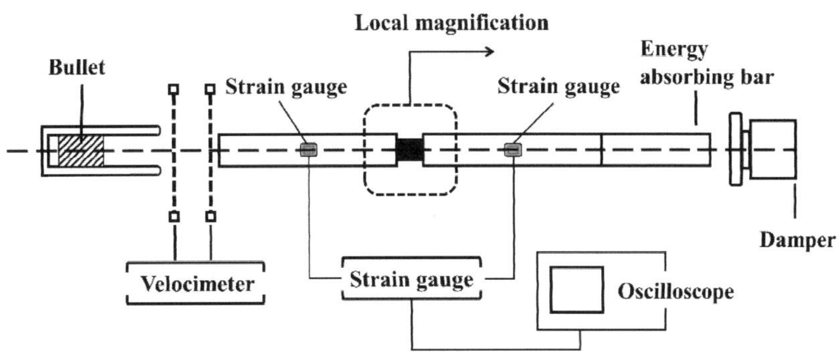


**Figure 1.** Schematic of split hopkinson pressure bar.

## 3. Results and discussion

The true stress-true strain curves and macroscopic morphology of VW75 and VW75+Ti alloy at high-speed impact are displayed in Figure 2. From Figure 2 (a), with increasing strain rate, the samples are continuously compressed and deformed, and the strain variable is continuously increased until the samples finally break into three pieces along the direction of 45°. The true stress-true strain curves of the two alloys' impacts at various strain rates are given in Figures 2 (b) and (c). The samples' maximum stress values gradually rise during the impact process as the strain rate rises, demonstrating both the positive strain effect and the strain strengthening effect. VW75 alloy specimens fracture at a strain rate of 2,700 $s^{-1}$, in which the fracture strain and fracture strength are 22% and 435 MPa, respectively. VW75+Ti alloy specimens fracture at a strain rate of 2,400 $s^{-1}$, in which the fracture strain and fracture strength are 17% and 449 MPa, respectively. According to Figure 2 (d)-Figure 2 (f), at the same strain rate, the true stress-true strain curves of the VW75+Ti alloys are higher than those of the VW75 alloys. VW75+Ti alloys break sooner than VW75 alloys, with higher breaking stress and smaller breaking strains. The addition of Ti particles increases fracture strength and decreases fracture plasticity of VW75 alloy.

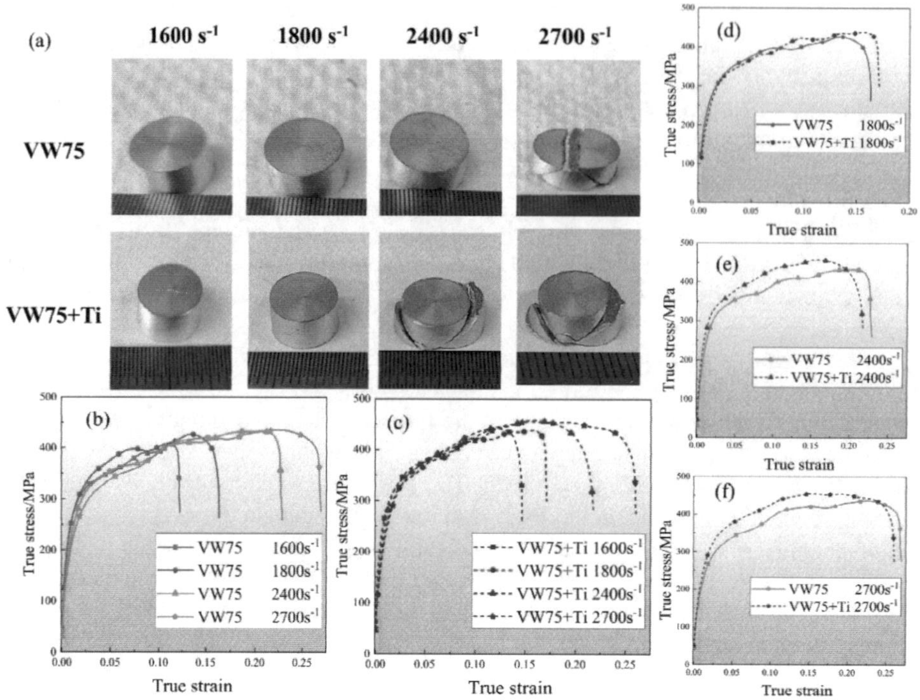


(a) Post-impact macroscopic morphology; (b) VW75; (c) VW75+Ti;
(d) 1,800 $s^{-1}$; (e) 2,400 $s^{-1}$; (f) 2,700 $s^{-1}$.

**Figure 2.** Macroscopic morphology and true stress-true strain curves of VW75 and VW75+Ti.

Figure 3 shows the longitudinal cross-section microstructure of VW75 and VW75+Ti alloys before and after dynamic deformation. The alloy microstructure after extrusion is equiaxial with no twins. The average grain sizes of VW75 alloy and VW75+Ti alloy are 46.6 μm and 24.6 μm respectively, as shown in Figures 3 (a) and (d). The incorporation of Ti particles significantly refined the alloy grains in Figures 3 (g) and (h). When the specimen was deformed according to the microstructure of high-speed impact, the number of twins increased with increasing strain rate in Figures 3 (b), (c), (e), and (f) and that of the VW75 + 2Ti alloys was less than that of the VW75 alloys under the same strain rate.

Slip and twinning play a decisive role in coordinated deformation at room temperature and high strain rate deformation [6]. When the sample is impacted by a high strain rate, twinning plays a major role, and twins are generated in the crystal. With the generation of twins, the mother crystal is divided into several pieces by twins, which increases the number of interfaces in the sample and becomes more serious. The interaction between twins and dislocation becomes more intense, and with the increase of the deformation rate, the percentage of twins increases.

The fracture morphology of the two alloys following impact is depicted in Figure 4, and the fracture characteristics of the alloy are mixed with toughness and brittleness. According to Figures 4 (b) and (e), the fracture surfaces of the two alloys are smooth and show a large number of step morphologies. There are some bright tearing ridges and tongue patterns on the step plane. There are small dimples in part of the fracture in Figures 4 (c) and (f), and in Figures 4 (a) and (d), there are fish scale structures. The fish scale structure appears on the edge of the smooth platform region, because under the condition of a high strain rate, the deformation of the material generates an enormous quantity of heat, causing the adiabatic temperature inside the material to rise and causing the temperature in the local area to exceed the melting point instantaneously, resulting in a melting phenomenon, and forming a heat softening zone. At the same high strain rate, the VW75+Ti alloy has more fish scale structure than the VW75 alloy. Cracks start at the edge of the thermal softening zone, expand rapidly, and finally lead to sample fracture. Therefore, the VW75+Ti alloy breaks before the VW75 alloy.

IOP Publishing
2720 (2024) 012059 doi:10.1088/1742-6596/2720/1/012059

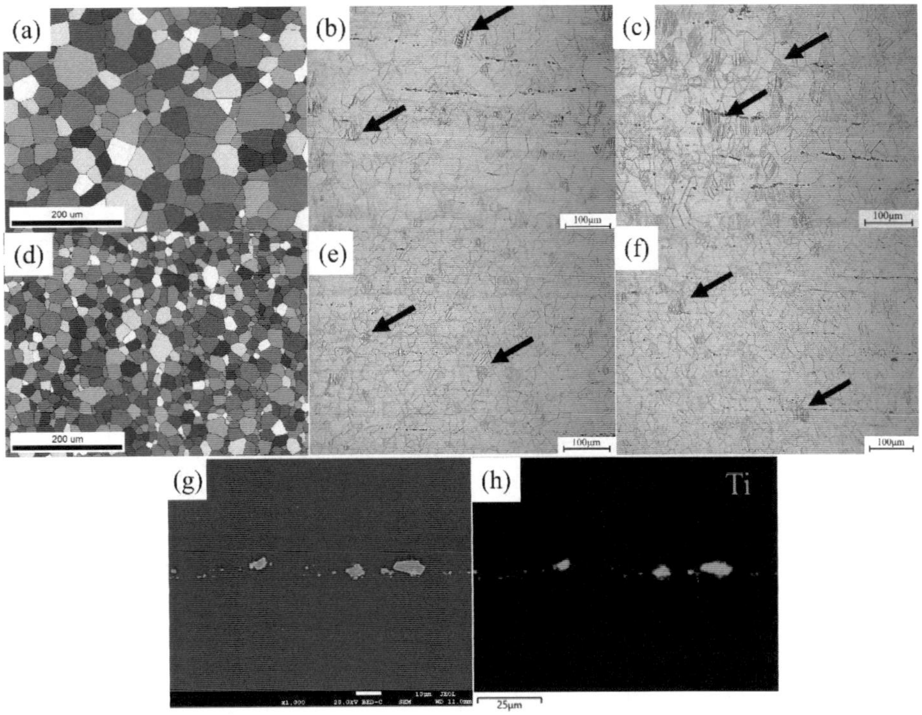


VW75 alloy: (a) Original microstructure; (b) 1, 600 $s^{-1}$; (c) 1, 800 $s^{-1}$;
VW75+Ti alloy: (d) Original structure; (e) 1, 600 $s^{-1}$; (f) 1, 800 $s^{-1}$; (g-h) Ti particles.

**Figure 3.** Microstructure of extruded VW75 and VW75+Ti alloy.

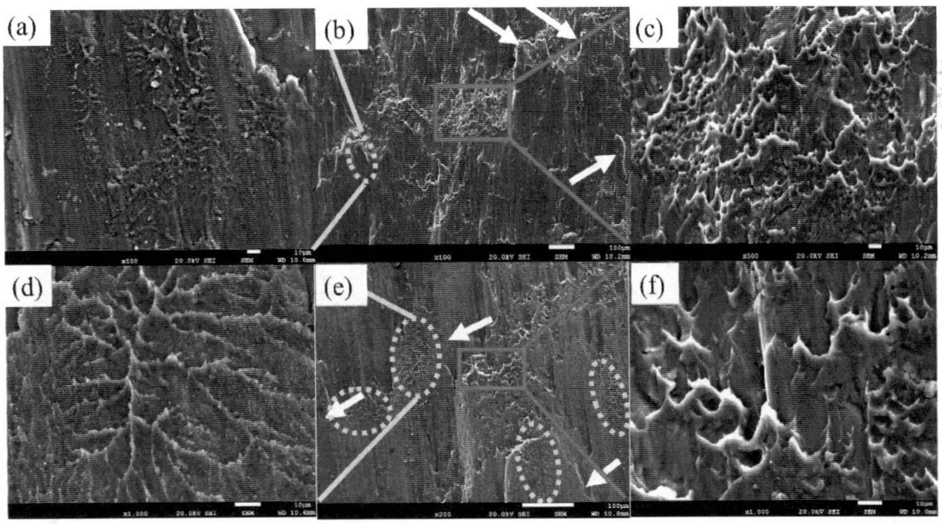


VW75 2,700 $s^{-1}$: (a) Fish scale structure; (b) Overall fracture morphology; (c) Dimples;
VW75+Ti 2,700 $s^{-1}$: (d) Fish scale structure; (e) Overall fracture morphology; (f) Dimples.

**Figure 4.** Fracture morphology after impact of extruded VW75 and VW75+Ti alloy.

The mechanism by which Ti particles affect the high-speed impact resistance of VW75 alloys is depicted in Figure 5. The VW75 alloys showed a 47% refinement in grain size during high-temperature extrusion with the Ti particle addition, and the grain boundaries increased. When VW75+Ti alloy is impacted by a high strain rate, the grain boundary obstructs the dislocation movement and increases the work-hardening ability, and the fracture strength of the alloy increases, but the fracture strain decreases. At the same time, during the impact process, with the grain boundary area growing, the coordination deformation ability between grains is improved, which can partially absorb energy. When subjected to high strain rate impacts, the grains are prone to rotation and the proportion of twinning decreases. Therefore, there are few twins in the VW75+Ti alloy, and the addition of Ti particles can reduce the twinning to some extent.

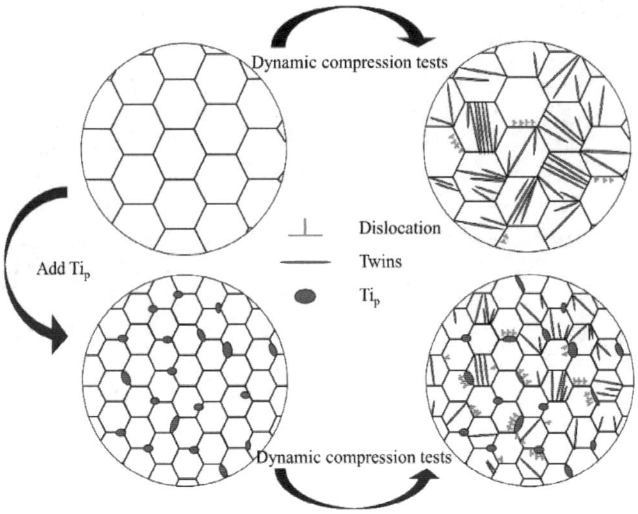


**Figure 5.** Effect of Ti particles on VW75 alloy.

## 4. Conclusions

In this paper, the effect of Ti particle addition on the high-speed impact resistance of VW75 alloy was studied by using SHPB.

(1). The true stress-true strain curves of VW75 and VW75+Ti alloys have no obvious yield point and show strain hardening and normal strain effects. At the same strain rate, true stress-true strain curves of the VW75+Ti alloy are higher than those of the VW75 alloy.

(2). Ti granules on the VW75 alloy had a grain refinement effect, and the grain size average dropped from 46.6 μm to 24.6 μm. After high-speed impact, twins are formed inside the grain, the quantity of twins in VW75+Ti alloy is small, and Ti particles reduce the formation of twins.

(3). The two kinds of alloy fracture present ductile and brittle mixed fracture characteristics. The addition of Ti particles to refined alloy grains increases the dynamic fracture strength of the alloy.

## Acknowledgments

This research was supported by the National Key Research and Development Program of China (Grant No. 2022YFB3709300).

## References

[1] You S., Huang Y., Kainer K. U., and Hort N. (2017). Recent research and developments on wrought magnesium alloys. Journal of Magnesium and Alloys, 5 (3): 239-253.

[2] Li Q. (2012). Mechanical properties and microscopic deformation mechanism of polycrystalline magnesium under high-strain-rate compressive loadings. Materials Science and Engineering, 540: 130-134.

[3] Sasaki G., Wang W. G., Hasegawa Y., Choi Y. B., Fuyama N., and Yanagisawa O. (2007). Surface treatment of $Al_{18}B_4O_{33}$ whisker and development of $Al_{18}B_4O_{33}$/ZK60 magnesium alloy matrix composite. Journal of Materials Processing Technology, 187: 429-432.

[4] Kondoh K., Kawakami M., Imai H., Umeda J., and Fujii H. (2010). Wettability of pure Ti by molten pure Mg droplets. Acta Materialia, 58: 606-614.
[5] Kurt P. and Kurnaz S. C. (2022). Investigation of the Effect of La and Ti Addition to Mg-4Sn-2Al Alloy on Microstructure, Mechanics, and Corrosion Properties. International Journal of Metalcasting, 17: 1, 388-1, 401.
[6] Wu W. X., Jin L., Dong J., and Ding W. J. (2014). Deformation behavior and texture evolution in an extruded Mg-1Gd alloy during uniaxial compression. Materials Science and Engineering, 593: 48-54.

# Real-time temperature monitoring and alarm system for biological samples storage based on FPGA

**Chennong Li [1], Yanxia Chen [1, 2,3], Xingping Zhong [1], Youchun Xu [2], and Lei Qin [1]**

[1]Beijing Sensor Key Laboratory, Beijing Information Science & Technology University, Beijing 100101, China

[2]School of Medicine, Tsinghua University, Beijing 100084, China

[3]Corresponding author's e-mail: chenyanxia@bistu.edu.cn

**Abstract**: In the field of biosensing, high-quality biological samples are a prerequisite for precise analysis and detection. Strictly controlled storage temperature is crucial to ensure the stability of biological samples and achieve effective detection. Different biological samples require different temperature ranges for storage. Therefore, it is essential to design a temperature monitoring and alarm system capable of alarming based on different needs. This work presents a real-time temperature monitoring and alarm system based on the DS18B20 temperature sensor and FPGA chip. The system with simple structures has a wide temperature measurement range, an adjustable temperature alarm range, and a timely alarm function for various scenarios monitoring. This work demonstrates the temperature monitoring and alarm system for simple, accurate, and low-cost modes of temperature monitoring. The real-time temperature monitoring and alarm system will provide an effective means to strictly control the storage temperature of different biological samples, ensuring the stability and effectiveness of biological samples and the accuracy and sensitivity of biosensing and analysis.

## 1. Introduction

The accurate measurement and control of temperature by using a suitable sensor system is not only important in the industry [1-2], agriculture [3-4], and environmental temperature monitoring [5-6] but also is crucial for the field of biosensing. Accurate and sensitive detections of biological samples have always been the goal pursued in the biosensing technology [7-9]. The high quality of biological samples is a prerequisite for achieving accurate analysis and detection. Maintaining strict and accurate storage temperatures is a critical element in ensuring the stability of biological samples. The storage temperature can have a significant impact on the quality of the samples. So, it is essential to monitor the temperature during storage [10].

Cryoprecipitates (additional Fresh Frozen plasma, FFP) and Molecular DNA/RNA should be stored in the range of -20°C ~ -25°C. Some body fluids, such as saliva, urine, and blood samples, are stored in the range of 2 ~ 8°C [11]. Storage of sweat samples at 25°C for more than 7 days affects sweat lactate and ammonia concentrations but can be extended to 28 days at -20°C [12]. The expression level of the Late Cornified Envelope 1D (LCE1D) gene in human skin swabs stored at 25°C is significantly higher than those stored at 40°C [13]. To prevent storage temperature from affecting the quality of biological samples, it is crucial to design a temperature monitoring and alarm system.

Recently, a temperature detection and tracking system was designed for storing COVID-19 vaccines. The analog signal output of this system needs modulation and conversion for temperature data reading, increasing the complexity of the system hardware structure [14]. A real-time monitoring system with ultra-low temperature freezers was exploited to store biomarkers at -80°C. When the temperature exceeds the set value, the abnormal information is alarmed via SMS or email. This system uses a thermocouple circuit for temperature measurement and an additional pre-calibrated resistance temperature detector for calibrating the circuit's deviation [15]. These may lead to complications of the system hardware structure and increased costs.

We develop a real-time temperature monitoring and alarm system by utilizing the DS18B20 as the temperature sensor and an FPGA chip as the central controller. This system has a wide temperature measurement range from -55°C to 125°C and has three selectable modes for different application scenarios. This system, based on three modes, is simple, low-cost, and suitable for temperature monitoring and alarming in various scenarios.

## 2. System design

### *2.1 System hardware design*

The overall hardware design of the system uses a modular approach. The system is divided into three modules, including a temperature measurement module, a control module, and a display and alarm module, as shown in Figure 1 (a).

The temperature measurement module contains the DS18B20 temperature sensor (oval wireframe in Figure 1 (a)). DS18B20 can measure temperature in 750 ms at the most and accurately in the range of - 55°C ~ 125°C. Meanwhile, it has excellent resistance to interference and stability.

The control module contains the FPGA main control chip (Cyclone IV E: EP4CE10F17C8), MP2359 chip, XTAL, standard JTAG debug ports, M5P16 flash storage chip, and four physical buttons (rectangular wireframe in Figure 1 (a)). The FPGA EP4CE10F17C8 chip is a controller and processes data transmission between modules. The MP2359 chip is a power supply chip and provides a stable 5 V operating voltage for the system. XTAL supplies a 50 MHz clock pulse to the system. The JTAG ports connect directly to the FPGA downloader for downloading programs or online debugging. The M5P16 Flash chip with a capacity of 16 Mbit can store the written FPGA program. Four physical buttons contain two switching buttons and two adjusting buttons.

The display and alarm module contains a digital tube, buzzer, and 7 LED lights (hexagonal wireframe in Figure 1 (a)). For the 7 LED lights (each row in Figure 1 (b)), 3 LED lights are alarm indicator lights (Warn 1_led, Warn 2_led, Warn 3_led in Figure 1 (b)), 3 LED lights are working mode indicator lights (Mode 1_led, Mode 2_led, Mode 3_led in Figure 1 (b)), and 1 LED light is adjusting indicator light (Adjust_led in Figure 1 (b)). The key to the alarm function is the adjustment of the temperature limit range according to the switching of three working modes (Modes 1, 2, and 3). The difference between these modes is the maximum and minimum of the temperature limits. In Mode 1, both the maximum and minimum limits are above zero degrees Celsius. In Mode 2, the maximum is above zero degrees Celsius, and the minimum is below zero degrees Celsius. In Mode 3, both maximum and minimum are below zero degrees Celsius. The maximum and minimum temperature limits are adjusted by using buttons. If the measured temperature exceeds the limits in different modes, the buzzer and the corresponding alarm indicator light will start working.

Taking Mode 1 as an example (Figure 1 (b)), the Mode 1_led indicator is lit (green), indicating that the system is working in Mode 1. When the temperature measured exceeds the limits in Mode 1, the Warn 1_led indicator will light up (red), and the buzzer will sound (the second row in Figure 1 (b)). When the Adjust_led is lit (blue), it indicates that the maximum and minimum values of the temperature limits of Mode 1 could be adjusted (the third row in Figure 1 (b)). If the temperature measured exceeds the limits when Adjust_led has lighted up (blue), the Warn 1_led indicator will light up (red), and the buzzer will sound as well (the fourth row in Figure 1 (b)). Other working modes are similar to Mode 1, so we can switch working modes by switching buttons.

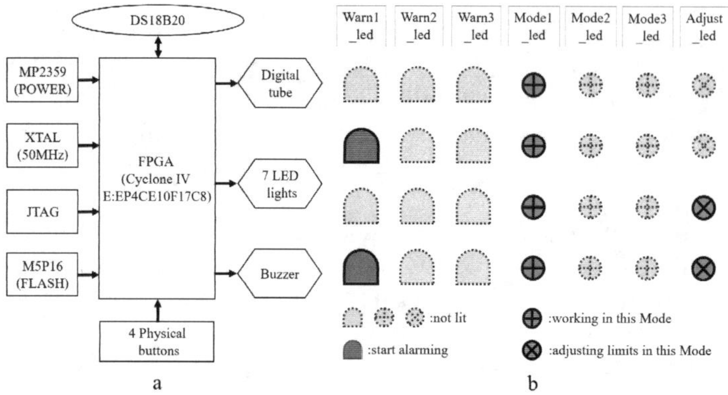


**Figure 1**. The hardware structure of the system and indicator light information in Mode 1.

*2.2 System software design*

The overall software design of the system uses the Top-down design method. Functions implemented by FPGA will be divided into six modules (Figure 2): the Jitter eliminating module, the DS18B20 controlling module, a data processing module, a display module, a buzzer driver module, and a counter module. A jitter-eliminating module will be used to eliminate the jitters of physical buttons and generate useful keystroke signals. DS18B20 controlling module will realize correct communication between the FPGA chip and DS18B20 for temperature acquisition. The data processing module will realize to switch working modes and adjust the maximum and minimum values of the temperature limits according to keystroke signals, and it will determine whether the temperature measured is abnormal or not. The display module will be used to drive a digital tube to display the temperature values measured, the maximum values, or the minimum values of the current working mode and will drive all indicator lights. The Buzzer driver module will realize to drive the buzzer. The counter module will be used to count the number of clock pulse signals and provide a working clock.

We use Verilog HDL as the FPGA programming language to design the system in the Quartus II integrated environment. Verilog HDL allows the modeling and describes digital systems at multiple levels. It not only simplifies the design task but also increases efficiency.

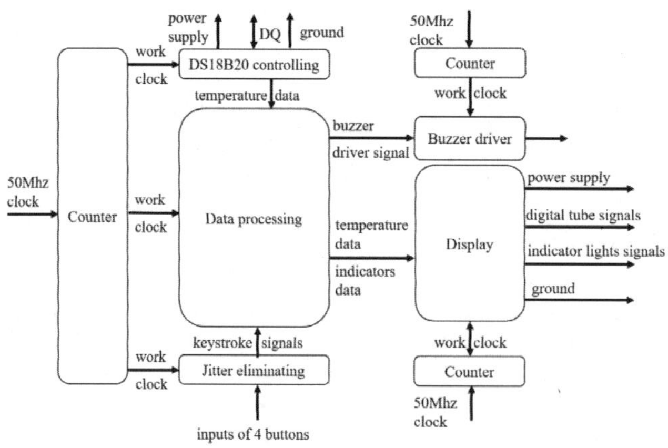


**Figure 2**. The software architecture of the system.

## 3. System synthesis and debugging

After completing the system design, we compiled and synthesized the system in the Quartus II environment. Figure 3 is the RTL-level view of the top module of the system. The system is divided into four functional modules after synthesis (m1, m2, m3, m4 in Figure 3). Module m1 realizes temperature acquisition. Module m2 not only realizes the elimination of jitters of physical buttons and receives temperature data but also generates alarm data and display data. Module m3 realizes the buzzer setting according to the alarm data received. Module m4 realizes digital data display according to the display data received. Inputting and outputting pins are used for pin constraints (dotted box in Figure 3). 50 MHz work clock and button signals are input to the system by inputting pins. Data in the process of communication between DS18B20 and FPGA chip, signals of ground, signals of power supply, signals of all indicator lights, and display data are output by outputting pins.

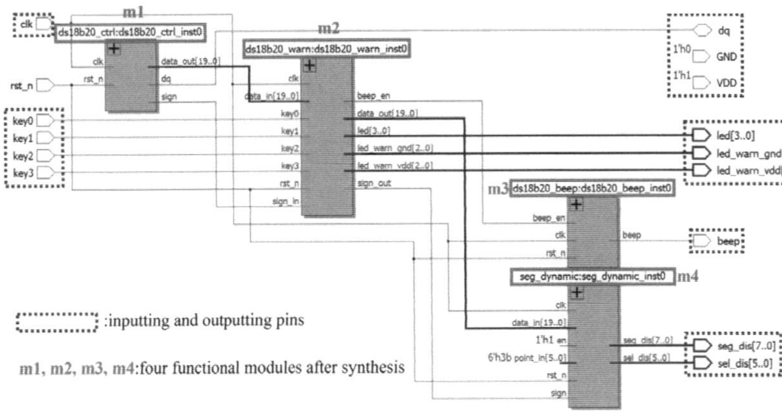


**Figure 3**. RTL view of the top module.

Under cooperating with the hardware system, the program debugging diagram of the top module is shown in Figure 4. When the measured temperature of the system working in Mode 1 (led was 1, 000 b) in Figure 4 (a) was at 25.312°C (within the normal limit of Mode 1: 10.0 ~ 81.0°C), the system did not alarm (beep was 0, led_warn vdd was 000 b) in Figure 4 (a). We adjusted the maximum of limits from 81°C (led was 1, 000 b) in Figure 4 (b) to 80°C (led became 1, 001 b) in Figure 4 (c) by pressing the switching button (Key 3 went from 1 to 0) in Figure 4 (b), and the system did not alarm as well in Figure 4 (c). Then we switched to Mode 2 (Key 2 went from 1 to 0 and led to 0100 b) in Figure 4 (d). When the measured temperature of the system working in Mode 2 (-5.0 ~ 8.0°C) was at 25.5°C (abnormal for Mode 2), the system started to alarm (beep became 1 and led_warn_vdd became 010 b) in Figure 4 (d). The predetermined target functions have been realized based on the real-time temperature monitoring and alarm system.

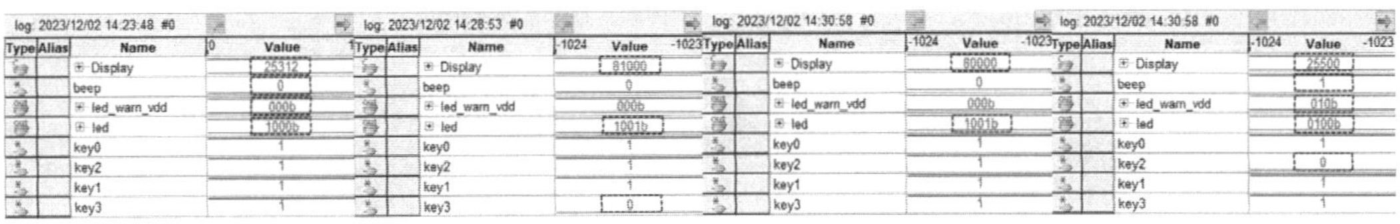


Display : current temperature data or maximum or minimum of limits
beep : buzzer starts to work at 1.
key0, key2 : two switching buttons and effective at 0.
key1, key3 : two adjusting buttons and effective at 0.
led_warn_vdd[2:0] : three alarm indicators
led[3:1] : three working mode indicators
led[0] : one adjustment indicator

**Figure 4**. Program debugging diagram.

## 4. Performances of the real-time temperature monitoring and alarm system

The performances of the real-time temperature monitoring and alarm system were investigated. The system working in Mode 1 was verified in an adjustable temperature ranging from 29.5°C to 30.5°C. The results show that the maximum error between the temperature measurement values and the preset temperature values did not exceed 0.55°C (Figure 5 (a)), and the maximum fluctuation range of multiple sets of temperature data did not exceed 0.3°C (Figure 5 (b)) at the same preset temperature. It demonstrates the accuracy and stability of temperature measurement by real-time temperature monitoring and alarm systems, respectively.

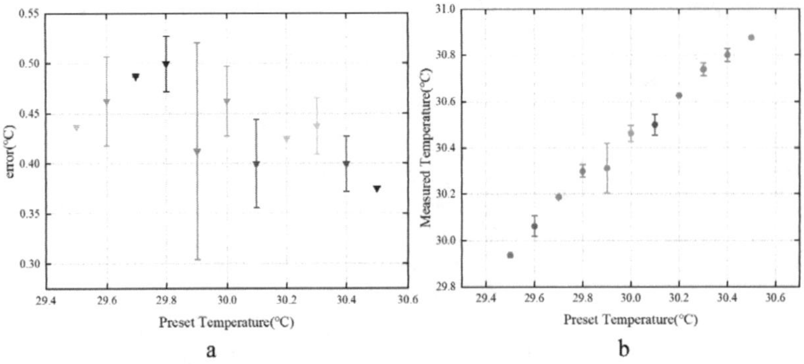


**Figure 5**. The accuracy and stability of system temperature measurement.

The real-time temperature monitoring and alarm system offers a simple and effective approach to measure and control storage temperature, which is a prerequisite for ensuring high-quality biological samples. As proof of concept, saliva and RNA were used as biological samples to investigate the adaptability of the real-time temperature monitoring and alarm system in Mode 1 and Mode 3 (Figure 6). As shown in Figure 6 (a), when the storage temperature of the saliva sample was at 4.187°C (within 2 ~ 8°C, we inset 1 in Figure 6 (a)), the system did not alarm. When the storage temperature of the saliva sample was at 10.125°C (outside 2 ~ 8°C, we inset 2 in Figure 6 (a)) or 1.375°C (outside 2 ~ 8°C, we inset 3 in Figure 6 (a)), the Warn 1_led was lit (we inset 2 and 3 in Figure 6 (a)), the buzzer sounded, and the system in Mode 1 started to alarm. Then, the system in Mode 1 stopped alarming when the minimum was adjusted to 1°C (we inset 4 in Figure 6 (a)). It demonstrates that the temperature alarm function and the adjustment of temperature limits are correct.

Additionally, we used the system to monitor the storage temperature of RNA after switching to Mode 3. As shown in Figure 6 (b), when the storage temperature of the RNA sample was at -22.624°C (within -25 ~ -20°C, we inset 5 in Figure 6 (b)), the system did not alarm. When the storage temperature of the saliva sample was at -18.25°C (outside -25 ~ -20°C, we inset 6 in Figure 6 (b)) or -27.75°C (outside -25 ~ -20°C, we inset 7 in Figure 6 (b)), the Warn 3_led was lit (we inset 6 and 7 in Figure 6 (b)), the buzzer sounded, and the system in Mode 3 started to alarm. All functions worked correctly, showing excellent adaptability to the real-time temperature monitoring and alarm system.

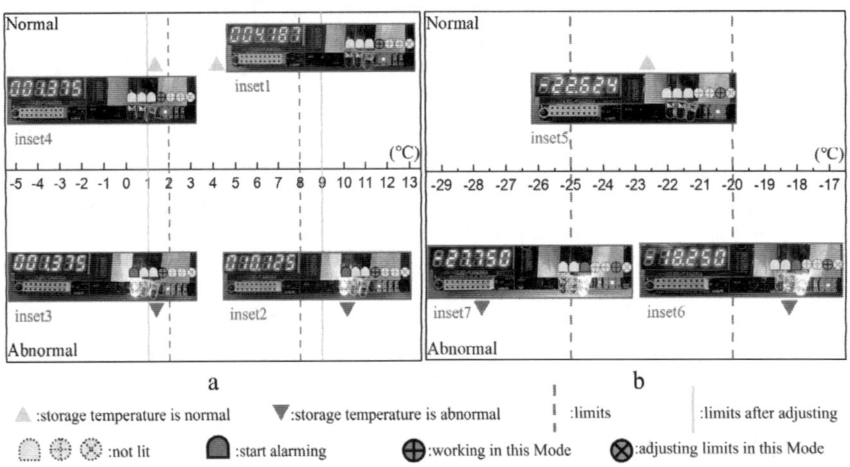


**Figure 6**. Saliva and RNA storage in mode 1 and mode 3.

## 5. Conclusion

In conclusion, we have developed a real-time temperature monitoring and alarm system for biological sample storage based on DS18B20 and FPGA chips. The system is simple, low-cost, and multi-mode. Synthesis and debugging verify the correctness of the system design. Meanwhile, we confirm the accuracy and stability of the temperature measurement of the system. The system realizes the storage temperature monitoring of the saliva and RNA samples and alarming. The real-time temperature monitoring and alarm system provides an effective approach to monitoring and controlling the storage temperature of biological samples. This system design has the potential to ensure the high quality of biological samples and the accuracy and sensitivity of biosensing and analysis. This design will have a significant application promise in the field of temperature monitoring and related research areas.

**Acknowledgments**

The work is supported by the National Natural Science Foundation of China (22004009), Beijing Natural Science Foundation (2214057), Classification Development of Universities-Key Research and Cultivation Projects (2021YJPY240), General Projects of Beijing Science and Technology Plan (KM202111232005), the National Natural Science Foundation of China (U2006218), and Classification Development of Universities-Scientific Research Platform Condition Construction Projects (2021KYPT117).

## References

[1] Usamentiaga R., Fernandez M. A., Villan A. F., and Carus J. L. (2018). *IEEE Trans. Industr. Inform.* 14 5, 667-77.
[2] Liu X., Du J., and Ye Z. S. (2021). *IEEE Trans. Industr. Inform.* 18 986-95.
[3] Hu X., Ren H., Tansey K., Zheng Y. T., Ghent D., Liu X. F., and Yan L. (2019). *Agric. For. Meteorol.* 279 107, 707.
[4] Zhang Z., Xu W., Shi Z., and Qin Q. M. (2021). *IEEE J. Sel. Topics Appl. Earth Observ. Remote Sens.* 14 2, 113-26.
[5] Hu T., Renzullo L. J., van Dijk A. I. J. M., He J., Tian S. Y., Xu Z. H., Zhou J., Liu T. J., and Liu Q. H. (2020). *Remote Sens. Environ.* 236 111, 419.
[6] Aneja B., Singh S., Chandna U., and Maheshwari V. (2011). *Int. J. Electr. Electron. Eng.* 3 29-37.
[7] Chen Y. X., Li K., Zhang S. D., Qin L., Deng S. H., Ge L. Y., Xu L. P., Ma L. L., Wang S. T., and Zhang X. J. (2020). *ACS Nano.* 14 4, 654-61.
[8] Chen Y. X., Min X. H., Zhang X. Q., Zhang F. L., Lu S. M., Xu L. P., Lou X. D., Xia F., Zhang X. J., and Wang S. T. (2018). *Biosens. Bioelectron.* 111 124-30.
[9] Chen Y. X., Xu L. P., Meng J. X., Deng S. H., Ma L. L., Zhang S. D., Zhang X. J., and Wang S. T. (2018). *Biosens. Bioelectron.* 102 418-24.
[10] Yin Z., Dong X., Xue L., and Dong E. (2021). *Chinese Sci. Bull.* 66 1, 812-20.
[11] Seneviratne V., Bae S., Jeon S. B., Won Y., and Yoon G. (2021). *J. Med. Syst.* 45 1-11.
[12] Klous L., Folkerts M., Daanen H., and Gerrett N. (2021). *Temperature.* 8 254-61.
[13] Ibrahim S. F. (2018). *Egypt. J. Forensic Sci.* 8 1-5.
[14] Muriyan A. S. V., Alshehhi S. A., Alblooshi M. A., Saeed A. N., and Alshehhi M. E. (2021). *Int. J. Eng. App. Sci. Technol.* 6 5-12.
[15] Powell S., Molinolo A., Masmila E., and Kaushal S. (2019). *Biopreserv. Biobank.* 17 139-42.

# Imprinted CNTs field emission cathode for electron brushing

**Zhanliang Wang[1,2,4], Zhanpeng Lin[2], Lei Yan[1], Zhifeng Xu[1], Jianlong Liu[2], Zhiyuan He[3], Dongzhi Jianmucuo[2], Ruirui Jiang[2] and Yubin Gong[2]**

[1]Science and Technology on Low-Light-Level Night Vision Laboratory, Xi'an 710000, China
[2]School of Electronic Science and Engineering, University Electronic Science and Technology of China, Chengdu 610054, China
[3]Shenzhen Institute for Advanced Study, University Electronic Science and Technology of China, Shenzhen 518000, China

[4]Corresponding author's e-mail: 46969449@qq.com

**Abstract:** Microchannel plates (MCPs) are the key components for low-level light night vision systems. Electron brushing is a crucial processing for MCPs manufacturing. In this paper, an imprinted carbon nanotubes (CNTs) field emission cathode is fabricated and tested. And the electron brushing experiment based on this cathode is carried out. The result shows that the imprinted CNTs cathode has some advantages and can be used in electron brushing. This research lays the foundation for massive, low cost and highly efficient electron brushing applications in the future.

## 1. Introduction

Night vision technology originated from the discovery of photoelectric emission by Hertz in 1887 and the publication of the photoelectric effect by Einstein in 1905. It is a high-tech that converts light to electrons, electrons to image and enhances the image under weak night light at last. The instruments based on night vision technology can greatly improve people's ability in weak light. So, the instruments can be applied to various tasks such as wide-spectrum tracking, all-weather observation, aiming and guidance. Nowadays, night vision technology has wider and more important applications in many fields, such as astronomy, aerospace, shipping, biology, medicine, nuclear physics, satellite monitoring and high-speed photography [1-8].

Researchers discovered the photocathode on Bi-Ag-O-Cs in 1938, on Na-K-Sb and Na-K-Sb-Cs in 1955 and on K-Cs-Sb in 1963. Based on the above photoelectric cathode materials and the fiber optic panel which was invented by Kapany, the first-generation (1G) of night vision technology was developed. The extremely weak target image is transferred to a photoelectric cathode and excites electrons in the 1G night vision system. The excited electrons are accelerated and focused on a fluorescent screen, then the enhanced image appears. The 1G system is passive with high gain because it has three sections of enhancement. However, 1G systems are destined for some shortcomings, such as bulky volume, heavy weight, and poor imaging effect in bright light.

In 1965, the GaAs activated by Cs were used as a high-quantum efficiency photoelectric cathode, which was discovered by J. J. Scheer and J. Vanlaar. In 1968, A. A. Turnbull and G. B. Evans adopted GaAs activated by Cs-O to achieve a better negative electron affinity (NEA) material, which was the basis of a higher-efficiency photoelectric cathode. The combination of NEA materials and microchannel plates (MCP), which were successfully developed in the early 1970s, gave birth to the second-generation (2G) night vision system [1-4]. The MCPs with high secondary electron emission channels ensure tens of thousands of electron multiplication and keep a significant gain in 1/3 length of the 1G tubes.

In 1989, J. Dupuy achieved a significantly improved on photoelectric current sensitivity of 600uA/lm. With MCPs noise factor reduced, the third-generation (3G) night vision system was successfully developed, which extended the viewing distance by more than 50% [1-4,9-11].

As a key part of the night vision systems, MCPs are composed of millions of 2 um to 5 um hollow channels, as shown in Figure 1. The inner wall of the channels is a high secondary electron emission material. The surface impurity will block the electrons and reduce the secondary electron multiplication. Electron scrubbing uses electron bombardment to remove surface impurities and improves the stability of MCPs' channels. So, it is a key step in the production process of MCPs. The quality of electron scrubbing dominates the performance, stability and life of night vision instruments.

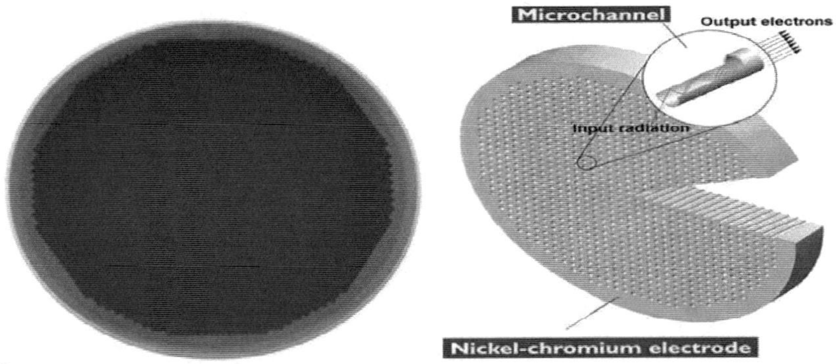


**Figure 1.** Real view and detailed schematic diagram of MCPs.

In this article, we first develop imprinted CNTs field emission cathode and then apply this cathode to electron brushing. The result verifies that the cathode can be used for MCP electron brushing.

## 2. The imprinted CNTs field emission cathode

Field emission is usually represented by the F-N formula [12-15]:

$$J = A\frac{(\beta E)^2}{\varphi}e^{-\frac{B\phi^{3/2}}{\beta E}} \tag{1}$$

Here, J is the current density, the unit is $A/cm^2$. E is the electric field strength on the emitter (V/cm). φ is the work function that determines the height of the potential barrier on the surface of the material (eV). A and B are constants:

$$A = 1.54\times10\text{-}6\ (A\cdot eV\cdot V\text{-}2) \tag{2}$$

$$B = 6.83\times107\ (V\cdot(eV)\text{-}3/2cm\text{-}1) \tag{3}$$

β is the field enhancement factor, which depends on the geometric shape of the cathode. Here, β is dominated by the aspect ratio of the emitter tip. Simple simplification of Formula (1) leads to:

$$\ln\left(\frac{J}{E^2}\right) = \frac{-B\phi^{3/2}}{\beta E} + \ln\left(\frac{A}{\phi}\beta^2\right) \tag{4}$$

Therefore, $\ln(J/E^2)$ is proportional to 1/E. The relationship curve is approximately linear known as the F-N curve, which can determine whether the emitter material belongs to field emission.

The imprinted CNTs field emission cathode is fabricated as follows [16-20]. Firstly, single-walled CNTs are mixed with absolute alcohol and sonicated for 2 hours. Then the CNTs ethanol solution is dried to a paste at 60°C. Third, the imprinted mold is manufactured, which is used to print a silver paste layer onto a clean graphite substrate as a transition layer. Fourth, the CNTs paste is further imprinted onto the silver paste transition layer. Fifth, the printed silver paste and CNTs are dried in Ar atmosphere at 200°C for 20 minutes. Sixth, the imprinted CNTs cathode is annealed at 500°C for 30 minutes to remove organic impurities. At last, the imprinted CNTs cathode is obtained and the process is shown in Figure 2.

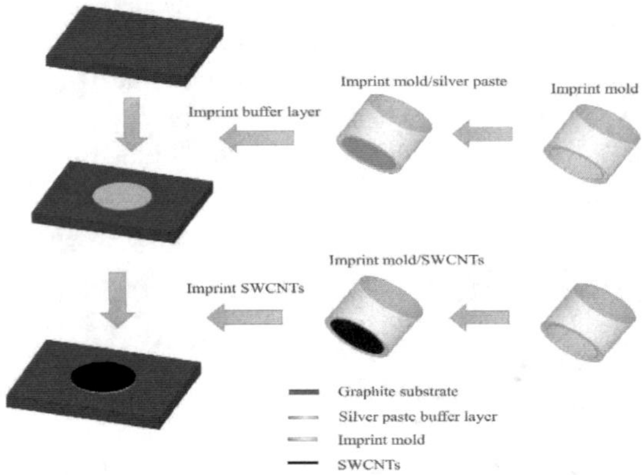


**Figure 2.** The process of the imprinted CNTs field emission cathode.

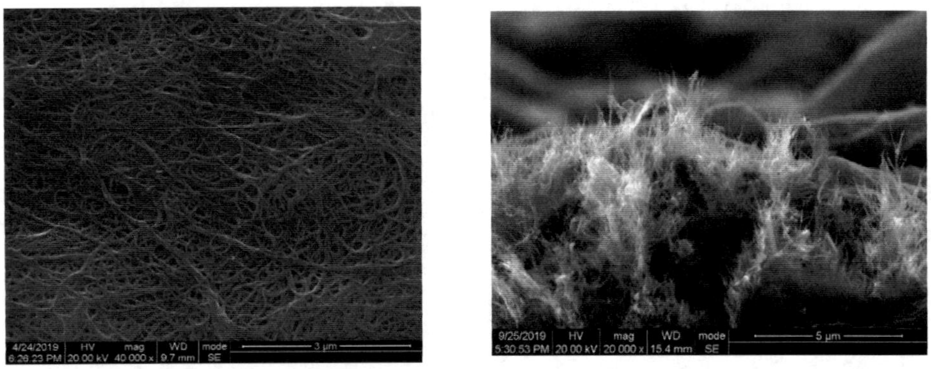

(a) Before testing (b) After testing

**Figure 3**. The SEM image of the imprinted CNTs cathode.

Figure 3 (a) shows the SEM image of the imprinted CNTs cathode, where the surface is flat and clean without impurities on it. Therefore, the emission uniformity is good. Figure 3 (b) is the SEM image after field emission testing. It can be seen that the CNTs also keep clean without impurities. A large number of CNT tips are perpendicular to the emission surface and become effective emission points, which reduce the turn-on field and the threshold field and keep the cathode emitting stably.

Figure 4 shows the field emission results of the imprinted CNTs cathode, where the turn-on field and the threshold field are low at 0.62 V/μm and 1.04 V/μm respectively [19-21]. The field enhancement factor β reaches 15800.

Figure 5 shows the long-term field emission stability test of the cathode. When the current density keeps 3.5 mA/cm$^2$, the cathode is continuously emitted for 140 minutes without significant fluctuations, which indicates that the imprinted CNTs cathode has good field emission stability and can operate stably for a long time [16, 18-20].

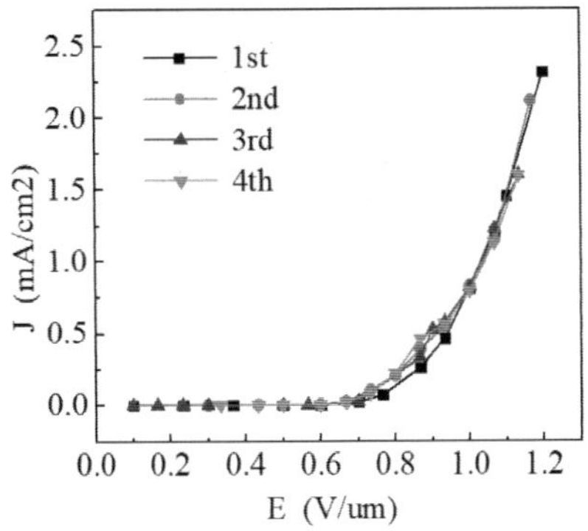


(a) J V.S. field

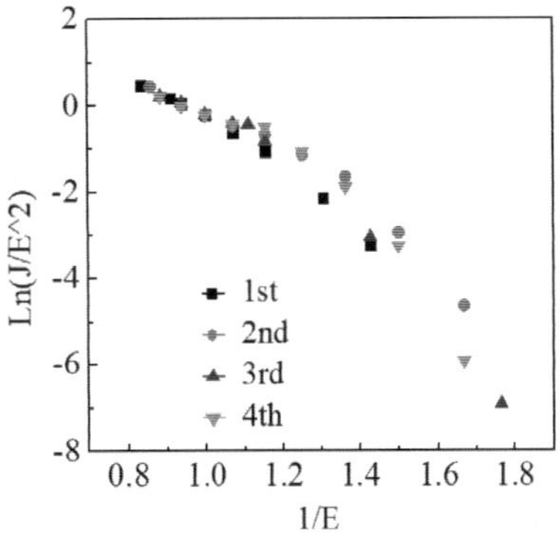


(b) F-N curve

**Figure 4**. The field emission testing of the imprinted CNTs cathode.

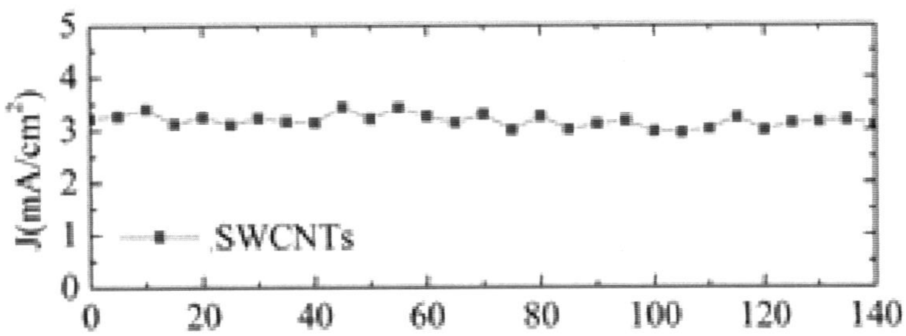


**Figure 5**. The long-term field emission test result.

### 3. Electron brush experimental

The electronic brushing platform is shown in the following Figure 6. It consists of the electron gun, the MCP and a display screen. The electron beam is amplified through the MCP and then projects onto the display screen [2-4]. In general, a thermic electron gun including a cathode, a focusing electrode, a heating, an anode and some support structures is embedded in the electron gun sleeve. The imprinted CNTs cathode does not require heating or controlling structure to form an electron beam shape. As shown in Figure 7, the imprinted CNTs cathode is fixed on a quartz sheet with a stainless grid fixed 0.5 mm away from its surface. The grid is made of stainless lines with a diameter of 0.25 mm and the grid mesh size is 1 mm * 1 mm. The imprinted CNTs field emission cathode is connected to a negative high-voltage power, the grid is grounded. The MCP and the display screen are fixed at a distance of 100 mm from the grid respectively which is connected to a voltage of + 800 V and + 3200 V.

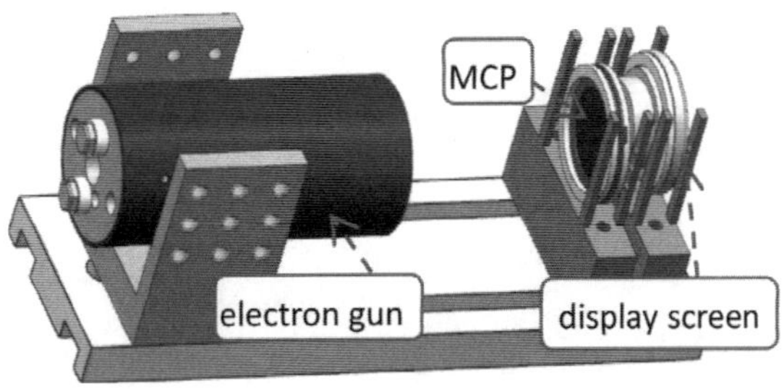


**Figure 6.** The electronic brushing platform.

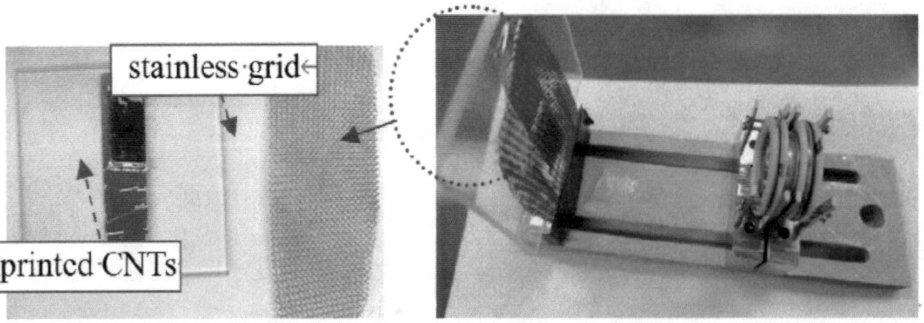


**Figure 7.** The imprinted CNTs cathode, the stainless grid & the platform.

When the negative high voltage is -800 V and applied to the imprinted cathode, preliminary lighting can be observed as shown in Figure 8 (a). Due to the unevenness of the grid, part of the electron exceeds the emission threshold and emits. With the continuous increase of negative high voltage, the brightness range of the display screen gradually expands. When it reaches -1000 V, the entire screen is lit up, indicating that all areas exceed the threshold and start to emit electrons, as shown in Figure 8 (b). With the increase of the negative voltage, the brightness of the display screen also increases. When the voltage reaches -1500 V, the display screen shows a strong brightness level, as shown in Figure 8 (d). To avoid damaging the MCP and display screen, the voltage increase is stopped. So, the electron brushing process can be operated at 1000 V - 1500 V.

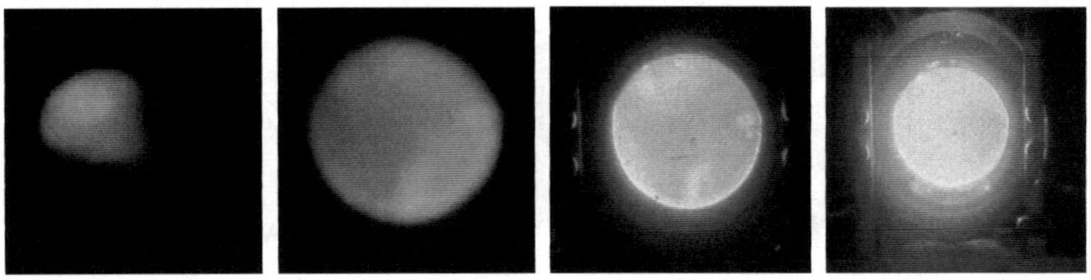

(a) -800 V (b) -1000 V (c) -1200 V (d) -1500 V

**Figure 8.** The brightness of the screen.

## 4. Conclusion

The imprinted CNTs cathode has some virtues, such as easy to manufacture, high emission efficiency, no heating requirement, stable performance and easy atmospheric maintenance. This paper has verified that the field emission cathode has the potential to apply electron brushing. This work provides a new method for realizing large-scale electron brushing based on the imprinted CNTs field emission cathode. In addition, it also lays a foundation for further research on electron brushing technology in the future.

## Acknowledgments

This work is supported by the State Key Laboratory Foundation (J20190103) and Key R&D Program of Guangdong (Grant No 2021B0101300003) and National Key Research and Development Program (Grant No G072022YFF0707602).

## References

[1] Cheng Hongchang, Shi Feng, Li Zhoukui, et al. Preliminary study on distinguishment method of low-level-light night vision devices, *Journal of Applied Optics*, 2021, 42 (6): 1092-1101.

[2] Xiang Shiming, Ni Guoqiang, 1999, *The Principle of Photoelectronic Imaging Devices,* (Beijing: National Defence Industry Press).

[3] Shi Feng, Cheng Hongchang, Yan Lei, 2017, *UV detection technology,* (Beijing: National Defence Industry Press).

[4] Chang Benkang, 2017, *GaAs Photoelectronic Cathode*, (Beijing: Science Press).
[5] Pan Jingsheng, Image Intensifier Upgraded Performance and Evaluation Standard, *Infrared Technology*, 2020, 42 (6): 509-518.
[6] Aebi V W, Costello K A, Arcuni P W, et al. EBAPS: Next generation, low power, digital night vision, *OPTRO 2005 International Symposium*, 2005: 1-10.
[7] Holst G, Electro-Optical Imaging System Performance, *Edition Jed Publishing Winter Park*, 2000, 18 (20): 123-125.
[8] Honggang Wang, Jing Gong, et al. Effect of Pro-Proximity Pulse Voltage on the Noise Characteristics of Generation III Low-Light-Level Image Intensifiers. *Optik*, 2018, 154.
[9] Dang Tao, 2010, *Dynamic (fast motion) low-light-level image enhancement.* (Chengdu: University Electronic Science and Technology of China).
[10] Donal J, Denvir, Emer Conroy, Electron Multiplying CCD Technology: the New ICCD, *Proc. SPIE,* 2002, 4796: 167-174.
[11] Lin Zhulun, Wang Xiaoju, 2013, *Cathode Electronics* (Beijing: National Defence Industry Press).
[12] Chen Qingyun, 2020, *Research on Electronic Optics System and Backward Wave Oscillator Based on Carbon Nanotube Cold Cathode*, (Chengdu: University Electronic Science and Technology of China).
[13] Mai Yapan, 2007, *CNTs-Science and applications*. (Beijing: Science Press).
[14] Iijima, S. Helical microtubules of graphitic carbon. *nature,* 1991, 354 (6348): 56-58.
[15] M Cole, K Teo, O Gröning, et al. Deterministic cold cathode electron emission from carbon nanofiber arrays. *Scientific Reports*, 2014, 4: 4840.
[16] X. Yan, Y. Wu, B. Wang, et al. Fabrication of carbon nanotube on nickel–chromium alloy wire for high-current field emission. *Applied Surface Science*, 2018, 450: 38-45.
[17] Tang, Z., Li, S., Zhu, Z., et al. Electron field emission from graphene nanosheets grown on Si nonporous pillar array. *Materials Science in Semiconductor Processing*, 2019, 89: 105-109.
[18] H. J. Kim, J. J. Choi, J. Han, et al. Design and field emission test of carbon nanotube pasted cathodes for traveling-wave tube applications. *IEEE Transactions on Electron Devices*, 2006, 53 (11): 2674-268.
[19] G. Ulisse, F. Brunetti, E. Tamburri, et al. Carbon nanotube cathodes for electron gun. *IEEE Electron Device Letters,* 2013, 34 (5): 698-700.
[20] Kajiwaraa, K., Saitob, Y. 2006, *Carbon nanotubes for electron emitting devices*. (Carbon Nanotechnology. Elsevier).
[21] Bonard, J.-M., Croci, M., Klinke, C., et al. Carbon nanotube films as electron field emitters. *Carbon*, 2002, 40 (10): 1715-1728.

# Formation and prevention of turning crack of screw for electronic parking brake system

**Dong Liu[1], Liejun Li[2], Hu Long[1,4], Shoujun Zhong[1], Fuqiang Li[1], Jixiang Gao[3]**

[1] Guangdong Shaoguan Zhongnan Iron and Steel Co., Ltd., Shaoguan, 512123, P.R. China

[2] National Engineering Research Center for Near-net-shape for Metallic Materials, South China University of Technology, Guangzhou, 510640, P. R. China

[3] Guangdong Polytechnic Normal University, Guangzhou 510635, China

[4]Corresponding author's e-mail: longhu042@126.com

**Abstract.** The cracking of screws made from medium-carbon cold heading steel for the electronic parking brake system was studied through the analysis of the microstructure and the systematic investigation of the whole process, and the corresponding improvement measures were proposed at last. Results showed that the cracks initiated at the root of the threads due to insufficient plasticity generated by the large pressure during thread rolling were the resource of the fracture, which was further extruded during the turning process, while the centering condition was not well. It was necessary to adjust the appropriate parameters of rolling pressure and rolling time to avoid heavy work hardening and also pay attention to the cleanness on the surface of the rolling plate or the wheel.

## 1. Introduction

With the fast industrial upgrading and development of automobiles, electrification, and intelligent technology, the Electronic Parking Brake (EPB) as a safety configuration has been widely used in vehicles [1,2]. When the vehicle is parked, the motor is turning forward, the parking torque is increased through the double-stage gear reduction mechanism, and then the rotary motion of the motor is converted into the linear motion of the lead screw through the screw nut mechanism, thus pulling the parking cable [3-5]. The structure is shown in Figure 1. The complex manufacturing process of lead screw nuts for electronic parking brake systems is as follows. Hot rolled wire → sandblasting → spheroidizing annealing → cold drawing → thread rolling → turning. The processing deformation is large and the deformation speed is fast, which puts forward higher requirements on the mechanical properties, formability, surface quality, fatigue life, and processing adaptability of raw materials. Medium carbon cold-heading steel is usually used as the raw material of such products to meet the composite mechanical properties and processability.

Material researchers always pay attention to fracture in the manufacturing process of the lead screw nut, and it is necessary to find out the resource to evaluate the usability of the rest batch of materials. The impact factors are complex. On the one hand, the internal defects of raw materials such as inclusions, porosity, shrinkage, carbide net, and other segregation [6-8]; on the other hand, the external factors such as the high rotation speed used in the drawing process, the large compression rate of some passes and the working cone angle, etc. [9-11].

In this paper, a sample of processing crack defects was inspected and analyzed. The direct cause of the cracking was found through the analysis of the microstructure and the systematic investigation of the whole process. The corresponding improvement measures were proposed in combination with the production process.

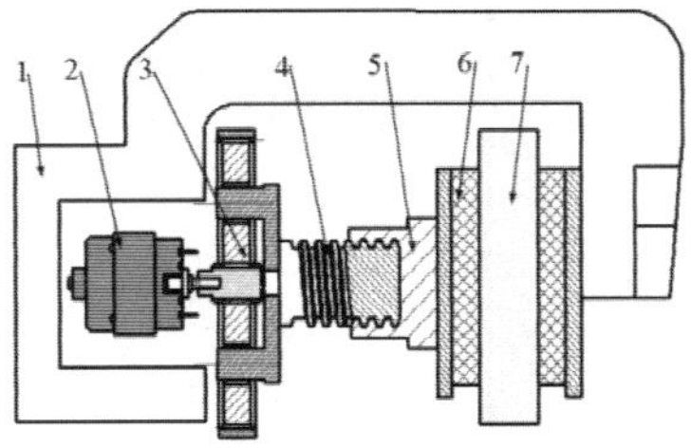


1-Brake caliper, 2-motor, 3-planetary gear reduction mechanism, 4-screw, 5-nut, 6-friction block, 7-brake disc

**Figure 1.** Electronic parking system structure diagram.

## 2. Fracture morphology and analysis

The chemical composition of cold-heading steel samples is shown in Table 1. The hot rolled wire with a diameter of 10 mm was drawn to the refined wire with a diameter of 9 mm after blasting and spheroidization annealing. The rolling process was normal, but there was a fracture in the turning process. First, a LEICA DVM6 3D stereo microscope was used to observe the macro morphology of the fracture after turning. Secondly, EVO18 Zeiss scanning electron microscope and EDS spectroscopy were used to examine and analyze the crack source.

**Table 1.** Chemical composition of steel (weight percentage %).

| Element | C | Si | Mn | P | S | Als |
|---|---|---|---|---|---|---|
| Percentage | 0.46 | 0.22 | 0.72 | 0.011 | 0.005 | 0.004 |

### *2.1. The fracture morphology*

The macro morphology of fracture on the screw was observed by stereo microscope. The top view of the fracture was shown on the left, and the lateral view was shown on the right. From the observation of radial cracks, it can be seen that the fracture source is at the crack of the tooth base, as shown in Figure 2.

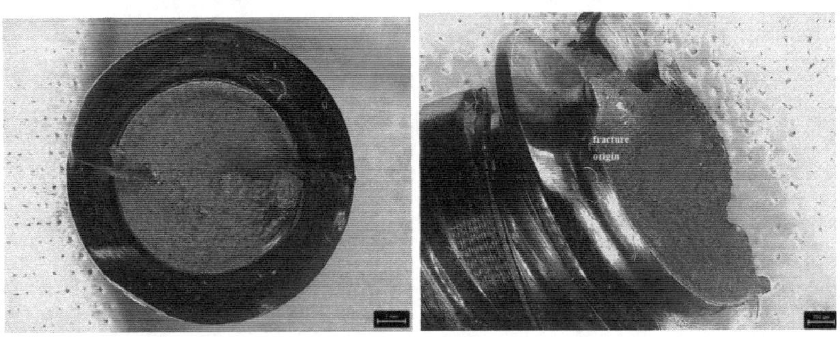


**Figure 2.** Fracture morphology of screw (30×).

### *2.2. The longitudinal morphology of the fracture*

The longitudinal metallographic samples were further prepared and observed. Cracks or peeling were found at the root of the upper R corner on both sides of the threads, which were symmetrically in the same direction with an angle of 15-30° from the surface to the inner, with the depth in the range of 10-125 microns, as shown in Figure 3.

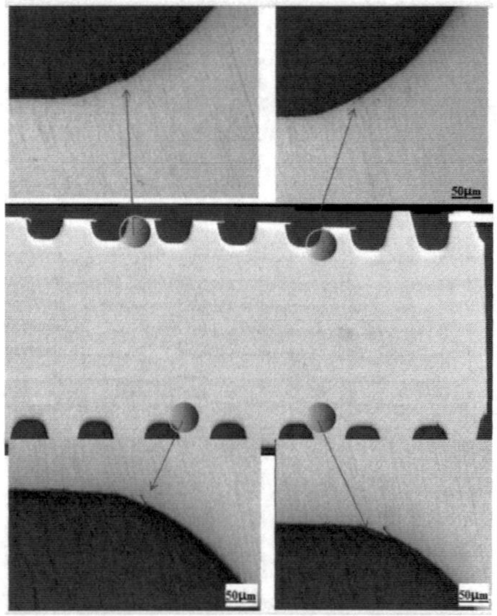


**Figure 3.** Distribution of cracks in tooth base of screw.

Figure 4 shows the morphology of the cracks before and after corrosion at the root of the threads. It was observed that the cracks are symmetrical, with depths of 102 microns and 125 microns, respectively. The microstructure after corrosion was spheroidized, the deformation of the structure at the crack was large, the crack defect on one side was thicker than that on the other side, and abnormal inclusion or structure was not found.

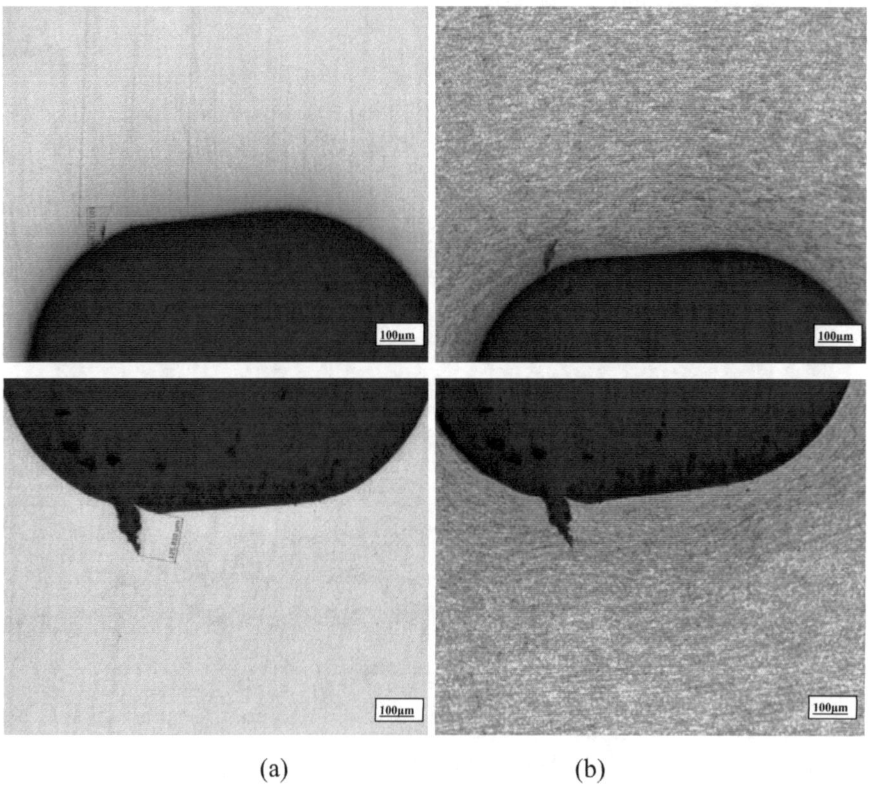


**Figure 4.** Microstructure at the crack of tooth base: (a) No corrosion; (b) Corrosion.

After etching with a nitric acid solution, each crack was observed along the longitudinal section under a scanning electron microscope, as shown in Figure 5. No abnormal inclusions and abnormal structures were found near the crack, granular cementites were uniformly distributed, matrix metal rheology was regular, and metal streamline bending could be observed at the crack.

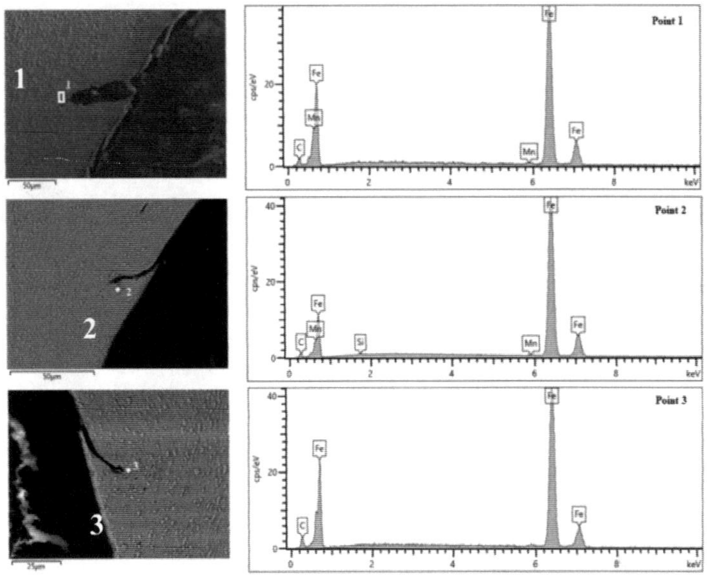


**Figure 5** EDS analysis results of large-size inclusions.

The linear distribution analysis of elements was further carried out in the cracked and non-cracked areas. As shown in Figure 6, the cracked area was filled with cold mounting material, so the content of carbon was high, and the distribution of other components was not significantly different from that in the non-cracked area. The matrix was Fe, containing trace amounts of C, and there was a layer of FeO with a height of 10 μm at the edge.

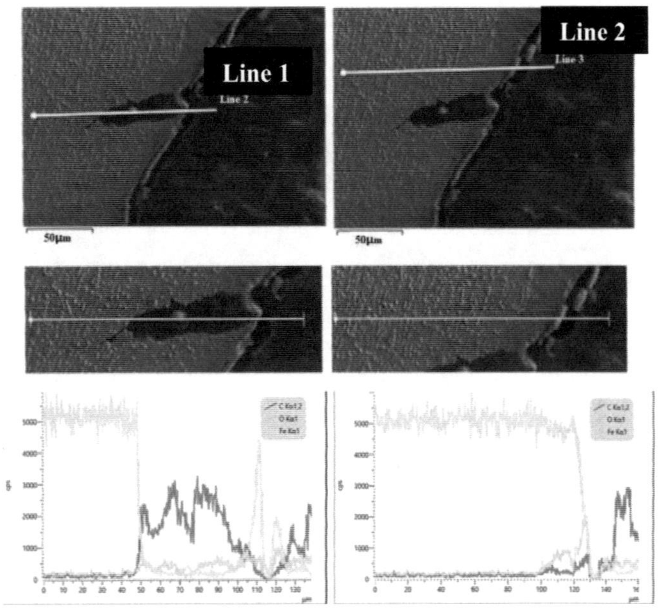


(a) (b)

**Figure 6** The linear distribution of elements at crack and non-crack: (a) Crack; (b) Non-crack.

### 2.3. Internal defect analysis

Further analysis of the internal defects of the material was studied and it was shown that the size of inclusions was limited to 10 μm, with the components of $CaO$-$Al_2O_3$-$SiO_2$-$MgO$, as shown in Figure 7. There was no direct relationship with edge defects. The microstructure and mechanical properties of raw materials were also in the normal range, with a yield strength of 386-387 MPa, a tensile strength of 670-682 MPa, an elongation of 30%, and a section shrinkage of 56%.

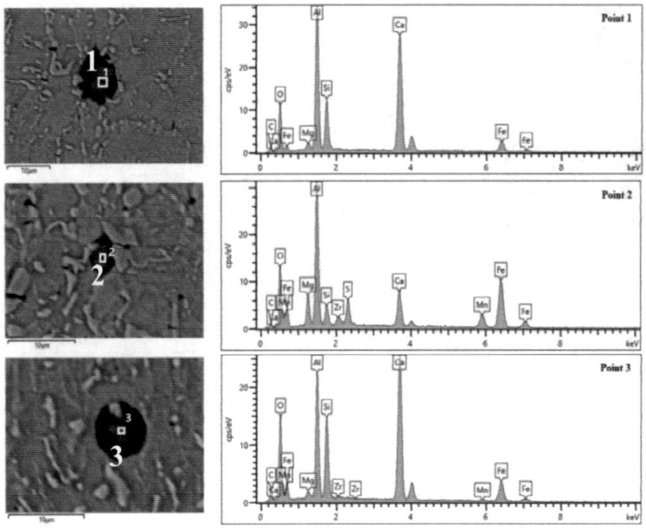


**Figure 7** EDS analysis results of internal inclusions.

## 3. Analysis and Discussion

### 3.1. Causes of cracks at the root of the threads

Cracks took place during the cold work such as thread rolling. At first, the cold work hardening of the thread surface was caused by the strength increased with the fibrous organization during the thread rolling. The plastic deformation was obvious, and the surface hardness increased, and then, the cracks were initiated due to insufficient plasticity under the too-large rolling pressure or the long rolling time. The surface folding cracks with large openings were found on the side of some screws, and deformed fibrous structures were found around the surface folding cracks, indicating that the cracks generated during the process of thread rolling were further extruded and deformed to form folding cracks, which may be related to contaminants on the surface of the rolling plate or the rolling wheel.

### 3.2. Mechanism of cracking

After the threads were rolled, the surface hardness of the screw increased, with a large residual stress. The thread at both ends of the screw needed to be removed turning by the tool with strong impact resistance. During the turning process, the poor centering condition leads to the transverse strike between the tool and the sample, which leads to the crack at the root of the threads being further expanded by the external force.

### 3.3. Analysis of improvement measures

In the rolling process, it is necessary to adjust the appropriate rolling pressure and rolling time to avoid heavy work hardening; Given the partial folding crack generated after plastic deformation, attention should be paid to the cleaning on the surface of the rolling plate or the wheel. In the subsequent turning process, the alignment accuracy should be ensured to avoid the transverse impact between the turning tool and the sample, which will lead to the further expansion of the crack at the root of the threads.

## 4. Conclusion

The resource of the fracture was the crack at the root of the threads, which was distributed along the circumference, with the depth in the range of 101 μm to 125 μm. There was no decarburization, abnormal inclusion, or abnormal organization at the crack. No cracks were found on the surface of the screw or threads. From the review of the whole production process, and the analysis of mechanical properties and inclusion control, it was concluded that the machining fracture was caused by the widespread surface cracks at the root of the threads:

(1) Composition and mechanical properties met the requirements of the raw materials with good surface quality, no circumferential or transverse defects, and the hardness was appropriate and uniform after annealing;

(2) In consideration of rolling processing, it was necessary to further adjust the appropriate rolling pressure and rolling time to avoid too strong work hardening;

(3) For some folding cracks generated after plastic deformation, attention should be paid to cleaning the surface of the rolling plate or wheel.

(4) In the subsequent turning process, the alignment accuracy should be ensured to avoid the transverse impact between the turning tool and the sample, which will lead to the further expansion of the crack of the tooth base.

**Acknowledgments**

The authors gratefully acknowledge financial support from the Guangdong Science and Technology Project (2017B090907015);

## References

[1] Zhang Wenjie, Zhao Huan, HU Weitan. Automobile electronic parking brake system analysis[J]. Internal combustion engine and accessories, 2018(22):40-41.

[2] Min Qing Fu, Ruan Wen Long. Research on parking force of automotive Electronic Parking Brake System (EPB)[J]. Transmission technology, 2019, 33(02):25-30.

[3] Zhuang L I, Di W U, Wei L. Effects of Rolling and Cooling Conditions on Microstructure and Mechanical Properties of Low Carbon Heading Steel[J]. Journal of Iron and Steel Research, 2012.

[4] Li M, Xu Q, Li J. Evolution of Nonmetallic Inclusion during Steelmaking Process of Cold Heading Steel SWRCH35K[J]. Mineral and Material Properties and Engineering, 2022(005):010.

[5] Ramesha C M, Rajendra P, Anilkumar T, et al. Fatigue Analysis of a Forged Medium Carbon Low Alloy Steel Quenched in a Polymer for Vehicle Structures Applications[J]. International Journal of Vehicle Structures and Systems, 2022.

[6] Su Zhenwei, Zhao Yun, Tu Xingeng. Cracking Analysis and technological improvement of cold heading steel of SWRCH35K[J]. Special steel, 2022, 43(04).

[7] Zhong S, Li L, Long H, et al. Prevention of Sharp Fracture Caused by Large Size Inclusion in Cold Heading Steel[J]. Journal of Physics: Conference Series, 2021.

[8] Sabih A, Wanjara P, Nemes J. Characterization of Internal Voids and Cracks in Cold Heading of Dual Phase Steel[J]. ISIJ International, 2005, 45(8):p.1179-1186.

[9] Wang Zhiyi, Wang Lijun, Chen Jilin, et al. Cause analysis of cold heading cracking of steel ML45n2 for gear shaft[J]. Heat treatment of metals, 2016, 41(10):198-201.

[10] Ping M X, Jun H W, Ming L C, et al. Bauschinger Effect of Micro alloyed Steel for Cold Heading Bolt in Cold Drawing Process[J]. Journal of Iron & Steel Research, 2007, 19(1):39-40.

[11] Nunes R M, Rocha A D S, Hirsch T G K. Residual stress states in cementite and ferrite in a combined cold drawing process of AISI 1045 steel using neutron diffraction and synchrotron radiation[J]. Editora Cubo, 2021.

# AUTHOR INDEX